Renaud Gauthier

Derived Algebraic Geometry

De Gruyter Expositions in Mathematics

Volume 75

Renaud Gauthier

Derived Algebraic Geometry

An Elemental Approach

2nd edition

DE GRUYTER

Mathematics Subject Classification 2020
Primary: 14-02, 14A30, 14A20, 18F20, 18N40, 18N60; Secondary: 81-10, 81T30

Author
Renaud Gauthier
University of Mary
Bismarck, ND, USA

ISBN 978-3-11-133366-3
e-ISBN (PDF) 978-3-11-133407-3
e-ISBN (EPUB) 978-3-11-133421-9
ISSN 0938-6572

Library of Congress Control Number: 2023946883

Bibliographic information published by the Deutsche Nationalbibliothek
The Deutsche Nationalbibliothek lists this publication in the Deutsche Nationalbibliografie; detailed bibliographic data are available on the Internet at http://dnb.dnb.de.

Typesetting: VTeX UAB, Lithuania
Printing and binding: CPI books GmbH, Leck

www.degruyter.com

Ad Majorem Dei Gloriam

Preface to the second edition

What prompted a second edition to this volume is a fairly comprehensive rewriting of the chapter on Physical Mathematics, where notations about Segal categories of higher states have been simplified to emphasize the iterative nature of the construction of such states, which led to further work on what makes states what they are. In the present edition, states are referred to as deformations, in order not to confuse them with physical states. It is actually a more apt description of what they are. All of this the reader will find in this extended chapter. Speaking of extensions, some introductory material about ∞-categories has been added to smoothen the introduction to that concept. At about the same time that this occurred, some work had been done on derived motivic spectra, something that found its rightful place in this volume, so it is been added as well. Note that there is also a chapter about motifs in derived algebraic geometry, not to be confused with motives. Motives are cohomological in nature, motifs are patterns, or variables in a sense. Another motivation for releasing a second edition is that in the first edition a few links for notations and the index were broken and had to be fixed. Last but not least, the title of this volume has been slightly changed at my publisher's request: it reads "DAG: an Elemental Approach", as opposed to the original "Elements of DAG".

https://doi.org/10.1515/9783111334073-201

Introduction

We should start by saying what this book is all about and what it is not. First and foremost, this is not a textbook, which is clear from its size. A textbook on Derived Algebraic Geometry would have an introduction the size of the present book. This does not mean however that the present work contains research material only; a substantial part of this volume is devoted to foundational material, as we will explain below. However, we provide only the results that we need, in contrast to a textbook which would give a comprehensive coverage of the material, along with proofs. Aside from the introductory material, our aim in the rest of the book is to present ideas that one would characterize as being off-the-beaten path, yet ideas that we consider to be vitally important from our perspective, in the hope that the interested reader will benefit from being exposed to such work.

There are essentially two parts to this volume then, foundations and applications. The foundational material, whether it be in the prerequisites, in the part on Derived Algebraic Geometry (DAG) or the one on Spectral Algebraic Geometry (SAG), is presented in a utilitarian fashion; there are no proofs as mentioned above, but only that which we need later is presented, and moreover key concepts are given in a top down fashion, since we regard this part to be a compendium of results. From that perspective, the readers should be able to quickly identify what they need and may keep reading on for clarifications if necessary. We did not want them to get bogged down in some thick text, most of which they may not need. This is once again because all of that material is standard and our exposition is not meant to be a progressive introduction to the subject. Key references are always mentioned in that regard, and in those the readers will find all results fully fleshed out.

Since it is no textbook, and the points of view therein are rather new, this text is intended for graduate students who would like to get a bird's eye view of some topics such as homotopical Algebraic Geometry and Higher Category Theory, with a few interesting applications. In the prerequisites, the student will find the basics of those theories, which are typically scattered amongst multiple sources, so in that regard the first part of this text is useful. The parts about DAG and SAG collect together what we regard as essential results for our purposes, and there also the student will find those parts to be a convenient summary of the basics of those theories. Researchers may find in the applications some new avenues of research.

Regarding the title, Elements of DAG does not mean Elementary DAG, because once again that would correspond to a textbook, which was not our intention of writing. The word Elements is chosen because it is an apt description of the contents of this book. You have a first part that is very much foundational and gives you the rudiments of certain fields, for the sake of understanding what comes thereafter. However, that will not give you a working knowledge of the subject matter; it is really meant for you to get acclimated to certain constructs. In that regard, we were very careful to reference all results

https://doi.org/10.1515/9783111334073-202

used, and in the first foundational part, there is no claim to originality, save for the remarks we made. References are built in the text or put down as footnotes if possible in order not to clutter the text. The second part consists of applications, which are reflective of our choice, and for that reason, the word elements again is a good description of what we produced. Essentially then, our aim was to quickly provide some foundations and some applications. One can regard this work as relatively self-contained, as most concepts needed are introduced as they surface.

At the heart of this work is the idea that many mathematical constructs can be iterated in a simplicial fashion for generalizing certain classical results (something we refer to as "simplicialization"), a move prompted by certain theories used in conjunction with those results as this book will illustrate, most notably when dealing with singularities. Perhaps one first example mentioned by many, and by Toen [73] and Lurie [54] in particular, is the following problem in intersection theory. If X and Y are smooth algebraic curves in $\mathbb{CP}^2$ of degrees m and n, respectively, then assuming that they intersect transversely, $X \cap Y$ has mn points. One can make this statement precise algebraically using fundamental classes; X and Y have fundamental classes in $H^2(\mathbb{CP}^2;\mathbb{Z})$, which satisfy $[X] \cup [Y] = [X \cap Y]$, where $[X \cap Y] \in H^4(\mathbb{CP}^2;\mathbb{Z}) \cong \mathbb{Z}$ counts exactly the number of intersection points. More generally though, if X and Y do not intersect transversally, then one only has a strict inequality $[X] \cup [Y] > [X \cap Y]$. Now it seems the intersection number is not enough, and we need something else to fix this apparent discrepancy. It turns out that it just suffices to consider intersections with their multiplicities. If $p \in X \cap Y$ is one such intersection, then its multiplicity is defined as $\dim_{\mathbb{C}} \mathcal{O}_{X,p} \otimes_{\mathcal{O}_{\mathbb{P}^2,p}} \mathcal{O}_{Y,p}$. Note that this object is the local ring of $X \times_{\mathbb{P}^2} Y$, a scheme-theoretic intersection. Thus if we focus on the intersection of X and Y, and we have to consider a local ring of this type at every intersection point, then this means we are no longer dealing with subvarieties, but with schemes. Counting multiplicities in this manner, the equality $[X] \cup [Y] = [X \cap Y]$ becomes correct. Worse still, suppose now X and Y are singular subvarieties of complementary dimension in a smooth algebraic variety Z but still intersect in a finite number of points. For the fundamental classes equation to hold, one still has to consider intersection multiplicities, but in this case, $\dim_{\mathbb{C}}(\mathcal{O}_{X,p} \otimes_{\mathcal{O}_{Z,p}} \mathcal{O}_{Y,p})$ is no longer enough. It turns out that Serre's intersection formula [69] tells us one has to consider the Euler characteristic instead to account for the right intersection multiplicity $\sum(-1)^n \dim_{\mathbb{C}} \operatorname{Tor}_n^{\mathcal{O}_{Z,p}}(\mathcal{O}_{X,p}, \mathcal{O}_{Y,p})$, where the first term is just the dimension of $\mathcal{O}_{X,p} \otimes_{\mathcal{O}_{Z,p}} \mathcal{O}_{Y,p}$, which was sufficient before. This already tells us that schemes are not sophisticated enough with generic situations and that more is needed. Now the question is whether there be an algebraic object that accounts for all those extra terms? Recall that if R is a commutative ring and A and B are R-modules, then $\operatorname{Tor}_*(A, B)$ is given by the homology of $\cdots \to P_2 \otimes_R B \to P_1 \otimes_R B \to P_0 \otimes_R B$, where $\cdots P_2 \to P_1 \to P_0 \to A$ is a projective resolution of A, a "simplicialization" of sort. Since for our purposes, A and B are commutative R-algebras, one has the tensor product over R of B with the projective resolution of A to be a commutative dg algebra over R, which one denotes by $A \otimes_R^{\mathbb{L}} B$ and which one refers to as the derived tensor product of A and B. Now this being defined, the intersection multiplicity reads $\sum(-1)^n \dim_{\mathbb{C}} H_n(\mathcal{O}_{X,p} \otimes^{\mathbb{L}}_{\mathcal{O}_{Z,p}} \mathcal{O}_{Y,p})$. Thus we are looking at something slightly more general than a scheme; we want on the intersection $X \cap Y$ a sheaf not of rings, but of commutative dg algebras, and this is a first move towards Derived Algebraic Geometry or Spectral Algebraic Geometry. Strictly speaking, we have at present a derived scheme. However, still more generally, as argued in [54], if we take a generic field k in lieu of the field of complex numbers as in Complex Algebraic Geometry, then the above theory is not well-behaved in positive characteristic, and one has to use $\mathbb{E}_\infty$-rings instead of commutative dg algebras. This is where Spectral Algebraic Geometry comes in. The resulting objects are called spectral schemes, pairs $(X, \mathcal{O}_X)$ consisting of a topological space X along with a sheaf $\mathcal{O}_X$ of $\mathbb{E}_\infty$-rings on it. To argue that taking $\mathbb{E}_\infty$-rings is not simply a side issue but is rather a generalization of commutative dg algebras, observe that they produce the same results in characteristic zero, but there are $\mathbb{E}_\infty$-algebras that are not quasi-isomorphic to commutative dg algebras in positive characteristic.

On the DAG side of things, first note that one can formalize the theory of schemes by seeing it as one facet of a still more general theory, namely relative Algebraic Geometry [37]. To see how this comes about, recall the classical formalism as expounded in [78]. One has the category of affine schemes $\mathrm{Aff} = \mathrm{CRing}^{\mathrm{op}}$, CRing being the category of commutative unital rings, which itself is the category of commutative unital monoids in the symmetric monoidal category $(\mathbb{Z}\text{-Mod}, \otimes)$: $\mathrm{CRing} = \mathrm{Comm}(\mathbb{Z}\text{-Mod}, \otimes)$. From there we consider the Yoneda embedding

$$\begin{aligned} h : \mathrm{Aff} &\to \mathrm{sPr}(\mathrm{Aff}), \\ X &\mapsto h_X = \mathrm{Hom}(-, X) \end{aligned}$$

where sPr refers to simplicial presheaves. This first move for us is of the utmost importance, since with an "inert" object X one associates the "dynamic" object $\mathrm{Hom}(-, X) \in \mathrm{Set}_\Delta$. We refer to $\mathrm{Hom}(-, X)$ as the perception of X, since it effectively corresponds functorially to all the possible ways of looking at X. Thus from simple schemes we are naturally led to considering simplicial presheaves. So far, Aff just corresponds to affine schemes: if $A \in \mathrm{CRing}$, then $A^{\mathrm{op}} =: \mathrm{Spec}\, A \in \mathrm{Aff}$. Schemes crop up if we have a Grothendieck topology on Aff to glue affine schemes. If (Aff, τ) is now a Grothendieck site and τ a Grothendieck topology on Aff, then for the sake of coherence, one may want functors $F : \mathrm{Aff}^{\mathrm{op}} \to \mathrm{Set}_\Delta$ that preserve this topology in some sense. This corresponds to asking that they satisfy τ-hyperdescent, and such functors correspond to stacks.

Having this elementary construction of affine schemes in mind, relative Algebraic Geometry consists in abstracting this formalism by considering any symmetric monoidal category $(\mathcal{C}, \otimes)$ in lieu of $(\mathbb{Z}\text{-Mod}, \otimes)$, considering $\mathrm{Comm}(\mathcal{C})$, whose opposite category, denoted $\mathcal{A}\mathrm{ff}_{\mathcal{C}}$, will be our category of $\mathcal{C}$-schemes, or stacks, depending on the context.

At this point, one would like to infuse our current theory with some homotopical flavor. Thus we consider base symmetric monoidal model categories instead, and this further enlarges our theory to that of Homotopical Algebraic Geometry. All of this can be found in the seminal papers of Toen and Vezzosi, which one can find in the references. This move is important since from our perspective, this allows us to have homotopical constructs supple enough to allow for some manipulations that would not be possible otherwise.

Note that Comm($\mathcal{C}$) in this situation does not automatically have a model structure, so we are really only focusing on those base model categories $\mathcal{C}$ for which Comm($\mathcal{C}$) is also a model category. This produces another model category $\mathcal{A}\mathrm{ff}_{\mathcal{C}} = \mathrm{Comm}(\mathcal{C})^{\mathrm{op}}$ of affine stacks on $\mathcal{C}$. Now one faces the problem of constructing a homotopical Yoneda embedding $\mathcal{A}\mathrm{ff}_{\mathcal{C}} \to \mathrm{sPr}(\mathcal{A}\mathrm{ff}_{\mathcal{C}})$. Naively, this means that objects in the image of this functor are objects $F : \mathcal{A}\mathrm{ff}_{\mathcal{C}}^{\mathrm{op}} \to \mathrm{Set}_{\Delta}$ that preserve weak equivalences. One way to enforce this is by considering $\mathcal{A}\mathrm{ff}_{\mathcal{C}}^{\wedge} = L_{W(\mathcal{A}\mathrm{ff}_{\mathcal{C}})}\mathrm{sPr}(\mathcal{A}\mathrm{ff}_{\mathcal{C}})$, the left Bousfield localization of $\mathrm{sPr}(\mathcal{A}\mathrm{ff}_{\mathcal{C}})$ (with its objectwise model structure) with respect to weak equivalences in the base $\mathcal{A}\mathrm{ff}_{\mathcal{C}}$. One can construct a fully faithful functor $\mathrm{Ho}\,(\mathcal{A}\mathrm{ff}_{\mathcal{C}}) \to \mathrm{Ho}\,(\mathcal{A}\mathrm{ff}_{\mathcal{C}}^{\wedge})$ of homotopy categories with the flavor of a Yoneda embedding, which is therefore regarded as the correct homotopical generalization of the Yoneda embedding for our purposes.

Next comes the problem of putting a topology on the model category $\mathcal{A}\mathrm{ff}_{\mathcal{C}}$. Following [78], [76], and [79], this means defining a homotopical version of a Grothendieck topology, making $(\mathcal{A}\mathrm{ff}_{\mathcal{C}}, \tau)$ into a model site. Then objects of $\mathcal{A}\mathrm{ff}_{\mathcal{C}}^{\wedge}$ that satisfy an appropriate τ-hyperdescent condition are referred to as stacks.

At this point, it is worth making the following observation: if $\mathcal{A}\mathrm{ff}_{\mathcal{C}}^{\sim,\tau} \subset \mathcal{A}\mathrm{ff}_{\mathcal{C}}^{\wedge}$ denotes the model category of stacks on $\mathcal{A}\mathrm{ff}_{\mathcal{C}}$, then one typically considers topologies that are subcanonical, meaning those for which the homotopical Yoneda embedding factors through $\mathrm{Ho}\,(\mathcal{A}\mathrm{ff}_{\mathcal{C}}^{\sim,\tau})$. Then objects in its essential image are referred to as affine stacks, or representable stacks, and geometric stacks are built from those by gluing. The point of view we follow however is the following: once one has a homotopical Yoneda embedding, $\mathcal{A}\mathrm{ff}_{\mathcal{C}}$ is forgotten, and the new model category to focus on is now $\mathcal{A}\mathrm{ff}_{\mathcal{C}}^{\wedge} \subset \mathrm{sPr}(\mathcal{A}\mathrm{ff}_{\mathcal{C}})$, since the homotopy Yoneda embedding allowed us to enlarge our horizons. Then $\mathcal{A}\mathrm{ff}_{\mathcal{C}}^{\sim,\tau} \subset \mathcal{A}\mathrm{ff}_{\mathcal{C}}^{\wedge}$ is the model category of stacks, some of its objects are geometric stacks in particular, but it is really the whole model category we are interested in, insofar as it is a larger environment so to speak within which one can talk about homotopical stacks, and if one is interested in stacks proper, then it suffices to work in the homotopy category $\mathrm{Ho}\,(\mathcal{A}\mathrm{ff}_{\mathcal{C}}^{\sim,\tau})$.

Now comes the important step of spectralization, or simplicialization: if $\mathcal{C} = k$-Mod, $k \in \mathrm{CRing}$, then one gets a theory of higher Artin stacks if we follow the above construction. If one takes $\mathcal{C} = sk$-Mod instead, whose objects are simplicial commutative k-modules, then one gets derived algebraic stacks, whose model category is denoted by k-D^{-}Aff. It turns out there is a fully faithful embedding of the category of higher Artin stacks into the category of derived algebraic stacks by Lemma 2.2.4.1 of [79]. This shows us that in some respect spectralization further enlarges our field of operation. This would correspond to working with $s\mathcal{C}$ instead of $\mathcal{C}$ as our base model category. It turns out that some constructs using stacks on $\mathcal{C}$ are illuminated when viewed as derived objects. Some classical objects in Algebraic Geometry displaying some incongruities, or downright ill-defined, are resolved in the derived setting [46]. Some of the manipulations we do in the classical setting turn out to already be simplicial anyway. Thus it seems spectralizing our objects is a natural thing to do and stacks on such base model categories $s\mathcal{C}$ are part of DAG, the subject of this book. To summarize then, there are two essential moves we operate whenever possible: using larger spaces for more flexibility and a better vantage point, and spectralization/simplicialization.

With regards to the second part of this volume, Applications, there the reader will find a few topics on the subject matter. The chapter about dual representation in SAG is strictly about SAG, whereas the others deal with DAG. In the SAG chapter, one relates the perception $\mathrm{Hom}(-,X)$ of an object X to its manifestation, or its representation theory. That we have such a relation is naively self-evident but far from obvious. In that chapter, we lay down the conditions under which such a relation exists. We have incorporated this chapter in the present volume mainly for two reasons; SAG provides an alternative treatment to DAG, and the notion of perception is something we use throughout. All the other chapters are all contributing in one way or another to the very last chapter on Physical Mathematics, which is really the primary reason why we worked with DAG and Segal topoi in the first place. The series of chapters about Segal topoi and ∞-topoi show that under the assumption that physical laws be modeled by simplicial algebras, one can construct topoi with internal homs within which phenomena take place, internal objects that are not visibly apparent at the level of objects. One can show there is a universal object out of which all phenomena are generated, and one can show we have a hierarchy of phenomena as we go higher in the tower of internal hom objects in such a manner that the higher we go, the more abstract are those phenomena, but also the more complete. Interestingly, in lower levels, those phenomena cannot be analyzed with the tools available at that particular level. The other chapters are more ancillary; the one about supersymmetric stacks is written with a view to Theoretical Physics. The one about motifs is meant to investigate formal constructions involving derived stacks. The chapter about internal Bousfield localizations is a technical one delving into the problem of localizing in the event that we have enriched categories, which is pertinent since stacks arise from localizations. The chapter about schematization is a nice application of the formalism of derived stacks, though this particular chapter is fairly disjoint from the philosophy adopted in the rest of the book. The one about Goodwillie Calculus is really the only chapter that deals with geometric stacks and is meant to draw similarities between the theory of the calculus of homotopy functors as developed by Goodwillie and certain aspects of derived geometric stacks.

Regarding notations, we use abbreviations throughout for ease of reading. In particular, in footnotes, Theorem, Proposition, Notation, Definition, Example, Remark, and Section will be abbreviated by Thm, Prop, Not, Def, Ex, Rmrk, and Sec, respectively. We also use L for left, R for right, whenever we find it necessary. We write RLP and LLP for right/left lifting property, which is fairly standard. Spectral Algebraic Geometry is referred to as SAG, and Derived Algebraic Geometry is referred to as DAG, something we have been doing already. CD denotes a commutative diagram. We reserve the notation $\cong$ for isomorphism, whereas $\simeq$ will denote an equivalence. At the end of the book the reader will find an index, along with a convenient list of all notations used in this book.

Last but not least, we would like to warmly thank all those persons without whom this book would not have seen the light of day: J. Bergner, A. Blaer, O. Caramello, L. Gabet, J. Gemmer, J. Gutierrez, M. Hoyois, T. Pilling, F. Vouzellaud, D. White, and D. Yetter.

Contents

Part II: **Derived algebraic geometry**

Part III: **Spectral algebraic geometry**

Part I: **Prerequisites**

1 Schemes

As a first introduction to schemes, recall how varieties were defined. In Algebraic Geometry, we work with zero sets of polynomials. To formalize this, fix an algebraically closed field k and define affine n-space over k by $\mathbb{A}^n = \{(a_1, \dots, a_n) : a_i \in k, 1 \le i \le n\}$. For $f \in k[x_1, \dots, x_n]$, define its zero set by $Z(f) = \{P \in \mathbb{A}^n : f(P) = 0\}$. More generally, if S is a subset of $k[x_1, \dots, x_n]$, then define its zero set by $Z(S) = \{P \in \mathbb{A}^n : f(P) = 0, \forall f \in S\}$. Topologize $\mathbb{A}^n$ by taking those zero sets as closed sets. One then defines an affine algebraic variety[1] as an irreducible closed subset of $\mathbb{A}^n$. Observe that maximal ideals of $k[x_1, \dots, x_n]$ correspond to points of $\mathbb{A}^n$. When we define schemes, we will replace the polynomial algebra $k[x_1, \dots, x_n]$ by any commutative ring, and the points of our space will be all prime ideals, not just maximal ideals. From our perspective, sufficiently enlarging a theory gives ample room to develop new Mathematics. Thus the present generalization from varieties to schemes is a strong motivator alone for developing a theory of schemes.

One can nevertheless mention some more down to earth classical motivations for the existence of schemes, such as problems in intersection theory for instance; if one has nontransverse intersections, then one has to also consider intersections with multiplicities as discussed in the introduction. Those multiplicities are given by the dimension of local rings on intersections.

Thus one has to consider pairs consisting of topological spaces and sheaves of rings on them. If one does this carefully, then schemes appear to be needed.

Schemes are a certain type of ringed spaces, pairs consisting of a topological space and a sheaf of rings on them. Thus we first discuss sheaves before we introduce ringed spaces. Then we can define schemes.

1.1 Sheaves

Sheaves are typically introduced at the same time that schemes are discussed, so we first briefly explain what sheaves are. In this regard, we can mention the excellent reference [33] on sheaves. We mainly follow [39].

Heuristically, a sheaf on a space is a functorialization of a structure defined locally on that space. If for spaces we take topological spaces, locality is defined in terms of open sets. For structure, we can take differentiable functions defined on open sets of a given space X for instance. Doing this globally, so that definitions agree on overlapping open sets, and doing this in a coherent fashion, meaning functorially, one arrives at the sheaf of differentiable functions on X. If one takes differential operators defined on X instead, one would arrive at the definition of another sheaf on X, namely the sheaf of differential

1 Sec. I.1 [39].

https://doi.org/10.1515/9783111334073-001

operators on X. We regard sheaves then as manifestations of a given space. For instance, from the perspective of differentiable functions, the sheaf of differentiable functions on a space X tells us what kind of such functions can be defined on X.

Let X be a topological space, and let $\mathcal{U}(X)$ be its category of open sets with inclusions as morphisms. Let $\mathcal{C}$ be a category. Then a *$\mathcal{C}$-valued presheaf on X*[2] is an object $\mathcal{F}$ of $\mathrm{Fun}(\mathcal{U}(X)^{\mathrm{op}}, \mathcal{C})$ such that for any inclusion $U \hookrightarrow V$ in $\mathcal{U}(X)$, we have a *restriction map* $\rho_{VU} = \mathcal{F}(U \hookrightarrow V) : \mathcal{F}(V) \to \mathcal{F}(U)$ such that $\mathcal{F}(\emptyset) = *$, $\rho_{UU} = \mathrm{id}$, and for inclusions $U \hookrightarrow V \hookrightarrow W$, we have $\rho_{WV} \circ \rho_{VU} = \rho_{WU}$. Morphisms of presheaves are natural transformations in $\mathrm{Fun}(\mathcal{U}(X)^{\mathrm{op}}, \mathcal{C})$. We denote by $\mathrm{Pr}_{\mathcal{C}}(X)$ the category of $\mathcal{C}$-valued presheaves on X. If $\mathcal{C}$ is the category Ab of abelian groups, then we use the notation $\mathrm{Pr}(X)$ for the corresponding category, which we just refer to as the category of presheaves on X by abuse of language. Quite often in classical Algebraic Geometry, $\mathcal{C}$ will be the category of abelian groups; for $U \in \mathcal{U}(X)$, $\mathcal{F}(U)$ is an abelian group. In this section, we will focus on this case, but our constructions also hold for sheaves valued in any other category $\mathcal{C}$, be it rings, simplicial sets, etc.

Note that abelian groups have an underlying set, so we can talk about the elements $s \in \mathcal{F}(V)$ for $V \in \mathcal{U}(X)$. For the sake of notation, if $U \hookrightarrow V$, then we denote $\rho_{VU}(s)$ by $s|_U$. A presheaf $\mathcal{F}$ on a topological space X is a *sheaf*[3] if whenever $\bigcup U_i = U$ is an open cover of some object $U \in \mathcal{U}(X)$ such that $s|_{U_i} = 0$ for all i, then $s \cong 0$, and whenever $\bigcup U_i = U$ is an open cover of some object $U \in \mathcal{U}(X)$, if we have some $s_i \in \mathcal{F}(U_i)$ for all i that agree on overlaps: $s_i|_{U_i \cap U_j} = s_j|_{U_i \cap U_j}$, then one can find a global $s \in \mathcal{F}(U)$ such that $s|_{U_i} = s_i$ for all i. We define a morphism of sheaves to be a morphism of their underlying presheaves. We denote by $\mathrm{Sh}(X)$ the category of sheaves on X, a notation we also use for $\mathcal{C}$-valued sheaves if it is clear from the context what kind of sheaves we are working with.

With this notion of morphism, we can relate sheaves and presheaves. Since sheaves are just presheaves with just added conditions, one may imagine that there is a way to "sheafify" a presheaf into a sheaf. There is indeed such a move: for a presheaf $\mathcal{F}$, there are a sheaf $\mathcal{F}^+$ and a morphism of presheaves $a : \mathcal{F} \to \mathcal{F}^+$, the *sheafifying functor*, universal among all morphisms from $\mathcal{F}$ into a sheaf.[4] We refer to $\mathcal{F}^+$ as the *sheaf associated with $\mathcal{F}$*.

A trivial, but very convenient, example of a sheaf is the *constant sheaf*:[5] let X be a topological space, $R \in \mathrm{Ab}$. We denote by the same letter R the constant sheaf on X taking the value R. On $U \in \mathcal{U}(X)$, $R(U)$ is defined to be the group of continuous maps $U \to R$. For this to be well-defined, we have to topologize R. Just pick the discrete topology on R. One can verify that this produces a sheaf. Further, if U is connected, then $R(U) \cong R$.

---

2 Def. II.1.- [39].

3 Def. II.1.- [39].

4 Prop. II.1.2 [39].

5 Ex. 1.0.3 [39].

An important construct for sheaves is that of a *stalk*:[6] for a presheaf $\mathcal{F}$ on a topological space X, $P \in X$, since $\mathcal{U}(X)$ is a directed system, we can compute the direct limit of objects $\mathcal{F}(U) \in \mathrm{Ab}$ over those $U \ni P$ in $\mathcal{U}(X)$. We call this direct limit the stalk $\mathcal{F}_P$ of $\mathcal{F}$ at the point P:

$$\mathcal{F}_P = \varinjlim \mathcal{F}(U).$$

To illustrate what stalks tell us about the behavior of a sheaf, note that[7] if $\mathcal{F}, \mathcal{G} \in \mathrm{Sh}(X)$, X a topological space, then $f : \mathcal{F} \to \mathcal{G}$ is an isomorphism iff $f_P : \mathcal{F}_P \to \mathcal{G}_P$ is an isomorphism for all $P \in X$.

We have functorial constructions on sheaves: if $f : X \to Y$ is a morphism of topological spaces, i. e., a continuous map, then we have a pair of adjoint functors: $f_* : \mathrm{Sh}(X) \rightleftarrows \mathrm{Sh}(Y) : f^{-1}$, where f_* is called the *direct image functor*[8] and is defined by

$$(f_*\mathcal{F})(U) = \mathcal{F}(f^{-1}U)$$

for $U \in \mathcal{U}(Y)$, and f^{-1} is the *inverse image functor*, where $f^{-1}\mathcal{G}$ is the sheaf associated with the presheaf $U \mapsto \lim \mathcal{G}(V)$, where the limit is taken over all open sets V containing $f(U)$. We have $f^{-1} \dashv f_*$, which we formulate as follows:

$$\mathrm{Hom}_{\mathrm{Sh}(X)}(f^{-1}\mathcal{G}, \mathcal{F}) \cong \mathrm{Hom}_{\mathrm{Sh}(Y)}(\mathcal{G}, f_*\mathcal{F}).$$

1.2 Ringed spaces

Once we have the notion of sheaf, we can make a first move towards defining schemes: one defines a *ringed space*[9] to be a pair $(X, \mathcal{O}_X)$ consisting of a topological space X and a sheaf of rings $\mathcal{O}_X$ on it. Morphisms $(X, \mathcal{O}_X) \to (Y, \mathcal{O}_Y)$ of ringed spaces are given by pairs $(f, f^\#)$ consisting of a continuous map $f : X \to Y$ and of an induced map $f^\# : \mathcal{O}_Y \to f_*\mathcal{O}_X$ in $\mathrm{Sh}(Y)$.

More specifically, we will be interested in *locally ringed spaces* (affine schemes are locally ringed spaces), ringed spaces $(X, \mathcal{O}_X)$ such that the structure sheaf $\mathcal{O}_X$ is such that its stalks are local rings at all points of X. A morphism of locally ringed spaces $(X, \mathcal{O}_X) \to (Y, \mathcal{O}_Y)$ is a morphism of ringed spaces preserving their local nature, that is, such that for all points $P \in X$, the induced map of locally ringed spaces $f_P^\# : \mathcal{O}_{Y,f(P)} \to \mathcal{O}_{X,P}$ is a local homomorphism, that is, the preimage under $f_p^\#$ of the maximal ideal of $\mathcal{O}_{X,P}$ is the maximal ideal of $\mathcal{O}_{Y,f(P)}$.

6 Def. II.1.- [39].

7 Prop. II.1.1 [39].

8 Def. II.1.- [39].

9 Def.II.2.- [39].

Now of course once one has ringed spaces, one can consider sheaves of modules. To be precise, for a ringed space $(X, \mathcal{O}_X)$, a *sheaf of $\mathcal{O}_X$-modules*[10] is a sheaf $\mathcal{F}$ on X that is objectwise a module on X, that is, for all objects $U \in \mathcal{U}(X)$, $\mathcal{F}(U)$ is an $\mathcal{O}_X(U)$-module, consistent with the restriction maps. This means we have a commutative diagram in $\mathrm{Sh}(X)$ for every inclusion $U \hookrightarrow V$ in $\mathcal{U}(X)$:

$$\begin{array}{ccc} \mathcal{F}(V) & \xrightarrow{\mathcal{F}(U \hookrightarrow V)} & \mathcal{F}(U) \\ \uparrow & & \uparrow \\ \mathcal{O}_X(V) & \xrightarrow[\mathcal{O}_X(U \hookrightarrow V)]{} & \mathcal{O}_X(U). \end{array}$$

For $\mathcal{O}_X$-modules $\mathcal{F}, \mathcal{G}$, a morphism $\mathcal{F} \to \mathcal{G}$ in $\mathrm{Sh}(X)$ is a morphism of $\mathcal{O}_X$-modules if it is objectwise on $\mathcal{U}(X)$ a module homomorphism, that is, for all $U \in \mathcal{U}(X)$, $\mathcal{F}(U) \to \mathcal{G}(U)$ is a morphism of $\mathcal{O}_X(U)$-modules. We denote by $\mathrm{Mod}_{\mathcal{O}_X}$ the category of sheaves of $\mathcal{O}_X$-modules. Among common examples, we can name *sheaves of ideals* $\mathcal{I}$ that are objectwise ideals of $\mathcal{O}_X$, that is, for all $U \in \mathcal{U}(X)$, $\mathcal{I}(U)$ is an ideal of $\mathcal{O}_X(U)$. Also, for a morphism $(X, \mathcal{O}_X) \to (Y, \mathcal{O}_Y)$ of ringed spaces and $\mathcal{F} \in \mathrm{Mod}_{\mathcal{O}_X}$, observe that $f_*\mathcal{F} \in \mathrm{Mod}_{\mathcal{O}_Y}$. Indeed, $f_*\mathcal{F}$ is an $f_*\mathcal{O}_X$ module, but by the definition of morphism of ringed spaces, we have an induced map $f^\# : \mathcal{O}_Y \to f_*\mathcal{O}_X$, and thus $\mathrm{Mod}_{f_*\mathcal{O}_X} \subset \mathrm{Mod}_{\mathcal{O}_Y}$. Thus we have a direct image functor $f_* : \mathrm{Mod}_{\mathcal{O}_X} \to \mathrm{Mod}_{\mathcal{O}_Y}$. In the same manner, if $\mathcal{G} \in \mathrm{Mod}_{\mathcal{O}_Y}$, then $f^{-1}\mathcal{G}$ is an $f^{-1}\mathcal{O}_Y$-module. On the other hand, since $f^{-1} \dashv f_*$, we have a map $f^{-1}\mathcal{O}_Y \to \mathcal{O}_X$ corresponding to the map $\mathcal{O}_Y \to f_*\mathcal{O}_X$. In this manner, $\mathcal{O}_X$ can be regarded as an $f^{-1}\mathcal{O}_Y$-module as well. Note incidentally that this morphism also shows why we cannot simply take $f^{-1}\mathcal{G}$ as an object of $\mathrm{Mod}_{\mathcal{O}_X}$. Thus we fix this problem with the following definition:

$$f^*\mathcal{G} = f^{-1}\mathcal{G} \otimes_{f^{-1}\mathcal{O}_Y} \mathcal{O}_X$$

where a general definition of the tensor product in $\mathrm{Mod}_{\mathcal{O}_X}$ is as a sheafification of the objectwise tensor product functor $U \mapsto \mathcal{F}(U) \otimes_{\mathcal{O}_X(U)} \mathcal{G}(U)$ on $\mathcal{U}(X)$ for $\mathcal{F}, \mathcal{G} \in \mathrm{Mod}_{\mathcal{O}_X}$.

Thus we have a functor $f^* : \mathrm{Mod}_{\mathcal{O}_Y} \to \mathrm{Mod}_{\mathcal{O}_X}$ called the *inverse image functor*. Observe that $f^* \dashv f_*$, and we have

$$\mathrm{Hom}_{\mathcal{O}_X}(f^*\mathcal{G}, \mathcal{F}) \cong \mathrm{Hom}_{\mathcal{O}_Y}(\mathcal{G}, f_*\mathcal{F}).$$

1.3 Schemes

We can now define schemes. The reason for introducing them, aside from the fact that they will be used later, albeit sparingly, is that the constructions are very standard, and

10 Def. II.5.- [39].

will be just generalized for higher objects, such as spectral schemes or spectral DM stacks. A scheme is essentially a ringed space that locally looks like a special kind of ringed space, whose base topological space (called spectrum) is built from a commutative ring. Our presentation follows [39].

We start with a commutative ring R, and we define

$$|\mathrm{Spec}\, R| = \{\mathfrak{p} \, : \, \mathfrak{p} \text{ prime ideal of } R\}.$$

Such a set is called the spectrum of R. We will see below that one can topologize this set and put a sheaf of rings on it, resulting in a ringed space that we also call the spectrum of R. To avoid any ambiguity, one can refer to $|\mathrm{Spec}\, R|$ as the base spectrum.

Since objects of the form $|\mathrm{Spec}\, R|$ will be the spaces we deal with, note the general departure from the simple notion of affine variety, whose points simply correspond to maximal ideals. Now define

$$V(\mathfrak{a}) = \{\mathfrak{p} \supseteq \mathfrak{a} \, : \, \mathfrak{p} \in |\mathrm{Spec}\, R|\}$$

for any ideal $\mathfrak{a}$ of R. We topologize $|\mathrm{Spec}\, R|$ by taking as open subsets the complements of the sets $V(\mathfrak{a})$. This defines what we call the *Zariski topology* on $|\mathrm{Spec}\, R|$. This set thus topologized will be the topological space used in the definition of an affine scheme. We now need to define a sheaf of rings on it to make it into a ringed space. For a prime ideal $\mathfrak{p}$ in R, let $R_\mathfrak{p}$ be the localization of R at $\mathfrak{p}$. Recall how this works: one localizes a commutative ring with respect to a multiplicative system, that is, a subset S of R closed under the product and containing the unit for that product, which we denote by 1. In this case, the localization $S^{-1}R$ of R by S is the ring of equivalence classes of fractions r/s for $r \in R$ and $s \in S$, where the equivalence is defined as follows: $s/r \sim s'/r'$ iff $\exists s'' \in S : s''(s'r - sr') = 0$. For our purposes, if $\mathfrak{p}$ is a prime ideal of a commutative ring R, then $S = R - \mathfrak{p}$ is a multiplicative system, and the corresponding localization is denoted $R_\mathfrak{p}$.

We construct the *structure sheaf* $\mathcal{O}_R$ on $|\mathrm{Spec}\, R|$ objectwise: let $U \in \mathcal{U}(|\mathrm{Spec}\, R|)$. Note that points of U are ideals of R as well. Define

$$\mathcal{O}_R(U) = \{s : U \to \coprod_{\mathfrak{p} \in U} R_\mathfrak{p}\}$$

where $s(\mathfrak{p}) \in R_\mathfrak{p}$. We need a local condition however as this is too rigid. We ask that this last condition on s be true in a neighborhood of $\mathfrak{p}$, that is, for $\mathfrak{q}$ sufficiently close to $\mathfrak{p}$, we still have $s(\mathfrak{q}) \in R_\mathfrak{q}$, in a uniform manner. So we have $s = a/f$ with $f \in R - \mathfrak{q}$, that is, $f \notin \mathfrak{q}$. We formalize this: for $\mathfrak{p}$ fixed, one can find a neighborhood V of $\mathfrak{p}$ in U such that over V, there are $a, f \in R$ such that for all $\mathfrak{q} \in V, f \notin \mathfrak{q}$, and $s(\mathfrak{q}) = a/f$. Thus defined $\mathcal{O}_R(U)$ is a commutative ring. It is not hard to see that $\mathcal{O}_R$ is a sheaf of rings on $|\mathrm{Spec}\, R|$. Finally one defines the *spectrum*[11] of the commutative ring R by

$$\operatorname{Spec} R = (|\operatorname{Spec} R|, \mathcal{O}_R).$$

Thus defined the spectrum $\operatorname{Spec} R$ of R is a locally ringed space.[12] Observe that functorially a morphism of commutative rings $A \to B$ produces a morphism of spectra $\operatorname{Spec} B \to \operatorname{Spec} A$. For simplicity, we will just use the notation $\mathcal{O}$ for the structure sheaf $\mathcal{O}_R$ of $\operatorname{Spec} R$. Observe[13] that $\mathcal{O}_\mathfrak{p} \cong R_\mathfrak{p}$ and $\Gamma(|\operatorname{Spec} R|, \mathcal{O}) \cong R$, where we use the notation $\Gamma(U, \mathcal{F}) = \mathcal{F}(U)$, which we call the group of sections of $\mathcal{F}$ over U.

Next, we define an *affine scheme* to be a locally ringed space that is isomorphic to the spectrum of a ring[14] (it being understood that it is commutative, since we do commutative Algebra). From there one defines a *scheme* to be a locally ringed space that locally looks like an affine scheme or, equivalently, the spectrum of a ring. To be more precise, a locally ringed space $(X, \mathcal{O}_X)$ is a scheme if $\forall p \in X$, $\exists U$ open in X containing p such that $(U, \mathcal{O}_X|_U) \cong \operatorname{Spec} R_U$ for some commutative ring R_U. Here we use the following notation: if $\iota : U \hookrightarrow X$ is the inclusion map and $\mathcal{F}$ is a sheaf on X, then one calls the inverse image sheaf $\iota^{-1}\mathcal{F}$ on U the restriction of $\mathcal{F}$ to U, denoted by $\mathcal{F}|_U$ for simplicity. Morphisms of schemes $X \to Y$ are morphisms $X \to Y$ in the category of locally ringed spaces. We denote by Sch the category of schemes.

Observe that Spec is a contravariant functor $\operatorname{Spec} : \mathrm{CRing}^{\mathrm{op}} \to \mathrm{Sch}$.

In what follows, we will quite often define morphisms of schemes over a base scheme. We say[15] that an S-scheme X, or a scheme over a fixed base scheme S, is just given by a morphism $X \to S$, and morphisms of S-schemes are morphisms of schemes $X \to Y$ over S that make the following diagram commute:

$$\begin{array}{ccc} X & \longrightarrow & Y \\ & \searrow \quad \swarrow & \\ & S. & \end{array}$$

We denote by Sch_S the category of schemes over S. Observe that $\operatorname{Spec} \mathbb{Z}$ is the final object in the category Sch, and thus $\mathrm{Sch}_{\operatorname{Spec} \mathbb{Z}} = \mathrm{Sch}$. When we talk of a scheme over an algebraically closed field k, we really mean an object of the form $X \to \operatorname{Spec} k$.

Armed with those concepts, one can relate varieties over an algebraically closed field k and schemes over k. It turns out[16] that we have a fully faithful functor from the

11 Def. II.2.- [39].

12 Prop. II.2.3 [39].

13 Prop. II.2.2 [39].

14 Def. II.2.- [39].

15 Def. II.2.- [39].

16 Prop. II.2.6 [39].

category of varieties over k to the category of schemes over k, thereby showing that schemes are a generalization of varieties.

One construct on schemes we will use quite a bit, albeit in a general setting, is that of a *fiber product*,[17] the categorical notion of intersection. That we can do so is not surprising as the theory of schemes can be formalized using Category Theory, as we will see later in relative Algebraic Geometry. Fix a base scheme S and consider $X, Y \in \mathrm{Sch}_S$. We simply define the fiber product of X and Y over S to be the pullback $X \times_S Y$ in the category of schemes over S. If $S = \mathrm{Spec}\,\mathbb{Z}$, then we simply write $X \times Y$ for this fiber product. That we can form a pullback in the category of schemes is a theorem, namely Theorem II.3.3 of [39]. Observe that if $X = \mathrm{Spec}\,A$, $Y = \mathrm{Spec}\,B$, and $S = \mathrm{Spec}\,R$, then we have $X \times_S Y = \mathrm{Spec}\,A \times_{\mathrm{Spec}\,R} \mathrm{Spec}\,B = \mathrm{Spec}\,(A \otimes_R B)$.

Once this is defined, one may want to change base schemes, since schemes over S have to be put in perspective relative to schemes over other schemes S', or in other terms one ought to have objects defined in relation to other such objects. Once $X \times_S Y$ is defined, suppose we have a morphism of schemes $S' \to S$. Then $S' \times_S (X \times_S Y)$ is now an object of $\mathrm{Sch}_{S'}$.

Now that we have defined fiber products of schemes, fibers come next. To do this, we need a definition: for a scheme X, a point $x \in X$, and the local ring $\mathcal{O}_x$ at x with maximal ideal $\mathfrak{m}_x$, we call $k(x) := \mathcal{O}_x/\mathfrak{m}_x$ the *residue field of X at x*.[18] Then for a field κ, a morphism $\mathrm{Spec}\,\kappa \to X$ is given by a choice of a point $x \in X$ and an inclusion $k(x) \hookrightarrow \kappa$. In particular, for $\kappa = k(x)$, to $k(x) = k(x)$ we have an associated morphism $\mathrm{Spec}\,k(x) \to X$. We apply this to the notion of fibers: for a morphism $X \to Y$ of schemes, $y \in Y$, and its residue field $k(y)$, we have a morphism $\mathrm{Spec}\,k(y) \to Y$ as just pointed out, so we can form the fiber product

$$X \times_Y \mathrm{Spec}\,k(y) := X_y$$

which we call the fiber of $X \to Y$ over y. This is the correct generalization of the usual notion of fiber $X \times_Y \{y\} = X_y$ in Topology, since a point y of a scheme Y corresponds to a morphism $\mathrm{Spec}\,k(y) \to Y$ as explained above, and hence $X \times_Y \mathrm{Spec}\,k(y)$ does correspond to a fiber of X at y for schemes.

With the notion of sheaves of modules that we have seen above and the notion of schemes just introduced, one can discuss quasi-coherent sheaves. For a commutative ring R and a left R-module M, one can define a sheaf of modules $\tilde{M}$[19] on the topological space $|\mathrm{Spec}\,R|$ in exactly the same way that we defined a structure sheaf $\mathcal{O}$ on $|\mathrm{Spec}\,R|$. First, if $\mathfrak{p} \subset R$ is a prime ideal, M being a R-module, then the localization $M_{\mathfrak{p}}$ is well-defined. Then in defining $\tilde{M}$ objectwise, sections will be of the form m/f with $m \in M$

17 Def. II.3.- [39].

18 Ex. II.2.7 [39].

19 Def. II.5.- [39].

and $f \in R$, whereas for $\mathcal{O}$, sections were of the form r/f with $r, f \in R$. To be precise, for $U \in \mathcal{U}(|\mathrm{Spec}\, R|)$, let $\tilde{M}(U)$ be the set of functions $s : U \to \coprod_{\mathfrak{p} \in U} M_{\mathfrak{p}}$ in such a manner that $s(\mathfrak{p}) \in M_{\mathfrak{p}}$, and locally, one can find a neighborhood V of $\mathfrak{p}$ in U such that over U, there are $m \in M$ and $f \in R$ such that for all $\mathfrak{q} \in V, f \notin \mathfrak{q}$ and $s(\mathfrak{q}) = m/f$. This defines a sheaf $\tilde{M}$ on $|\mathrm{Spec}\, R|$. As one would expect, $\tilde{M}$ is an $\mathcal{O}_X$-module.[20] We also have results that held true in form for $\mathcal{O}$, that is, $(\tilde{M})_{\mathfrak{p}} \cong M_{\mathfrak{p}}$ and $\Gamma(|\mathrm{Spec}\, R|, \tilde{M}) = M$. Note also that by construction, if $M = R$, then $\tilde{M} = \mathcal{O}_R$ is the structure sheaf of $\mathrm{Spec}\, R$.

Now for a scheme $(X, \mathcal{O}_X)$, a sheaf $\mathcal{F} \in \mathrm{Mod}_{\mathcal{O}_X}$ is said to be *quasi-coherent*[21] if we have an open covering $X = \bigcup U_i$ where each $U_i = |\mathrm{Spec}\, A_i|$ is affine, and for each i, one can find some $M_i \in \mathrm{Mod}_{A_i}$ such that $\mathcal{F}|_{U_i} \cong \tilde{M}_i$. If further those modules are finitely generated, then we say $\mathcal{F}$ is a *coherent sheaf*. We denote by $\mathrm{QCoh}(X)$ the category of quasi-coherent sheaves on a scheme X. As an example, observe that the structure sheaf of a scheme is a quasi-coherent sheaf.[22]

Note however that it is really quasi-coherent sheaves we use most often, since to use coherent sheaves, one wants the schemes to be Noetherian. By definition quasi-coherent sheaves are sheaves that locally are constructed from modules over rings in the same fashion that structure sheaves are constructed. They can be regarded as a generalization of the notion of vector bundle, a pervasive object in Theoretical Physics. Since most of our applications are derived from physical phenomena, quasi-coherent sheaves ought to be considered, aside from the fact that they are a robust tool in Sheaf Theory. Quasi-coherent sheaves are especially pertinent since vector bundles are not very well-behaved under some operations that we routinely perform on schemes, and later on stacks. Quasi-coherent sheaves however offer a more "stable", optimal generalization of the notion of vector bundle.

There is one type of morphism that one uses quite a bit: *flat morphisms*. First recall that $M \in \mathrm{Mod}_R$ is said to be flat if the functor $M \otimes_R -$ is exact. We apply this to morphisms of commutative rings: if $A \to B$ is a morphism of commutative rings, then we can regard B as an A-module, and we say this morphism is flat if $B \in \mathrm{Mod}_A$ is flat. We can generalize that to sheaves in a pointwise fashion; namely, suppose we have a morphism of schemes $f : X \to Y$ and we have a sheaf of $\mathcal{O}_X$-modules $\mathcal{F}$ on X. We have an induced map in $\mathrm{Sh}(Y)$ $f^{\#} : \mathcal{O}_Y \to f_* \mathcal{O}_X$ that pointwise reads as follows: $f_x^{\#} : \mathcal{O}_{Y, f(x)} \to \mathcal{O}_{X,x}$. Letting $y = f(x)$, we can therefore regard $\mathcal{F}_x$ as an $\mathcal{O}_{Y,y}$-module. If $f_x^{\#}$ is flat for all $x \in X$, then we say $\mathcal{F}$ is flat over Y. In particular, if $\mathcal{F} = \mathcal{O}_X$ is flat, then we say f is a flat morphism.[23] It then follows from Proposition III.9.2 of [39] that $A \to B$ is a flat morphism of commutative rings iff $\mathrm{Spec}\, B \to \mathrm{Spec}\, A$ is a flat morphism of affine schemes.

---

20 Prop. II.5.1 [39].

21 Def. II.5.- [39].

22 Ex. II.5.2.1 [39].

23 Def. III.9.- [39].

In [39], it is observed that the fibers of a morphism of schemes $X \to Y$ essentially give us a family of schemes $\{X_y\}$ parameterized by Y. From a deformation theoretic perspective, one may then want to study how that family varies and in what fashion. For instance, suppose one has a scheme X_0 over a field k, and consider a morphism $X \to Y$ of schemes such that there is $y \in Y$ for which $k(y) \cong k$ and $X_y \cong X_0$. Then the other fibers of X over Y can be regarded as deformations of X_0. Now to have this family vary smoothly in Y, we need to take $X \to Y$ to be a flat morphism. This is one example among many, but there are numerous situations where flatness produces nice results.

There is also the notion of étale morphism, which is pervasive in Algebraic Geometry. However, the machinery needed to introduce such a class of morphisms dwarfs the interest of introducing this concept at this point. Our objective throughout is to introduce themes that are contextually self-evident and that will later be generalized to higher objects. One would not find it very illuminating at this point to introduce étale morphisms of schemes. It is better to introduce them later when they crop up.

2 Simplicial sets

As a first motivation for the notion of simplicial sets, observe that if X is a topological space with the homotopy type of a CW complex, then one has a homotopy equivalence $|\mathrm{Sing}_*(X)| \to X$ with $\mathrm{Sing}_*(X) \in \mathrm{Set}_\Delta$. Thus, homotopy-theoretically, it makes no difference whether we work with topological spaces or simplicial sets, and those latter are much more manageable.

Let $\mathbf{\Delta}$ denote the category of combinatorial simplices, whose objects are linearly ordered sets $[n] = \{0, 1, \dots, n\}$ and whose morphisms are nonstrictly order-preserving maps. One defines a *simplicial set* to be a functor $\mathbf{\Delta}^{\mathrm{op}} \to \mathrm{Set}$. A working definition goes as follows: a simplicial set K consists of sets K_n for $n \geq 0$ with a map $f^* : K_n \to K_m$ for each order-preserving map $f : [m] \to [n]$, with composition defined as follows: if $g : [n] \to [p]$ with a corresponding induced map $g^* : K_p \to K_n$, then one has $(g \circ f)^* = f^* \circ g^*$. In a functorial treatment, we have $K_n = K([n])$.

Among such maps, we distinguish the following: consider $p_j : [n-1] \to [n]$ defined for $0 \leq j \leq n$ by $p_j(i) = i$ if $i < j$ and $i+1$ otherwise. Then $p_j^* = d_j : K_n \to K_{n-1}$ is called the jth *face map*. If now $q_j : [n+1] \to [n]$ is defined by $q_j(i) = i$ if $i \leq j$ and $i-1$ otherwise, then $q_j^* = s_j : K_n \to K_{n+1}$ is called the jth *degeneracy map*.

The face and degeneracy maps satisfy the simplicial identities [28]

$$
\begin{aligned}
d_i d_j &= d_{j-1} d_i \quad \text{if } i < j, \\
d_i s_j &= s_{j-1} d_i \quad \text{if } i < j, \\
d_j s_j &= d_{j+1} s_j = \mathrm{id}, \\
d_i s_j &= s_j d_{i-1} \quad \text{if } i > j+1, \\
s_i s_j &= s_{j+1} s_i \quad \text{if } i \leq j.
\end{aligned}
$$

We denote by Set_Δ the category of simplicial sets. One pervasive simplicial set is the following. Define

$$
\begin{aligned}
\Delta^n : \mathbf{\Delta}^{\mathrm{op}} &\to \mathrm{Set}, \\
[p] &\mapsto \mathrm{Hom}([p], [n])
\end{aligned}
$$

that is, $\Delta^n = \mathrm{Hom}(-, [n]) \in \mathrm{Set}_\Delta$. Observe that if $K \in \mathrm{Set}_\Delta$, then $K_n \cong \mathrm{Hom}_{\mathrm{Set}_\Delta}(\Delta^n, K)$ for all $n \geq 0$.

For $0 \leq j \leq n$, one defines the *jth horn* $\Lambda_j^n \subsetneq \Delta^n$ to be the simplicial set Λ_j^n such that $\sigma \in (\Lambda_j^n)_m \subset \mathrm{Hom}([m], [n])$ is given by an order-preserving map such that $\{j\} \cup \sigma[m] \neq [n]$.

Another type of simplicial set that we will use extensively is that of a *Kan complex*.[1] $K \in \mathrm{Set}_\Delta$ is said to be a Kan complex if for all $0 \leq i \leq n$, any map $\Lambda_i^n \to K$ can be extended

1 Def. 1.1.2.1 [52].

https://doi.org/10.1515/9783111334073-002

to a map $\Delta^n \to K$. One denotes by Kan the full subcategory of Set_Δ spanned by Kan complexes, which we regard as a simplicial category (see chapter on model categories for a complete definition). By taking its simplicial nerve one gets the *∞-category of spaces*: $\mathcal{S} = \mathrm{N}(\mathrm{Kan})$.[2] The simplicial nerve is defined as follows: let J be a finite nonempty linearly ordered set. One defines a simplicial category $\mathfrak{C}[\Delta^J]$ as follows.[3] Its objects are those of J, and its morphisms are defined like so: $\mathrm{Map}(i,j) = \mathrm{N}(P_{ij})$ if $i \leq j$, $\emptyset$ otherwise, where $\mathrm{N}(P_{ij})$ is the nerve of the partially ordered set $P_{ij} = \{I \subseteq J : i,j \in I \text{ and } \forall\, k \in I, i \leq k \leq j\}$. For the composition, one takes the obvious map induced by the unions as follows. For $i_0 \leq i_1 \leq \cdots \leq i_n$, one has a product

$$P_{i_0,i_1} \times \cdots \times P_{i_{n-1},i_n} \to P_{i_0,i_n},$$
$$(I_1,\ldots,I_n) \mapsto I_1 \cup \cdots \cup I_n.$$

Then one defines the *simplicial nerve*[4] $\mathrm{N}(\mathcal{C})$ of a simplicial category $\mathcal{C}$ as the simplicial set satisfying

$$\mathrm{Hom}_{\mathrm{Set}_\Delta}(\Delta^n, \mathrm{N}(\mathcal{C})) = \mathrm{Hom}_{\mathrm{Cat}_\Delta}(\mathfrak{C}[\Delta^n], \mathcal{C})$$

where Cat_Δ denotes the category of simplicial categories.

A morphism $p : X \to S$ in Set_Δ is said to be an *inner fibration*[5] if it has the right lifting property(henceforward abbreviated as RLP) with respect to all horn inclusions $\Lambda^n_i \subsetneq \Delta^n$, $0 < i < n$.

A *trivial Kan fibration* is a trivial fibration, and a Kan fibration, where a morphism $p : X \to S$ in Set_Δ is called a *Kan fibration* if it has the RLP with respect to all horn inclusions, and it is said to be a *trivial fibration* if it has the RLP with respect to every inclusion $\partial\Delta^n \subseteq \Delta^n$.

One can put a model structure on Set_Δ called the *Joyal model structure*.[6] First, observe that the functor $\mathfrak{C} : \boldsymbol{\Delta} \to \mathrm{Cat}_\Delta$, $[n] \mapsto \mathfrak{C}[\Delta^n]$ has a unique extension to a functor $\mathfrak{C} : \mathrm{Set}_\Delta \to \mathrm{Cat}_\Delta$ that preserves colimits. We have a left proper, combinatorial model structure (see chapter on model categories) on Set_Δ for which a map $p : K \to S$ in Set_Δ is a cofibration if and only if it is a monomorphism, and it is a categorical equivalence if and only if the induced map of simplicial categories $\mathfrak{C}[K] \to \mathfrak{C}[S]$ is an equivalence in Cat_Δ, which by definition[7] is a map such that the induced map on homotopy categories is an equivalence. Recall how this works: one first defines the *homotopy category*

2 Def. 1.2.16.1 [52].

3 Def. 1.1.5.1. [52].

4 Def. 1.1.5.5 [52].

5 Def. 2.0.0.3 [52].

6 Sec. 2.2.5 [52].

7 Def. 1.1.4.4 [52].

of spaces[8] as the homotopy category $\mathcal{H} = h\mathcal{C}$ where $\mathcal{C}$ is the category of CW complexes and continuous maps between them, where the mapping spaces are endowed with the compact open topology. Independently, let $\mathcal{CG}$ be the category of compactly generated, weakly Hausdorff, topological spaces. For $X \in \mathcal{CG}$, one can find a CW complex Y, unique up to homotopy equivalence, along with a weak homotopy equivalence $Y \to X$. In other terms, we have a well-defined functor from $\mathcal{CG}$ to $\mathcal{H}$ that with X associates $[X] \ni Y$. Additionally, we have the geometric realization functor $\mathrm{Set}_\Delta \to \mathcal{CG}$.

Composing both functors, one gets a functor $\mathrm{Set}_\Delta \to \mathcal{H}$, and applying that functor to all the mapping spaces of a simplicial category $\mathcal{C}$, one gets an $\mathcal{H}$-enriched category $h\mathcal{C}$ called the homotopy category of $\mathcal{C}$.

One simple operation on simplicial sets we need is the join, which we will use quite a bit for the sake of extensions. It is defined like so: for $S, S' \in \mathrm{Set}_\Delta$, one defines the *join*[9] of S and S', denoted by $S \star S'$, as follows. For any nonempty finite linearly ordered set J,

$$S \star S'(J) = \coprod_{J = I \coprod I'} S(I) \times S'(I')$$

where one considers those decompositions of $J = I \coprod I'$ such that $i < i'$ for $i \in I$ and $i' \in I'$. Armed with this definition, we can define overcategories: by virtue of Proposition 1.2.9.2 of [52], for $S, K \in \mathrm{Set}_\Delta$ and a map $p : K \to S$, there exists some $S_{/p} \in \mathrm{Set}_\Delta$ with the following universal property:

$$\mathrm{Hom}_{\mathrm{Set}_\Delta}(Y, S_{/p}) = \mathrm{Hom}_{/p}(Y \star K, S)$$

where on the right-hand side one considers those morphisms whose restriction to K are p. In particular, if $p = x : [0] \to S$, then one denotes $S_{/p}$ by S_x.

Let $p : X \to S$ be an inner fibration in Set_Δ, and let $f : x \to y$ be an edge of X. One says f is *p-Cartesian*[10] if the induced map

$$X_{/f} \to X_{/y} \times_{S_{/py}} S_{/pf}$$

is a trivial Kan fibration. Later we will need the notion of being *p-coCartesian*: p is said to be coCartesian if it is Cartesian with respect to $p^{\mathrm{op}} : X^{\mathrm{op}} \to S^{\mathrm{op}}$. In other terms, for all $z \in X$, we have a homotopy pullback square

$$\begin{array}{ccc} \mathrm{Map}_X(y, z) & \longrightarrow & \mathrm{Map}_X(x, z) \\ \downarrow & & \downarrow \\ \mathrm{Map}_S(py, pz) & \longrightarrow & \mathrm{Map}_S(px, pz). \end{array}$$

8 Ex. 1.1.3.3. [52].

9 Section 1.2.8 [52].

10 Def. 2.4.1.1. [52].

The theory of p-Cartesian edges and Cartesian fibrations is motivated by the need to have functorial lifts in $X \in \mathrm{Set}_\Delta$ of maps in $S \in \mathrm{Set}_\Delta$ in the event that we have inner fibrations $p : X \to S$ for example. The fibers of such fibrations are ∞-categories, and sometimes one is interested in pullback lifts that are functors. The proper way to achieve this is to first define p-Cartesian edges, and if this can be done consistently, one can then define Cartesian fibrations, which are evidence of the existence of our desired functors.

A map $p : X \to S$ in Set_Δ is a *Cartesian fibration*[11] if it is an inner fibration, and for any edge $e : x \to y$ of S and any vertex $\tilde{y}$ of X over y, there is a p-coCartesian edge $\tilde{e} : \tilde{x} \to \tilde{y}$ of X that projects down to e. One says p is a *coCartesian fibration* if $p^{\mathrm{op}} : X^{\mathrm{op}} \to S^{\mathrm{op}}$ is a Cartesian fibration.

11 Def. 2.4.2.1 [52].

3 ∞-categories

3.1 Motivation

Before we delve into ∞-categories proper, we would like to motivate the introduction of such a notion. Suppose we start with a topological space X. Homotopically we can think of continuous paths between its points. But then one may consider paths between paths. As soon as we do that, we are led to defining first-order paths as 1-morphisms and paths between those as 2-morphisms. We can pursue in this fashion and therefore define higher order n-morphisms. If the points of X are regarded as objects, then what we have obtained is an n-category. Now the question is how far should we go? It turns out that if we do this indefinitely, then we end up with an ∞-category called the fundamental ∞-groupoid $\Pi_\infty X$ of X, an object we will introduce fully in the next section, which captures the whole homotopy type of X, as recalled in [52]. We can regard $\Pi_\infty X$ as a categorification of X for homotopical purposes. Observe that if all the intermediate n-categories $\Pi_{\leq n} X$ of points of X and k-morphisms between them for $k \leq n$ are higher categories, then in contrast X is not a category. Perhaps one can replace this by its associated locale to put all those objects on a same footing, and this is something we will indeed do below.

For now, we have obtained an obvious generalization of a classical object such as a topological space X, where all morphisms are really invertible. As pointed out in [52], invertibility is a feature we would like to have, at least at a higher order, since higher generalizations of classical notions do require that such higher morphisms be invertible. One then refers to higher categories for which higher k-morphisms for $k > n$ are invertible as (∞, n)-categories. The fundamental groupoid $\Pi_\infty X$ is an example of an $(\infty, 0)$-category. Amongst all such higher categories, one may ask which one is more central than the others?

Again as pointed out in [52], from the perspective of $(\infty, 1)$-category theory, there is an inductive construction of (∞, n)-categories using enriched category theory, which points to the fact that all we really need is a notion of $(\infty, 1)$-categories to start with.

Before formally defining $(\infty, 1)$-categories, we go back to the fundamental ∞-groupoid $\Pi_\infty X$ of a topological space and see how we can ease our way into $(\infty, 1)$-categories from there. By construction, $\Pi_\infty X$ is a higher category that encapsulates the homotopy type of the space X as pointed above, so it has a categorical flavor, as well as a homotopical one. Let us take the intermediate step of just working homotopically for the moment. Up to a weak homotopy equivalence, we can work with the singular complex $\mathrm{Sing}X$ of X. That we have such an equivalence is by virtue of the fact that one has inverse equivalences between the homotopy category of CW complexes and the homotopy category of Kan complexes, a special type of simplicial set that satisfies the

https://doi.org/10.1515/9783111334073-003

following extension property:[1] $K \in \mathrm{Set}_\Delta$ is a Kan complex if for all $0 \le i \le n$, one has a diagonal map as shown below making such a diagram commutative:

$$\begin{array}{ccc} \Lambda^n_i & \longrightarrow & K \\ \downarrow & \nearrow & \\ \Delta^n. & & \end{array}$$

It turns out that if $X \in \mathrm{Top}$, then Sing$X$ is a Kan complex. So homotopically we can work with Sing$X$ in lieu of just X. This is not categorical yet. Now by virtue of Proposition 1.1.2.2 of [52], those simplicial sets K such that $K \simeq N(\mathcal{C})$, the nerve of a small category $\mathcal{C}$, do satisfy an extension property:

$$\begin{array}{ccc} \Lambda^n_i & \longrightarrow & K \\ \downarrow & \nearrow & \\ \Delta^n & & \end{array}$$

for all $0 < i < n$, for a unique diagonal map, and vice-versa. The nerve being a simplicial deconstruction of a category, such simplicial sets consequently appear as having a categorical flavor. But Kan complexes satisfy this extension property. Thus starting from a topological space X, homotopically one can work with Sing$X$, a Kan complex, so it satisfies the above extension property, hence it is the nerve of a category, and is therefore regarded as having a categorical flavor. An $(\infty, 1)$-category is a slightly more general type of such simplicial sets, where one just drops the uniqueness condition.

An obvious question is why does that work so well? Among all the different notions of higher categories, why this one in particular? To come back to that unique extension property above, we mentioned that taking a nerve provides a simplicial, categorical deconstruction of a given category we started with. It turns out that for any category $\mathcal{C}$, its nerve $N(\mathcal{C})$ is an $(\infty, 1)$-category by definition. Thus taking the nerve allows us to move from classical category theory into the realm of $(\infty, 1)$-category theory. If Kan denotes the category of Kan complexes, which is still a classical notion, then taking its nerve gives us $\mathcal{S} = N(\mathrm{Kan})$, the $(\infty, 1)$-category of spaces. Quite often, we will focus on manifestations and perceptions of given objects. One dynamic manifestation in particular is given by the Yoneda lemma, which says that for any category $\mathcal{C}$, the map $j : \mathcal{C} \to \mathrm{Set}^{\mathcal{C}^{\mathrm{op}}}$ that with any object C of $\mathcal{C}$ associates $\mathrm{Hom}_{\mathcal{C}}(-, C)$ is fully faithful. Now replace $\mathcal{C}$ by $K = N(\mathcal{C})$, and replace Set by $\mathcal{S}$. In the formalism of $(\infty, 1)$-categories, $\mathcal{P}(K) = \mathrm{Fun}(K, \mathcal{S})$ is also an $(\infty, 1)$-category, and thus we have a very stable theory at our disposal. What is important is that according to Proposition 5.1.3.1 of [52], the resulting functor $j : K \to \mathcal{P}(K)$ is

1 Def. 1.1.2.1 of [52].

fully faithful as well and therefore provides us with a natural higher categorical generalization of the classical Yoneda lemma.

Another motivation for working with $(\infty,1)$-categories, albeit a roundabout way of motivating them, is as follows. Lurie [52] defines n-topoi, a higher generalization of the classical notion of Grothendieck topos. We can see that this is the right notion since by Remark 6.4.1.3 of [52], an $(\infty,1)$-category $\mathcal{X}$ is a 1-topos iff it is equivalent to the nerve of an ordinary Grothendieck topos, that is, it is its simplicial deconstruction, its $(\infty,1)$-categorical version as it were. Of interest also is the fact that an $(\infty,1)$-category $\mathcal{X}$ is a 0-topos iff it is equivalent to the nerve of a locale, by Corollary 6.4.2.6. Thus it seems a tower of n-topos constructed via truncation functors $\tau_{\leq n}$ would provide an $(\infty,1)$-categorical way of connecting a space, or a locale, with all its higher manifestations. It turns out from the perspective of an ∞-topos $\mathcal{X}$ in this formalism, truncations $\tau_{\leq n}\mathcal{X}$ thereof appear as Postnikov sections of the former, thereby emphasizing the holistic character of such a theory. More specifically, once one fleshes out the definition of an ∞-topos, it is defined in [52] as a localization of a category of $\mathcal{S}$-valued sheaves on a higher category $\mathcal{C}$, and the proper higher categorical notion we need here is that of an $(\infty,1)$-category.

3.2 Foundations

At a very elementary level, categories are used to formalize classical mathematical constructs. Very quickly, one encounters difficulties when classifying objects up to homotopies, rather than isomorphisms. At this point, higher categories are needed. ∞-categories is but one example of higher categories.

Before we formally define ∞-categories, perhaps we can give one very simple example of an ∞-category:[2] the fundamental groupoid $\Pi_\infty X$ of a topological space X, with objects the points of that space, a morphism (or 1-morphism) from a point x to a point y of X being a continuous path from x to y in X, a 2-morphism is a homotopy between paths, etc., an $(n+1)$-morphism is a homotopy between n-morphisms. This type of ∞-category is called a $(\infty,0)$-category (or ∞-groupoid), where one defines an (∞,n)-category to be an ∞-category for which k-morphisms are invertible if $k > n$. In the rest of this work, by an ∞-category we will implicitly mean an $(\infty,1)$-category.

A preliminary definition of an ∞-category would read like so: an ∞-category $\mathcal{C}$ consists of a collection of objects and ∞-groupoids $\mathrm{Map}_{\mathcal{C}}(X,Y)$ between any two objects $X, Y \in \mathrm{Ob}\,\mathcal{C}$, subject to an associative composition law. If we regard ∞-groupoids as topological spaces, then one can further refine this picture by considering weakly Hausdorff, compactly generated topological spaces for mapping objects, and this would yield the definition of a *topological category*. However, though simple enough to understand,

2 Ex. 1.1.1.4 [52].

this is not a working definition. There are various ways to define an ∞-category, and the definition that we will use is the following: an *∞-category* is an object $K \in \mathrm{Set}_\Delta$ such that for all $0 \le i \le n$, any map $f : \Lambda^n_i \to K$ can be extended to a map $\bar{f} : \Delta^n \to K$.

Before we pursue studying ∞-categories further, one can mention an issue one encounters when working with homotopy commutative diagrams, as expounded in Section 1.2.6 of [52]. The proper way to address this issue is to work with homotopy coherent diagrams instead; for an ordinary category $\mathcal{I}$, $\mathcal{C} \in \mathrm{Cat}_\infty$ (the ∞-category of ∞-categories), a homotopy coherent diagram in $\mathcal{C}$ is a functor $F : \mathcal{I} \to h\mathcal{C}$ along with all the information needed to obtain a lift of F to a functor $\tilde{F} : \mathcal{I} \to \mathcal{C}$. However, this information is unwieldy. This situation illustrates one of our motivations for considering spectralized objects in the present work. Clearly, here the problem stems from using an ordinary category as our index category. A quick fix should be replacing it with an ∞-category instead. That is exactly what Lurie does: he defines a *homotopy coherent diagram in an ∞-category $\mathcal{C}$* to be a map $N(\mathcal{I}) \to \mathcal{C}$ in Cat_∞ (the nerve $N(\mathcal{I}) \in \mathrm{Cat}_\infty$, as we will argue later), where the higher coherence is given by the restriction of F to higher simplices of $N(\mathcal{I})$, whereas its restriction to lower-degree simplices can be used to induce the map into $h\mathcal{C}$. Thus by doing an evident generalization whereby all objects are of a same nature, namely ∞-categories in this context, one easily solves a problem, which otherwise would have been intractable.

One defines objects and morphisms in an ∞-category the way they are defined for simplicial sets: for $S \in \mathrm{Set}_\Delta$, objects are elements of $\mathrm{Ob}\, S = \mathrm{Mor}\,(\Delta^0, S)$ and are referred to as vertices, and morphisms are elements of $\mathrm{Mor}\, S = \mathrm{Mor}\,(\Delta^1, S)$ and are referred to as edges. One says a morphism $f : X \to Y$ in $\mathcal{C} \in \mathrm{Cat}_\infty$ is an *equivalence* if it is an isomorphism in $h\mathcal{C} = h\mathfrak{C}[\mathcal{C}]$. Here f corresponds to $f : \Delta^1 \to \mathcal{C}$ with $f(0) = X$ and $f(1) = Y$, and we have used the fact that for a simplicial category $\mathcal{C}$, its homotopy category $h\mathcal{C}$ is obtained by applying the functor $\mathrm{Set}_\Delta \to \mathcal{H}$ to all its mapping spaces, something we have used above while defining the Joyal model structure.

We have already seen what a Kan complex is when we covered simplicial sets. A Kan complex is an example of an ∞-category.[3] Another important example, which will implicitly be used throughout, is the following: the nerve of an ordinary category is an ∞-category. In this manner, we will identify an ordinary category with its nerve, thereby viewing it as an ∞-category.[4]

We have briefly mentioned the space of morphisms $\mathrm{Map}_\mathcal{C}(X, Y)$ between two objects X and Y of a topological category $\mathcal{C}$. For an ∞-category, we have a similar concept, by way of simplicial sets:[5] if $S \in \mathrm{Set}_\Delta$ and $x, y \in S$, define

$$\mathrm{Map}_S(x,y) := \mathrm{Map}_{hS}(x,y) = \mathrm{Map}_{h\mathfrak{C}[S]}(x,y) = [\,|\mathrm{Map}_{\mathfrak{C}[S]}(x,y)|\,]$$

3 Ex. 1.1.2.5 [52].

4 Ex. 1.1.2.6 [52].

5 Def. 1.2.2.1 [52].

where in the second equality, we make use of Definition 1.1.5.14 of [52]. Observe that this is an object of $\mathcal{H}$, the homotopy category of spaces. Already the first equality showed that our definition would be at the level of homotopy categories. If one wants a representative for $\mathrm{Map}_S(x,y)$, then one has to define the space of *right morphisms*, but we will not need such a notion. The interested reader can find more details in Section 1.2.2 of [52].

Now that we have defined ∞-categories, we then define *functors of ∞-categories*; since ∞-categories are in particular simplicial sets, we just define functors between ∞-categories $\mathcal{C}$ and $\mathcal{D}$ to be maps on their underlying simplicial sets. For $S, \mathcal{C} \in \mathrm{Set}_\Delta$, we use the notation[6] $\mathrm{Fun}(S,\mathcal{C})$ to denote the simplicial set of mappings $\mathrm{Map}_{\mathrm{Set}_\Delta}(S,\mathcal{C})$ in the event that $\mathcal{C}$ is an ∞-category. It is also an ∞-category by Proposition 1.2.7.3 of [52]. Regarding functors, one says a functor $F : \mathcal{C} \to \mathcal{D}$ between ∞-categories is *essentially surjective*[7] if the induced functor on the respective homotopy categories is essentially surjective, and F is said to be fully faithful if hF is fully faithful. This can be generalized: one says F satisfies a property P if hF does. Finally, F is an equivalence iff it is fully faithful and essentially surjective.

To define a subcategory of an ∞-category, we use the same philosophy of defining things at the level of homotopy categories and lifting that up to the level of ∞-categories using a pullback. To be precise, if $\mathcal{C}$ is an ∞-category, $\mathcal{C}'$ is a subcategory of $h\mathcal{C}$, and $\mathcal{C}'_0$ is defined by the pullback

$$\begin{array}{ccc} \mathcal{C}'_0 & \longrightarrow & \mathcal{C} \\ \downarrow & & \downarrow \\ \mathcal{C}' & \longrightarrow & h\mathcal{C} \end{array}$$

then one says $\mathcal{C}'_0$ is a subcategory of $\mathcal{C}$, spanned by $\mathcal{C}'$. If the latter is a full subcategory of $h\mathcal{C}$, then one says $\mathcal{C}'_0$ is a full subcategory of $\mathcal{C}$.

One now defines the ∞-category of ∞-categories as follows: one first considers $\mathrm{Cat}^\Delta_\infty \in \mathrm{Cat}_\Delta$ as the category of small ∞-categories such that for small ∞-categories $\mathcal{C}$ and $\mathcal{D}$, $\mathrm{Map}_{\mathrm{Cat}^\Delta_\infty}(\mathcal{C},\mathcal{D})$ is the largest Kan complex in $\mathrm{Fun}(\mathcal{C},\mathcal{D})$. Then one defines the ∞-category Cat_∞ of ∞-categories as $\mathrm{Cat}_\infty = N(\mathrm{Cat}^\Delta_\infty)$, the simplicial nerve of $\mathrm{Cat}^\Delta_\infty$. In this structure, morphisms between ∞-categories are naturally given by functors.

We mentioned small ∞-categories. To be precise, if κ is a regular cardinal, a set S is κ-small if $\mathrm{Card}(S) < \kappa$. We generalize that to set-theoretic objects. For example, a category $\mathcal{C}$ is κ-small if $\mathrm{Ob}\,\mathcal{C}$ and $\mathrm{Mor}\,\mathcal{C}$ are κ-small. To distinguish between small and large objects, we fix a strongly inaccessible cardinal κ and say an object is small if it is κ-small. Objects that are not small are referred to as being large.

6 Not. 1.2.7.2 [52].

7 Def. 1.2.10.1 [52].

For later purposes, an ∞-category is said to be *presentable*[8] if it is accessible and admits small colimits. One can mention one application of the concept of presentability in the proof of the adjoint functor Theorem 5.5.2.9 of [52], which states that if $F : \mathcal{C} \to \mathcal{D}$ is a functor between presentable ∞-categories, then F has a right (abbr. R) adjoint iff it preserves small colimits, and it has a left (abbr. L) adjoint iff it is accessible and preserves small limits. For example, $\mathcal{S}$ is a presentable ∞-category, and so is $\mathrm{Pr}(K) = \mathrm{Fun}(K^{\mathrm{op}}, \mathcal{S})$ for some small simplicial set K.[9]

For a regular cardinal κ, an ∞-category $\mathcal{C}$ is κ-accessible if there exist $\mathcal{C}^0 \in \mathrm{Cat}_\infty$ and an equivalence $\mathrm{Ind}_\kappa(\mathcal{C}^0) \to \mathcal{C}$. One then says $\mathcal{C}$ is *accessible*[10] if it is κ-accessible for some κ. Accessible ∞-categories allow us to work with ∞-categories that are not necessarily small, but which are determined by small subcategories in such a manner that one does not run into set-theoretic difficulties. See discussion in Section 5.4 of [52]. Standard examples of accessible ∞-categories are the ∞-category $\mathcal{S}$ of spaces or even presheaf ∞-categories $\mathrm{Pr}(\mathcal{C})$ for an ∞-category $\mathcal{C}$.

For a small ∞-category $\mathcal{C}$ and a regular cardinal κ, $\mathrm{Ind}_\kappa(\mathcal{C})$ is the full subcategory of $\mathrm{Pr}(\mathcal{C})$ spanned by those functors $\mathcal{C}^{\mathrm{op}} \to \mathcal{S}$ that classify R fibrations $\tilde{\mathcal{C}} \to \mathcal{C}$ for some κ-filtered ∞-category $\tilde{\mathcal{C}}$. A map $p : K \to S$ in Set_Δ is a *right fibration*[11] if it has the RLP with respect to any inclusion $\Lambda^n_i \subseteq \Delta^n$ for $0 < i \leq n$.

For a regular cardinal κ and an ∞-category $\mathcal{C}$, $\mathcal{C}$ is κ*-filtered*[12] if for any κ-small $K \in \mathrm{Set}_\Delta$, for any map $f : K \to \mathcal{C}$, there exists a map $\overline{f} : K^{\triangleright} \to \mathcal{C}$ extending f, where $K^{\triangleright} = K \star \Delta^0$.

$\mathcal{C} \in \mathrm{Cat}_\infty$ is said to be *stable*[13] if it admits finite colimits, it has an object that is both initial and final, which we denote by 0, and the suspension functor $\Sigma X = 0 \coprod_X 0$ provides an equivalence of ∞-categories. Stable ∞-categories will be needed later in Spectral Algebraic Geometry.

3.3 ∞-topos

We now define the notion of ∞-topos. Those will replace topological spaces later when we define spectral Deligne–Mumford stacks, and for the following reason. If X is a topological space, then the ∞-category $\mathrm{Sh}(X)$ of sheaves on X is an ∞-topos. Additionally, if X is sober, then one can recover X from $\mathrm{Sh}(X)$. Thus they are interchangeable. This points to the fact that the theory of ∞-topos may be regarded as a generalization of the the-

8 Def. 5.5.0.1 [52].

9 Rmrk 5.5.3.7 [52].

10 Def. 5.4.2.1. [52].

11 Def. 2.0.0.3 [52].

12 Def. 5.3.1.7 [52].

13 Def. 0.2.3.5 [54].

ory of topological spaces. As a matter of fact, quite often in classical constructions that depend on a topological space X, one will replace this space by an ∞-topos. To see that we have such a generalization, one can define a whole spectrum of n-topoi, $0 \leq n \leq \infty$, as done in [52], the case $n = 0$ corresponding to the case of topological spaces. To be precise, Corollary 6.4.2.6 of [52] states that $\mathcal{X} \in \mathrm{Cat}_{\infty}$ is a 0-topos iff $\mathcal{X} \simeq N(\mathcal{V})$ where $\mathcal{V}$ is a locale.[14] Recall that if X is a topological space, then the collection $\mathcal{U}(X)$ of open sets of X is a locale.[15] Thus we can regard locales as generalizations of topological spaces and 0-topos as being equivalent to the nerve of locales. n-topoi are related in accordance to Proposition 6.4.5.7 of [52], which essentially shows that one has a localization functor from n-topoi to p-topoi for $p \leq n$. In particular, this is true for $n = \infty$, thereby presenting all the truncations of an ∞-topos as Postnikov sections thereof and a topological space as the ultimate one. Thus we can see ∞-topoi as generalizations of topological spaces or topological spaces as truncations of ∞-topoi.

An ∞-category $\mathcal{X}$ is an *∞-topos*[16] if there exist a small ∞-category $\mathcal{C}$ and an accessible L exact localization functor $\mathrm{Pr}(\mathcal{C}) \to \mathcal{X}$, where $\mathrm{Pr}(\mathcal{C}) = \mathrm{Fun}(\mathcal{C}^{\mathrm{op}}, \mathcal{S})$ is the ∞-category of presheaves on $\mathcal{C}$.

A functor between ∞-categories is a *localization*[17] if it has a fully faithful right adjoint.

If $\mathcal{C}$ is an accessible ∞-category, then a functor $F : \mathcal{C} \to \mathcal{C}'$ is said to be *accessible*[18] if it is κ-continuous for some regular cardinal κ. Note that if $\mathcal{C}$ is a small ∞-category, then $\mathrm{Pr}(\mathcal{C})$ in particular is accessible.

For an ∞-category $\mathcal{C}$ that admits small κ-filtered colimits, a functor $f : \mathcal{C} \to \mathcal{D}$ is said to be *κ-continuous*[19] if it preserves κ-filtered colimits.

Let $f : \mathcal{C} \to \mathcal{D}$ be a functor between ∞-categories, and let κ be a regular cardinal. One says f is κ-right exact[20] if for any fibration $\mathcal{D}' \to \mathcal{D}$ where $\mathcal{D}'$ is κ-filtered, $\mathcal{C}' = \mathcal{C} \times_{\mathcal{D}} \mathcal{D}'$ is also κ-filtered. One says f is R exact if it is ω-right exact. One has a dual notion of *left exactness*:[21] $f : \mathcal{C} \to \mathcal{D}$ is L exact if $f^{\mathrm{op}} : \mathcal{C}^{\mathrm{op}} \to \mathcal{D}^{\mathrm{op}}$ is R exact.

In the definition of ind-objects, to characterize accessibility, we mentioned a classifying map. The notion of classifying map is fairly technical, yet ubiquitous, so we cover it presently.

Before we discuss classifying maps, we have to discuss straightening and unstraightening functors in the unmarked case first, and then in the marked case.

---

14 Def. 6.4.2.3 [52].

15 Ex. 6.4.2.4 [52].

16 Def. 6.1.0.4 [52].

17 Def. 5.2.7.2 [52].

18 Def. 5.4.2.5 [52].

19 Def. 5.3.4.5 [52].

20 Def. 5.3.2.1 [52].

21 Rmrk. 5.3.2.10 [52].

Note that the reason why we invoked a classifying map for a right fibration in the definition of $\mathrm{Ind}_\kappa(\mathcal{C})$ is Proposition 3.3.2.5 of [52], which states that if $p : X \to S$ is a Cartesian fibration in Set_Δ, then it is a R fibration if and only if there exists $f : S \to \mathcal{S}^{\mathrm{op}}$ classifying p.

In a first time, let $S \in \mathrm{Set}_\Delta$, $\mathcal{C} \in \mathrm{Cat}_\Delta$, and $\phi : \mathfrak{C}[S] \to \mathcal{C}^{\mathrm{op}}$. Form a pushout

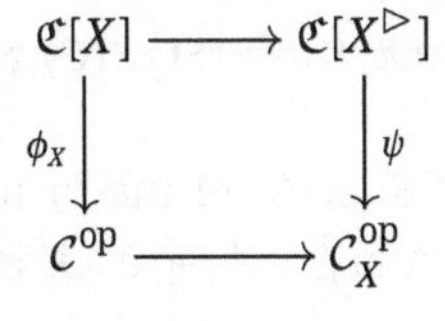

where $(X \to S) \rightsquigarrow (\mathfrak{C}[X] \to \mathfrak{C}[S] \xrightarrow{\phi} \mathcal{C}^{\mathrm{op}}) = \phi_X$. Observe that if c is a vertex of X, then $\phi_X(c) \in \mathcal{C}$, and we denote this point by C. One can identify c with a map $c : \Delta^0 \to X$, so that $c \star \mathrm{id}_{\Delta^0} : \Delta^1 \to X^{\triangleright}$, an edge of $X^{\triangleright}$, which under ψ maps to an edge $\tilde{c} : C \to *$ of $\mathcal{C}_X^{\mathrm{op}}$. Thus one defines the *straightening functor*[22] $\mathrm{St}_\phi : (\mathrm{Set}_\Delta)_{/S} \to (\mathrm{Set}_\Delta)^{\mathcal{C}}$ as follows: for $X \to S \in (\mathrm{Set}_\Delta)_{/S}$, $\mathrm{St}_\phi X : \mathcal{C} \to \mathrm{Set}_\Delta$ is defined by $\mathrm{St}_\phi X(C) = \mathrm{Map}_{\mathcal{C}_X^{\mathrm{op}}}(C, *)$, where $*$ is the image under ψ of the cone point of $X^{\triangleright}$.

For later purposes, if $X \to S$ and $\sigma \in \mathrm{Map}_{\mathcal{C}^{\mathrm{op}}}(C, D)_n$, then the latter induces $\sigma^* : \mathrm{St}_\phi X(D)_n \to \mathrm{St}_\phi X(C)_n$. Now if $f : c \to d$ is an edge of X, mapping under ϕ_X to an edge $F : C \to D$ of $\mathcal{C}^{\mathrm{op}}$:

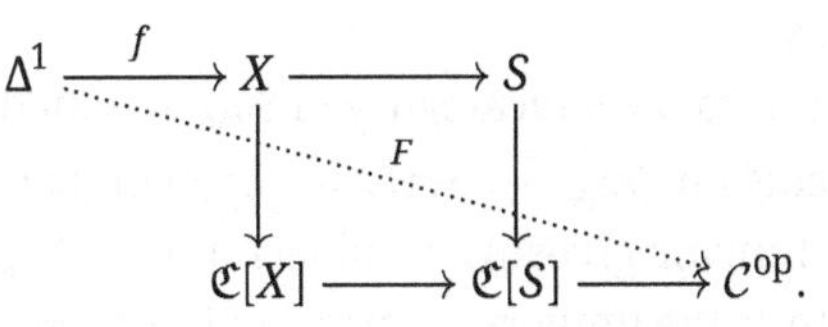

If one regards f as $f : \Delta^1 \to X$, then $f \star \mathrm{id}_{\Delta^0} : \Delta^2 \to X^{\triangleright}$ determines a map $\Delta^2 \to \mathcal{C}_X^{\mathrm{op}}$, a diagram

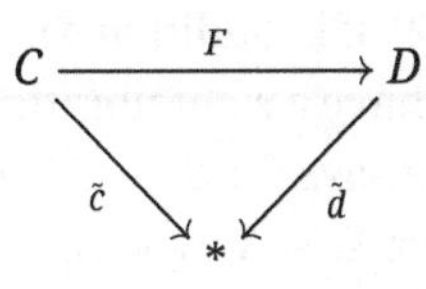

along with an edge $\tilde{c} \to \tilde{d} \circ F$ in $\mathrm{St}_\phi X(C)$.

One now defines a *marked simplicial set*[23] to be a pair $(X, \mathcal{E})$, with $X \in \mathrm{Set}_\Delta$ and a set $\mathcal{E}$ of edges of X containing all degenerate edges. A morphism $f : (X, \mathcal{E}) \to (X', \mathcal{E}')$ of

22 Sec. 2.2.1 [52].

23 Def. 3.1.0.1 [52].

marked simplicial sets is a map $f : X \to X'$ in Set_Δ such that $f(\mathcal{E}) \subseteq \mathcal{E}'$. One denotes by Set_Δ^+ the category of marked simplicial sets.[24]

Now for $S \in \mathrm{Set}_\Delta$, $\mathcal{C} \in \mathrm{Cat}_\Delta$, $\phi : \mathfrak{C}[S] \to \mathcal{C}^{\mathrm{op}}$, and $(X, \mathcal{E}) \in (\mathrm{Set}_\Delta^+)_{/S}$, one defines the straightening functor on marked simplicial sets as follows:

$$\begin{aligned}\mathrm{St}_\phi^+ : (\mathrm{Set}_\Delta^+)_{/S} &\to (\mathrm{Set}_\Delta^+)^{\mathcal{C}}, \\ (X, \mathcal{E}) &\mapsto (\mathrm{St}_\phi X(C), \mathcal{E}_\phi(C))\end{aligned}$$

where $\mathcal{E}_\phi(C)$ is the set of edges of $\mathrm{St}_\phi X(C)$ of the form $G^*\tilde{f}$ for $G \in \mathrm{Map}_{\mathcal{C}^{\mathrm{op}}}(C, D)_1$ and $f : d \to e$ in $\mathcal{E}$, with an induced map $\tilde{f} : \tilde{d} \to \tilde{e} \circ F$ in $\mathrm{St}_\phi X(D)_1$, $F : D \to E$, in such a manner that $G^*\tilde{f} \in \mathrm{St}_\phi X(C)_1$.

Then by Theorem 3.2.0.1 of [52], St_ϕ^+ has a right adjoint $\mathrm{Un}_\phi^+ : (\mathrm{Set}_\Delta^+)^{\mathcal{C}} \to (\mathrm{Set}_\Delta^+)_{/S}$, called the *unstraightening functor*.

Now recall that we have an adjunction (see [52]): $\mathfrak{C} : \mathrm{Set}_\Delta \rightleftarrows \mathrm{Cat}_\Delta : \mathrm{N}$, and $\mathrm{Cat}_\infty = \mathrm{N}(\mathrm{Cat}_\infty^\Delta)$, thus $\mathrm{Cat}_\infty^{\mathrm{op}} = \mathrm{N}((\mathrm{Cat}_\infty^\Delta)^{\mathrm{op}})$. It follows $\mathfrak{C}[\mathrm{Cat}_\infty^{\mathrm{op}}] = \mathfrak{C}(\mathrm{N}((\mathrm{Cat}_\infty^\Delta)^{\mathrm{op}})) \to (\mathrm{Cat}_\infty^\Delta)^{\mathrm{op}}$. If we let $S = \mathrm{Cat}_\infty^{\mathrm{op}} \in \mathrm{Set}_\Delta$ and $\mathcal{C} = \mathrm{Cat}_\infty^\Delta$, then we therefore have a morphism $\phi : \mathfrak{C}[S] \to \mathcal{C}^{\mathrm{op}}$, and the straightening and unstraightening functors provide an adjunction

$$(\mathrm{Set}_\Delta^+)_{/\mathrm{Cat}_\infty^{\mathrm{op}}} \rightleftarrows (\mathrm{Set}_\Delta^+)^{\mathrm{Cat}_\infty^\Delta}$$

where we trivially regard a simplicial set as being a marked simplicial set by taking all its edges as marked edges.

Now $\mathrm{Cat}_\infty^\Delta$ corresponds to the subcategory of Set_Δ^+ spanned by fibrant–cofibrant objects, thus one has an inclusion $\mathrm{Cat}_\infty^\Delta \hookrightarrow \mathrm{Set}_\Delta^+$, an object of $(\mathrm{Set}_\Delta^+)^{\mathrm{Cat}_\infty^\Delta}$, which after applying the unstraightening functor gives us an object of $(\mathrm{Set}_\Delta^+)_{/\mathrm{Cat}_\infty^{\mathrm{op}}}$, a Cartesian fibration $q : \mathcal{Z} \to \mathrm{Cat}_\infty^{\mathrm{op}}$, referred to as the universal Cartesian fibration, as discussed in [52]. Then one says a functor $f : S \to \mathrm{Cat}_\infty^{\mathrm{op}}$ *classifies*[25] a Cartesian fibration $p : X \to S$ in Set_Δ if one has an equivalence of Cartesian fibrations $X \to S \times_{\mathrm{Cat}_\infty^{\mathrm{op}}} \mathcal{Z}$. Here we use the fact that Cartesian fibrations are stable under base change: if q is Cartesian, so is $S \times_{\mathrm{Cat}_\infty^{\mathrm{op}}} \mathcal{Z} \to S$. Further by Proposition 3.3.2.5 of [52], if in addition p is a right fibration, f factors through $\mathcal{S}^{\mathrm{op}} \subseteq \mathrm{Cat}_\infty^{\mathrm{op}}$, so by base change we have a right fibration, called the universal R fibration: $\mathcal{Z}_0 = \mathcal{S}^{\mathrm{op}} \times_{\mathrm{Cat}_\infty^{\mathrm{op}}} \mathcal{Z} \to \mathcal{S}^{\mathrm{op}}$, and one says $f : S \to \mathcal{S}^{\mathrm{op}}$ classifies $p : X \to S$ if one has an equivalence of Cartesian fibrations: $X \to S \times_{\mathcal{S}^{\mathrm{op}}} \mathcal{Z}_0$.

We can mention in passing that marked simplicial sets can be used to show that Cat_∞ admits small limits and colimits by virtue of Proposition 4.2.4.8 of [52], which states that for a combinatorial simplicial model category $\mathcal{C}$, $N(\mathcal{C}^\circ)$ admits small limits and colimits, $\mathcal{C}^\circ$ being the simplicial subcategory of $\mathcal{C}$ spanned by fibrant–cofibrant objects; Set_Δ^+

24 Def. 3.1.0.1 [52].

25 Def. 3.3.2.2 [52].

is a simplicial model category (while Set_Δ is not), and one has an equivalence of simplicial categories $\mathrm{Cat}_\infty^\Delta \simeq (\mathrm{Set}_\Delta^+)^\circ$, which along with the fact that $\mathrm{Cat}_\infty = N(\mathrm{Cat}_\infty^\Delta)$ allows us to conclude.

The collection of ∞-topoi forms an ∞-category $\infty\mathcal{T}op$ with ∞-topoi as objects and *geometric morphisms*[26] as morphisms. If $\mathcal{X}$ and $\mathcal{Y}$ are ∞-topoi, then one defines a geometric morphism from $\mathcal{X}$ to $\mathcal{Y}$ to be given by an adjunction $f^* : \mathcal{X} \rightleftarrows \mathcal{Y} : f_*$ with $f^* \dashv f_*$ and f^* L exact.

3.4 ∞-categorical Grothendieck construction

One pervasive construction in our work is the Grothendieck construction (see [83] for example), adapted by Lurie to the ∞-categorical setting in [52]. It is given by a R Quillen functor $(\mathrm{Set}_\Delta)^{\mathfrak{C}[S]^{\mathrm{op}}} \to (\mathrm{Set}_\Delta)_{/S}$, which is actually a Quillen equivalence. This section might be perceived as being repetitious since it hinges on the straightening/unstraightening functors we have just defined. However, it is worth quickly going over this construction again and add a few more details.

Given a functor $\phi : \mathfrak{C}[S] \to \mathcal{C}^{\mathrm{op}}$ between simplicial categories with $S \in \mathrm{Set}_\Delta$ and $\mathcal{C} \in \mathrm{Cat}_\Delta$, if $X \in (\mathrm{Set}_\Delta)_{/S}$, then $f : X \to S$ induces $\mathfrak{C}[f] : \mathfrak{C}[X] \to \mathfrak{C}[S]$. We can form the pushout

$$\begin{array}{ccc} \mathfrak{C}[X] & \xrightarrow{\phi\circ\mathfrak{C}[f]} & \mathcal{C}^{\mathrm{op}} \\ \downarrow & & \downarrow \\ \mathfrak{C}[X^{\triangleright}] & \longrightarrow & \mathcal{M} \end{array}$$

and regard the simplicial category $\mathcal{M}$ as a correspondence from $\mathcal{C}^{\mathrm{op}}$ to $\{v\}$, v the cone point of $X^{\triangleright} = X \star \Delta^0$. One can identify $\mathcal{M}$ with a simplicial functor

$$\begin{aligned} \mathrm{St}_\phi(X) : \mathcal{C} &\to \mathrm{Set}_\Delta, \\ C &\mapsto \mathrm{Map}_{\mathcal{M}}(C, v) \end{aligned}$$

giving a functor $\mathrm{St}_\phi : (\mathrm{Set}_\Delta)_{/S} \to (\mathrm{Set}_\Delta)^{\mathcal{C}}$, which we defined earlier as the straightening functor [52]. If $\mathcal{C} = \mathfrak{C}[S]^{\mathrm{op}}$ and $\phi = \mathrm{id}$, then we just write St_S for St_ϕ. By the adjunction functor theorem there is a right adjoint $\mathrm{Un}_\phi : (\mathrm{Set}_\Delta)^{\mathcal{C}} \to (\mathrm{Set}_\Delta)_{/S}$, called the unstraightening functor, providing a Quillen adjunction. If ϕ is an equivalence, then $(\mathrm{St}_\phi, \mathrm{Un}_\phi)$ is a Quillen equivalence.

26 Def. 6.3.1.1 [52].

For a clearer definition of St_ϕ, we have a pushout

$$\begin{array}{ccccc} & & \overset{\phi_X}{\frown} & & \\ \mathfrak{C}[X] & \longrightarrow & \mathfrak{C}[S] & \xrightarrow{\ \phi\ } & C^{\mathrm{op}} \\ \big\downarrow & & & & \big\downarrow \\ \mathfrak{C}[X^{\triangleright}] & & \xrightarrow[\ \psi\ ]{} & & \mathcal{M} = \mathfrak{C}[X^{\triangleright}] \coprod_{\mathfrak{C}[X]} C^{\mathrm{op}}. \end{array}$$

If c is a vertex of X, then $C = \phi_X(c) \in C$, equivalently represented as $c : \Delta^0 \to X$, so $c \star \mathrm{id}_{\Delta^0} : \Delta^1 \to X^{\triangleright}$. In this manner, one has a morphism from c to the cone point of $X^{\triangleright}$. Now $c \star \mathrm{id}_{\Delta^0}$ edge of $X^{\triangleright}$ maps to an edge $C \to *$ of $\mathcal{M}$. To make this precise, $c \star \mathrm{id}_{\Delta^0} : \Delta^0 \star \Delta^0 \to X \star \Delta^0 = X^{\triangleright}$, so that a morphism in $X^{\triangleright}$ is between objects in $(X^{\triangleright})_0 = X_0 \cup \{v\}$. The induced morphism in $\mathcal{M}$ is of the form

$$\begin{gathered} \Big(\mathfrak{C}[X^{\triangleright}] \coprod_{\mathfrak{C}[X]} C^{\mathrm{op}}\Big)_0 \to \Big(\mathfrak{C}[X^{\triangleright}] \coprod_{\mathfrak{C}[X]} C^{\mathrm{op}}\Big)_0, \\ \Big(c \coprod_{c} C\Big) \mapsto \Big(v \coprod_{\emptyset} \emptyset\Big) \end{gathered}$$

which we represent by $C \to *$ or, more precisely, by $C \to v$.

4 Model categories

4.1 Model categories

In a model category, one essentially has three classes of morphisms: fibrations, cofibrations, and weak equivalences, all satisfying certain axioms that we will give below. As an illustration of the importance of model categories, if $\mathcal{M}$ is a simplicial model category for example and $\mathcal{M}^\circ$ is its subcategory of fibrant and cofibrant objects, then $\mathcal{M}^\circ$ is a fibrant simplicial category. It follows by Proposition 1.1.5.10 of [52] that $N(\mathcal{M}^\circ)$ is an ∞-category. Thus we can view model categories as ways of obtaining ∞-categories. Actually, we should regard them as deconstructions of ∞-categories. As another example of application, the localization of a model category with respect to its set of weak equivalences is defined to be its homotopy category, and it turns out that many categories we work with are homotopy categories. This again presents model categories as deconstructions of such working categories. One may argue that it just suffices to localize a category with respect to its equivalences. The problem with doing this though is that the resulting object may not be a set. Having fibrations and cofibrations along with the axioms satisfied by a model category ensures that the homotopy category is indeed a category.

One defines a *model structure*[1] on a category $\mathcal{C}$ to consist of subcategories of Mor $(\mathcal{C})$ called weak equivalences, fibrations and cofibrations, and functorial factorizations (α, β) and (γ, δ), subject to:

- if f and g are objects of Mor $(\mathcal{C})$ such that gf is defined, and if two of f, g, and gf are weak equivalences, so is the third (2–3 property).
- For $f, g \in$ Mor $(\mathcal{C})$, f a retract of g, then if g is a fibration, cofibration, or a weak equivalence, so is f.
- Trivial cofibrations (meaning cofibrations that are also weak equivalences) have the LLP with respect to all fibrations, and trivial fibrations (fibrations that are also weak equivalences) have the RLP with respect to all cofibrations.
- For all $f \in$ Mor $(\mathcal{C})$, $\alpha(f)$ is a cofibration, $\beta(f)$ is a trivial fibration, $\gamma(f)$ is a trivial cofibration, and $\delta(f)$ is a fibration.

One defines a *model category*[2] to be a category with all small limits and colimits that is endowed with a model structure.

Note that[3] the class of cofibrations is closed under pushouts, and so is the class of trivial cofibrations, and the class of fibrations is closed under pullbacks, and so is the class of trivial fibrations.

---

1 Def. 1.1.2 [42].

2 Def. 1.1.4 [42].

3 Prop. 7.2.5 [40].

https://doi.org/10.1515/9783111334073-004

Observe that if $\mathcal{C}$ is a model category, then it has an initial object 0 and a final object 1. One says $C \in \mathcal{C}$ is *cofibrant* if $0 \to C$ is a cofibration, and C is *fibrant* if $C \to 1$ is a fibration.

Now consider the unique map $0 \to X$. Apply the functorial factorization (α, β) to this map. One gets

$$0 \xrightarrow{\text{cof}} X^- \xrightarrow{\text{triv.fibr.}} X.$$

Denote X^- by QX; it is cofibrant. Thus $QX \xrightarrow{\text{triv.fibr.}} X$. This is functorial: $Q : \mathcal{C} \to \mathcal{C}$ associating QX as just defined with any object X is called the *cofibrant replacement functor*. This is the beginning of a long series of approximations of objects by fibrant or cofibrant objects. The reason for doing this of course is that those equivalences project to isomorphisms in the homotopy category, which allows us to replace objects by their fibrant or cofibrant approximations.

Likewise, applying the functorial factorization (γ, δ) to the map $X \to 1$, one gets

$$X \xrightarrow{\text{triv.cof}} X^+ \xrightarrow{\text{fibr}} 1.$$

Denote X^+ by RX. It is fibrant, and $X \to RX$ is a trivial cofibration. This is functorial: $R : \mathcal{C} \to \mathcal{C}$ assigning to any object X the fibrant object RX just defined is a functor, called the *fibrant replacement functor*.

If $\mathcal{M}$ is a model category, then a *cofibrant approximation*[4] to an object X is a pair $(\tilde{X}, i)$, where $\tilde{X}$ is cofibrant, and $i : \tilde{X} \to X$ is a weak equivalence. A *fibrant cofibrant approximation* is a cofibrant approximation for which i is a trivial fibration. A *fibrant approximation* to an object X is a pair $(\hat{X}, j)$, where $\hat{X}$ is fibrant, and $j : X \to \hat{X}$ is a weak equivalence. A *cofibrant fibrant approximation* is one for which j is a trivial cofibration.

In a model category, every object has a fibrant cofibrant approximation and a cofibrant fibrant approximation.[5]

For a model category $\mathcal{M}$ and $X \in \operatorname{Ob}\mathcal{M}$, if $(\tilde{X}, i)$ is a cofibrant approximation to X and $(\tilde{X}', i')$ is a fibrant cofibrant approximation to X, then by Proposition 8.1.7 of [40] there exists a map of cofibrant approximations $g : \tilde{X} \to \tilde{X}'$, unique up to homotopy over X, and it is a weak equivalence. Likewise, if $(\hat{X}, j)$ is a cofibrant fibrant approximation to X and $(\hat{X}', j')$ is a fibrant approximation to X, then there exists a map of fibrant approximations $g : \hat{X} \to \hat{X}'$, unique up to homotopy under X, and it is a weak equivalence.

In a model category $\mathcal{M}$, a *cofibrant approximation to a map* $g : X \to Y$[6] is provided by cofibrant approximations $(\tilde{X}, i_X)$ for X, $(\tilde{Y}, i_Y)$ for Y, and a map $\tilde{g} : \tilde{X} \to \tilde{Y}$ such that $i_Y \tilde{g} = g i_X$. In addition, one says we have a *fibrant cofibrant approximation to g* if both cofibrant approximations are fibrant cofibrant approximations.

---

4 Def. 8.1.2 [40].

5 Prop. 8.1.3 [40].

6 Def. 8.1.22 [40].

In a model category, by Proposition 8.1.23 of [40] every map $g : X \to Y$ has a fibrant cofibrant approximation $\tilde{g} : \tilde{X} \to \tilde{Y}$.

One defines the *homotopy category*[7] of a model category $\mathcal{C}$ by $\mathrm{Ho}\,(\mathcal{C}) = \mathcal{C}[W^{-1}]$, where W denotes the subcategory of weak equivalences. This is also denoted $h\mathcal{C}$. Note that there is a functor $\gamma : \mathcal{C} \to \mathrm{Ho}\,(\mathcal{C})$, the identity on objects, but which sends maps in W to isomorphisms in $\mathrm{Ho}\,\mathcal{C}$. Now by Proposition 1.2.3 of [42] one has equivalences of categories

$$\begin{array}{ccccc} \mathrm{Ho}\,\mathcal{C}_{cf} & \longrightarrow & \mathrm{Ho}\,\mathcal{C}_c & \longrightarrow & \mathrm{Ho}\,\mathcal{C} \\ & \searrow & & \nearrow & \\ & & \mathrm{Ho}\,\mathcal{C}_f & & \end{array}$$

where $\mathcal{C}_c$ is the subcategory of cofibrant objects of $\mathcal{C}$, $\mathcal{C}_f$ is the subcategory of fibrant objects, and $\mathcal{C}_{cf}$ is the subcategory of objects that are fibrant and cofibrant.

Practically, we can give another definition of a homotopy category. We need a few definitions for that purpose. Let $\mathcal{C}$ be a model category, and let $B, X \in \mathrm{Ob}\,\mathcal{C}$. A *cylinder object*[8] for B is given by a factorization of the fold map $\nabla : B \coprod B \xrightarrow{i_0+i_1} B' \xrightarrow{s} B$, where $i_0 + i_1$ is a cofibration, and s is a weak equivalence. A *path object* for X is given by a factorization of the diagonal map $\Delta : X \xrightarrow{r} X' \xrightarrow{(p_0,p_1)} X \times X$, where r is a weak equivalence, and (p_0, p_1) is a fibration. Now let f and g be two maps from B to X in $\mathcal{C}$. One defines a *left homotopy*: $f \to g$ to be a map $H : B' \to X$ for a cylinder object B' of B, such that $Hi_0 = f$ and $Hi_1 = g$. We write $f \sim^l g$. Similarly, a *right homotopy* from f to g is given by a map $K : B \to X'$, where X' is a path object of X, satisfying $p_0 K = f$ and $p_1 K = g$. We write $f \sim^r g$. We simply write $f \sim g$ if f and g are L and R homotopic. Now one says f is a *homotopy equivalence* if there exists $h : X \to B$ with $hf \sim 1_B$ and $fh \sim 1_X$. By Corollary 1.2.7 of [42] the relation of homotopy on $\mathcal{C}_{cf}$ is an equivalence relation, compatible with the composition, making $\mathcal{C}_{cf}/\sim$ well-defined. The functor $\mathcal{C}_{cf} \to \mathcal{C}_{cf}/\sim$ inverts homotopy equivalences in $\mathcal{C}_{cf}$. By Corollary 1.2.9 of [42], $\mathcal{C}$ being a model category, one has maps $\gamma : \mathcal{C}_{cf} \to \mathrm{Ho}\,(\mathcal{C}_{cf})$ and $\delta : \mathcal{C}_{cf} \to \mathcal{C}_{cf}/\sim$.

There is a unique isomorphism j of categories making the following diagram commutative:

$$\begin{array}{ccc} \mathcal{C}_{cf}/\sim & \xrightarrow{\;j\;} & \mathrm{Ho}\,(\mathcal{C}_{cf}) \\ {\scriptstyle\delta}\uparrow & \nearrow{\scriptstyle\gamma} & \\ \mathcal{C}_{cf} & & \end{array}$$

which is the identity on objects.

7 Def. 1.2.1 [42].

8 Def. 1.2.4 [42].

By Theorem 1.2.10 of [42], $\mathcal{C}$ being a model category, there are natural isomorphisms

$$\mathrm{Hom}_{\mathcal{C}_{cf}/\sim}(QRX, QRY) \cong \mathrm{Hom}_{\mathrm{Ho}\,\mathcal{C}}(\gamma X, \gamma Y) \cong \mathrm{Hom}_{\mathcal{C}_{cf}/\sim}(RQX, RQY).$$

This follows from Proposition 1.2.3, which states that $\mathrm{Ho}\,\mathcal{C}_{cf} \simeq \mathrm{Ho}\,\mathcal{C}$, and Corollary 1.2.9: $\mathcal{C}_{cf}/\sim \cong \mathrm{Ho}\,\mathcal{C}_{cf}$. We also have that

$$\mathrm{Hom}_{\mathrm{Ho}\,\mathcal{C}}(\gamma X, \gamma Y) \cong \mathrm{Hom}_{\mathcal{C}_{cf}/\sim}(QX, RY).$$

An important result is the following: if $f : A \to B$ is a morphism in $\mathcal{C}$, γf being an isomorphism in $\mathrm{Ho}\,\mathcal{C}$ implies it is a weak equivalence in $\mathcal{C}$. Those results allow us to have a working definition of the homotopy category and to use the concepts of fibrant and cofibrant approximations.

Note that the construction $\mathcal{C} \mapsto \mathrm{Ho}\,(\mathcal{C})$ is functorial in the following sense: one has a pseudo-2-functor $\mathrm{Ho} : \mathrm{Mod} \to \mathrm{Cat}_{\mathrm{ad}}$,[9] where Mod denotes the 2-category of model categories, Quillen adjunctions (see below) and natural transformations between them, and $\mathrm{Cat}_{\mathrm{ad}}$ denotes the 2-category of categories and adjunctions, and natural transformations. A *pseudo-2-functor*[10] between two categories $\mathcal{C}$ and $\mathcal{D}$ is given by three maps $F_i : \mathcal{C}_i \to \mathcal{D}_i$, $i = 0, 1, 2$, that we collectively denote by $F : \mathcal{C} \to \mathcal{D}$, a 2-isomorphism $\alpha : F(\mathrm{id}_X) \to \mathrm{id}_{FX}$ for all $X \in \mathrm{Ob}\,\mathcal{C}$, and a 2-isomorphism $m_{gf} : Fg \circ Ff \to F(g \circ f)$ for any two $f, g \in \mathrm{Mor}\,\mathcal{C}$ for which $g \circ f$ is defined, such that F preserves domains and codomains, and is functorial with respect to vertical composition: $F(\mathrm{id}_f) = \mathrm{id}_{Ff}$ for all $f \in \mathcal{C}_1$, and for $\tau, \sigma \in \mathcal{C}_2$ for which the vertical composition $\tau \bullet \sigma$ is well-defined, $F(\tau \bullet \sigma) = F\tau \bullet F\sigma$.

One has a coherently associative diagram: for all $f, g, h \in \mathcal{C}_1$ such that the composition $h \circ g \circ f$ is well-defined,

$$\begin{array}{ccccc}
(Fh \circ Fg) \circ Ff & \xrightarrow{m_{hg} \circ \mathrm{id}_{Ff}} & F(h \circ g) \circ Ff & \xrightarrow{m_{(h \circ g)f}} & F((h \circ g) \circ f) \\
\| & & & & \| \\
Fh \circ (Fg \circ Ff) & \xrightarrow[\mathrm{id}_{Fh} \circ m_{gf}]{} & Fh \circ F(g \circ f) & \xrightarrow[m_{h(g \circ f)}]{} & F(h \circ (g \circ f)).
\end{array}$$

There is also the following commutative diagrams for any $f : X \to Y$ in $\mathcal{C}$:

$$\begin{array}{ccc}
F\mathrm{id}_Y \circ Ff & \xrightarrow{m} & F(\mathrm{id}_Y \circ f) \\
{\scriptstyle \alpha \circ \mathrm{id}_{Ff}} \downarrow & & \| \\
(\mathrm{id}_{FY}) \circ Ff & = & Ff
\end{array}$$

9 Thm 1.4.3 [42].

10 Def. 1.4.2 [42].

and:

$$\begin{array}{ccc} Ff \circ F\mathrm{id}_X & \xrightarrow{m} & F(f \circ \mathrm{id}_X) \\ {\scriptstyle \mathrm{id}_{Ff}\circ\alpha}\downarrow & & \| \\ Ff \circ \mathrm{id}_{FX} & = & Ff. \end{array}$$

Further the map m is functorial for horizontal composition: if $\sigma : f \Rightarrow f'$ and $\tau : g \Rightarrow g'$ in $\mathcal{C}$ for which the composition $g \circ f$ is well-defined, then

$$\begin{array}{ccc} Fg \circ Ff & \xrightarrow{m_{gf}} & F(g \circ f) \\ {\scriptstyle F\tau\circ F\sigma}\downarrow & & \downarrow{\scriptstyle F(\tau\circ\sigma)} \\ Fg' \circ Ff' & \xrightarrow[m_{g'f'}]{} & F(g' \circ f'). \end{array}$$

Now let $\mathcal{C}$ and $\mathcal{D}$ be two model categories. A functor $F : \mathcal{C} \to \mathcal{D}$ is called a *left Quillen functor*[11] if it is a L adjoint and preserves cofibrations and trivial cofibrations. A functor $U : \mathcal{D} \to \mathcal{C}$ is called a *right Quillen functor* if it is a R adjoint and preserves fibrations and trivial fibrations. One defines a *Quillen adjunction* to be a pair (F, U, ϕ), where $F \dashv U$, F is a L Quillen functor, and $\phi : \mathrm{Hom}_{\mathcal{D}}(FA, B) \xrightarrow{\cong} \mathrm{Hom}_{\mathcal{C}}(A, UB)$. The reason for introducing that concept of L and R Quillen functors is that one would like functors between model categories to preserve their structure. It is however too much asking that they preserve fibrations and cofibrations.

Left and right Quillen functors preserve half that, but it turns out that they provide a very robust theory.

Now suppose $\mathcal{C}$ and $\mathcal{D}$ are two model categories and $F : \mathcal{C} \to \mathcal{D}$ is a L Quillen functor. One defines the *total left derived functor*[12] $\mathbb{L}F$ of F to be the functor $\mathbb{L}F : \mathrm{Ho}\,\mathcal{C} \to \mathrm{Ho}\,\mathcal{D}$ defined as the composition $\mathrm{Ho}\,\mathcal{C} \xrightarrow{\mathrm{Ho}\,Q} \mathrm{Ho}\,\mathcal{C}_c \xrightarrow{\mathrm{Ho}\,F} \mathrm{Ho}\,\mathcal{D}$. In other terms, $\mathbb{L}F(X) = F(QX)$. Here $\mathrm{Ho}\,Q$ is the only map that makes the following diagram commutative:

$$\begin{array}{ccccc} \mathcal{C} & \xrightarrow{Q} & \mathcal{C}_c & \xrightarrow{\gamma} & \mathrm{Ho}\,\mathcal{C}_c \\ & {\scriptstyle\gamma}\searrow & & \nearrow{\scriptstyle \mathrm{Ho}\,Q} & \\ & & \mathrm{Ho}\,\mathcal{C}. & & \end{array}$$

Now if $\tau : F \Rightarrow F'$ is a natural transformation of L Quillen functors, then one defines $\mathbb{L}\tau = \mathrm{Ho}\,\tau \circ \mathrm{Ho}\,Q$, that is, $(\mathbb{L}\tau)_X = \tau_{QX}$ for $X \in \mathrm{Ob}\,\mathcal{C}$. Indeed, consider

11 Def. 1.3.1 [42].

12 Def. 1.3.6 [42].

$$\begin{array}{ccccc} \mathrm{Ho}\,\mathcal{C} & \xrightarrow{\mathrm{Ho}\,Q} & \mathrm{Ho}\,\mathcal{C}_c & \xrightarrow{\mathrm{Ho}\,F} & \mathrm{Ho}\,\mathcal{D} \\ \| & & \| & & \| \\ \mathrm{Ho}\,\mathcal{C} & \xrightarrow[\mathrm{Ho}\,Q]{} & \mathrm{Ho}\,\mathcal{C}_c & \xrightarrow[\mathrm{Ho}\,F']{} & \mathrm{Ho}\,\mathcal{D} \end{array}$$

with $\mathrm{id} : \mathrm{Ho}\,Q \Longrightarrow \mathrm{Ho}\,Q$ and $\mathrm{Ho}\,\tau : \mathrm{Ho}\,F \Longrightarrow \mathrm{Ho}\,F'$. Thus we have a horizontal composite $\mathbb{L}\tau : \mathbb{L}F \Longrightarrow \mathbb{L}F'$ with $\mathbb{L}\tau = \mathrm{Ho}\,\tau \circ \mathrm{id}_{\mathrm{Ho}\,Q}$, so that $(\mathbb{L}\tau)_X = \mathrm{Ho}\,\tau(QX) = \tau_{QX}$.

In the same manner, if $U : \mathcal{D} \to \mathcal{C}$ is a R Quillen functor, then the *total right derived functor* $\mathbb{R}U$ of U is the functor $\mathbb{R}U : \mathrm{Ho}\,(\mathcal{D}) \to \mathrm{Ho}\,(\mathcal{C})$ defined as the composite $\mathrm{Ho}\,(\mathcal{D}) \xrightarrow{\mathrm{Ho}\,(R)} \mathrm{Ho}\,(\mathcal{D}_f) \xrightarrow{\mathrm{Ho}\,(U)} \mathrm{Ho}\,(\mathcal{C})$, that is, $\mathbb{R}U(Y) = U(RY)$. For a Quillen adjunction $(U, F, \phi) : \mathcal{C} \rightleftarrows \mathcal{D}$, by Lemma 1.3.10 of [42] one has

$$\mathrm{Ho}\,(U, F, \phi) = \mathbb{L}(U, F, \phi) = (\mathbb{L}U, \mathbb{R}F, \mathbb{R}\phi).$$

Finally, a Quillen adjunction $(F, U, \phi) : \mathcal{C} \to \mathcal{D}$ is called a *Quillen equivalence*[13] if for all $X \in \mathcal{C}_c$ and $Y \in \mathcal{D}_f$, $f : FX \to Y$ is an equivalence in $\mathcal{D}$ iff $\phi f : X \to UY$ is an equivalence in $\mathcal{C}$.

4.2 Cofibrantly generated model categories

One says a model category $\mathcal{C}$ is *cofibrantly generated*[14] if there are subsets I and J of $\mathrm{Mor}\,\mathcal{C}$ such that the domains of maps of I are small relative to I-cell, the domains of maps of J are small relative to J-cell, $\mathrm{Fibr}(\mathcal{C}) = J$-inj, and triv.Fibr.$(\mathcal{C}) = I$-inj. One calls I the set of *generating cofibrations* and J the set of *generating trivial cofibrations*. In this definition, if I is a class of maps in a category $\mathcal{C}$, then a map is *I-injective*[15] if it has the RLP with respect to all maps in I. Such maps form a class that we denote by I-inj. Similarly, a map is said to be *I-projective* if it has the LLP with respect to any map in I, and I-proj denotes the class of all such maps. A map is an *I-cofibration* if it has the LLP with respect to all maps in I-inj, and we denote by I-Cof the class of all such maps. A map is an *I-fibration* if it has the RLP with respect to I-projective maps. Cofibrantly generated model categories are very useful when it comes to localizing model categories, but one should point out however that combinatorial model categories, defined later, seem to be preferred in modern treatments.

If $\mathcal{C}$ contains all small colimits, then one defines a *relative I-cell complex*[16] to be a transfinite composition of pushouts of elements of I. We denote by I-cell the collection

13 Def. 1.3.12.

14 Def. 2.1.17 [42].

15 Def. 2.1.7 [42].

16 Def. 2.1.9 [42].

of relative I-cell complexes. $A \in \mathrm{Ob}\,\mathcal{C}$ is an I-cell complex if the map $0 \to A$ is a relative I-cell complex.

An important fact for our purposes: by Proposition 2.1.18 of [42], if $\mathcal{C}$ is a cofibrantly generated model category with I as a set of generating cofibrations and J as a set of generating trivial cofibrations, then Cof = I-Cof, and triv.Cof. = J-Cof.

An important example of cofibrantly generated model category is provided by Theorem 2.1.19 of [42], which states that if $\mathcal{C}$ is a model category with all small limits and colimits, $W \subset \mathcal{C}$ a subcategory, and $I, J \subset \mathrm{Mor}\,\mathcal{C}$, then there is a cofibrantly generated model structure on $\mathcal{C}$ with I as a set of generating cofibrations, J as a set of generating trivial cofibrations, and W as a set of weak equivalences if and only if:

1- W satisfies the 2–3 property and is closed under retracts;
2- The domains of elements of I are small relative to I-cell;
3- The domains of elements of J are small relative to J-cell;
4- J-cell $\subseteq W \cap (I\text{-Cof})$ and I-inj $\subseteq W \cap (J\text{-inj})$;
5- $W \cap I$-Cof $\subset J$-Cof, or $W \cap J$-inj $\subseteq I$-inj.

One can put a cofibrantly generated model structure on Set_Δ as follows.[17] Let $I = \{\partial\Delta^n \hookrightarrow \Delta^n \,:\, n \geq 0\}$ and $J = \{\Lambda^r_n \hookrightarrow \Delta^n \,:\, n > 0, 0 \leq r \leq n\}$. Then

$$\mathrm{Cof}(\mathrm{Set}_\Delta) = I\text{-Cof},$$
$$\mathrm{Fibr}(\mathrm{Set}_\Delta) = J\text{-inj}$$

where the latter are also referred to as *Kan fibrations*, and finally $f \in W(\mathrm{Set}_\Delta)$ iff $|f|$ is a weak equivalence in Top, the category of topological spaces. One refers to this model structure as the *Kan model structure* on Set_Δ. Observe that f is a cofibration in this model structure iff it is injective.[18] It follows that any $K \in \mathrm{Set}_\Delta$ is automatically cofibrant.

4.3 Combinatorial model categories

Another type of model categories that is widely used besides cofibrantly generated model categories is that of *combinatorial model categories*.[19] A model category $\mathcal{M}$ is said to be combinatorial if it is *presentable* (accessible with all small colimits [1]), there exists a set I of generating cofibrations such that Cof($\mathcal{M}$) is the smallest weakly saturated class of morphisms containing I, and there exists a set J of generating trivial cofibrations such that triv.Cof.($\mathcal{M}$) is the smallest weakly saturated class of morphisms containing J.

17 Def. 3.2.1 [42].
18 Prop. 3.2.2. [42].
19 Def. A.2.6.1 [52].

In a category $\mathcal{C}$ with all small colimits, one says that a class S of morphisms in $\mathcal{C}$ is *weakly saturated*[20] if it is closed under pushouts, retracts, and transfinite compositions.

Combinatorial model categories are used quite a bit in Derived Algebraic Geometry, where we principally work with functor categories. If those are of the form $\mathrm{Fun}(\mathcal{C}, \mathcal{M})$ for a small category $\mathcal{C}$ and a combinatorial model category $\mathcal{M}$, then one can put two combinatorial model structures on them, the *projective model structure*[21] with objectwise fibrations and weak equivalences and the *injective model structure* with objectwise cofibrations and weak equivalences.

One should point out that it seems combinatorial model categories are favored over cofibrantly generated model categories since results that make use of the combinatorial model structure are cleaner in a sense. Here is an example of such a result: by Proposition A.3.7.6 of [52], if $\mathcal{C}$ is an ∞-category, then $\mathcal{C}$ is presentable as an ∞-category iff there exists a combinatorial simplicial model category $\mathcal{M}$ such that $\mathcal{C} \simeq N(\mathcal{M})$.

4.4 Monoidal model categories

A *monoidal model category*[22] is a closed monoidal category with a model structure such that the tensor product for the monoidal structure $\otimes : \mathcal{C} \times \mathcal{C} \to \mathcal{C}$ is a Quillen bifunctor, and if $QS \xrightarrow{q} S$ is a cofibrant replacement for the unit S of $\otimes$, then $QS \otimes X \xrightarrow{q\otimes 1} S \otimes X$ is a weak equivalence for all $X \in \mathrm{Ob}\, \mathcal{C}_c$, and $X \otimes QS \xrightarrow{1\otimes q} X \otimes S$ is a weak equivalence as well. One can regard monoidal categories as the categorical analog of rings.

For $\mathcal{C}, \mathcal{D}, \mathcal{E}$ model categories, an adjunction of two variables

$$(\otimes, \mathrm{Hom}_r, \mathrm{Hom}_l, \phi_r, \phi_l) : \mathcal{C} \times \mathcal{D} \to \mathcal{E}$$

is called a *Quillen adjunction of two variables*[23] if for any cofibrations $f : U \to V$ in $\mathcal{C}$ and $g : X \to Y$ in $\mathcal{D}$, the induced map $f \square g : (V \otimes X) \coprod_{U \otimes X} (U \otimes Y) \to V \otimes Y$ is a cofibration in $\mathcal{E}$, trivial if either of f or g is. One typically refers to the tensor product of a Quillen adjunction of two variables as a *Quillen bifunctor*, and $f \square g$ is referred to as the pushout product of f and g.

A *closed monoidal structure*[24] on a model category $\mathcal{C}$ is given by $(\otimes, \alpha, \lambda, \rho, \mathrm{Hom}_r, \mathrm{Hom}_l, \phi_r, \phi_l)$, where $(\otimes, \alpha, \lambda, \rho)$ corresponds to a monoidal structure on $\mathcal{C}$, and $(\otimes, \mathrm{Hom}_r, \mathrm{Hom}_l, \phi_r, \phi_l)$ is an adjunction of two variables on $\mathcal{C}$.

20 Def. A.1.2.2 [52].

21 Prop. A.2.8.2 [52].

22 Def. 4.2.6 [42].

23 Def. 4.2.1 [42].

24 Def. 4.1.13 [42].

An *adjunction of two variables*[25] from $\mathcal{C}\times\mathcal{D}\to\mathcal{E}$ is given by $(\otimes,\mathrm{Hom}_r,\mathrm{Hom}_l,\phi_r,\phi_l)$, where $\otimes:\mathcal{C}\times\mathcal{D}\to\mathcal{E}$, $\mathrm{Hom}_r:\mathcal{D}^{\mathrm{op}}\times\mathcal{E}\to\mathcal{C}$, $\mathrm{Hom}_l:\mathcal{C}^{\mathrm{op}}\times\mathcal{E}\to\mathcal{D}$, and ϕ_r and ϕ_l are natural isomorphisms:

$$\mathrm{Hom}_{\mathcal{C}}(C,\mathrm{Hom}_r(D,E)) \xleftarrow[\phi_r]{\cong} \mathrm{Hom}_{\mathcal{E}}(C\otimes D,E) \xrightarrow[\phi_l]{\cong} \mathrm{Hom}_{\mathcal{D}}(D,\mathrm{Hom}_l(C,E)).$$

Note that Hom_r and Hom_l are internal Homs, whence the name "closed". Those are both right adjoints to the tensor product $\otimes$, which explains why one is calling this an adjunction of two variables.

Observe that by Proposition 4.2.8 of [42], Set_Δ is a symmetric monoidal model category with the usual product as tensor product, with adjoint the function complex functor $(K,L)\mapsto\mathrm{Map}(K,L)$, defined by $\mathrm{Map}(K,L)_n=\mathrm{Hom}_{\mathrm{Set}_\Delta}(K\times\Delta^n,L)$.

One thing that we will use repeatedly is the following concept: if $\mathcal{C}$ is a monoidal model category, a *$\mathcal{C}$-model category*[26] is a $\mathcal{C}$-module $\mathcal{D}$ with a model structure such that $\otimes:\mathcal{D}\times\mathcal{C}\to\mathcal{D}$ is a Quillen bifunctor, and if $QS\xrightarrow{q}S$ is a cofibrant replacement for the unit S of $\otimes$, then $X\otimes QS\to X\otimes S$ is a weak equivalence for all $X\in\mathcal{D}_c$. In particular, a Set_Δ-model category is referred to as a *simplicial model category*.

By Proposition 4.3.1 [42], if $\mathcal{C},\mathcal{D},\mathcal{E}$ are model categories and $(\otimes,\mathrm{Hom}_r,\mathrm{Hom}_l,\phi_r,\phi_l):\mathcal{C}\times\mathcal{D}\to\mathcal{E}$ is a Quillen adjunction of two variables, then the total derived functor of this adjunction is given by

$$(\otimes^{\mathbb{L}},\mathbb{R}\underline{\mathrm{Hom}}_r,\mathbb{R}\underline{\mathrm{Hom}}_l,\mathbb{R}\phi_r,\mathbb{R}\phi_l):\mathrm{Ho}(\mathcal{C})\times\mathrm{Ho}(\mathcal{D})\to\mathrm{Ho}(\mathcal{E}).$$

It follows that if $\mathcal{C}$ is a (symmetric) monoidal category, then $\mathrm{Ho}(\mathcal{C})$ inherits the structure of a closed (symmetric) monoidal category.[27]

4.5 Simplicial model categories

One can also define a simplicial model category to be a simplicial category with a compatible model structure. First, a *simplicial category*[28] is a category $\mathcal{C}$ together with a mapping space $\mathrm{Map}(X,Y)\in\mathrm{Set}_\Delta$ for all $X,Y\in\mathrm{Ob}\,\mathcal{C}$, with a composition for all $X,Y,Z\in\mathrm{Ob}\,\mathcal{C}$:

$$c_{XYZ}:\mathrm{Map}(Y,Z)\times\mathrm{Map}(X,Y)\to\mathrm{Map}(X,Z).$$

---

25 Def. 4.1.12 [42].
26 Def. 4.2.18 [42].
27 Thm. 4.3.2 [42].
28 Def. 9.1.2 [40].

We also have, $\forall X \in \operatorname{Ob}\mathcal{C}$, a map $i_X : * \to \operatorname{Map}(X,X)$ in Set_Δ, and $\forall X, Y \in \operatorname{Ob}\mathcal{C}$, $\operatorname{Map}(X,Y)_0 \cong \operatorname{Hom}_{\mathcal{C}}(X,Y)$, compatible with composition, such that we have the associativity and unit diagrams

$$
\begin{array}{ccc}
((\operatorname{Map}(Y,Z) \times \operatorname{Map}(X,Y)) \times \operatorname{Map}(W,X) & \longrightarrow & \operatorname{Map}(X,Z) \times \operatorname{Map}(W,X) \\
\downarrow \cong & & \\
\operatorname{Map}(Y,Z) \times (\operatorname{Map}(X,Y) \times \operatorname{Map}(W,X)) & & \downarrow \\
\downarrow \cong & & \\
\operatorname{Map}(Y,Z) \times \operatorname{Map}(W,Y) & \longrightarrow & \operatorname{Map}(W,Z)
\end{array}
$$

and

$$
\begin{array}{ccc}
* \times \operatorname{Map}(X,Y) & \longrightarrow & \operatorname{Map}(Y,Y) \times \operatorname{Map}(X,Y) \\
& \searrow^{\cong} \quad \swarrow & \\
& \operatorname{Map}(X,Y) &
\end{array}
$$

as well as

$$
\begin{array}{ccc}
\operatorname{Map}(X,Y) \times * & \longrightarrow & \operatorname{Map}(X,Y) \times \operatorname{Map}(X,X) \\
& \searrow^{\cong} \quad \swarrow & \\
& \operatorname{Map}(X,Y). &
\end{array}
$$

For instance, Set_Δ can be made into a simplicial category by taking the mapping space $\operatorname{Map}(X,Y)$ between $X, Y \in \mathrm{Set}_\Delta$ defined by $\operatorname{Map}(X,Y)_n = \operatorname{Hom}_{\mathrm{Set}_\Delta}(X \times \Delta^n, Y)$.

One defines a *simplicial model category*[29] to be a model category $\mathcal{C}$ that is also a simplicial category such that for all $X, Y \in \operatorname{Ob}\mathcal{C}$ and $K \in \mathrm{Set}_\Delta$, there are objects $X \otimes K$ and Y^K of $\mathcal{C}$ along with isomorphisms

$$\operatorname{Map}(X \otimes K, Y) \cong \operatorname{Map}(K, \operatorname{Map}(X,Y)) \cong \operatorname{Map}(X, Y^K)$$

and if $i : A \to B$ is a cofibration in $\mathcal{C}$ and $p : X \to Y$ is a fibration in $\mathcal{C}$, then the map in Set_Δ

$$\operatorname{Map}(B,X) \xrightarrow{i^* \times p_*} \operatorname{Map}(A,X) \times_{\operatorname{Map}(A,Y)} \operatorname{Map}(B,Y)$$

29 Def. 9.1.6 [40].

is a fibration, trivial if either of i or p is. For instance, for $X \in \mathrm{Set}_\Delta$ and $K \in \mathrm{Set}_\Delta$, defining $X \otimes K = X \times K$ and $X^K = \mathrm{Map}(K, X)$ makes Set_Δ (along with its Kan model structure) into a simplicial model category.

When we do localizations, we will use the following result quite a bit:[30] if $\mathcal{M}$ is a simplicial model category and $X, Y, Z \in \mathrm{Ob}\,\mathcal{M}$, then:

1- If $X \in \mathcal{M}_c$ and $g : Y \to Z$ is a weak equivalence in $\mathcal{M}_f$, then g induces a weak equivalence in Set_Δ: $g_* : \mathrm{Map}(X, Y) \to \mathrm{Map}(X, Z)$.

2- If $Z \in \mathcal{M}_f$ and $h : X \to Y$ is a weak equivalence in $\mathcal{M}_c$, then h induces a weak equivalence in Set_Δ: $h^* : \mathrm{Map}(Y, Z) \to \mathrm{Map}(X, Z)$.

We also have Corollary 9.7.5 of [40], which states that given $g : X \to Y$ in $\mathcal{M}$, still assumed to be a simplicial model category, then:

1- If $X, Y \in \mathcal{M}_f$, then g is an equivalence iff for all $W \in \mathcal{M}_c$, the map $g_* : \mathrm{Map}(W, X) \to \mathrm{Map}(W, Y)$ is an equivalence in Set_Δ.

2- $X, Y \in \mathcal{M}_c$, then g is an equivalence if for all $Z \in \mathcal{M}_f$, the map $g^* : \mathrm{Map}(Y, Z) \to \mathrm{Map}(X, Z)$ is an equivalence in Set_Δ.

4.6 Localization of model categories

Generally speaking, if $\mathcal{M}$ is a model category, and S is a class of maps in $\mathcal{M}$, then a *localization* of $\mathcal{M}$ with respect to S as defined in [40] is a map of model categories $\mathcal{M} \xrightarrow{L} L\mathcal{M}$ such that projections in $\mathrm{Ho}\,\mathcal{M}$ of elements of S map to isomorphisms in $\mathrm{Ho}\,L\mathcal{M}$, and L is initial among all such maps. Now since we have L and R Quillen functors between model categories, one can therefore distinguish L and R localizations. One defines a *left localization*[31] of $\mathcal{M}$ with respect to S to be a model category $L_S\mathcal{M}$, along with a L Quillen functor $j : \mathcal{M} \to L_S\mathcal{M}$ such that $\mathbb{L}j : \mathrm{Ho}\,\mathcal{M} \to \mathrm{Ho}\,L_S\mathcal{M}$ maps projections in $\mathrm{Ho}\,\mathcal{M}$ of elements of S to isomorphisms in $\mathrm{Ho}\,L_S\mathcal{M}$, and j is initial among all such maps.

Let $\mathcal{M}$ be a model category, and let C be a class of maps in $\mathcal{M}$. One defines the *left Bousfield localization*[32] of $\mathcal{M}$ with respect to C to be a model category structure on $\mathcal{M}$, denoted by $L_C\mathcal{M}$, such that $W(L_C\mathcal{M})$ is the set of C-local equivalences, $\mathrm{Cof}(L_C\mathcal{M}) = \mathrm{Cof}(\mathcal{M})$, and $\mathrm{Fibr}(L_C\mathcal{M})$ is the set of maps that have the RLP with respect to maps that are cofibrations and C-local equivalences.

For a model category $\mathcal{M}$ and a class C of maps in $\mathcal{M}$, an object W of $\mathcal{M}$ is said to be *C-local*[33] if it is fibrant and if for all $f : A \to B$ in C, the induced map of homotopy function complexes $f^* : \mathrm{map}(B, W) \to \mathrm{map}(A, W)$ is a weak equivalence. Note: if this is true for

---

30 Cor. 9.3.3 [40].

31 Def. 3.1.1 [40].

32 Def. 3.3.1 [40].

33 Def. 3.1.4 [40].

any homotopy function complex, then it is true of all homotopy function complexes. Now one says $g : X \to Y$ is a *C-local equivalence* if for all C-local objects W, the induced map of homotopy function complexes $g^* : \mathrm{map}(Y, W) \to \mathrm{map}(X, W)$ is a weak equivalence in Set_Δ.

A convenient way to define the notion of space of maps between two objects of a given category is by using homotopy function complexes, which we define presently. A *homotopy function complex* from X to Y is used generically to refer to a L homotopy function complex, a R homotopy function complex, or a two-sided homotopy function complex.[34]

For a model category $\mathcal{M}$ and $X, Y \in \mathrm{Ob}(\mathcal{M})$, we use the notation $\mathrm{map}(X, Y)$ to represent a simplicial set that is a homotopy function complex from X to Y.[35] This means $\mathrm{map}(X, Y)$ refers to either $\mathcal{M}(\tilde{\mathbf{X}}, \hat{Y})$, as part of a left homotopy function complex $(\tilde{\mathbf{X}}, \hat{Y}, \mathcal{M}(\tilde{\mathbf{X}}, \hat{Y}))$, or $\mathcal{M}(\tilde{X}, \hat{\mathbf{Y}})$, as part of a R homotopy function complex $(\tilde{X}, \hat{\mathbf{Y}}, \mathcal{M}(\tilde{X}, \hat{\mathbf{Y}}))$, or even $\mathrm{diag}\mathcal{M}(\tilde{\mathbf{X}}, \hat{\mathbf{Y}})$, as part of a two-sided homotopy function complex $(\tilde{\mathbf{X}}, \hat{\mathbf{Y}}, \mathcal{M}(\tilde{\mathbf{X}}, \hat{\mathbf{Y}}))$.

For a model category $\mathcal{M}$ and $X, Y \in \mathrm{Ob}(\mathcal{M})$, a *L homotopy function complex*[36] from X to Y is a triple $(\tilde{\mathbf{X}}, \hat{Y}, \mathcal{M}(\tilde{\mathbf{X}}, \hat{Y}))$, where $\tilde{\mathbf{X}}$ is a cosimplicial resolution of X, $\hat{Y}$ is a fibrant approximation to Y, and $\mathcal{M}(\tilde{\mathbf{X}}, \hat{Y})$ is defined by $\mathcal{M}(\tilde{\mathbf{X}}, \hat{Y})_n = \mathcal{M}(\tilde{\mathbf{X}}^n, \hat{Y})$.

For a model category $\mathcal{M}$ and $X \in \mathrm{Ob}(\mathcal{M})$, a *cosimplicial resolution*[37] of X is a cofibrant approximation $\tilde{X} \to \mathrm{cc}_* X$ in the Reedy model category structure on $\mathcal{M}^\Delta$, where $\mathrm{cc}_* X$ is a constant cosimplicial object with value X.

By virtue of Theorem 3.3.19 of [40], if $L_S\mathcal{M}$ is a L Bousfield localization of $\mathcal{M}$ with respect to S, then the identify functor $\mathcal{M} \to L_S\mathcal{M}$ in particular is a L localization of $\mathcal{M}$ with respect to S.

Left Bousfield localizations exist, and one way to obtain those is via Theorem 4.1.1 of [40], which states that if $\mathcal{M}$ is a L proper cellular model category and S is a set of maps in $\mathcal{M}$, then there is a L Bousfield localization of $\mathcal{M}$ with respect to S, and $L_S\mathcal{M}$ is also a L proper cellular model category. Observe that if $\mathcal{M}$ is a simplicial model category, then it induces a simplicial model structure on $L_S\mathcal{M}$.

For later purposes, note that if $\mathcal{M}$ is a L proper cellular model category and C is a small category, then $\mathcal{M}^C$ is also a L proper cellular model category.

We now define what are L proper and cellular model categories. An example is provided by Set_Δ.[38]

A model category $\mathcal{M}$ is said to be *left proper*[39] if any pushout of a weak equivalence along a cofibration is still a weak equivalence.

---

34 Def. 17.4.1 [40].

35 Def. 17.4.2 [40].

36 Def. 17.1.1 [40].

37 Def. 16.1.2 [40].

38 Prop. 4.1.4 [40].

39 Def. 13.1.1 [40].

A cofibrantly generated model category $\mathcal{M}$ is a *cellular model category*[40] if the domains and codomains of I are compact, the domains of J are small relative to I, and cofibrations are effective monomorphisms.

For a cofibrantly generated model category $\mathcal{M}$ with generating cofibrations I and for a cardinal γ, $W \in \operatorname{Ob}\mathcal{M}$ is said to be *γ-compact*[41] if it is γ-compact relative to I, and one says W is *compact* if it is γ-compact for some cardinal γ.

If C is a cocomplete category, meaning it has all small colimits, I is a set of maps in C, γ is a cardinal, then $W \in \operatorname{Ob} C$ is said to be *γ-compact relative to I*[42] if for any presented relative I-cell complex $f : X \to Y$, every map $W \to Y$ factors through a subcomplex of f of size at most γ.

The complex f above is said to be a *presented relative I-cell complex*[43] if it is a relative I-cell complex, together with a particular presentation $(X = X_0 \to X_1 \to \cdots \to X_\beta \to \cdots (\beta < \lambda), \{S^\beta, \sigma^\beta, p^\beta\}_{\beta<\lambda})$.

Now if

$$(f : X \to Y, X = X_0 \to X_1 \to \cdots \to X_\beta \to \cdots (\beta < \lambda), \{S^\beta, \sigma^\beta, p^\beta\}_{\beta<\lambda})$$

is a presented relative I-cell complex, then the *set of cells of f*[44] is $\coprod_{\beta<\lambda} S^\beta$, and the *size of f* is the cardinal thereof.

For a relative I-cell complex $f : X \to Y$, a *presentation of f*[45] is given by a λ-sequence $X = X_0 \to X_1 \to \cdots \to X_\beta \to \cdots$ for $\beta < \lambda$, for some ordinal λ, and a sequence $\{(S^\beta, \sigma^\beta, p^\beta)\}_{\beta<\lambda}$, such that for all $\beta < \lambda$, S^β is a set, there is a map $\sigma^\beta : S^\beta \to I$, such that if $i \in S^\beta$, then $\sigma_i^\beta : C_i \to D_i$ is a map in I, and there is a map $p_i^\beta : C_i \to X_\beta$ such that one has a pushout

$$\begin{array}{ccc} \coprod_{S^\beta} C_i & \longrightarrow & \coprod_{S^\beta} D_i \\ {\scriptstyle \coprod p_i^\beta}\downarrow & & \downarrow \\ X_\beta & \longrightarrow & X_{\beta+1}. \end{array}$$

Let C be a category closed under small colimits, and let I be a set of maps in C. The subcategory of *relative I-cell complexes*[46] is the subcategory of maps obtained from transfinite compositions of pushouts of elements of I.

40 Def. 12.1.1 [40].

41 Def. 11.4.1 [40].

42 Def. 10.8.1 [40].

43 Def. 10.6.3 [40].

44 Def. 10.6.4 [40].

45 Def. 10.6.2 [40].

46 Def. 10.5.8 [40].

If $\mathcal{M}$ is a cocomplete category, I is a set of maps in $\mathcal{M}$, and κ is a cardinal, then an object is said to be *κ-small relative to I*[47] if it is κ-small relative to the subcategory of relative I-cell complexes. An object is *small relative to I* if it is κ-small relative to I for some κ.

Let $\mathcal{C}$ be a category closed under pushouts, and let $f : A \to B$ be a map in $\mathcal{C}$. It is said to be an *effective monomorphism*[48] if it is the equalizer of the pair of natural inclusions $B \rightrightarrows B \coprod_A B$.

4.7 A result on functor categories

Since we have been discussing cellular model categories, for later purposes, we can mention the following result about functor categories built from such categories. For a cellular model category $\mathcal{M}$ and a small category $\mathcal{C}$, $\mathcal{M}^{\mathcal{C}}$ can be endowed with a cellular model structure as follows:[49] by virtue of Theorem 11.6.1 of [40], if $\mathcal{M}$ is a cofibrantly generated model category, then $\mathcal{M}^{\mathcal{C}}$ is also a cofibrantly generated model category with generating cofibrations $I = F_I^{\mathcal{C}}$ and generating trivial cofibrations $J = F_J^{\mathcal{C}}$. A map $X \to Y$ is a weak equivalence if $X_\alpha \to Y_\alpha$ is a weak equivalence in $\mathcal{M}$ for all $\alpha \in \mathcal{C}$, it is a fibration if $X_\alpha \to Y_\alpha$ is a fibration in $\mathcal{M}$ for all $\alpha \in \mathcal{C}$, and it is a cofibration if it is a retract of a transfinite composition of pushouts of elements of I.

Let $\mathcal{M}$ be a model category and $\mathcal{C}$ a small category, and let K be a set of maps in $\mathcal{M}$. Then one defines[50] $F_K^{\mathcal{C}} \subset \operatorname{Mor} \mathcal{M}^{\mathcal{C}}$ as the set $\{F_{A_k}^\alpha \to F_{B_k}^\alpha\}$ where $A_k \to B_k$ is a map in K and $\alpha \in \mathcal{C}$.

For a small category $\mathcal{C}$, $\alpha \in \operatorname{Ob} \mathcal{C}$, a cocomplete category $\mathcal{M}$, and $X \in \operatorname{Ob} \mathcal{M}$, the *free diagram on X generated by α*[51] is the $\mathcal{C}$-diagram in $\mathcal{M}$ defined by $F_X^\alpha = X \otimes F_*^\alpha$ defined pointwise by $F_X^\alpha(\beta) = \coprod_{\operatorname{Hom}_{\mathcal{C}}(\alpha,\beta)} X$.

For a small category $\mathcal{C}$ and $\alpha \in \operatorname{Ob} \mathcal{C}$, the *free $\mathcal{C}$-diagram*[52] of sets generated at position α is the $\mathcal{C}$-diagram of a set F_*^α such that $F_*^\alpha(\beta) = \operatorname{Hom}_{\mathcal{C}}(\alpha, \beta)$ for $\beta \in \operatorname{Ob} \mathcal{C}$ and $(F_*^\alpha(g))(h) = gh$ if $h \in F_*^\alpha(\beta)$ and $g : \beta \to \gamma$ is in $\mathcal{C}_1$.

4.8 Homotopy limits and colimits

Two ubiquitous notions in relative Algebraic Geometry are those of homotopy limit and homotopy colimit. First, if $\mathcal{M}$ is a model category, then it is said to be *right*

47 Def. 105.12 [40].

48 Def. 10.9.1 [40].

49 Prop. 12.1.5 [40].

50 Def. 11.5.33 [40].

51 Def. 111.5.25 [40].

52 Def. 11.5.7 [40].

proper[53] if any pullback of a weak equivalence along a fibration is still a weak equivalence.

Now let $\mathcal{M}$ be a right proper model category, and let E be any fixed functorial factorization of maps: if $g : X \to Y$ is any map, then it factors as $X \xrightarrow{i_g} E(g) \xrightarrow{p_g} Y$, where i_g is a trivial cofibration, and p_g is a fibration. Then the *homotopy pullback* of the diagram $X \xrightarrow{g} Z \xleftarrow{h} Y$ is the pullback of $E(g) \xrightarrow{p_g} Z \xleftarrow{p_h} E(h)$.

A very useful result is provided by Proposition 13.5.3 of [40], which states that for a R proper model category $\mathcal{M}$, given a diagram

$$\begin{array}{ccccc} X & \xrightarrow{g} & Z & \xleftarrow{h} & Y \\ \downarrow & & \downarrow & & \downarrow \\ \tilde{X} & \xrightarrow[\tilde{g}]{} & \tilde{Z} & \xleftarrow[\tilde{h}]{} & \tilde{Y} \end{array}$$

where the vertical maps are equivalences, the induced map of homotopy pullbacks

$$E(g) \times_Z E(h) \to E(\tilde{g}) \times_{\tilde{Z}} E(\tilde{h})$$

is a weak equivalence.

4.9 Reedy structures

One important type of category we use for localizations of functor categories in the context of stacks is that of a Reedy category. By definition a *Reedy category*[54] is a small category $\mathcal{C}$ along with two subcategories $\mathcal{C}^+$ and $\mathcal{C}^-$ called the *direct subcategory* and the *inverse subcategory*, respectively, both containing all objects of $\mathcal{C}$, but such that their objects can be assigned a degree in such a manner that every nonidentity morphism of $\mathcal{C}^+$ raises the degree, whereas every nonidentity morphism of $\mathcal{C}^-$ lowers the degree, and we ask also that every morphism f of $\mathcal{C}$ has a unique factorization as $f = f^+ \circ f^-$, where $f^+ \in \operatorname{Mor} \mathcal{C}^+$ and $f^- \in \operatorname{Mor} \mathcal{C}^-$.

Let $\mathcal{C}$ be a Reedy category, and let $\alpha \in \operatorname{Ob} \mathcal{C}$. One defines the *latching category*[55] $\partial \mathcal{C}^+_{/\alpha}$ of $\mathcal{C}$ at α to be the full subcategory of $\mathcal{C}^+_{/\alpha}$ containing all objects but id_α. In the same manner, one defines the *matching category* $\partial \mathcal{C}^-_{\alpha/}$ of $\mathcal{C}$ at α to be the full subcategory of $\mathcal{C}^-_{\alpha/}$ containing all objects save id_α.

If $\mathcal{C}$ is a Reedy category and $\mathcal{M}$ is a model category, then we consider the functor category $\mathcal{M}^{\mathcal{C}}$. Let X be an object thereof, and let $\alpha \in \operatorname{Ob} \mathcal{C}$. We abuse notation and also

53 Def. 13.1.1 [40].

54 Def. 15.1.2 [40].

55 Def. 15.2.3 [40].

denote by X the induced $\partial\mathcal{C}^{+}_{/\alpha}$-diagram defined pointwise by $X_{(\beta\to\alpha)} = X_\beta$ and the induced $\partial\mathcal{C}^{-}_{\alpha/}$-diagram defined by $X_{(\alpha\to\beta)} = X_\beta$. One defines the *latching object*[56] of X at α to be $L_\alpha X = \operatorname{colim}_{\partial\mathcal{C}^{+}_{/\alpha}} X$, and one defines the *matching object* of X at α to be given by $M_\alpha X = \lim_{\partial\mathcal{C}^{-}_{\alpha/}} X$.

Let now $\mathcal{C}$ be a Reedy category, let $\mathcal{M}$ be a model category, and let $X, Y \in \mathcal{M}^{\mathcal{C}}$. Following Definition 15.3.3 of [40], one defines a map $f : X \to Y$ to be a *Reedy weak equivalence* if it is a pointwise weak equivalence in $\mathcal{M}$, that is, $f_\alpha : X_\alpha \to Y_\alpha$ is a weak equivalence for all $\alpha \in \operatorname{Ob}\mathcal{C}$, and one says f is a *Reedy cofibration* if the *relative latching map* $X_\alpha \coprod_{L_\alpha X} L_\alpha Y \to Y_\alpha$ is a cofibration in $\mathcal{M}$ for all α, and finally one says f is a *Reedy fibration* if the *relative matching map* $X_\alpha \to Y_\alpha \times_{M_\alpha Y} M_\alpha X$ is a fibration in $\mathcal{M}$ for all α.

By Theorem 15.3.4 of [40], the Reedy equivalences, Reedy fibrations, and Reedy cofibrations as defined above give a model structure $\mathcal{M}^{\mathcal{C}}$, which one calls the *Reedy model structure*. Additionally, if $\mathcal{M}$ is a left/right proper model category, then so is $\mathcal{M}^{\mathcal{C}}$.

Observe that if $\mathcal{C}$ is a Reedy category, $\mathcal{M}$ is a model category, and $f : X \to Y$ is a morphism in $\mathcal{M}^{\mathcal{C}}$, supposing f is a Reedy fibration, then Proposition 15.3.11 of [40] tells us that for all $\alpha \in \operatorname{Ob}\mathcal{C}$, $f_\alpha : X_\alpha \to Y_\alpha$ and $M_\alpha X \to M_\alpha Y$ are fibrations in $\mathcal{M}$, and if X is Reedy fibrant, then X_α and $M_\alpha X$ are fibrant as well in $\mathcal{M}$ for all α.

Suppose now that $\mathcal{M}$ is a simplicial model category and that $\mathcal{C}$ is a small category. Let $X \in \mathcal{M}^{\mathcal{C}}$ and $K \in \mathrm{Set}_\Delta$. Following Definition 11.7.1 of [40], we define a simplicial structure on $\mathcal{M}^{\mathcal{C}}$ as follows: one defines $X \otimes K \in \mathcal{M}^{\mathcal{C}}$ pointwise by $(X \otimes K)_\alpha = X_\alpha \otimes K$ for $\alpha \in \operatorname{Ob}\mathcal{C}$, and one defines $Y^K \in \mathcal{M}^{\mathcal{C}}$ pointwise by $(Y^K)_\alpha = (Y_\alpha)^K$. The mapping space is then provided by $\operatorname{Map}(X, Y)_n = \operatorname{Hom}_{\mathcal{M}^{\mathcal{C}}}(X \otimes \Delta^n, Y)$.

If $\mathcal{M}$ is a simplicial model category and $\mathcal{C}$ is a Reedy category, then the above simplicial structure on the Reedy model category $\mathcal{M}^{\mathcal{C}}$ makes it into a simplicial model category.[57]

Now let $\mathcal{C}$ be a Reedy category, let $\mathcal{M}$ be a cofibrantly generated model category, and let $X, Y \in \mathcal{M}^{\mathcal{C}}$. According to Proposition 15.6.3 of [40], if $f : X \to Y$ is a Reedy fibration, then it is also a fibration in the cofibrantly generated model structure on $\mathcal{M}^{\mathcal{C}}$, and if f is a cofibration in the cofibrantly generated model structure of $\mathcal{M}^{\mathcal{C}}$, then it is a Reedy cofibration.

Now let $\mathcal{M}$ be a cellular model category, and let $\mathcal{C}$ be a Reedy category. Then by Theorem 15.7.6 of [40] the Reedy model structure on $\mathcal{M}^{\mathcal{C}}$ is cellular.

A useful result is the following: by Theorem 15.8.7 of [40], in the Reedy model structure of $\mathrm{Set}_\Delta^{\Delta^{\mathrm{op}}}$, a morphism of bisimplicial sets is a cofibration iff it is a monomorphism.

By Corollary 15.8.8 of [40] a simplicial object in Set_Δ is always Reedy cofibrant.

56 Def. 15.2.5 [40].

57 Thm 15.3.4 [40].

Let $\mathcal{M}$ be a model category. Consider

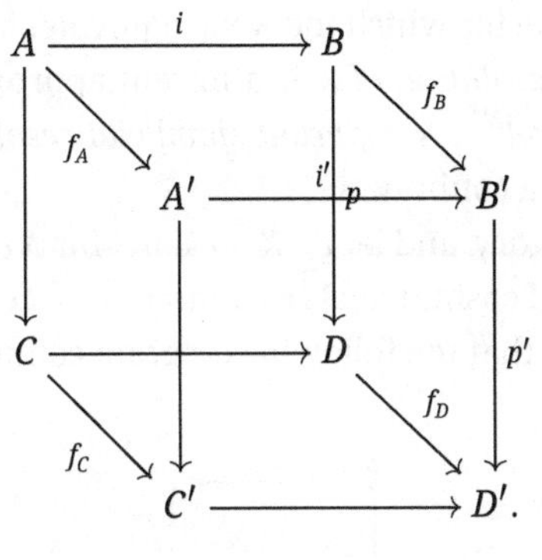

If all objects are fibrant, the squares in the front and back are pullbacks, p and p' are fibrations, and f_B, f_C, and f_D are weak equivalences, then by Proposition 15.10.10 of [40], f_A is a weak equivalence as well, a very useful result.

Let $X \in \mathrm{Set}_\Delta^{\Delta^{\mathrm{op}}} = \mathrm{sSet}_\Delta$. The *geometric realization*[58] $|X|$ of X is the object of Set_Δ defined as the coequalizer of

$$\coprod_{\alpha:[n]\to[k]} X_n \times \Delta^k \underset{\psi}{\overset{\phi}{\rightrightarrows}} \coprod_{n\geq 0} X_n \times \Delta^n$$

where for $\alpha : [n] \to [k]$ in Δ^{op}, $\phi|_\alpha = \mathrm{id}_{X_n} \times \alpha^*$, and $\psi|_\alpha = \alpha_* \times \mathrm{id}_{\Delta^k} : X_n \times \Delta^k \to X_k \times \Delta^k$.

For a bisimplicial set X, its *diagonal*[59] is the object of Set_Δ defined by $(\mathrm{diag}X)_n = X_{n,n}$ for $n \geq 0$.

By Theorem 15.11.6 of [40], if $X \in \mathrm{sSet}_\Delta$, then $|X| \cong \mathrm{diag}X$.

By Theorem 15.11.9 of [40], if $X \in \mathrm{sSet}_\Delta$, $Y : \Delta \to \mathrm{Set}_\Delta$, and $Z \in \mathrm{Set}_\Delta$, then one has a natural isomorphism in Set_Δ:

$$\mathrm{Map}(X \otimes_\Delta Y, Z) \cong \mathrm{Map}(Y, \mathrm{Map}(X, Z)).$$

A convenient result is the following: if $\mathcal{C}$ is a Reedy category, $\mathcal{M}$ is a simplicial model category, X is Reedy cofibrant in $\mathcal{M}^{\mathcal{C}}$, and Y is fibrant in $\mathcal{M}$, then by Lemma 15.11.10 of [40], $\mathrm{Map}(X, Y)$ is Reedy fibrant.

We now apply this formalism of having a Reedy model structure to the need of having resolutions, whether they be fibrant or cofibrant. For a model category $\mathcal{M}$ and $X \in \mathrm{Ob}\,\mathcal{M}$, let cc_*X denote the *constant cosimplicial object at* X, and let cs_*X denote

58 Def. 15.11.1 [40].

59 Def. 15.11.3 [40].

the *constant simplicial object at* X.[60] One defines a *cosimplicial resolution of* X[61] to be a cofibrant approximation $\tilde{X} \to \mathrm{cc}_* X$ in the Reedy model structure of $\mathcal{M}^{\Delta}$. A *fibrant cosimplicial resolution* is one for which the weak equivalence $\tilde{X} \to \mathrm{cc}_* X$ is a Reedy trivial fibration. A *simplicial resolution of* X is a fibrant approximation $\mathrm{cs}_* X \to \hat{X}$ in the Reedy model structure of $\mathcal{M}^{\Delta^{\mathrm{op}}}$. A *cofibrant simplicial resolution* is one for which the weak equivalence above is a cofibration.

Let $\mathcal{M}$ be a model category, and let $g : X \to Y$ in Mor $\mathcal{M}$. One defines a *cosimplicial resolution of* g[62] to consist of cosimplicial resolutions $\tilde{X} \to \mathrm{cc}_* X$ and $\tilde{Y} \to \mathrm{cc}_* Y$, together with a map $\tilde{g} : \tilde{X} \to \tilde{Y}$ such that the following diagram commutes:

$$
\begin{array}{ccc}
\tilde{X} & \xrightarrow{\tilde{g}} & \tilde{Y} \\
\downarrow & & \downarrow \\
\mathrm{cc}_* X & \longrightarrow & \mathrm{cc}_* Y
\end{array}
$$

and one defines a *simplicial resolution of* g to consist of simplicial resolutions $\mathrm{cs}_* X \to \hat{X}$ and $\mathrm{cs}_* Y \to \hat{Y}$ along with a map $\hat{g} : \hat{X} \to \hat{Y}$ such that the following diagram commutes:

$$
\begin{array}{ccc}
\mathrm{cs}_* X & \longrightarrow & \mathrm{cs}_* Y \\
\downarrow & & \downarrow \\
\hat{X} & \xrightarrow[\hat{g}]{} & \hat{Y}.
\end{array}
$$

By Proposition 16.1.22 of [40], for a model category $\mathcal{M}$ and a morphism $g : X \to Y$ in $\mathcal{M}$, there exists a natural cosimplicial resolution $\tilde{g} : \tilde{X} \to \tilde{Y}$ of g such that $\tilde{X}$ and $\tilde{Y}$ are fibrant cosimplicial resolutions and $\tilde{g}$ is a Reedy cofibration, and there exists a natural simplicial resolution $\hat{g} : \hat{X} \to \hat{Y}$ of g such that $\hat{X}$ and $\hat{Y}$ are cofibrant simplicial resolutions and $\hat{g}$ is a Reedy fibration.

Now let $\mathcal{M}$ be a model category, and let $g : X \to Y$ be a weak equivalence in $\mathcal{M}$. Then by Proposition 16.1.24 of [40] any (co)simplicial resolution of g is a Reedy weak equivalence.

If $\mathcal{M}$ is a model category and $\tilde{X}$ is a cosimplicial object of $\mathcal{M}$, then one says $\tilde{X}$ is a *cosimplicial resolution*[63] if there is $X \in \mathcal{M}$ such that $\tilde{X}$ is a cosimplicial resolution of X. In the same manner, a simplicial object of $\mathcal{M}$ is said to be a *simplicial resolution* if it is the simplicial resolution of some object of $\mathcal{M}$.

60 Not. 16.1.1 [40].

61 Def. 16.1.2 [40].

62 Def. 16.1.20 [40].

63 Def. 16.1.26 [40].

Let $\mathcal{M}$ be a model category, let $\mathbf{X}$ be a cosimplicial object of $\mathcal{M}$, and let $Y \in \operatorname{Ob}\mathcal{M}$. Then $\operatorname{Hom}_{\mathcal{M}}(\mathbf{X}, Y) \in \mathrm{Set}_\Delta$ is defined by $\operatorname{Hom}_{\mathcal{M}}(\mathbf{X}, Y)_n = \operatorname{Hom}_{\mathcal{M}}(\mathbf{X}^n, Y)$. If in addition $\mathbf{Y}$ is a simplicial object in $\mathcal{M}$, then one defines the bisimplicial set $\operatorname{Hom}_{\mathcal{M}}(\mathbf{X}, \mathbf{Y})$ by $\operatorname{Hom}_{\mathcal{M}}(\mathbf{X}, \mathbf{Y})_{n,p} = \operatorname{Hom}_{\mathcal{M}}(\mathbf{X}^p, \mathbf{Y}_n)$. Now if X is just an object of $\mathcal{M}$ but $\mathbf{Y}$ is a simplicial object of $\mathcal{M}$, then one defines $\operatorname{Hom}_{\mathcal{M}}(X, \mathbf{Y}) \in \mathrm{Set}_\Delta$ by $\operatorname{Hom}_{\mathcal{M}}(X, \mathbf{Y})_n = \operatorname{Hom}_{\mathcal{M}}(X, \mathbf{Y}_n)$. If in addition $\mathbf{X}$ is a cosimplicial object of $\mathcal{M}$, then one defines $\operatorname{diagHom}_{\mathcal{M}}(\mathbf{X}, \mathbf{Y}) \in \mathrm{Set}_\Delta$ by $(\operatorname{diagHom}_{\mathcal{M}}(\mathbf{X}, \mathbf{Y}))_n = \operatorname{Hom}_{\mathcal{M}}(\mathbf{X}^n, \mathbf{Y}_n)$.

By Theorem 16.4.2 of [40], for a model category $\mathcal{M}$, a cosimplicial object $\mathbf{A}$ of $\mathcal{M}$, $X \in \operatorname{Ob}(\mathcal{M})$, and $K \in \mathrm{Set}_\Delta$, one has a natural isomorphism in Set_Δ:

$$\operatorname{Hom}_{\mathrm{Set}_\Delta}(K, \operatorname{Hom}_{\mathcal{M}}(\mathbf{A}, X)) \cong \operatorname{Hom}_{\mathcal{M}}(\mathbf{A} \otimes K, X).$$

If $\mathbf{Y}$ is a simplicial object of $\mathcal{M}$, $B \in \operatorname{Ob}(\mathcal{M})$, and $K \in \mathrm{Set}_\Delta$, then there is a natural isomorphism in Set_Δ:

$$\operatorname{Hom}_{\mathrm{Set}_\Delta}(K, \operatorname{Hom}_{\mathcal{M}}(B, \mathbf{Y})) \cong \operatorname{Hom}_{\mathcal{M}}(B, \mathbf{Y}^K).$$

Here are a few important results to deal with homotopy function complexes as objects: if $\mathcal{M}$ is a model category, $\mathbf{B}$ is a cosimplicial resolution in $\mathcal{M}$, and $X \in \mathcal{M}_f$, then by Corollary 16.5.3 of [40], $\operatorname{Hom}_{\mathcal{M}}(\mathbf{B}, X)$ is a fibrant simplicial set.

In a model category $\mathcal{M}$, if $i : \mathbf{A} \to \mathbf{B}$ is a Reedy cofibration of cosimplicial resolutions and $X \in \mathcal{M}_f$, then $i^* : \operatorname{Hom}_{\mathcal{M}}(\mathbf{B}, X) \to \operatorname{Hom}_{\mathcal{M}}(\mathbf{A}, X)$ is a fibration in Set_Δ, trivial if i is a Reedy trivial cofibration.[64]

If now i is a Reedy weak equivalence of cosimplicial resolutions, then i^* is a weak equivalence of fibrant simplicial sets.[65]

For $B \in \mathcal{M}_c$ and a Reedy weak equivalence $p : \mathbf{X} \to \mathbf{Y}$ of simplicial resolutions in $\mathcal{M}$, $p_* : \operatorname{Hom}_{\mathcal{M}}(B, \mathbf{X}) \to \operatorname{Hom}_{\mathcal{M}}(B, \mathbf{Y})$ is a weak equivalence of fibrant objects in Set_Δ.

To argue that all homotopy function complexes are the same for practical purposes, we use Theorem 17.4.14 of [40], which states that if $\mathcal{M}$ is a model category and $X, Y \in \operatorname{Ob}\mathcal{M}$, then any two fixed homotopy function complexes between two objects are connected by an essentially unique zig-zag of change of homotopy function complexes.

If $\mathcal{M}$ is a model category and $X, Y \in \operatorname{Ob}\mathcal{M}$, a *change of homotopy function complexes*[66] is a finite composition in $(\mathcal{M}^{\Delta}_{\mathrm{cc}_* X/})^{\mathrm{op}} \times (\mathcal{M}^{\Delta^{\mathrm{op}}}_{\mathrm{cs}_* Y/})$ consisting of a combination of changes of L/R/2-sided homotopy function complex maps or of L/R to 2-sided change of homotopy function complex maps.

One defines a *change of L homotopy function complex map*[67] $(\tilde{\mathbf{X}}, \hat{Y}, \operatorname{Hom}_{\mathcal{M}}(\tilde{\mathbf{X}}, \hat{Y})) \to (\tilde{\mathbf{X}}', \hat{Y}', \operatorname{Hom}_{\mathcal{M}}(\tilde{\mathbf{X}}', \hat{Y}'))$ to be given by a map of cosimplicial resolutions in $\mathcal{M}$, $f : \tilde{\mathbf{X}}' \to \tilde{\mathbf{X}}$,

---

64 Cor. 16.5.4 [40].

65 Cor. 16.5.5 [40].

66 Def. 17.4.7 [40].

67 Def. 17.1.5 [40].

a map of fibrant approximations in $\mathcal{M}$, $g : \hat{Y} \to \hat{Y}'$, and a map $h : \mathrm{Hom}_{\mathcal{M}}(\tilde{\mathbf{X}}, \hat{Y}) \to \mathrm{Hom}_{\mathcal{M}}(\tilde{\mathbf{X}}', \hat{Y}')$ in Set_Δ, induced by f and g.

In the same manner, one defines a *change of R homotopy function complex map*[68] $(\tilde{X}, \hat{\mathbf{Y}}, \mathrm{Hom}_{\mathcal{M}}(\tilde{X}, \hat{\mathbf{Y}})) \to (\tilde{X}', \hat{\mathbf{Y}}', \mathrm{Hom}_{\mathcal{M}}(\tilde{X}', \hat{\mathbf{Y}}'))$ to be given by a map of cofibrant approximations to X: $f : \tilde{X}' \to \tilde{X}$, a map of simplicial resolutions in $\mathcal{M}$, $g : \hat{\mathbf{Y}} \to \hat{\mathbf{Y}}'$, and a map $h : \mathrm{Hom}_{\mathcal{M}}(\tilde{X}, \hat{\mathbf{Y}}) \to \mathrm{Hom}_{\mathcal{M}}(\tilde{X}', \hat{\mathbf{Y}}'))$ in Set_Δ, induced by f and g.

A *change of 2-sided homotopy function complex map*[69] $(\tilde{\mathbf{X}}, \hat{\mathbf{Y}}, \mathrm{diagHom}_{\mathcal{M}}(\tilde{\mathbf{X}}, \hat{\mathbf{Y}})) \to (\tilde{\mathbf{X}}', \hat{\mathbf{Y}}', \mathrm{diagHom}_{\mathcal{M}}(\tilde{\mathbf{X}}', \hat{\mathbf{Y}}'))$ is given by a map of cosimplicial resolutions of X, $f : \tilde{\mathbf{X}}' \to \tilde{\mathbf{X}}$, a map of simplicial resolutions of Y, $g : \hat{\mathbf{Y}} \to \hat{\mathbf{Y}}'$, and a map $h : \mathrm{diagHom}_{\mathcal{M}}(\tilde{\mathbf{X}}, \hat{\mathbf{Y}}) \to \mathrm{diagHom}_{\mathcal{M}}(\tilde{\mathbf{X}}', \hat{\mathbf{Y}}')$ in Set_Δ, induced by f and g.

Finally, a *L to 2-sided change of homotopy function complex map*[70] is a map

$$(\tilde{\mathbf{X}}, \mathrm{cs}_*\hat{\mathbf{Y}}_0, \mathrm{diagHom}_{\mathcal{M}}(\tilde{\mathbf{X}}, \mathrm{cs}_*\hat{\mathbf{Y}}_0)) \to (\tilde{\mathbf{X}}, \hat{\mathbf{Y}}, \mathrm{diagHom}_{\mathcal{M}}(\tilde{\mathbf{X}}, \hat{\mathbf{Y}}))$$

the identity on $\tilde{\mathbf{X}}$, and a *R to 2-sided change of homotopy function complex map* is a map

$$(\mathrm{cc}_*\tilde{\mathbf{X}}_0, \hat{\mathbf{Y}}, \mathrm{diagHom}_{\mathcal{M}}(\mathrm{cc}_*\tilde{\mathbf{X}}_0, \hat{\mathbf{Y}})) \to (\tilde{\mathbf{X}}, \hat{\mathbf{Y}}, \mathrm{diagHom}_{\mathcal{M}}(\tilde{\mathbf{X}}, \hat{\mathbf{Y}}))$$

that is the identity on $\mathbf{Y}$, where in both instances, we used the embedding $\mathcal{M} \hookrightarrow \mathcal{M}^\Delta$ (or $\mathcal{M}^{\Delta^{\mathrm{op}}}$) to see an object of $\mathcal{M}$ as a constant simplicial or cosimplicial object.

By Theorem 17.6.3 of [40], for a model category $\mathcal{M}$, $W, X, Y, Z \in \mathrm{Ob}\,\mathcal{M}$, and an equivalence $g : X \to Y$, any map of homotopy function complexes $g_* : \mathrm{map}(W, X) \to \mathrm{map}(W, Y)$ induced by g is a weak equivalence of fibrant simplicial sets. Likewise, any map of homotopy function complexes $g^* : \mathrm{map}(Y, Z) \to \mathrm{map}(X, Z)$ induced by g is a weak equivalence of fibrant simplicial sets.

By virtue of Proposition 17.7.6 of [40], for a model category $\mathcal{M}$, $g : X \to Y$ is a weak equivalence iff either g induces weak equivalences of homotopy function complexes

$$g_* : \mathrm{map}(X, X) \cong \mathrm{map}(X, Y),$$
$$g_* : \mathrm{map}(Y, X) \cong \mathrm{map}(Y, Y)$$

or g induces weak equivalences of homotopy function complexes

$$g^* : \mathrm{map}(Y, X) \cong \mathrm{map}(X, X),$$
$$g^* : \mathrm{map}(Y, Y) \cong \mathrm{map}(X, Y).$$

68 Def. 17.2.5 [40].

69 Def. 17.3.3 [40].

70 Def. 17.4.5 [40].

Now let $\mathcal{M}$ be a model category, and let $g : X \to Y$ be a morphism in $\mathcal{M}$. By Theorem 17.7.7 of [40] the following are equivalent:

1- g is a weak equivalence.

2- For all $W \in \mathcal{M}$, we have a weak equivalence $g_* : \mathrm{map}(W, X) \simeq \mathrm{map}(W, Y)$.

3- For all $W \in \mathcal{M}_c$, we have a weak equivalence $g_* : \mathrm{map}(W, X) \simeq \mathrm{map}(W, Y)$.

4- For all $Z \in \mathcal{M}$, we have a weak equivalence $g^* : \mathrm{map}(Y, Z) \simeq \mathrm{map}(X, Z)$.

5- For all $Z \in \mathcal{M}_f$, we have a weak equivalence $g^* : \mathrm{map}(Y, Z) \simeq \mathrm{map}(X, Z)$.

Part II: **Derived algebraic geometry**

5 Introduction

One can view Derived Algebraic Geometry as a generalization of classical Algebraic Geometry; it is a theory deep enough to resolve some classical constructions that would otherwise remain obscure. We are in particular thinking of what happens when working with some singular moduli spaces. In a situation such as this one, we are proponents of the hidden smoothness argument [49], according to which singular objects are merely truncations of smooth higher objects. Derived Algebraic Geometry provides the correct setting for defining such higher objects.

Our interest is to develop a general theory of Algebraic Geometry, as done in [76], [78], [79], an elementary subcase thereof being that of classical Algebraic Geometry, obtained by first considering $\mathbb{Z}$-Mod, regarded as a symmetric monoidal category. Taking Comm($\mathbb{Z}$-Mod) = CRing and then its opposite category, one gets a category Aff of affine schemes. Putting a topology on this category, one can study various presheaves, prestacks, and their accompanying notions on it. The idea then is to just generalize this formalism by starting with a base symmetric monoidal category, and this would place us within the realm of relative Algebraic Geometry [37]. More specifically, we consider symmetric monoidal model categories as this will allow us to consider useful localization techniques. At this point, one is working in Homotopical Algebraic Geometry. Abstractly, by loosening some structures one gains in flexibility, which is something we strive to achieve throughout this work. Thus our approach is purely homotopical.

Following this prescription, one starts with a symmetric monoidal model category $(\mathcal{C}, \otimes)$ in lieu of the usual category $(\mathbb{Z}\text{-Mod}, \otimes)$. We consider its category of commutative monoid objects Comm($\mathcal{C}$), itself a model category by assumption (see [79]), and one denotes its opposite model category by $\mathcal{C}$-Aff or $\mathcal{A}\mathit{ff}_{\mathcal{C}}$. This will be our model category of relative schemes, or stacks, depending on the context. At this point, this language can be further abstracted by considering prestacks and stacks on model categories that we generically denote by M, an example being $M = \mathcal{A}\mathit{ff}_{\mathcal{C}}$. The model category $M^{\wedge}$ of prestacks on M is the left Bousfield localization of sPr(M) with respect to equivalences in M, and if one puts a topology τ on M making it into a model site, then one can define a notion of local equivalence in sPr(M), which allows us to define stacks on the model site (M, τ) as being those objects of $M^{\wedge}$ that are local with respect to local equivalences. It turns out one can also obtain $M^{\sim,\tau}$, the model category of stacks on (M, τ), as a left Bousfield localization of $M^{\wedge}$ with respect to a certain class of homotopy hypercovers. That construction itself follows from the construction of stacks on simplicial sites that consists in starting with simplicial presheaves on simplicial categories. This will dictate how we introduce stacks; we will first start by defining stacks on simplicial sites, following which we will use this material to define stacks over model sites (M, τ), to finally apply this to the case $M = \mathcal{A}\mathit{ff}_{\mathcal{C}}$.

Many constructions surrounding stacks will use the fact that for a given category $\mathcal{C}$, the category of simplicial presheaves sPr($\mathcal{C}$) on $\mathcal{C}$ is a simplicial model category. Thus presently we show how this comes about, as shown in [78]. Denote by $\underline{\mathrm{Hom}}$ the simplicial

https://doi.org/10.1515/9783111334073-005

Hom of the simplicial category $sPr(\mathcal{C})$. The simplicial model structure is given by having this simplicial category tensored and cotensored over Set_Δ: for $K \in Set_\Delta$ and $F \in sPr(\mathcal{C})$, one defines $(K \times F)(x) = K \times F(x)$ for $x \in \mathcal{C}$, and $(F^K)(x) = \underline{Hom}_{Set_\Delta}(K, F(x))$, where $\underline{Hom}_{Set_\Delta}$ is the simplicial Hom for the simplicial model category Set_Δ. The simplicial set of morphisms between objects F and G of $sPr(\mathcal{C})$ is given by $\underline{Hom}(F, G)_n = Hom(\Delta^n \times F, G)$. There is an adjunction isomorphism

$$Hom_{sPr(\mathcal{C})}(K \times F, G) \cong \underline{Hom}_{Set_\Delta}(K, \underline{Hom}(F, G)) \cong Hom_{sPr(\mathcal{C})}(F, G^K)$$

showing that $K \times - \dashv \exp_K(-)$.

6 Stacks on simplicial sites

We are ultimately interested in defining derived stacks, which consists in working with $\mathcal{A}\!f\!f_{\mathcal{C}} = \mathrm{Comm}(\mathcal{C})^{\mathrm{op}}$ as our base model category, with $\mathcal{C} = sk\text{-Mod}$, on which we define stacks. Abstracting this, it suffices to define stacks over model sites (M, τ). We will see later that this is Quillen equivalent to defining stacks over their simplicial localizations LM, which are simplicial categories. Thus one can equivalently focus on developing a theory of stacks over simplicial categories. Since this is a ubiquitous theory, we start with this formalism. It turns out that if T is a simplicial category with an appropriately chosen topology τ, then one can define a model category of stacks on (T, τ), which is typically denoted by $\mathrm{sPr}_{\tau}(T)$. If we have $T = LM$ for a model category M, then Theorem 4.7.1 of [76] ensures that there is a Quillen equivalence $M^{\sim,\tau} \simeq \mathrm{sPr}_{\tau}(LM)$, where $M^{\sim,\tau}$ is the model category of stacks over (M, τ), thereby showing that we get equivalent theories of stacks.

The material is fairly lengthy, so we will present it in a top-down fashion. All definitions and accompanying constructions can be found in [76].

We have the following result:[1] if (T, τ) is an S-site, that is, a simplicial category with a topology, then one has a model structure on $\mathrm{sPr}(T)$ called the *local projective model structure*,[2] with π_*-equivalences as weak equivalences, defined using homotopy sheaves and cofibrations for the projective model structure on $\mathrm{sPr}(T)$, the projective model structure having pointwise equivalences and fibrations. The resulting model category is denoted $\mathrm{sPr}_{\tau}(T)$ and is called the *model category of stacks* on the S-site (T, τ). One can also represent $\mathrm{sPr}_{\tau}(T)$ as a left Bousfield localization of $\mathrm{sPr}(T)$ with respect to a certain class of morphisms.[3] The construction of this class is fairly technical but useful for the following reason; if the local projective model structure allows one to obtain the model category of stacks on the S-site (T, τ), then this does not tell us what a stack, an object of $\mathrm{sPr}_{\tau}(T)$, looks like. Using the class used in the construction of the left Bousfield localization of the model category of prestacks $\mathrm{sPr}(T)$ to obtain the model category of stacks, one can define a hyperdescent condition that clearly characterizes objects of $\mathrm{sPr}_{\tau}(T)$. We will come back to this construction after we discuss the local projective model structure, which we do presently, by first introducing all the requisite notions.

Following [76], one defines[4] an *S-category* to be a category enriched over Set_{Δ}, that is, a simplicial category. We denote by $\underline{\mathrm{Hom}}_T$ the simplicial set of morphisms in T. An *S-topology* on an S-category T is a Grothendieck topology on $\mathrm{Ho}\,(T)$,[5] and one calls such a pair (T, τ) an *S-site*.

1 Thm. 3.4.1 [76].

2 Thm. 3.4.1 [76].

3 Cor. 3.4.5 [76].

4 Def. 2.1.1 [76].

5 Def. 3.1.1 [76].

https://doi.org/10.1515/9783111334073-006

The homotopy category Ho (T) of T is defined as follows: Ob Ho (T) = Ob (T), and $\mathrm{Hom}_{\mathrm{Ho}\,(T)} = \pi_0\underline{\mathrm{Hom}}_T$.

We now define π_*-equivalences. By Definition 3.2.2 of [76] a morphism $F \to G$ in sPr(T) is a *π_*-equivalence* if the induced morphism $\pi_0(F) \to \pi_0(G)$ is an isomorphism of sheaves on Ho (T), and for all $x \in$ Ob (T), $s \in \pi_0(F(x))$, and $n > 0$, one has an isomorphism of sheaves $\pi_n(F,s) \to \pi_n(G, f(s))$ on Ho (T/x).

We can mention a convenient characterization of π_*-equivalences: by Lemma 3.3.3 of [76], $F \to G$ a morphism of sPr(T) is a π_*-equivalence if for all $n \geq 0$, the induced morphism $F^{\mathbb{R}\Delta^n} \to F^{\mathbb{R}\partial\Delta^n} \times_{G^{\mathbb{R}\partial\Delta^n}} G^{\mathbb{R}\Delta^n}$ is a covering. By Definition 3.1.3 of [76] a morphism $F \to G$ in Ho (sPr(T)) is a *covering* if the induced morphism of sheaves $\pi_0(F) \to \pi_0(G)$ is an epi. We can mention at this point is that if one naively looks at coverings as being epimorphisms, or more abstractly fibrations, then with this definition we achieve is characterizing coverings in Ho (sPr(T)) as being global epi insofar as they are so at the level of connected components, or homotopy like fibrations, in a pedestrian sense. To understand what type of covers π_*-equivalences are as defined above, we have to put this construction of stacks over an S-site (T,τ) in perspective. This construction hinges on first considering sSPr(T), the category of simplicial objects in sPr(T), on which we put a Reedy model structure, for which fibrations $F_* \to G_*$ are those maps such that for all n, the maps $F_*^{\Delta^n} \cong F_n \to F_*^{\partial\Delta^n} \times_{G_*^{\partial\Delta^n}} G_*^{\Delta^n}$ are fibrations in sPr(T). Thus we can see fibrations in sSPr(T) as hyperfibrations. Now by Definition 3.2.3 of [76] a morphism $F_* \to G_*$ in sSPr(T) is a *hypercover* if for all n, the induced morphism $F_*^{\mathbb{R}\Delta^n} \to F_*^{\mathbb{R}\partial\Delta^n} \times^h_{G_*^{\mathbb{R}\partial\Delta^n}} G_*^{\mathbb{R}\Delta^n}$ is a covering in Ho (sPr(T)), and coverings being homotopy like fibrations, a hypercover therefore appears as a homotopy like hyperfibration, which is further illustrated by the very definition of fibrations in sSPr(T). That we have such a transparent interpretation of a hypercover is by virtue of the fact that we put the Reedy model structure on sSPr(T), and hypercovers are defined as they are. Now coming back to π_*-equivalences, given their above characterizations, we see that they appear as hypercovers for constant simplicial objects in sSPr(T).

Thus one thing we can say at this point is that hypercovers correspond to generalizations of π_*-equivalences.

Focusing now on the notion of π_*-equivalences, there are multiple notions to be discussed to fully make sense of them; the sheaves $\pi_0(F)$ and higher sheaves $\pi_n(F,s)$, and the homotopy category Ho (T/x). T being an S-category, $F \in$ sPr(T), one can form a presheaf of connected components $\pi_0 F$ in Fun$(T^{\mathrm{op}}, \mathrm{Set})$ defined pointwise. By universality of the homotopy category this factors through Ho $(T)^{\mathrm{op}}$. This is functorial in F, so we lift that to a functor sPr$(T) \to \mathrm{Set}^{\mathrm{Ho}\,(T)^{\mathrm{op}}}$, which itself factors through Ho (sPr(T)), since equivalences in sPr(T) are defined objectwise, so equivalent simplicial presheaves have the same presheaves of connected components. One denotes the resulting functor by π_0^{pr}, and by $\pi_0(F)$ the sheaf associated with $\pi_0^{\mathrm{pr}}(F)$ on the site (Ho (T), τ).

Higher sheaves are defined as follows: for $F \in \mathrm{sPr}(T)$, $x \in \mathrm{Ob}\,(T)$, $s \in \pi_0(F(x))$, and $n > 0$,

$$\pi_n(F,s) = \pi_0(j_x^*(F)^{\mathbb{R}\Delta^n} \times_{j_x^*(F)^{\mathbb{R}\partial\Delta^n}} *).$$

We define the different constituents of this definition. First for $x \in \mathrm{Ob}\,(T)$, (T,τ) being an S-site, by definition τ is a topology on $\mathrm{Ho}\,(T)$, which induces one on $\mathrm{Ho}\,(T)/x$, and the map $\mathrm{Ho}\,(T/x) \to \mathrm{Ho}\,(T)/x$ allows one to pull this topology back to $\mathrm{Ho}\,(T/x)$. Now consider the natural projection $j_x : T/x \to T$. This induces $j_x^* : \mathrm{sPr}(T) \to \mathrm{sPr}(T/x)$ and hence a left derived functor $\mathbb{L}j_x^* : \mathrm{Ho}\,(\mathrm{sPr}(T)) \to \mathrm{Ho}\,(\mathrm{sPr}(T/x))$, which we still denote by j_x^* to follow [76]. Now let $s \in \pi_0(F(x))$. Recall that $F(x) \cong \mathbb{R}\underline{\mathrm{Hom}}(\underline{h}_x, F)$,[6] where $\underline{h}_x = \underline{\mathrm{Hom}}_T(-,x)$. This is an isomorphism in $\mathrm{Ho}\,(\mathrm{Set}_\Delta)$. Thus F here is seen as an object of $\mathrm{Ho}\,(\mathrm{sPr}(T))$. This isomorphism gives us $\pi_0(F(x)) \cong \pi_0\mathbb{R}\underline{\mathrm{Hom}}(\underline{h}_x, F)$. Thus $s \in \pi_0(F(x))$ corresponds to a class in $\pi_0\mathbb{R}\underline{\mathrm{Hom}}(\underline{h}_x, F)$. Pick a representative in $\mathbb{R}\underline{\mathrm{Hom}}(\underline{h}_x, F)$. By adjointness, for $K \in \mathrm{Set}_\Delta$, $\mathrm{Hom}_{\mathrm{sPr}(T)}(K \otimes \underline{h}_x, F) \cong \mathrm{Hom}_{\mathrm{Set}_\Delta}(K, \underline{\mathrm{Hom}}(\underline{h}_x, F))$, where $(K \otimes F)(x) = K \times F(x)$. Thus the corresponding representative by adjointness produces a map $\underline{h}_x \to F$, and we take its homotopy class in $\mathrm{Ho}\,(\mathrm{sPr}(T))$. Under j_x^*, this induces a morphism in $\mathrm{Ho}\,(\mathrm{sPr}(T/x))$: $j_x^*(\underline{h}_x) \to j_x^*(F)$, which we still denote by s. Now $j_x^*(\underline{h}_x)$ has a global point, and composing this with s, one gets a global point $* \to j_x^*(F)$.

We now discuss the exponentiation notation. First, $\mathrm{sSPr}(T)$ is tensored and cotensored over Set_Δ. Let $F_* \in \mathrm{sSPr}(T)$, $K \in \mathrm{Set}_\Delta$. For the sake of notation, a choice that will become clear below, the tensor product of K with F_* is denoted $\underline{K} \otimes F_*$ and is the simplicial object of $\mathrm{sPr}(T)$, which with $[n] \in \Delta^{\mathrm{op}}$ associates $\coprod_{K_n} F_n$. Exponentiation is defined by the adjunction formula $\mathrm{Hom}_{\mathrm{sSPr}(T)}(\underline{K} \otimes F_*, G_*) \cong \mathrm{Hom}_{\mathrm{sSPr}(T)}(F_*, G_*^{\underline{K}})$. Now one typically writes F_*^K for $(F_*^{\underline{K}})_0$, which explains now the choice of notation. This produces a functor

$$\begin{aligned} \exp_K : \mathrm{sSPr}(T) &\to \mathrm{sPr}(T), \\ F_* &\mapsto F_*^K \end{aligned}$$

with a right derived functor

$$\begin{aligned} \mathbb{R}\exp_K : \mathrm{Ho}\,(\mathrm{sSPr}(T)) &\to \mathrm{Ho}\,(\mathrm{sPr}(T)), \\ F_* &\mapsto F_*^{\mathbb{R}K}. \end{aligned}$$

Now if we just have $F \in \mathrm{sPr}(T)$, then we regard it as a constant simplicial object of $\mathrm{sPr}(T)$ via the constant simplicial functor $F \rightsquigarrow c(F)_*$, which is identically F. Using the above formalism, one can make sense of the object $[c(F)_*]^{\mathbb{R}K}$. There is another way to represent this exponent: pick a fibrant replacement RF for F. We can exponentiate it using

6 Prop. 2.4.2 [76].

the simplicial structure of sPr(T): $(RF)^{\Delta^n}$. Collectively, $(RF)^{\Delta^*}$ is a fibrant replacement for $c(F)_*$ in sSPr(T). It follows that in Ho (sPr(T)),

$$c(F)_*^{\mathbb{R}K} = [(RF)^{\Delta^*}]^{\mathbb{R}K} = ([(RF)^{\Delta^*}]^{\underline{\mathbb{R}K}})_0 \cong (RF)^{\mathbb{R}K}.$$

One denotes this object by $F^{\mathbb{R}K}$. Thus the notation $j_x^*(F)^{\mathbb{R}K}$ is well-defined.

To understand the form given for the higher homotopy groups, for a simplicial presheaf F, $K \in \mathrm{Set}_\Delta$, one has $F^K(x) = \underline{\mathrm{Hom}}_{\mathrm{Set}_\Delta}(K, F(x))$ in Set_Δ, and $F^K(x) = \mathbb{R}\underline{\mathrm{Hom}}(K, F(x))$ in Ho (Set_Δ). Observe that this corresponds to $F^{\mathbb{R}K}(x)$. Now $\pi_n(F(x))$ should be a simplicial homotopy group [28], the set of homotopy classes of maps $\Delta^n \to F(x)$ mapping $\partial\Delta^n$ to a point of $F(x)$. In other terms,

$$\pi_n(F(x)) = \pi_0\big(\mathbb{R}\underline{\mathrm{Hom}}(\Delta^n, F(x)) \times^h_{\mathbb{R}\underline{\mathrm{Hom}}(\partial\Delta^n, F(x))} F(x)_0\big)$$

which can be rewritten in the desired form

$$\pi_n(F(x)) = \pi_0\big((F^{\mathbb{R}\Delta^n})(x) \times^h_{F^{\mathbb{R}\partial\Delta^n}(x)} F(x)_0\big).$$

With regards to the model structure on sPr(T), one defines the *projective model structure* to be the model structure for which equivalences and fibrations are defined pointwise, that is, $F \to G$ is a weak equivalence (resp., fibration) iff for all $x \in T$, $F(x) \to G(x)$ is a weak equivalence in Set_Δ (resp., a fibration).

We now come back to the claim that $\mathrm{sPr}_\tau(T)$ can be obtained from a left Bousfield localization of sPr(T). For this, one needs to introduce a certain class of morphisms in the latter model category. One first considers $E(T) = \coprod_{n\geq 0} \coprod_{x,y\in \mathrm{Ob}\,T} \underline{\mathrm{Hom}}_T(x,y)_n$. Let α be a cardinal, $\alpha > \mathrm{Card}E(T)$, and $\alpha > \aleph_0$. Let β be a cardinal such that $\beta > 2^\alpha$. For $x \in \mathrm{Ob}\,T$, consider a fibrant replacement $\underline{h}_x \to R(\underline{h}_x)$. Let $H_\beta(x)$ for $x \in \mathrm{Ob}\,T$ be a set of representatives of isomorphism classes of objects $F_* \in \mathrm{sSPr}(T)_{/R(\underline{h}_x)^{\Delta^*}}$ satisfying two conditions. The first condition is $F_* \to R(\underline{h}_x)^{\Delta^*}$ is a hypercover in Ho (sSPr(T)). Recall that a morphism $F_* \to G_*$ in sSPr(T) is a hypercover if for all n, the induced morphism $F_*^{\mathbb{R}\Delta^n} \to F_*^{\mathbb{R}\partial\Delta^n} \times^h_{G_*^{\mathbb{R}\partial\Delta^n}} G_*^{\mathbb{R}\Delta^n}$ is a covering in Ho (sPr(T)), and we simply say it is a hypercover as a morphism of Ho (sSPr(T)) if one of its representatives in sSPr(T) is a hypercover. The second condition is that for all $n \geq 0$, $\mathrm{Card}F_n = \coprod_{n\geq 0}\coprod_{x\in \mathrm{Ob}\,T} F_n(x) < \beta$, and $F_n \cong \coprod_{u\in I_n} \underline{h}_u$ in Ho (sPr(T)), where I_n is an indexing set of cofibrant objects of T. Then by Corollary 3.4.5 of [76], $\mathrm{sPr}_\tau(T)$ is the left Bousfield localization of sPr(T) with respect to the set $H = \{|F_*| \to \underline{h}_x \mid x \in \mathrm{Ob}\,T, F_* \in H_\beta(x)\}$, $|F_*|$ the geometric realization in sPr(T) of the object $F_* \in$ sSPr(T). Formally,

$$\mathrm{sPr}_\tau(T) = L_H \mathrm{sPr}(T).$$

Using this class $H_\beta(x)$, we have a clear characterization of stacks: it is an object of $\mathrm{sPr}(T)$ that satisfies *hyperdescent*,[7] that is, for all $x \in \mathrm{Ob}\,T$ and for any semirepresentable hypercover $H_* \cong \coprod \underline{h}_u \to \underline{h}_x$, the induced morphism $F(x) \cong \mathbb{R}\underline{\mathrm{Hom}}(\underline{h}_x, F) \to \mathbb{R}\underline{\mathrm{Hom}}(|H_*|, F)$ is an isomorphism in $\mathrm{Ho}\,(\mathrm{Set}_\Delta)$, or, in other terms, F_* is local with respect to the class of semirepresentable hypercovers. Let us emphasize the mechanics we have here. We do a left Bousfield localization with respect to H. Fibrant objects of $L_H\mathrm{sPr}(T)$ are H-local objects, and being H-local means exactly that such objects satisfy hyperdescent. Now every object of $\mathrm{sPr}(T)$ has a cofibrant fibrant approximation $F \to \hat{F}$ by a fibrant object $\hat{F}$ of $\mathrm{sPr}(T)$ via a trivial cofibration. Thus F and $\hat{F}$ are indistinguishable in $\mathrm{Ho}\,(L_H\mathrm{sPr}(T))$, where all objects therefore satisfy hyperdescent, and if we call such objects stacks, then this means $L_H\mathrm{sPr}(T)$ is a model category of stacks over T.

One last observation about $\mathrm{sPr}_\tau(T)$: as we said in our introductory remarks, it can be obtained from $\mathrm{sPr}(T)$ as a left Bousfield localization with respect to π_*-equivalences or as a left Bousfield localization with respect to hypercovers, as we just did above. We also mentioned at some point that hypercovers in $\mathrm{sSPr}(T)$ provide generalizations of the notion of π_*-equivalences. So left Bousfield localizations with respect to them both produce the same result, namely $\mathrm{sPr}_\tau(T)$, which says exactly that π_*-equivalences are actually sufficient to define stacks over simplicial categories.

7 Def. 3.4.8 [76].

7 Stacks on model sites

From [76], we can put various model structures on $\mathrm{sPr}(C)$, giving rise to model categories of prestacks and stacks. It turns out that those can be expressed as left Bousfield localizations, as we argued in the preceding chapter already. For instance, the model category of prestacks $M^{\wedge}$ on a model category M is the left Bousfield localization of $\mathrm{sPr}(M)$ with respect to equivalences in M in a sense to be precised below. Here $\mathrm{sPr}(M)$ is given its projective model structure. We have that $\mathrm{Ho}\,(M^{\wedge})$ corresponds to the full subcategory of $\mathrm{Ho}\,(\mathrm{sPr}(M))$ spanned by functors $F : M^{\mathrm{op}} \to \mathrm{Set}_{\Delta}$ preserving equivalences. Now if in addition (M, τ) is a model site, then one can define a notion of local equivalence on $M^{\wedge}$, giving rise to a local model structure on $M^{\wedge}$. This yields a new model category $M^{\sim,\tau}$, which can also be obtained as a left Bousfield localization of $M^{\wedge}$ with respect to a class of τ-hypercovers in a sense to be precised below. We follow [76].

We first define the model category of prestacks on a model category M. This is done via the notion of *restricted diagrams*.[1] For that purpose, we just need a category M along with a subcategory W of morphisms in it, which for a model category will be its subcategory of weak equivalences. We put the projective model structure on $\mathrm{sPr}(M)$. Now for any $x \in M$, one has an evaluation functor $i_x^* : \mathrm{sPr}(M) \to \mathrm{Set}_{\Delta}$, which with F associates $F(x)$. It has a left adjoint $(i_x)_! : \mathrm{Set}_{\Delta} \to \mathrm{sPr}(M)$, a left Quillen functor. Let I be a set of generating cofibrations in Set_{Δ}. For any $f : A \to B$ in I, for any morphism $u : x \to y$ in W, one has natural morphisms in $\mathrm{sPr}(M)$

$$f\Box u : (i_y)_! A \coprod_{(i_x)_! A} (i_x)_! B \to (i_y)_! B.$$

One defines the model category of restricted diagrams from $(M^{\mathrm{op}}, W^{\mathrm{op}})$ to Set_{Δ}[2] to be the left Bousfield localization of $\mathrm{sPr}(M)$ (with its projective model structure) along the class of morphisms of the form $f\Box u, f \in I, u \in W$. One denotes this model category by $\mathrm{Set}_{\Delta}^{M^{\mathrm{op}},W^{\mathrm{op}}}$ and takes this to be our model category of prestacks on M. Thus

$$M^{\wedge} = \mathrm{Set}_{\Delta}^{M^{\mathrm{op}},W^{\mathrm{op}}} = L_{\{f\Box u\}}\mathrm{sPr}(M)$$

and this is what is meant by saying that $M^{\wedge}$ is the left Bousfield localization of $\mathrm{sPr}(M)$ with respect to equivalences in M.

Next, suppose that (M, τ) is a model site. By Theorem 4.6.1 of [76] there is a closed model structure on $M^{\wedge}$, the local projective model structure, for which the weak equivalences are the local equivalences, to be defined below, and for which the cofibrations are the cofibrations for the projective model structure on $\mathrm{sPr}(M)$. One denotes by $M^{\sim,\tau}$ this model structure on $M^{\wedge}$, and one refers to it as the model category of stacks on (M, τ).

1 Def. 4.1.4 [76].

2 Def. 2.3.3 [76].

https://doi.org/10.1515/9783111334073-007

One can use some of the formalism of stacks on simplicial sites we have already covered for the simple reason that by Proposition 4.3.5 of [76] we have a one-to-one correspondence between model topologies on model categories M and S-topologies on their Dwyer–Kan simplicial localizations LM. Further, by Theorem 2.3.5 of the same reference the model categories $\mathrm{Set}_\Delta^{LM^{\mathrm{op}}}$ and $M^\wedge$ are Quillen equivalent. The former is of the form $\mathrm{sPr}(T)$ with $T = LM$, at which point we can use our construction of stacks over simplicial categories. Nevertheless, even if we do use some results from the simplicial case, there are developments that are simpler if we just work with model categories, and also our set of hypercovers is slightly different for the sake of left Bousfield localization, so it is worth going over that construction again for illustrative purposes.

To define local equivalences, one needs a notion of homotopy group on functors $F : M^{\mathrm{op}} \to \mathrm{Set}_\Delta$. First, consider $F_* \in \mathrm{sSPr}(M) = \mathrm{sPr}(M)^{\Delta^{\mathrm{op}}}$. Such objects are naturally tensored and cotensored over Set_Δ: for $K \in \mathrm{Set}_\Delta$, one defines

$$\begin{aligned} \underline{K} \otimes F_* : \Delta^{\mathrm{op}} &\to \mathrm{sPr}(M), \\ [n] &\mapsto \coprod_{K_n} F_n \end{aligned}$$

as well as $F_*^{\underline{K}}$ by $\mathrm{Hom}(\underline{K} \otimes F_*, G_*) \cong \mathrm{Hom}(F_*, G_*^{\underline{K}})$. Let $F_*^K = (F_*^{\underline{K}})_0$.

We put the Reedy model structure on $\mathrm{sSPr}(M)$: the equivalences are the degreewise equivalences, and fibrations $F_* \to G_*$ are those maps such that for all $[n] \in \Delta$, the induced map

$$F_*^{\Delta^n} \simeq F_n \to F_*^{\partial\Delta^n} \times_{G_*^{\partial\Delta^n}} G_*^{\Delta^n}$$

is a fibration in $\mathrm{sPr}(M)$. Armed with these notations, consider

$$\begin{aligned} \pi_0^{\mathrm{pr}}(F) : M^{\mathrm{op}} &\to \mathrm{Set}, \\ x &\mapsto \pi_0(F(x)) \end{aligned}$$

which defines a functor

$$\begin{aligned} \pi_0^{\mathrm{eq}} : M^\wedge &\to \mathrm{Pr}(M), \\ F &\mapsto \pi_0^{\mathrm{pr}}(RF). \end{aligned}$$

Since RF is fibrant, this functor preserves equivalences, so that π_0^{eq} factors through $\mathrm{Ho}\,(M)$:

$$\begin{aligned} \pi_0^{\mathrm{eq}} : M^\wedge &\to \mathrm{Pr}(\mathrm{Ho}\,(M)), \\ F &\mapsto \pi_0^{\mathrm{eq}}(F) \end{aligned}$$

which finally factors as

$$\pi_0^{\mathrm{eq}} : \mathrm{Ho}\,(M^\wedge) \to \mathrm{Pr}(\mathrm{Ho}\,(M)).$$

Now on the model site (M, τ), denote by $\pi_0(F)$ the sheaf associated with $\pi_0^{\mathrm{eq}}(F)$. One says a morphism $F \to G$ in $\mathrm{Ho}\,(M^\wedge)$ is a *τ-covering*[3] if $(\pi_0^{\mathrm{eq}}(F) \to \pi_0^{\mathrm{eq}}(G)) \rightsquigarrow (\pi_0(F) \twoheadrightarrow \pi_0(G))$. One says that such a morphism $F \to G$ is a τ-covering in $M^\wedge$ if it induces a τ-covering in its homotopy category.

Now consider $j_x : M_{/x} \to M$, which induces $j_x^* : M^\wedge \to (M_{/x}, W)^\wedge$. Let $F \in M^\wedge$, let $x \in M^f$ be fibrant, and let $s \in \pi_0^{\mathrm{eq}}(F(x))$ be represented by $s : h_x \to F$ in $\mathrm{Ho}\,(M^\wedge)$. The derived functor for j_x^* is denoted $\mathbb{R}j_x^* : \mathrm{Ho}\,(M^\wedge) \to \mathrm{Ho}\,((M_{/x}, W)^\wedge)$. We need to define its exponentiation in the derived setting. Recall that for $K \in \mathrm{Set}_\Delta$, one has a functor $\mathrm{sSPr}(M) \to \mathrm{sPr}(M)$, which with F_* associates F_*^K. Its right derived functor $\mathrm{Ho}\,(\mathrm{sSPr}(M)) \to \mathrm{Ho}\,(\mathrm{sPr}(M))$ is denoted $F_* \to F_*^{\mathbb{R}K}$. With this notation, for any $n > 0$, one defines[4]

$$\pi_n(F, s) = \pi_0\big(\mathbb{R}j_x^*(F)^{\mathbb{R}\Delta^n} \times_{\mathbb{R}j_x^*(F)^{\mathbb{R}\partial\Delta^n}} *\big)$$

where $\mathbb{R}j_x^* s$ has a global point:

$$* \to \mathbb{R}j_x^*(h_x) \xrightarrow{\mathbb{R}j_x^* s} \mathbb{R}j_x^* F.$$

Thus defined $\pi_n(F, s)$ defines a sheaf on $(\mathrm{Ho}\,(M_{/x}), \tau)$. Finally, one says a morphism $F \to G$ in $M^\wedge$ is a *π_*-local equivalence* or simply a *local equivalence* if $\pi_0(F) \xrightarrow{\cong} \pi_0(G)$ in $\mathrm{Pr}(\mathrm{Ho}\,(M))$, and if for all $x \in M^f$, $s \in \pi_0^{\mathrm{eq}}(F(x))$, and $n > 0$, the morphism $\pi_n(F, s) \to \pi_n(G, fs)$ is an isomorphism of sheaves on $\mathrm{Ho}\,(M_{/x})$. Observe that[5] $f : F \to G$ is a local equivalence if for all $n \geq 0$, the morphism

$$F^{\mathbb{R}\Delta^n} \to F^{\mathbb{R}\partial\Delta^n} \times^h_{G^{\mathbb{R}\partial\Delta^n}} G^{\mathbb{R}\Delta^n}$$

is a covering, where we regard $F \in \mathrm{sPr}(M)$ as a constant simplicial object of $\mathrm{sSPr}(M)$. These then are the local equivalences, which are used to define the *local projective model structure* on $M^\wedge$, giving rise to a model category of stacks $M^{\sim,\tau}$. It turns out that one can express this latter model category as a left Bousfield localization of $M^\wedge$ with respect to a certain class of morphisms, which we show presently.

Indeed, one can write $M^{\sim,\tau} = L_T M^\wedge$,[6] where $T = \{|F_*| \to h_x \,|\, x \in M^f,\ F_* \in \mathcal{H}_\beta(x)\}$ with $|F_*| = \mathrm{hcolim}_{[n]\in\Delta^{\mathrm{op}}} F_n$, F_* a simplicial object in $M^\wedge$, and $\beta > 2^\alpha$, α being greater

3 Def. 4.3.7 [76].

4 Def. 4.5.3 [76].

5 Lemma 4.5.4 [76].

6 Cor. 4.6.2 [76].

than the cardinality of the set of morphisms in M, or $\aleph_0$. Thus β is meant to deal with size issues. However, in this work, we do not worry about universe considerations. The reader who may want to keep track of all such issues will find all the relevant universes being considered in the work of Toen and Vezzosi. The set $\mathcal{H}_\beta(x)$ is defined as follows. It is the set of representatives of isomorphism classes of objects $F_* \to h_x$ in $sM^\wedge_{/h_x}$; $F_* \to h_x$ a pseudo-representable hypercover, and for all $n \geq 0$, $\mathrm{Card}(F_n) < \beta$. Here for $x \in M^f$, a pseudo-representable hypercover of x[7] is an object $F_* \to h_x$ of $sM^\wedge_{/h_x}$ such that for all $n \geq 0$, the map $F_n \to F_*^{\partial\Delta^n} \times_{h_x^{\partial\Delta^n}} h_x^{\Delta^n}$ is a pseudo-fibration and a pseudo-covering between pseudo-representable objects. An object $F \in M^\wedge$ is said to be pseudo-representable if it is a small disjoint union of representable presheaves: $F \simeq \coprod_{u \in I} h_u$. A morphism between pseudo-representable objects $f : \coprod_I h_u \to \coprod_J h_v$ is said to be a pseudo-fibration if the corresponding projection for all $u \in I$

$$f \in \prod_{u \in I} \coprod_{v \in J} \mathrm{Hom}(h_u, h_v) \to \coprod_J \mathrm{Hom}(h_u, h_v) \simeq \coprod_J \mathrm{Hom}(u, v)$$

is represented by a fibration in M. Finally, for f as above and for $v \in J$, let $I_v = \{u \in I : h_u \to h_v\}$. Then f is said to be a pseudo-covering if for all $v \in J$, the set $\{h_u \to h_v\}_{u \in I_v}$ corresponds to a covering family in M. Observe that by Lemma 4.4.2 of [76] a pseudo-representable hypercover $F_* \to h_x$ is a τ-hypercover, and given their definition, they are therefore referred to as homotopy τ-hypercovers. Thus localizing $M^\wedge$ with respect to T as defined above amounts to localizing with respect to homotopy τ-hypercovers. Here we referenced τ-hypercovers; by Definition 3.2.3 of [76] a morphism $F_* \to G_*$ in $\mathrm{sSPr}(M)$ is called a *τ-hypercover* if for all n, the induced morphism $F_*^{\mathbb{R}\Delta^n} \to F_*^{\mathbb{R}\partial\Delta^n} \times^h_{G_*^{\mathbb{R}\partial\Delta^n}} G_*^{\mathbb{R}\Delta^n}$ is a covering in $\mathrm{Ho}\,(\mathrm{sPr}(M))$.

Writing $M^{\sim,\tau} = L_T M^\wedge$ is formal however. To see what this means at the level of functors, we invoke Corollary 4.6.3 of [76], which states that $F \in M^{\sim,\tau}$ is fibrant iff it is objectwise fibrant, preserves equivalences, and satisfies the hyperdescent condition: for all $x \in M^f$ and $H_* \in \mathcal{H}_\beta(x)$, the natural morphism $F(x) \simeq \mathbb{R}_\omega\underline{\mathrm{Hom}}(h_x, F) \to \mathbb{R}_\omega\underline{\mathrm{Hom}}(|H_*|, F)$ is an isomorphism in $\mathrm{Ho}\,(\mathrm{Set}_\Delta)$, since fibrant objects are T-local, where $\mathbb{R}_\omega\underline{\mathrm{Hom}}$ denotes the derived simplicial Hom in $\mathrm{Ho}\,(M^\wedge)$. Then we use the fact that any object of $M^{\sim,\tau}$ can be approximated by a fibrant object.

We would like to point out at this point something important. One could define stacks over a model site (M, τ) by working with hypercovers on M, as done in [78], or work with hypercovers on $\mathrm{sPr}(M)$, as done in [76]. We briefly explain how this is done. Essentially the connection, as the reader can imagine, is done using the functor $h : M \to \mathrm{sPr}(M)$, which with any $x \in M$ associates $\mathrm{Hom}_M(-, x)$, viewed as a constant simplicial object. Write h_S for the set $\{h_f \mid f \in S \subset \mathrm{Mor}\,(M)\}$. Then define $M^\wedge = L_{h_W}\mathrm{sPr}(M)$, the left Bousfield localization of the simplicial category $\mathrm{sPr}(M)$ of simplicial presheaves

7 Def. 4.4.1 [76].

on M with respect to its set of weak equivalences W. Now put a topology τ on M to make it into a model site (M, τ). Consider

$$S_X = \{\coprod_{i \in I} h_{x_i} \to h_{\coprod x_i} \mid x_i \in M^c\}.$$

For $x \in M^f$, let

$$h\mathbb{H}\mathrm{cov}(x) = \{u_* \in s(M_{/x}) \mid \gamma u_* \in \mathrm{Ho}\,(sM)_{/x} \text{ htpy hypercover of } x\}.$$

Let

$$S_{h\mathbb{H}\mathrm{cov}} = \{h_{u_*} \to h_x \mid x \in M^f,\ u_* \in h\mathbb{H}\mathrm{cov}(x)\}.$$

Then the model category of stacks on M is defined as

$$M^{\sim,\tau} = L_{S_X \cup S_{h\mathbb{H}\mathrm{cov}}} M^{\wedge}.$$

Now it turns out, as shown in [76], that one can more compactly represent that as a left Bousfield localization with respect to a set of semirepresentable hypercovers in $M^{\wedge}$. In the proof of such a statement, it is also proved that this corresponds to having a local projective model structure on $M^{\wedge}$ whereby equivalences are local equivalences, or π_*-equivalences, maps $F \to G$ in $M^{\wedge}$ such that for all $n \geq 0$,

$$F^{\mathbb{R}\Delta^n} \to F^{\mathbb{R}\partial\Delta^n} \times^h_{G^{\mathbb{R}\partial\Delta^n}} G^{\mathbb{R}\Delta^n} \tag{7.1}$$

is a τ-hypercover, where $F \to G$ is regarded as a morphism of constant simplicial objects of $M^{\wedge}$. Here we have now made the transition from having hypercovers in the base model category M to working with hypercovers in the functor category $M^{\wedge}$. Going over the definition of π_*-equivalence, which makes use of the notion of higher homotopy sheaves on sPr(M), the connection with covers in M is made by virtue of the following fact: $F \to G$ being a local equivalence means in particular that (7.1) is a covering. However, those have the local homotopy lifting property: if $F \to G$ in $(M^{\wedge})^f$ is a τ-covering, then for all $x \in M$ and for any morphism $h_x \to G$ in $M^{\wedge}$, there is a covering family $\{u_i \to x\}$ in M such that for all i, there is a commutative diagram in $\mathrm{Ho}\,(M^{\wedge})$

$$\begin{array}{ccc} F & \longrightarrow & G \\ \uparrow & & \uparrow \\ h_{u_i} & \longrightarrow & h_x \end{array}$$

thereby establishing the connection between representable functors and covers in M. One can say more regarding this commutative diagram: if $u_i \to x$ is a covering map, then $h_{u_i} \to h_x$ is a covering. Thus $F \to G$ is a covering if it is locally so. Now if we place ourselves in the Quillen equivalent category sPr(LM), then one can imagine that if $F \to G$ is a π_*-equivalence, which we argued are constant hypercovers, then one should have a commutative diagram of the form

$$\begin{array}{ccc} F & \longrightarrow & G \\ \uparrow & & \uparrow \\ |H_*| & \longrightarrow & R\underline{h}_X \end{array}$$

for some hypercover $H_* \to R\underline{h}_X$, thereby showing that $F \to G$ is a hypercover if it is so locally. We do have this result: this is Lemma 3.4.3 of [76].

8 Stacks on Segal sites

Yet another way to obtain a category of stacks is by using Segal categories. For this, we follow [77]. Segal categories have the advantage over model categories of providing an internal theory, i. e., they have internal hom objects. We can also mention one reason why one would want to work with Segal categories; as pointed out in [76], one cannot develop a notion of geometric morphism on model topoi (model categories Quillen equivalent to some $\mathrm{sPr}_\tau(T)$, $T \in \mathrm{Cat}_\Delta$) or simplicial topoi (S-topoi) of the form LM for a model topos M. But on Segal topoi, we have such a notion, since we have internal homs.

We first define, following [77], a *Segal pre-category* to be given by a functor $A : \Delta^{\mathrm{op}} \to \mathrm{Set}_\Delta$ such that $A_0 \in \mathrm{Set}$. One denotes by SePC the category of such objects. For $n \geq 0$, the morphisms $[0] \to [n]$ in Δ induce morphisms in Set_Δ: $A_n \to A_0^{n+1}$, whose fibers over a point $(a_0, \dots, a_n) \in A_0^{n+1}$ are denoted $A_{(a_0 \cdots a_n)}$. Observe that one has $A_n \cong \coprod_{(a_0,\dots,a_n) \in A_0^{n+1}} A_{(a_0 \cdots a_n)}$ in Set_Δ. If we fix a set S, then denote by $\mathrm{SePC}[S] \subset \mathrm{SePC}$ the subcategory of objects A with $A_0 = S$, with morphisms inducing the identity on S. One puts a model structure on $\mathrm{SePC}[S]$ that is called the *isoprojective model structure*, with equivalences being isoequivalences (resp., fibrations being isofibrations), maps $f : A \to B$ in $\mathrm{SePC}[S]$ such that for all $(a_0 \cdots a_n) \in S^{n+1}$, $f_{(a_0 \cdots a_n)} : A_{(a_0 \cdots a_n)} \to B_{(a_0 \cdots a_n)}$ is an equivalence (resp., a fibration) in Set_Δ. In a sense, this model structure is a very natural thing to have for Segal categories with the same set of objects. Naturally, for Segal categories with differing sets of objects, one will have to use pullbacks to define a notion of model structure, something we will see later.

Observe that for $n \geq 2$ and $0 \leq i < n$, one also defines the maps

$$\begin{aligned} h_i : [1] &\to [n], \\ 0 &\mapsto i, \\ 1 &\mapsto i+1. \end{aligned}$$

For $A \in \mathrm{SePC}$, those induce morphisms $A_n \to A_1^n$ in Set_Δ called *Segal morphisms*. Objectwise, $A_{(a_0 \cdots a_n)} \to A_{(a_0 a_1)} \times A_{(a_1 a_2)} \times \cdots \times A_{(a_{n-1} a_n)}$. It turns out that one can corepresent $A_{(a_0 \cdots a_n)}$ by some object $h_{(a_0 \cdots a_n)} \in \mathrm{SePC}$. Thus the objectwise presentation of the Segal morphisms induces morphisms

$$h(a_0 \cdots a_n) : h_{(a_0 a_1)} \coprod \cdots \coprod h_{(a_{n-1} a_n)} \to h_{(a_0 \cdots a_n)}.$$

Let $H = \{h(a_0 \cdots a_n) \mid (a_0 \cdots a_n) \in A_0^{n+1}\}$. One defines the *model category of Segal categories over S* to be the left Bousfield localization of the isoprojective model structure on $\mathrm{SePC}[S]$ with respect to H: $\mathrm{SeCat}[S] = L_H \mathrm{SePC}[S]$. Note that the inclusion $\mathrm{Ho}\,(\mathrm{SeCat}[S]) \hookrightarrow \mathrm{Ho}\,(\mathrm{SePC}[S])$ has a left adjoint $\mathrm{SeCat} : \mathrm{Ho}\,(\mathrm{SePC}[S]) \to \mathrm{Ho}\,(\mathrm{SeCat}[S])$. By the definition of Bousfield localizations, Segal morphisms are equivalences in

https://doi.org/10.1515/9783111334073-008

L_HSePC$[S]$. In other terms,[1] a Segal pre-category A is a *Segal category* if for all $n \geq 2$ and $(a_0 \cdots a_n) \in A_0^{n+1}$,

$$A_{(a_0 \cdots a_n)} \to A_{(a_0 a_1)} \times A_{(a_1 a_2)} \times \cdots \times A_{(a_{n-1} a_n)}$$

is an equivalence in Set_Δ.

Thus we have defined Segal categories objectwise in SePC using localization, and we will need this construction later to define an internal model structure on SePC.

Note that if one regards $A_{(a_0 \cdots a_n)}$ as the simplicial set of composable morphisms $a_0 \to a_1 \to \cdots \to a_n$, then for $A \in \mathrm{Cat}_\Delta$, such a morphism is an isomorphism, and thus Cat_Δ is a full subcategory of SeCat, the category of Segal categories. One can regard this as saying that Segal categories are loose simplicial categories, and this is the point of view we adopt. By relaxing some conditions we get a more versatile formalism.

Along the same lines, observe that there is a weak composition law in SeCat:

$$\begin{array}{ccc} A_{(a,b,c)} & \longrightarrow & A_{(a,c)} \\ \downarrow \simeq & \nearrow & \\ A_{(a,b)} \times A_{(b,c)}. & & \end{array}$$

For $A \in$ SeCat, one defines its homotopy category Ho (A) as follows: Ob Ho (A) = Ob (A), and for all $a, b \in A_0$, one has $\mathrm{Hom}_{\mathrm{Ho}\,(A)}(a,b) = \pi_0 A_{(a,b)}$. The composition law in Ho (A) is obtained by applying π_0 to the weak composition diagram above: $\pi_0(A_{(a,b)}) \times \pi_0(A_{(b,c)}) \to \pi_0(A_{(a,c)})$.

Now note that $f : A \to B$ in SePC induces

$$\begin{aligned} f^* : \mathrm{SePC}[B_0] &\to \mathrm{SePC}[A_0], \\ B &\mapsto f^* B \end{aligned}$$

defined objectwise by $f^* B_{(a_0 \cdots a_n)} = B_{(fa_0 \cdots fa_n)}$ for $(a_0 \cdots a_n) \in A_n$. One has a CD in Ho (SePC$[A_0]$):

$$\begin{array}{ccccc} A & \longrightarrow & f^* B & \longrightarrow & f^* \mathrm{SeCat}(B) \\ \vdots & & & \nearrow & \\ \mathrm{SeCat}(A) & & & & \end{array}$$

where the diagonal arrow is defined objectwise by

$$\mathrm{SeCat}(A)_{(a,b)} \to \mathrm{SeCat}(B)_{(fa,fb)}$$

1 Def. 2.0.2 [77].

for $a, b \in \mathrm{Ob}\, A = A_0$. Along with the obvious morphism $A_0 \to B_0$, this defines a morphism $\mathrm{Ho}\,(\mathrm{SeCat}(A)) \xrightarrow{\mathrm{Ho}\, f} \mathrm{Ho}\,(\mathrm{SeCat}(B))$. To be precise, the morphism $f_{1,(a,b)}$: $\mathrm{SeCat}(A)_{(a,b)} \to \mathrm{SeCat}(B)_{(fa,fb)}$ is a morphism in $\mathrm{SePC}[A_0]$. In $\mathrm{SePC}[A_0]$, equivalences are pointwise equivalences of simplicial sets, so $f_{1,(a,b)} : \mathrm{SeCat}(A)_{(a,b)} \to \mathrm{SeCat}(B)_{(fa,fb)}$ is well-defined up to equivalences of simplicial sets, i. e., $\mathrm{Ho}\,(f_{1,(a,b)})$ is a well-defined morphism of simplicial sets up to isomorphisms. In particular, taking connected components, this gives us a well-defined morphism $\pi_0(\mathrm{SeCat}(A)_{(a,b)}) \to \pi_0(\mathrm{SeCat}(B)_{(fa,fb)})$. Thus together with the map $A_0 \to B_0$ above, this produces the desired map of categories $\mathrm{Ho}\,(f) : \mathrm{Ho}\,(\mathrm{SeCat}(A)) \to \mathrm{Ho}\,(\mathrm{SeCat}(B))$.

Now to come back to our original morphism $f : A \to B$ above, one says it is fully faithful if $\mathrm{SeCat}(A)_{(a,b)} \to \mathrm{SeCat}(B)_{(fa,fb)}$ is an isomorphism in $\mathrm{Ho}\,(\mathrm{Set}_\Delta)$, and one says f is essentially surjective if $\mathrm{Ho}\,(f)$ is. Finally, f is defined to be an equivalence in SePC if it is fully faithful and essentially surjective.[2] If we also define a class of cofibrations in SePC to be monomorphisms, then this defines a combinatorial simplicial model structure on SePC,[3] which is internal.[4] One denotes by $\underline{\mathrm{Hom}}$ the internal Hom of SePC, and by $\mathbb{R}\underline{\mathrm{Hom}}$ its derived Hom in Ho (SePC). With this definition, one defines morphisms $A \to B$ in SeCat to be objects of the Segal category $\mathbb{R}\underline{\mathrm{Hom}}(A, B)$. Note also that $\mathrm{Ho}\,(\mathrm{Cat}_\Delta) \simeq \mathrm{Ho}\,(\mathrm{SeCat})$.

One defines the *Segal category of simplicial sets*[5] by

$$\mathrm{Top} = L(\mathrm{Set}_\Delta)$$

where we have used a Dwyer–Kan localization, giving a simplicial category, which we regard as a Segal category. One defines the *Segal category of prestacks over* T,[6] a small Segal category, by

$$\hat{T} = \mathbb{R}\underline{\mathrm{Hom}}(T^{\mathrm{op}}, \mathrm{Top}).$$

That this is a Segal category will be clear later when we see the strictification theorem.

To define stacks on T, we need a notion of Segal site. A *Segal topology*[7] τ on a Segal category T is just a Grothendieck topology τ on its homotopy category $\mathrm{Ho}\,(T)$,[8] and (T, τ) is called a *Segal site.*

Let $\pi_0^{\mathrm{pr}} : \hat{T} \to \mathbb{R}\underline{\mathrm{Hom}}(\mathrm{Ho}\,(T)^{\mathrm{op}}, \mathrm{Set})$ be the presheaf of sets functor defined objectwise by $\pi_0^{\mathrm{pr}}(F)(x) = \pi_0(F(x))$ for $x \in T$. For $K \in \mathrm{Set}_\Delta$, the exponential map $\exp_K : \mathrm{Set}_\Delta \to$

2 Def. 2.0.3 [77].

3 Thm 2.0.4 [77].

4 Cor. 2.0.5 [77].

5 Def. 2.0.6 [77].

6 Def. 2.0.6 [77].

7 Def. 3.3.1 [77].

8 Def. 3.3.4 [77].

Set_Δ induces a localized map $L\exp_K : \mathrm{Top} \to \mathrm{Top}$. Upon taking the restriction to fibrant objects, which we denote by $L^f \exp_K$, one has an induced direct image functor

$$(-)^{\mathbb{R}K} = (L^f \exp_K)_* : \hat{T} \to \hat{T}.$$

For a small Segal site (T, τ), $F \to G$ in $\hat{T}$ is called a *local equivalence* if for all $n \geq 0$,

$$\pi_0^{\mathrm{pr}}(F^{\mathbb{R}\Delta^n}) \to \pi_0^{\mathrm{pr}}(F^{\mathbb{R}\partial\Delta^n} \times_{G^{\mathbb{R}\partial\Delta^n}} G^{\mathbb{R}\Delta^n})$$

is an epi. Denote by LE the set of local equivalences. Then $F \in \hat{T}$ is called a *stack*[9] if it is LE-local, that is for all $f : G \to H$ in LE, the induced morphism

$$f^* : \mathrm{Hom}_{\mathrm{Ho}(\hat{T})}(H, F) \to \mathrm{Hom}_{\mathrm{Ho}(\hat{T})}(G, F)$$

is a bijection. One denotes by $T^{\sim,\tau}$ the *Segal category of stacks on* T,[10] it is a full subcategory of $\hat{T}$.

We have the following connection between stacks on simplicial sites and Segal sites: if $T \in \mathrm{Cat}_\Delta$ and M is a cofibrantly generated simplicial model category, M^T being given the projective model structure, then by Theorem 4.2.1 of [77] one has an isomorphism in Ho (SeCat):

$$L(M^T) \to \mathbb{R}\underline{\mathrm{Hom}}(T, LM).$$

This is the so-called *strictification theorem.*[11] Note that it follows that for $M = \mathrm{Set}_\Delta$, we have $\hat{T} = \mathbb{R}\underline{\mathrm{Hom}}(T^{\mathrm{op}}, \mathrm{Top}) \simeq L(\mathrm{Set}_\Delta^{T^{\mathrm{op}}}) \in \mathrm{Cat}_\Delta \subset \mathrm{SeCat}$, thereby showing that $\hat{T}$ is a Segal category.

Consider now a small simplicial site (T, τ). Then one has

$$L(\mathrm{sPr}(T)) \cong \hat{T},$$
$$L(\mathrm{sPr}_\tau(T)) \cong T^{\sim,\tau}$$

where the above isomorphisms are in Ho (SeCat), and we have used the fact that $\mathrm{Cat}_\Delta \subset$ SeCat and the definition of small sites and Segal sites, which are formally the same.

We can apply this formalism to define derived stacks following [74]. One has the following localized theories for $\mathcal{C} = sk$-Mod:

$$L(\mathcal{C}) = L(sk\text{-Mod}) = dk\text{-Mod},$$
$$L(\mathrm{Comm}(\mathcal{C})) = L(sk\text{-Alg}) = dk\text{-Alg},$$

9 Def. 3.3.2 [77].

10 Def. 3.3.2 [77].

11 Thm. 4.2.1 [77].

$$L(sk\text{-Aff}) = dk\text{-Aff}$$

where $sk\text{-Aff} = (sk\text{-Alg})^{\mathrm{op}}$. We have seen that derived stacks correspond to functors $sk\text{-Alg} \to \mathrm{Set}_{\Delta}$. After localization, this produces functors $dk\text{-Alg} \to \mathrm{Top}$. One has $\widehat{dk\text{-Aff}} = \mathbb{R}\underline{\mathrm{Hom}}(dk\text{-Aff}^{\mathrm{op}}, \mathrm{Top}) \cong L(\mathrm{sPr}(dk\text{-Aff}))$. Then one defines the Segal category of stacks[12] to be the full subcategory of $\widehat{dk\text{-Aff}}$ spanned by functors that satisfy hyperdescent.

We also have the model category of derived stacks $\text{k-D}^{-}\mathrm{Aff}^{\sim,\tau}$. If one denotes by $\mathrm{dSt}(k)$ the *Segal category of derived stacks*, then one has

$$\mathrm{dSt}(k) \simeq L(\text{k-D}^{-}\mathrm{Aff}^{\sim,\tau}).$$

More generally, if H denotes the class of τ-hypercovers, then one has

$$\mathrm{dSt}(k) = L_H\widehat{dk\text{-Aff}} \cong L(\mathrm{sPr}_{\tau}(dk\text{-Aff}))$$

and $\mathrm{sPr}_{\tau}(LM) \cong M^{\sim,\tau}$, and thus

$$\mathrm{dSt}(k) \cong L(sk\text{-Aff}^{\sim,\tau}) = L(\text{k-D}^{-}\mathrm{Aff}^{\sim,\tau}).$$

One may wonder what need do we have for introducing different notions of stacks, whether it be those on simplicial sites, model sites, or Segal sites. It turns out that Segal categories are in a sense better suited for abstract developments, and this is very much our approach, whereas model categories are better suited for computations. To appreciate such a statement, one would have to read the work of Toen and Vezzosi on Homotopical Algebraic Geometry. Simplicial categories are simply mentioned here because of their link with model categories via the Dwyer–Kan localization, their link with Segal categories using the same type of localization, and the fact that they are used to formulate some strictification theorems.

12 Def. 4.2 [74].

9 Application: étale derived stacks

Now we follow [79] and apply this formalism to the setting where one starts with a base symmetric monoidal model category $\mathcal{C}$. One then considers the category of commutative monoid objects $\mathrm{Comm}(\mathcal{C})$ of $\mathcal{C}$, which we assume is also a model category; this is one of the assumptions of [79] to have a context within which one can do Homotopical Algebraic Geometry. Its opposite category is denoted $\mathcal{A}\mathrm{ff}_{\mathcal{C}}$ and is referred to as the model category of affine objects on $\mathcal{C}$. If in addition we have a topology τ on this model category, then one can define stacks on $(\mathcal{A}\mathrm{ff}_{\mathcal{C}}, \tau)$ using the formalism of Chapter 7, which are elements of the model category $\mathcal{A}\mathrm{ff}_{\mathcal{C}}^{\sim,\tau}$, whose homotopy category is denoted $\mathrm{St}(\mathcal{C}, \tau)$.

In Chapter 7, we briefly discussed the transition from having hypercovers in the base to having hypercovers in the functor category itself. If $M = \mathcal{A}\mathrm{ff}_{\mathcal{C}}$, then we can show how having hypercovers in the base gives a particularly nice characterization of stacks. By Corollary 1.3.2.4 of [79], if $F : \mathrm{Comm}(\mathcal{C}) \to \mathrm{Set}_\Delta$ models a stack on $\mathcal{A}\mathrm{ff}_{\mathcal{C}}$, then for any cosimplicial commutative A-algebra $A \to B_*$ corresponding to some hypercover $\mathrm{Spec}\, B_* \to \mathrm{Spec}\, A$, one has $F(A) \xrightarrow{\cong} \mathrm{holim}\, F(B_n)$ in $\mathrm{Ho}\,(\mathrm{Set}_\Delta)$. This is a reformulation of Lemma 1.3.2.3 of [79], which itself follows from Assumption 1.3.2.2, which states that if $\mathrm{Spec}\, B_* \to \mathrm{Spec}\, A$ is such a hypercover, then $A \to B_*$ satisfies the *descent condition* in the sense of Definition 1.2.12.1. The definition is not particularly enlightening for our purposes, and the reader is referred to [79] for details, but suffices it to say that if $A \to B_*$ satisfies descent, then by Lemma 1.2.12.3, $A \to \mathrm{holim}\, B_n$ is an isomorphism in $\mathrm{Ho}\,(\mathrm{Comm}(\mathcal{C}))$. Collecting things, a stack F is one functor for which we have, among other things, that $A \xrightarrow{\cong} \mathrm{holim}\, B_n$ implies $F(A) \xrightarrow{\cong} \mathrm{holim}\, F(B_n)$ in $\mathrm{Ho}\,(\mathrm{Set}_\Delta)$, which really says that stacks preserve the shape of hypercover. Note that we did have something of the same form already at the level of stacks on simplicial categories. If T is a simplicial category and $H_* \to \underline{h}_x$ are the hypercovers used in the left Bousfield localization to define the model category of stacks on (T, τ), then by [76] one can write $H_n \simeq \coprod \underline{h}_{u_i}$, so that fibrant objects of $\mathrm{sPr}(T)$ satisfy $F(x) \xrightarrow{\simeq} \mathrm{holim} \prod F(u_i)$ in Set_Δ.

We deem it important at this point to digress slightly and discuss the concept of commutative monoids. Strictly speaking, the correct generalization of commutative monoid for a very general base symmetric monoidal model category is the notion of $\mathbb{E}_\infty$-algebra [19, 78]. In other terms, $\mathbb{E}_\infty\text{-Alg}(\mathcal{C}) = \mathrm{Alg}_{\mathbb{E}_\infty}(\mathcal{C})$ plays the role of $\mathrm{Comm}(\mathcal{C})$ in more general homotopical contexts, and if $\mathcal{C}$ has the trivial model structure, then $\mathbb{E}_\infty\text{-Alg}(\mathcal{C})$ does produce $\mathrm{Comm}(\mathcal{C})$. This former category is naturally endowed with a model structure [72], whereas if we were to just consider the category of commutative monoids on $\mathcal{C}$, $\mathrm{Comm}(\mathcal{C})$, then it certainly exists [79], but we would have to make the assumption that it has a model structure. Note however that in most applications, $\mathrm{Comm}(\mathcal{C})$ does have a model structure, and in particular for the case of interest to us, $sk\text{-Mod} = \mathcal{C}$, $\mathrm{Comm}(\mathcal{C}) = sk\text{-Alg}$ has a model structure. The reason why we do not work with $\mathbb{E}_\infty$-algebras is that they make the work heavy on notations [79], and they do not form a universal derived form of classical Algebraic Geometry [73]. Thus in spite of the fact that, strictly speaking,

https://doi.org/10.1515/9783111334073-009

$\mathbb{E}_\infty$-algebras provide the correct homotopical generalization of commutative monoid object in a generic symmetric monoidal model category, we work with commutative monoid objects instead, a theory that we regard as a direct generalization of classical Algebraic Geometry. Note also that the homotopy theory of sk-Alg in characteristic different from zero is not the same as the homotopy theory of $\mathrm{Alg}_{\mathbb{E}_\infty}(sk\text{-Mod})$ [79], and since this is our primary focus, this further justifies our choice of simply working with commutative monoids.

In derived algebraic geometry, one considers $C = sk$-Mod, the symmetric monoidal model category of simplicial commutative k-modules, for some base commutative ring k. One puts the standard model structure on sk-Mod by taking fibrations and equivalences as being those defined on the underlying simplicial sets. Its model structure is combinatorial, proper, and it is in fact a simplicial model category structure. One denotes Comm(sk-Mod) = sk-Alg, with fibrations and equivalences still being defined as being those on the underlying simplicial sets, and $\mathcal{A}\mathrm{ff}_C$ =: k-D$^-$Aff, giving rise to a model category of stacks k-D$^-$Aff$^{\sim,\tau}$, with homotopy category denoted D$^-$St(k). If X is the object of $\mathcal{A}\mathrm{ff}_C$ corresponding to some $A \in sk$-Alg, then one denotes it by $X = \operatorname{Spec} A$. As a general notation, denote by $\mathbb{R}\underline{\mathrm{Hom}}$ the derived simplicial Hom in $\mathrm{Ho}(\mathcal{A}\mathrm{ff}_C^{\wedge})$ and by $\mathbb{R}_\tau\underline{\mathrm{Hom}}$ the derived simplicial Hom in $\mathrm{Ho}(\mathcal{A}\mathrm{ff}_C^{\sim,\tau})$.

We now put the *étale topology* on k-D$^-$Aff: a family of morphisms $\{\operatorname{Spec} A_i \to \operatorname{Spec} A\}$ in k-D$^-$Aff = sk-Alg^{op} is said to be an étale covering family if for all i, the morphisms $A \to A_i$ are étale, and if there exists a finite subset $J \subset I$ such that $\{A \to A_i\}_{i \in J}$ is a *formal covering family*.[1] For our purposes, a convenient characterization of étale morphisms in sk-Alg goes as follows: by Theorem 2.2.2.6 of [79] a morphism in sk-Alg is étale iff it is *strongly étale*. By definition a morphism $A \to B$ in sk-Alg is said to be strongly étale[2] if it is *strong* and if the morphism of affine schemes $\operatorname{Spec} \pi_0 B \to \operatorname{Spec} \pi_0 A$ is étale. A morphism $A \to B$ is said to be strong if the natural morphism $\pi_*(A) \otimes_{\pi_0 A} \pi_0 B \to \pi_* B$ is an isomorphism. Here for $A \in sk$-Alg, we define its underlying space $|A| = \mathrm{Map}_{sk\text{-Mod}}(1, A) \in \mathrm{Ho}(\mathrm{Set}_\Delta)$, which has a natural base point $* \in |A|$, which allows us to define $\pi_i A = \pi_i(|A|, *)$.

We also need to define formal coverings: a family of morphisms of commutative monoids $\{f_i : A \to A_i\}$ is said to be a formal covering family[3] if the family of functors $\{\mathbb{L}f_i^* : \mathrm{Ho}(A\text{-Mod}) \to \mathrm{Ho}(A_i\text{-Mod})\}$ is conservative . Recall that $f : A \to B$ induces $f^* : A\text{-Mod} \to B\text{-Mod}$, which with M associates $M \otimes_A B$. This functor is a left adjoint, with derived functor $\mathbb{L}f^* : \mathrm{Ho}(A\text{-Mod}) \to \mathrm{Ho}(B\text{-Mod})$, and we denote $\mathbb{L}f^* M = M \otimes_A^{\mathbb{L}} B$.

Note that with those characterizations of étale morphisms, we have the following result: a family of morphisms $\{\operatorname{Spec} A_i \to \operatorname{Spec} A\}$ is an étale covering family iff there exists a finite subset $J \subset I$ such that for all i, the natural morphism $\pi_* A \otimes_{\pi_0 B} \pi_0 A_i \to \pi_* A_i$

1 Def. 2.2.2.12 [79].

2 Def. 2.2.2.3 [79].

3 Def. 1.2.5.1 [79].

is an isomorphism, and $\coprod_{i\in J} \operatorname{Spec} \pi_0 A_i \twoheadrightarrow \operatorname{Spec} \pi_0 A$ is étale. Fact: étale covering families form a model topology.[4] Thus we can talk about the model category of stacks k-D$^-$Aff$^{\sim,\text{ét}}$.

4 Lemma 2.2.2.13 [79].

Part III: **Spectral algebraic geometry**

10 Introduction

In the Introduction, we saw that in intersection theory, one has to consider multiplicities given by the dimension of local rings, thereby motivating the notion of schemes. However, for very generic situations, that is no longer sufficient, and in accordance with Serre's intersection formula, one has to consider commutative dg algebras instead of just commutative rings, thus giving rise to the existence of dg schemes. More generally still, if one works in positive characteristic, then it seems $\mathbb{E}_\infty$-rings are better suited than commutative dg algebras, and this paves the way for spectral schemes, themselves part of Spectral Algebraic Geometry (abbr. SAG). Thus it seems SAG was introduced by the necessity to clarify certain situations by first considering ringed objects and then by spectralizing those objects, first, as dg objects and then as spectral objects, strictly speaking. Thus we have a combination of two sorts of moves: coming up with a higher concept to account for unexplained terms (such as in Serre's intersection formula) and collecting different concepts into one package. An example of this latter philosophy can be found in [54], Example 0.1.2.4, where one considers a pullback in the category of schemes:

$$\begin{array}{ccc} X' & \xrightarrow{g'} & X \\ \Big\downarrow{\scriptstyle f'} & & {\scriptstyle f}\Big\downarrow \\ Y' & \xrightarrow[g]{} & Y \end{array}$$

for quasi-compact separated schemes. Refer to this diagram as σ. The category $\mathrm{QCoh}(X)$ of quasi-coherent sheaves on a scheme X is an abelian category, so one can consider its derived category, which we denote by $\mathcal{D}(X)$. One has derived functors $\mathbb{R}f_* : \mathcal{D}(X) \to \mathcal{D}(Y)$, $\mathbb{R}f'_* : \mathcal{D}(X') \to \mathcal{D}(Y')$, $\mathbb{L}g^* : \mathcal{D}(Y) \to \mathcal{D}(Y')$, and $\mathbb{L}g'^* : \mathcal{D}(X) \to \mathcal{D}(X')$, along with a natural transformation $\alpha : \mathbb{L}g^* \circ \mathbb{R}f_* \Longrightarrow \mathbb{R}f'_* \circ \mathbb{L}g'^*$, which is an equivalence only if f or g is flat. Now it turns out that α is an equivalence when σ is a pullback square in the category of spectral schemes. Thus it seems that pullback in schemes + flatness = pullback in spectral schemes, or in other terms in this special instance, spectral algebraic geometry repackages some fiber products and some notion of flatness together. The idea is that when we go higher conceptually, and we do this correctly, we can collect certain notions together into one simple theory.

Coming back to the problem in intersection theory, the idea is that DAG, or SAG for that matter, are theories motivated by the necessity to deal with generic situations that cannot be dealt with classically. As pointed out in [54], one such example is provided by the following result: if X is a smooth algebraic variety over $\mathbb{C}$ and $\tilde{X}$ is a first-order deformation of X, then one has bijections between $\mathrm{Aut}_X(\tilde{X})$ and $H^0(X; T_X)$ and between isomorphism classes of first-order deformations of X and $H^1(X; T_X)$ as well. It is in an attempt to generalize this result to the nonsmooth case that SAG was initially developed. It turns out SAG is not only the proper venue for doing such a thing, but it also gives us

https://doi.org/10.1515/9783111334073-010

an interpretation of higher cohomology groups $H^n(X;TX)$ as well, whereas classically, for example, it is known that $H^2(X;T_X)$ is related to the problem of extending first-order deformations of X to second-order deformations, but that is all we can perceive; one does not know what $H^2(X;T_X)$ consists of. So in addition to motivating extra terms and to repackaging different objects together, it seems that SAG also sheds some light on concepts that were still classically opaque. We regard this as an important concept; if one can go to a higher theory that explains the presence of extra terms that combine loose concepts and provide a more complete picture, one should consider such a theory.

11 ∞-operads

$\mathbb{E}_\infty$*-rings* are central elements of Spectral Algebraic Geometry. They are commutative algebra objects of the ∞-category Sp of spectra. To be more precise, they are objects of CAlg := CAlg(Sp) := $\mathrm{Alg}_{\mathbb{E}_\infty}$(Sp), an example of ∞-category of the form $\mathrm{Alg}_{\mathcal{O}}(\mathcal{C})$, where $\mathcal{O}$ is an ∞-operad, and $\mathcal{C}$ is a symmetric monoidal ∞-category. Thus we need the notion of ∞-operad, on which the concept of $\mathbb{E}_\infty$-ring is based, which we develop presently. We follow [53].

Let $(\mathcal{C}, \otimes)$ be a symmetric monoidal category. One can define a new category $\mathcal{C}^\otimes$,[1] whose objects are finite, possibly empty, sequences of objects of $\mathcal{C}$ denoted by $[C_1, \dots, C_n]$. This category $\mathcal{C}^\otimes$ will act as a deconstruction of $(\mathcal{C}, \otimes)$. A morphism $\phi : [C_1, \dots, C_n] \to [C'_1, \dots, C'_m]$ is given by a triple $(S, \alpha, \{f_j\})$, where $S \subseteq \{1, \dots, n\}$, $\alpha : S \to \{1, \dots, m\}$, and one has a collection $\{f_j : \otimes_{\alpha(i)=j} C_i \to C'_j\}_{1 \le j \le m}$. Write f for such a collection. By abuse of language we will refer to such a triple (S, α, f) as simply f. The composition of such morphisms $f = (S, \alpha, f) : [C_1, \dots, C_n] \to [C'_1, \dots, C'_m]$ and $g = (T, \beta, g) : [C'_1, \dots, C'_m] \to [C''_1, \dots, C''_p]$ with $S \subseteq \{1, \dots, n\}$, $\alpha : S \to \{1, \dots, m\}$, $T \subseteq \{1, \dots, m\}$, and $\beta : T \to \{1, \dots, p\}$ is given by $(U, \beta \circ \alpha, \{g_j \circ f_i\})$, where $U = \alpha^{-1}T$, $\beta \circ \alpha : U \to \{1, \dots, p\}$, and for $1 \le k \le p$,

$$g_j \circ f_i : \otimes_{(\beta \circ \alpha)(i)=k} C_i \cong \otimes_{\beta(j)=k} \otimes_{\alpha(i)=j} C_i \xrightarrow{f_i} \otimes_{\beta(j)=k} C'_j \xrightarrow{g_j} C''_k.$$

One defines *Segal's category $\mathcal{F}\mathrm{in}_*$ of pointed finite sets*[2] as follows. For $n \ge 0$, let $\langle n \rangle = \{*, 1, 2, \dots, n\}$. Let those be objects of $\mathcal{F}\mathrm{in}_*$, and morphisms $\langle m \rangle \to \langle n \rangle$ are maps that keep $*$ fixed. Among such maps, we distinguish, for $1 \le i \le n$, maps $\rho^i : \langle n \rangle \to \langle 1 \rangle$ such that $\rho^i(j) = 1$ if $j = i$ and $*$ otherwise. Note that given a symmetric monoidal category $\mathcal{C}$, one has an obvious forgetful functor $\mathcal{C}^\otimes \xrightarrow{p} \mathcal{F}\mathrm{in}_*$ with fiber over $\langle n \rangle$ being denoted $\mathcal{C}^\otimes_{\langle n \rangle}$. Observe that $\mathcal{C}^\otimes_{\langle 1 \rangle} \simeq \mathcal{C}$. More generally, $\mathcal{C}^\otimes_{\langle n \rangle} \simeq \mathcal{C}^n$. There is another property enjoyed by the forgetful functor p:[3] for $C = [C_1, \dots, C_n] \in \mathrm{Ob}\, \mathcal{C}^\otimes$ and a morphism $f : \langle n \rangle \to \langle p \rangle$ in $\mathcal{F}\mathrm{in}_*$, there are an object $C' = [C'_1, \dots, C'_p]$ of $\mathcal{C}^\otimes$ and a universal map $\bar{f} : C \to C'$ over f. Observe, as pointed out in Remark 2.0.0.6 of [53], that a functor $p : \mathcal{C}^\otimes \to \mathcal{F}\mathrm{in}_*$ satisfying this lifting property and such that $\mathcal{C}^\otimes_{\langle n \rangle} \simeq \mathcal{C}^n$ is equivalent to putting a symmetric monoidal structure on $\mathcal{C}$. This is a technical advantage, as it is sometimes easier to define $\mathcal{C}^\otimes$, and those two properties we just discussed are easier to work with than the axioms for a symmetric monoidal structure.

One defines a *symmetric monoidal ∞-category*[4] to be a coCartesian fibration $p : \mathcal{C}^\otimes \to \mathrm{N}(\mathcal{F}\mathrm{in}_*)$ in Set_Δ such that for all $n \ge 0$, the maps ρ^i mentioned above induce

1 Constr. 2.0.0.1 [53].

2 Not. 2.0.0.2 [53].

3 Rmrk 2.0.0.5 [53].

4 Def. 2.0.0.7 [53].

https://doi.org/10.1515/9783111334073-011

functors $\rho^i_! : \mathcal{C}^{\otimes}_{\langle n\rangle} \to \mathcal{C}^{\otimes}_{\langle 1\rangle}$ that determine an equivalence $\mathcal{C}^{\otimes}_{\langle n\rangle} \simeq (\mathcal{C}^{\otimes}_{\langle 1\rangle})^n$. Thus we see this as being a straightforward ∞-categorical generalization of the construction of $\mathcal{C}^{\otimes}$ above for an ordinary symmetric monoidal category $\mathcal{C}$.

A morphism $f : \langle m\rangle \to \langle n\rangle$ in $\mathcal{F}\mathrm{in}_*$ is *inert*[5] if for all $i \in \langle n\rangle - \{*\}, f^{-1}(i)$ is a singleton.

One defines an ∞-*operad* as a further generalization of the concept of symmetric monoidal ∞-category,[6] to be a functor $p : \mathcal{O}^{\otimes} \to \mathrm{N}(\mathcal{F}\mathrm{in}_*)$ in Cat_∞ such that for every inert morphism $f : \langle m\rangle \to \langle n\rangle$ in $\mathrm{N}(\mathcal{F}\mathrm{in}_*)$ and every $C \in \mathrm{Ob}\,(\mathcal{O}^{\otimes}_{\langle m\rangle})$, there is a p-co-Cartesian morphism $\overline{f} : C \to C'$ in $\mathcal{O}^{\otimes}$ lifting f and inducing a functor $f_! : \mathcal{O}^{\otimes}_{\langle m\rangle} \to \mathcal{O}^{\otimes}_{\langle n\rangle}$. Further, for $C \in \mathcal{O}^{\otimes}_{\langle m\rangle}$, $C' \in \mathcal{O}^{\otimes}_{\langle n\rangle}$ and $f : \langle m\rangle \to \langle n\rangle$ in $\mathcal{F}\mathrm{in}_*$, denoting by $\pi_0^{/f}\mathrm{Map}(C, C')$ the union of connected components of $\mathrm{Map}(C, C')$ over f, if one chooses p-coCartesian morphisms $C' \to C'_i$ over the inert morphisms $\rho^i : \langle n\rangle \to \langle 1\rangle, 1 \le i \le n$, then the induced map

$$\pi_0^{/f}(C, C') \to \prod_{1\le i\le n} \pi_0^{/\rho^i\circ f}\mathrm{Map}(C, C'_i)$$

is a homotopy equivalence, and for any finite collection of objects $C_1, \ldots, C_n$ of $\mathcal{O}^{\otimes}_{\langle 1\rangle}$, there is $C \in \mathcal{O}^{\otimes}_{\langle n\rangle}$ along with a collection of p-coCartesian morphisms $C \to C_i$ over the morphisms ρ^i. One calls $\mathcal{O} = \mathcal{O}^{\otimes}_{\langle 1\rangle}$ the *underlying* ∞-*category of* $\mathcal{O}^{\otimes}$.[7]

One may object that this definition is not manifestly operadic in nature in the classical sense. One can remedy this by starting with an ∞-operad $p : \mathcal{O}^{\otimes} \to N(\mathcal{F}\mathrm{in}_*)$ as just defined, and let $\mathcal{O} = \mathcal{O}^{\otimes}_{\langle 1\rangle}$. One has a canonical equivalence $\mathcal{O}^{\otimes}_{\langle n\rangle} \simeq \mathcal{O}^n$,[8] which allows us to identify objects C of $\mathcal{O}^{\otimes}_{\langle n\rangle}$ with finite sequences $(C_1, \ldots, C_n)$ of objects of $\mathcal{O}$. For objects $X_1, \ldots, X_n, Y$ of $\mathcal{O}$, define

$$\mathrm{Mul}_{\mathcal{O}}(\{X_i\}, Y) = \bigcup \mathrm{Map}_{\mathcal{O}^{\otimes}}((X_1, \ldots, X_n), Y)$$

where the union is taken over morphisms $\beta : \langle n\rangle \to \langle 1\rangle$ such that $\beta^{-1}(*) = *$.[9] Then arguing as in Remark 2.1.1.17 of [53], we get that $\mathcal{O}$, together with the spaces of multiplications as just defined, assuming that they satisfy the axioms of a colored operad up to coherent homotopy, provide an alternative, manifestly operadic definition for $\mathcal{O}^{\otimes}$.

One example of ∞-operad that we will use extensively is the *commutative* ∞-*operad*[10] denoted $\mathrm{Comm}^{\otimes} = \mathrm{N}(\mathcal{F}\mathrm{in}_*)$.

---

5 Def. 2.1.1.8 [53].

6 Def. 1.2.1.10 [53].

7 Rmrk 2.1.1.12 [53].

8 Rmrk 2.1.1.15 [53].

9 Rmrk 2.1.1.16 [53].

10 Ex. 2.1.1.18 [53].

An important simple fact we will implicitly use to define algebra objects of ∞-operads is that if $p : \mathcal{O}^{\otimes} \to N(\mathcal{F}\mathrm{in}_*)$ is an ∞-operad, then p is a categorical fibration.[11]

Observe that in a sense an ∞-operad is a generalization of a symmetric monoidal ∞-category, where the fibration $p : \mathcal{O}^{\otimes} \to \mathrm{N}(\mathcal{F}\mathrm{in}_*)$ is not required to be a coCartesian fibration.

To define ∞-operad maps, we need the notion of inert morphism. Let $p : \mathcal{O}^{\otimes} \to \mathrm{N}(\mathcal{F}\mathrm{in}_*)$ be an ∞-operad. One says a morphism f in $\mathcal{O}^{\otimes}$ is *inert*[12] if $p(f)$ is inert and f is p-coCartesian. Then for two ∞-operads $\mathcal{O}^{\otimes}$ and $\mathcal{O}'^{\otimes}$, an *∞-operad map*[13] $\mathcal{O}^{\otimes} \to \mathcal{O}'^{\otimes}$ is given by a map $f : \mathcal{O}^{\otimes} \to \mathcal{O}'^{\otimes}$ over $\mathrm{N}(\mathcal{F}\mathrm{in}_*)$ preserving inert morphisms. One denotes by $\mathrm{Alg}_{\mathcal{O}}(\mathcal{O}')$ the full subcategory of $\mathrm{Fun}(\mathcal{O}^{\otimes}, \mathcal{O}'^{\otimes})$ spanned by the ∞-operad maps, whose objects are referred to as $\mathcal{O}$-algebra objects of $\mathcal{O}'$.

$\mathbb{E}_{\infty}$-rings are defined from the *little k-cubes operad*, which we define presently. From [53], for $k \geq 0$, denote $(-1,1)^k$ by $\square^k$. A map $f : \square^k \to \square^k$ is called a rectilinear embedding if it is linear in each argument, with a positive slope. More generally, S being a finite set, a map $\square^k \times S \to \square^k$ is a rectilinear embedding if it is an open embedding and its restriction to each of its connected components is rectilinear. One denotes by $\mathrm{Rect}(\square^k \times S, \square^k)$ the collection of all such maps. Then one defines a topological category ${}^t\mathbb{E}^{\otimes}{}_k$[14] with the objects $\langle n \rangle$ of $\mathcal{F}\mathrm{in}_*$ as its objects, and morphisms as maps $\alpha : \langle m \rangle \to \langle n \rangle$ such that for all $j \in \langle n \rangle, j \neq *$, one has a rectilinear embedding $\square^k \times \alpha^{-1}(j) \to \square^k$. One topologizes the mapping spaces $\mathrm{Hom}_{{}^t\mathbb{E}^{\otimes}{}_k}(\langle m \rangle, \langle n \rangle)$ in accordance with the presentation of such a space as the disjoint union over all maps $\alpha : \langle m \rangle \to \langle n \rangle$ of the product of the spaces $\mathrm{Rect}(\square^k \times \alpha^{-1}(j), \square^k)$. The composition is the obvious one. Let $\mathbb{E}_k^{\otimes} = \mathrm{N}({}^t\mathbb{E}^{\otimes}{}_k)$. Observe that $\mathbb{E}_k^{\otimes} \to \mathrm{N}(\mathcal{F}\mathrm{in}_*)$ is an ∞-operad, called the ∞-operad of little k-cubes.

One has a stabilization functor ${}^t\mathbb{E}_k \to {}^t\mathbb{E}_{k+1}$, which is the identity on objects and is simply given on morphisms by taking the product with the interval $(-1, 1)$. This functor induces a map of ∞-operads $\mathbb{E}_k^{\otimes} \to \mathbb{E}_{k+1}^{\otimes}$. The colimit of the sequence of ∞-operads $\mathbb{E}_0^{\otimes} \to \mathbb{E}_1^{\otimes} \to \cdots$ is denoted $\mathbb{E}_{\infty}^{\otimes}$ and is equivalent to the commutative ∞-operad $\mathrm{Comm}^{\otimes} = \mathrm{N}(\mathcal{F}\mathrm{in}_*)$.

Perhaps it is not entirely clear at this point why we use little k-cubes operad. Eventually, $\mathbb{E}_{\infty}$-ring spectra will be defined to be $\mathbb{E}_{\infty}$-algebra objects of the ∞-category Sp of spectra, where $\mathbb{E}_{\infty}^{\otimes}$ is an ∞-operad. $\mathbb{E}_{\infty}$-ring spectra are known to be infinite loop spaces. If they arise from an operadic construction, then one would expect that characterization as infinite loop space to be algebraically carried by $\mathbb{E}_{\infty}^{\otimes}$, which is the colimit of the sequence of ∞-operads we saw above. Now Theorem 5.1.0.8 of [53] shows that this is the case; if X is a topological space on which there is an action of the little k-cubes

11 Rmrk 2.1.1.13 [53].

12 Def. 2.1.2.3 [53].

13 Def. 2.1.2.7 [53].

14 Def. 5.1.0.2 [53].

operad ${}^t\mathbb{E}_k$ for $k > 0$, and $\pi_0 X$ is a group for the induced monoid structure, then X is weakly homotopy equivalent to $\Omega^k X_0$ for some pointed topological space X_0.

12 Spectra

Let $\mathcal{S}$ denote the ∞-category of spaces defined as the simplicial nerve of the full subcategory Kan of Set_Δ spanned by Kan complexes: $\mathcal{S} = \mathrm{N}(\mathrm{Kan})$.[1] We have seen already in the chapter on simplicial sets that one can use topological spaces and simplicial sets interchangeably, and that is why we refer to Kan complexes as spaces. Taking the nerve here allows us to obtain an ∞-category.

One defines $\mathcal{S}_*^{\mathrm{fin}}$ as the smallest full subcategory of $\mathcal{S}_*$, the ∞-category of pointed objects of $\mathcal{S}$, which is stable under finite colimits. We will see below why this particular category is of interest to us. Again, the reader can find all this material in [53].

A *spectrum*[2] is a *reduced*[3] (maps pushouts to pullbacks) and *excisive*[4] (preserves final objects) functor $X : \mathcal{S}_*^{\mathrm{fin}} \to \mathcal{S}$. We denote by Sp the ∞-category of spectra. Classically, the construction of the ∞-category $\mathrm{Sp}(\mathcal{C})$ of spectrum objects of an ∞-category $\mathcal{C}$ is associated with cohomology theories on $\mathcal{C}$. By this we mean that if one considers a generalized cohomology theory on $\mathcal{C}$ – satisfying certain conditions – given by a sequence of functors $\{h_n\}$, then by Brown representability each functor h_n can be represented by an object $E_n \in \mathcal{C}$ satisfying $E_n \cong \Omega E_{n+1}$, and the collection of all such objects assemble into a spectrum object of $\mathcal{C}$.[5] Now one may adopt a dual approach to constructing $\mathrm{Sp}(\mathcal{C})$ by considering functors from pointed spaces into $\mathcal{C}$, which morally corresponds to the functor-of-points approach to defining schemes. This would correspond to providing a theory with a homological flavor. If we take the simple example of the singular homology of topological spaces, then for our homological construction to be well-defined, one needs our functors defining spectra to be excisive and reduced as well, which justifies our definition.

Another (rather indirect way) of defining spectra goes as follows, as covered in [54]. If $\mathcal{H}$ denotes the category of pointed finite CW complexes with homotopy classes of pointed maps between them as morphisms, then one has a suspension functor $\Sigma : \mathcal{H} \to \mathcal{H}$. Observing that $\mathcal{H} = h\mathcal{S}_*^{\mathrm{fin}}$, the suspension functor lifts at the level of $\mathcal{S}_*^{\mathrm{fin}}$ to a functor, which we denote by the same letter, giving rise to a sequence $\cdots \to \mathcal{S}_*^{\mathrm{fin}} \xrightarrow{\Sigma} \mathcal{S}_*^{\mathrm{fin}} \cdots$, whose limit is denoted $\mathrm{Sp}^{\mathrm{fin}}$, and the ∞-category of spectra is defined as $\mathrm{Sp} = \mathrm{Ind}(\mathrm{Sp}^{\mathrm{fin}})$, the ∞-category of Ind objects of $\mathrm{Sp}^{\mathrm{fin}}$ obtained by formally adding filtered colimits.[6] The reason why we take the ∞-category of Ind-objects of $\mathrm{Sp}^{\mathrm{fin}}$ as our ∞-category of spectra and not just $\mathrm{Sp}^{\mathrm{fin}}$ is that the latter is missing some of the categorical constructs we will

---

1 Def. 1.2.16.1 [52].

2 Def. 1.4.2.8 [53].

3 Def. 1.4.2.1 [53].

4 Def. 1.4.2.1 [53].

5 Cor. 1.4.1.10 [54].

6 Constr. 0.2.3.10 [54].

https://doi.org/10.1515/9783111334073-012

need, hence the present enlargement by considering Ind-objects. In this setting a spectrum X is also identified with an infinite sequence of pointed spaces $\{X_n\}$ along with equivalences $X_n \simeq \Omega X_{n+1}$, Ω the loop space functor, thereby reproducing the classical definition of spectra.

The motivation for considering $\mathcal{H}$ in the first place is that CW complexes provide a convenient, combinatorial way of dealing with spaces. Additionally, the homotopy category of CW complexes is equivalent to the homotopy category of Kan complexes, so working with CW complexes as approximation to spaces, one can equivalently work with Kan complexes, which is purely combinatorial. Now typically in homotopy theory, as argued in [54], one is interested in homotopy classes of pointed maps between finite, pointed CW complexes. However, those classes are hard to describe, so we can limit ourselves to just asking what the stable homotopy classes of maps $[X, Y]_s = \lim_{\to}[\Sigma^n X, \Sigma^n Y]$ between two finite, pointed CW complexes X and Y look like. From there one defines the *Spanier–Whitehead category*[7] whose objects are pairs (X, n), where X is a finite, pointed CW complex, $n \in \mathbb{Z}$, and morphisms $(X, n) \to (Y, p)$ are defined by $\lim_{\to, k}[\Sigma^{n+k}X, \Sigma^{p+k}Y]$. It turns out that the Spanier–Whitehead category is the direct limit of the sequence $\cdots \to \mathcal{H} \xrightarrow{\Sigma} \mathcal{H} \to \cdots$ and therefore appears as the homotopy category of $\mathrm{Sp}^{\mathrm{fin}}$. Note however that this just provides us with the homotopy category of an ∞-category of finite spectra, and spectra are obtained by considering Ind-objects of $\mathrm{Sp}^{\mathrm{fin}}$, whence the necessity to lift $\Sigma : \mathcal{H} \to \mathcal{H}$ to the level of $\mathrm{Sp}^{\mathrm{fin}}$ to obtain our desired ∞-category of spectra Sp.

For our immediate purposes, observe that one has an essentially unique functor $\otimes : \mathrm{Sp} \times \mathrm{Sp} \to \mathrm{Sp}$ with the sphere spectrum $S = (S^0, 0)$ as unit: $S \otimes S = S$. One refers to this product as the smash product, and it makes $(\mathrm{Sp}, \otimes)$ into a symmetric monoidal ∞-category. This is a first step toward defining the ∞-category $\mathrm{Alg}_{\mathbb{E}_\infty}(\mathrm{Sp})$, of the form $\mathrm{Alg}_{\mathcal{O}}(\mathcal{C})$, well-defined for an ∞-operad $\mathcal{O}$ and a symmetric monoidal ∞-category $\mathcal{C}$.

7 Def. 0.2.3.1 [54].

13 $\mathbb{E}_\infty$-ring spectra

$\mathbb{E}_\infty$-rings are objects of the ∞-category CAlg = CAlg(Sp) of commutative algebra objects of Sp, where Sp is regarded as a symmetric monoidal ∞-category for the smash product monoidal structure. We have seen already that $\mathbb{E}_\infty$-rings are a natural improvement upon the use of commutative dg algebras in algebraic geometry, but from a purely algebraic perspective, recall that one has an equivalence of ∞-categories between $\mathrm{CAlg}^\heartsuit$, the ∞-category of discrete $\mathbb{E}_\infty$-rings, and the category of commutative rings, seen as an ∞-category through the nerve functor. Since $\mathrm{CAlg}^\heartsuit$ is a full subcategory of CAlg, in this manner, one can regard $\mathbb{E}_\infty$-rings as generalized commutative rings. We can make this idea more precise following [54]: $\mathbb{E}_\infty$-rings are in particular spectra. Let E be a spectrum, and consider its 0th-space $\Omega^\infty E$. On the latter, one can define a unital commutative additive law that is associative up to coherent homotopy. It turns out that the construction $E \mapsto \Omega^\infty E$ provides an equivalence of ∞-categories $\mathrm{Sp}^{\mathrm{cn}} \to \mathrm{CAlg}^{\mathrm{gp}}(\mathcal{S})$ into the ∞-category of grouplike $\mathbb{E}_\infty$-spaces, where we say, for instance, the $\mathbb{E}_\infty$-space $\Omega^\infty E$ is grouplike if the addition defined on it turns its set of connected components into an abelian group. In this manner, we can see that connective spectra have the flavor of a commutative group in a generalized sense, which prompts us to regard spectra as generalizations of commutative rings.

For $k \geq 0$, one defines a $\mathbb{E}_k$*-ring*[1] to be an object of the ∞-category $\mathrm{Alg}_{\mathbb{E}_k}(\mathrm{Sp})$. Independently, objects of the ∞-category $\mathrm{Alg}_{\mathrm{N}(\mathcal{F}\mathrm{in}_*)}(\mathrm{Sp}) = \mathrm{CAlg}(\mathrm{Sp}) = \mathrm{CAlg}$ are referred to as commutative algebra objects of Sp, and this ∞-category CAlg can be identified with the homotopy limit $\mathrm{hlimAlg}_{\mathbb{E}_k}(\mathrm{Sp})$. As such, objects thereof are called $\mathbb{E}_\infty$*-rings*. For $k \geq 0$, let R be an $\mathbb{E}_{k+1}$-ring. Let $\mathrm{Alg}_R^{(k)} = \mathrm{Alg}_{\mathbb{E}_k}(\mathrm{LMod}_R)$ be the *∞-category of $\mathbb{E}_k$-algebras over R*,[2] where $\mathrm{LMod}_R = \mathrm{LMod}_R(\mathrm{Sp})$ is regarded as an $\mathbb{E}_{k+1}$-monoidal ∞-category. One also denotes by $\mathrm{CAlg}_R = \mathrm{CAlg}(\mathrm{LMod}_R) = \mathrm{Alg}_{\mathbb{E}_\infty}(\mathrm{LMod}_R)$ the *∞-category of $\mathbb{E}_\infty$-algebras over R*.[3] Note that an $\mathbb{E}_1$-ring can be regarded as a space with an associative structure, up to coherent homotopy, and an $\mathbb{E}_\infty$-ring can be viewed as a space with a commutative structure up to homotopy. Thus the $\mathbb{E}_k$-rings in between give us a whole spectrum of notions of structures in between associativity and commutativity.

We now go over a few fundamental definitions revolving around the notion of $\mathbb{E}_\infty$-ring, following [53]. For $n \in \mathbb{Z}$, define the *n-sphere* to be $S^n = (S^0, n)$,[4] where S^0 is the 0-sphere. Recall that as such it is an object of the Spanier–Whitehead category. For $E \in \mathrm{Sp}$ and $n \in \mathbb{Z}$, define the *nth space of E* to be $\Omega^{\infty-n}E = \mathrm{Map}_{\mathrm{Sp}}(S^{-n}, E)$. This is well-defined, since the Spanier–Whitehead category is the homotopy category of $\mathrm{Sp}^{\mathrm{fin}}$, and taking the

1 Def. 7.1.0.1. [53].

2 Def. 7.1.3.5 [53].

3 Variant 7.1.3.8 [53].

4 Def. 0.2.3.2. [54].

https://doi.org/10.1515/9783111334073-013

Ind-objects of the latter category, we get Sp. Equivalently, one can regard $\Omega^\infty : \mathrm{Sp} \to \mathcal{S}$ as a functor given by evaluation at $S^0 \in \mathcal{S}_*^{\mathrm{fin}}$ and $\Omega^{\infty-n} : \mathrm{Sp} \to \mathcal{S}$ for $n \in \mathbb{Z}$ as the composite of Ω^∞ with the shift functor $E \mapsto E[n]$.[5] For X a space, the nth cohomology group of X with coefficients in $E \in \mathrm{Sp}$ is defined as $E^n(X) = \pi_0 \mathrm{Map}_{\mathcal{S}}(X, \Omega^{\infty-n}E)$. Finally, for $E \in \mathrm{CAlg}$, define $\pi_n E = E^{-n}(\{x\})$. Armed with this definition, one says E is *connective* if $\pi_n E = 0$ for $n < 0$. Let $\mathrm{CAlg}^{\mathrm{cn}}$ be the full subcategory of CAlg spanned by the connective $\mathbb{E}_\infty$-rings.

An $\mathbb{E}_\infty$-ring E is said to be *discrete* if $\pi_n E = 0$ for $n \neq 0$. Let $\mathrm{CAlg}^\heartsuit$ be the full subcategory of CAlg spanned by those objects. The construction $A \mapsto \pi_0 A$ gives an equivalence of ∞-categories between $\mathrm{CAlg}^\heartsuit$ and CRing (where we take the usual stance as in [52] of viewing ordinary categories as ∞-categories by taking their nerve). In this manner, we can identify a commutative ring with an $\mathbb{E}_\infty$-ring via this equivalence. Thus for a commutative ring k, one can define CAlg_k and the *∞-category of augmented $\mathbb{E}_\infty$-algebras* $(\mathrm{CAlg}_k)_{/k} = \mathrm{CAlg}_k^{\mathrm{aug}}$. $E \in \mathrm{CAlg}_k^{\mathrm{aug}}$ is said to be *artinian* if it is connective, $\pi_* E$ is a finite-dimensional vector space over k, and $\pi_0 E$ is a local ring. One denotes by $\mathrm{CAlg}_k^{\mathrm{art}}$ the ∞-category of artinian algebras.

One can see the need for artinian algebras when dealing with moduli problems. As discussed in [54], one can recover the formal completion of an affine algebraic variety over $\mathbb{C}$ from its restriction to local artinian $\mathbb{C}$-algebras, and this is why in particular one defines classical $\mathcal{C}$-valued formal moduli problems, $\mathcal{C}$ being an ∞-category, as given by functors $\mathrm{CAlg}_{\mathbb{C}}^{\heartsuit\mathrm{art}} \to \mathcal{C}$. At a much deeper level, the class of artinian objects of $\mathrm{CAlg}_k^{\mathrm{aug}}$ can be recovered from a certain deformation context, from which one can also have a deformation theory (a Koszul duality functor), which is instrumental in establishing an equivalence of ∞-categories between Lie_k, the ∞-category of dg Lie algebras, and Moduli_k, the ∞-category of formal moduli problems over a field k. This equivalence will be fundamental for us later when we relate perceptions to manifestations in SAG, whence the necessity to work with artinian objects.

For more detail on the deformation theoretical aspects of formal moduli problems, the reader is referred to Chapter 12 of [54] and to Example 11.5.0.7 in particular.

5 Not. 1.4.2.20 [53].

14 Spectral Deligne–Mumford stacks

The fundamental geometric object of interest to us in Spectral Algebraic Geometry is that of a *Spectral Deligne–Mumford stack*,[1] which we will just refer to as a spectral DM stack. Those are in particular spectrally ringed ∞-topos, which we first define. Before we do so, one should explain how spectral DM stacks come about. The reader can find all this material in [54].

From our perspective, generalizing commutative rings to commutative dg algebras is a natural thing to do, in the spirit of "spectralizing" our theory. Thus, following this logic, one goes from schemes to dg schemes, and as argued before, it is better in some respect to use $\mathbb{E}_\infty$-rings instead of commutative dg algebras, which would produce *spectral schemes*[2](a certain type of spectrally ringed spaces), which form an ∞-category SpSch containing the category Sch of schemes as a full subcategory. Morally, we have started from a ringed space and generalized it to an $\mathbb{E}_\infty$-ringed space. Suppose now that instead of modifying our sheaves of rings, we modify the base of our ringed spaces and use not topological spaces, but ∞-topoi instead, the reason being that those are more amenable to using categorical constructions. There is one type of ringed topos whose category contains Sch as a full subcategory, Deligne–Mumford stacks. Those form a 2-category DM that contains Sch as a full subcategory, though DM stacks and schemes are very different objects. The reason why DM stacks are defined the way they are is because they were introduced in particular to deal with elliptic curves, and it turns out that the 2-category DM contains the moduli stack of elliptic curves as an object. DM stacks though are developed retaining the philosophical construction of schemes; a DM stack is locally isomorphic to the étale spectrum of a commutative ring, whereas a scheme is locally isomorphic to the spectrum of a commutative ring. A DM stack is in particular a *ringed topos*,[3] a pair $(\mathcal{X}, \mathcal{O}_\mathcal{X})$ consisting of a Grothendieck topos $\mathcal{X}$ and a commutative ring object $\mathcal{O}_\mathcal{X}$ of $\mathcal{X}$. In particular, for a commutative ring R, if one lets $\mathcal{X} = \mathrm{Sh}_{\mathrm{Set}}(\mathrm{CAlg}_R^{\text{ét}})$ and $\mathcal{O}_\mathcal{X} = \mathcal{O} : \mathrm{CAlg}_R^{\text{ét}} \to \mathrm{Set}$ be the forgetful functor, then the corresponding ringed topos is denoted SpétR and is referred to as the *étale spectrum of* R. One has a fully faithful embedding Sch $\to$ DM provided by the construction Spec $R \to$ SpétR for a commutative ring R. Now if one combines spectral schemes and DM stacks, then one gets a so-called spectral DM stack. On the one hand, as argued in [54], spectral schemes were an ∞-categorical generalization of the notion of scheme, a sort of left derived version of a scheme, in a sense. DM stacks, thinking of derived objects, is morally a right derived generalization of schemes, in the sense that the embedding of Sch into DM does not work well with some colimits.

---

1 Def. 1.4.4.2. [54].

2 Def. 1.1.2.8 [54].

3 Def. 1.2.1.1 [54].

https://doi.org/10.1515/9783111334073-014

A spectral DM stack combines both pictures, providing a fully derived generalization of a scheme; one can think of a spectral DM stack as a pair consisting of an ∞-topos and a sheaf of $\mathbb{E}_\infty$-rings on it, which locally looks like the étale spectrum of some object, which, as we will see, has to be an $\mathbb{E}_\infty$-ring. We have a fully faithful embedding DM $\to$ SpDM provided by the construction $(\mathcal{X}, \mathcal{O}_\mathcal{X}) \to (\mathrm{Sh}(\mathcal{X}), \mathcal{O})$,[4] where $\mathrm{Sh}(\mathcal{X})$ is the 1-localic ∞-topos associated with the Grothendieck topos $\mathcal{X}$. Further, since we have an equivalence between the category of commutative ring objects of $\mathcal{X}$ and the ∞-category of connective 0-truncated sheaves of $\mathbb{E}_\infty$-rings on $\mathrm{Sh}(\mathcal{X})$, $\mathcal{O}$ above is the sheaf of $\mathbb{E}_\infty$-rings corresponding to $\mathcal{O}_\mathcal{X}$ under this equivalence.[5] Note that there is a fully faithful embedding SpSch $\to$ SpDM.[6]

We now define all the notions necessary to understand a spectral DM stack. A *spectrally ringed ∞-topos*[7] is a pair $(\mathcal{X}, \mathcal{O})$ consisting of an ∞-topos $\mathcal{X}$ and a sheaf $\mathcal{O}$ of $\mathbb{E}_\infty$-rings on it that is a sheaf valued in the ∞-category CAlg.

For an ∞-category $\mathcal{C}$, a *$\mathcal{C}$-valued sheaf*[8] on an ∞-topos $\mathcal{X}$ is a functor $\mathcal{X}^{\mathrm{op}} \to \mathcal{C}$ that preserves small limits. Observe that this is a convenient, site-independent definition of sheaf, in which the descent condition is no longer needed. However, in the event that $\mathcal{C}$ admits small limits, one has an equivalence of ∞-categories $\mathrm{Sh}_\mathcal{C}(\mathrm{Sh}(\mathcal{A})) \simeq \mathrm{Sh}_\mathcal{C}(\mathcal{A})$ for an essentially small ∞-category $\mathcal{A}$ by Proposition 1.3.1.7 of [54], which connects our present notion of sheaf on an ∞-topos with the standard notion of sheaf on an ∞-categorical site. This is something we will come back to below.

By definition a spectrally ringed ∞-topos $X = (\mathcal{X}, \mathcal{O}_\mathcal{X})$ is a *spectral DM stack* if it has a collection of objects $U_\alpha \in \mathcal{X}$ covering $\mathcal{X}$ such that for all α, there is $\mathrm{A}_\alpha \in \mathrm{CAlg}$ along with an equivalence of spectrally ringed ∞-topos $(\mathcal{X}_{/U_\alpha}, \mathcal{O}_\mathcal{X}|_{U_\alpha}) \simeq \mathrm{Sp\acute{e}tA}_\alpha$, and the structure sheaf $\mathcal{O}_\mathcal{X}$ is connective. We now define all the requisite notions.

If $\mathcal{O}_\mathcal{X}$ is a $\mathcal{C}$-valued sheaf on an ∞-topos $\mathcal{X}$, where $\mathcal{C}$ is an ∞-category, then for $U \in \mathrm{Ob}(\mathcal{X})$, one denotes by $\mathcal{O}_\mathcal{X}|_U$ the composite $(\mathcal{X}_{/U})^{\mathrm{op}} \to \mathcal{X}^{\mathrm{op}} \xrightarrow{\mathcal{O}_\mathcal{X}} \mathcal{C}$. Observe that it is also a $\mathcal{C}$-valued sheaf on $\mathcal{X}_{/U}$.[9]

For an $\mathbb{E}_\infty$-ring R, define the *étale spectrum of R*[10] by $\mathrm{Sp\acute{e}t}R = (\mathrm{Sh}_R^{\acute{e}t}, \mathcal{O})$, where $\mathrm{Sh}_R^{\acute{e}t}$ is the ∞-category $\mathrm{Sh}(\mathrm{CAlg}_R^{\acute{e}t})$ of sheaves of spaces on $\mathrm{CAlg}_R^{\acute{e}t}$ for the étale topology, and $\mathcal{O} : \mathrm{CAlg}_R^{\acute{e}t} \to \mathrm{CAlg}$ is the forgetful functor. Thus defined, $\mathrm{Sp\acute{e}t}R$ is a spectrally ringed ∞-topos. Observe that $\mathcal{O}$ is a strictly Henselian (see the next chapter for a definition) sheaf of $\mathbb{E}_\infty$-rings on the ∞-topos $\mathrm{Sh}_R^{\acute{e}t}$, and it is also a sheaf on $\mathrm{CAlg}_R^{\acute{e}t}$ with respect to the étale topology. Recall from [54] that we have two ways of defining sheaves. For a fixed

4 Rmrk 1.4.8.3 [54].

5 Rmrk 1.4.1.5 [54].

6 Rmrk 1.6.6.5 [54].

7 Def. 1.4.1.1 [54].

8 Def. 1.3.1.4 [54].

9 Not. 1.4.4.1 [54].

10 Def. 1.4.2.5. [54].

∞-category $\mathcal{C}$ and an ∞-topos $\mathcal{X}$, we have $\mathcal{C}$-valued sheaves on $\mathcal{X}$ as defined above, and they are objects of the ∞-category $\mathrm{Sh}_{\mathcal{C}}(\mathcal{X})$. But if $\mathcal{A}$ is an essentially small ∞-category with a Grothendieck topology, a $\mathcal{C}$-valued sheaf on $\mathcal{A}$[11] is a functor $\mathcal{F} : \mathcal{A}^{\mathrm{op}} \to \mathcal{C}$ such that for all $U \in \mathrm{Ob}\,(\mathcal{A})$ and for any covering sieve C of U, we have an equivalence in $\mathcal{C}$: $\mathcal{F}(U) \to \lim_{V \in C} \mathcal{F}(V)$. We let $\mathrm{Sh}_{\mathcal{C}}(\mathcal{A})$ be the full subcategory of $\mathrm{Fun}(\mathcal{A}^{\mathrm{op}}, \mathcal{C})$ spanned by the $\mathcal{C}$-valued sheaves. The connection between these two concepts is provided by Proposition 1.3.1.7 of [54], which states that one has an equivalence of ∞-categories: $\mathrm{Sh}_{\mathcal{C}}(\mathrm{Sh}(\mathcal{A})) \simeq \mathrm{Sh}_{\mathcal{C}}(\mathcal{A})$.

Letting $\mathcal{C} = \mathrm{CAlg}$, $\mathcal{A} = \mathrm{CAlg}_R^{\text{ét}}$, this reads $\mathrm{Sh}_{\mathrm{CAlg}}(\mathrm{Sh}_R^{\text{ét}}) \simeq \mathrm{Sh}_{\mathrm{CAlg}}(\mathrm{CAlg}_R^{\text{ét}})$, from which we see that $\mathcal{O}$ above can be regarded in two ways.

Recall from [54] that a morphism $\phi : R \to S$ in CAlg is *étale* if $\pi_0 R \to \pi_0 S$ is étale and in addition the morphism ϕ induces an isomorphism $\pi_0 S \otimes_{\pi_0 R} \pi_* R \to \pi_* S$. One denotes by $\mathrm{CAlg}_R^{\text{ét}}$ the full subcategory of CAlg_R consisting of *étale R-algebras*. To define $\mathrm{Sh}_R^{\text{ét}}$, we put a Grothendieck topology on $(\mathrm{CAlg}_R^{\text{ét}})^{\mathrm{op}}$ defined as follows:[12] given $A \in \mathrm{CAlg}_R$, a sieve C over A is a covering sieve iff it contains a finite family of étale morphisms $\{A \to A_i\}_{1 \le i \le n}$ with a faithfully flat induced map $A \to \prod_i A_i$. One calls this the *étale topology* on $(\mathrm{CAlg}_R^{\text{ét}})^{\mathrm{op}}$.[13]

11 Def. 1.3.1.1. [54].

12 Prop. B.6.2.1. [54].

13 Def. B.6.2.2 [54].

15 Spectral Artin representability

One key result of [54] for our purposes is the *Spectral Artin Representability Theorem*,[1] which states that if R is a Noetherian $\mathbb{E}_\infty$-ring such that $\pi_0 R$ is a Grothendieck ring, given a natural transformation $p : \mathbf{X} \to \operatorname{Spec} R$ in Fun(CAlg^{cn}, $\mathcal{S}$), given $n \geq 0$, if $\mathbf{X}$ is such that $\mathbf{X}(A)$ is n-truncated for all $A \in \text{CAlg}^{\heartsuit}$, if $\mathbf{X}$ is a sheaf for the étale topology, if it is nilcomplete, integrable, infinitesimally cohesive, admits a connective cotangent complex, and p is locally almost of finite presentation, then $\mathbf{X}$ is representable by a spectral DM n-stack X, locally almost of finite presentation over R.

Those are numerous conditions for having representability, but observe that many of them hold for a representable functor, that is, if $\mathbf{X} : \text{CAlg}^{\text{cn}} \to \mathcal{S}$ is representable, then it has a cotangent complex and is integrable, infinitesimally cohesive, and nilcomplete. Now if we want X (the object that represents $\mathbf{X}$) to be a spectral DM n-stack, then we may want $\mathbf{X}(A)$ to be n-truncated for some objects A to be specified, and if X is locally almost of finite presentation, then this is over some $\mathbb{E}_\infty$-ring R, and thus we want an analogous statement for $\mathbf{X}$. Since one has a morphism $X \to \text{Spét}R$, this induces the natural transformation $\mathbf{X} \to \text{Hom}(-, \text{Spét}R) = \operatorname{Spec} R$ mentioned above. The conditions on R will be spelled out below. We also want $\mathbf{X}$ to be a sheaf for the étale topology, which was to be expected if it is to be representable. Thus there is nothing too surprising in the statement of the representability theorem above considering some of the properties of a representable functor $\mathbf{X}$ we already know about and the conditions put on the representing object $X \in \text{SpDM}$. It turns out that as expounded in Section 18.3.3 of [54], if X is as presented in the statement of the spectral Artin representability theorem and if we just suppose $R \in \text{CAlg}^{\text{cn}}$, then all the conditions on $\mathbf{X}$ are met. It suffices then to check that the converse holds.

Representability for us is important, since writing $\mathbf{X} = \text{Hom}(-, X)$ gives us a functorial "perception" of X, a term we will define later. Presently, we can take this in the colloquial sense; $\text{Hom}(-, X)$ corresponds to all the possible ways to have morphisms into X. We will see later that this can be connected to the representation theory of X.

First, let us mention that asking $\mathbf{X}(A)$ to be n-truncated for every discrete commutative ring A ensures that X is a spectral DM n-stack if it exists.

We now go over the concepts used in the statement of the spectral Artin representability theorem, following [54].

15.1 Noetherian $\mathbb{E}_\infty$-rings

An $\mathbb{E}_\infty$-ring R is *Noetherian*[2] if it is coherent and $\pi_0 R$ is Noetherian, and R is *coherent* if it is left coherent as an $\mathbb{E}_1$-ring. Here we use Remark 7.1.0.4 of [53]: if R is an $\mathbb{E}_k$-ring, then it

1 Thm 18.3.0.1 [54].

2 Def. 7.2.4.30 [53].

https://doi.org/10.1515/9783111334073-015

can be regarded as an $\mathbb{E}_l$-ring for $l \leq k$. By definition an $\mathbb{E}_1$-ring R is *left coherent*[3] if it is connective, $\pi_0 R$ is left-coherent,[4] that is, every finitely generated left ideal of R is finitely presented in LMod_R, and for all $n \geq 0$, $\pi_n R$ is finitely presented as a module over $\pi_0 R$. Now a commutative ring R is said to be a *Grothendieck ring* [55] if it is Noetherian and for any prime ideal $\mathfrak{p}$ of R, the map $R_{\mathfrak{p}} \to \hat{R}$ is geometrically regular. Here $\hat{R}$ is the completion of $R_{\mathfrak{p}}$ with respect to its maximal ideal. From the same reference,[5] a morphism $f : R \to S$ of Noetherian commutative rings is *geometrically regular* if it is flat and for any prime ideal $\mathfrak{p} \subseteq R$ and any finite extension k of the residue field $k(\mathfrak{p})$, $k \otimes_R S$ is regular. It is expected that R is Noetherian since in the original paper [3], where we have Artin's representability theorem, we work with Noetherian rings. It is also expected that $\pi_0 R$ is a Grothendieck ring since in the classical representability theorem, one works with a Grothendieck ring R. In the proof of the spectral generalization of this theorem in [54], one needs those assumptions in two key, highly technical results Proposition 18.3.2.1 and Proposition 18.3.1.1, and one cannot help but think that one could circumvent these hypotheses of being a Noetherian and Grothendieck ring by perhaps doing something wholly different. Already in the classical case, this spurred a discussion on that topic [10].

15.2 Spec *R*

For $R \in \mathrm{CAlg}^{\mathrm{cn}}$, one defines $\mathrm{Spec}\, R = \mathrm{Hom}_{\mathrm{CAlg}}(R, -) : \mathrm{CAlg}^{\mathrm{cn}} \to \mathcal{S}$.[6] If $\infty\mathcal{T}\mathrm{op}^{\mathrm{sHen}}_{\mathrm{CAlg}}$ denotes the ∞-category of spectrally ringed ∞-topos with strictly Henselian structure sheaves, then one also has

$$\mathrm{Spec}\, R = \mathrm{Hom}_{\mathrm{CAlg}}(R, -) = \mathrm{Hom}_{\infty\mathcal{T}\mathrm{op}^{\mathrm{sHen}}_{\mathrm{CAlg}}}(\mathrm{Sp\acute{e}t}(-), \mathrm{Sp\acute{e}t} R).$$

We can also mention in passing[7] that if SpDM denotes the ∞-category of spectral DM stacks, then one has a fully faithful embedding $h : \mathrm{SpDM} \to \mathrm{Fun}(\mathrm{CAlg}^{\mathrm{cn}}, \mathcal{S})$, mapping $\mathrm{Sp\acute{e}t} R$ to $\mathrm{Spec}\, R$. Having such an embedding will allow us later to identify $X \in \mathrm{SpDM}$ with the functor $\mathbf{X} = \mathrm{Hom}(\mathrm{Sp\acute{e}t}(-), X)$ it represents. From that perspective, we will regard the *manifestation* $\mathbf{QCoh}(\mathbf{X})$ as the manifestation of X through its functor of points.

The choice of word "manifestation" comes from the fact that if $\mathbf{X} : \mathrm{CAlg}^{\mathrm{art}}_k \to \mathcal{S}$ is a formal moduli problem over a field k of characteristic zero and $\mathfrak{g}_*$ is the dg Lie algebra associated with k, then by Theorem 13.4.0.1 of [54] we have a fully faithful monoidal

3 Def. 7.2.4.16 [53].

4 Def. 7.2.4.13 [53].

5 Def. 0.0.4 [55].

6 Not. 6.2.2.3 [54].

7 Rmrk 6.2.2.4 [54].

embedding $\mathrm{QCoh}_{\mathrm{art}}(\mathbf{X}) \hookrightarrow \mathrm{Rep}_{\mathfrak{g}_*}$. We will later see how $\mathrm{QCoh}(\mathbf{X})$ is related to $\mathrm{QCoh}_{\mathrm{art}}(\mathbf{X})$. Thus $\mathrm{QCoh}(\mathbf{X})$ has a representation-theoretic interpretation. Since it is also connected to $\mathrm{Hom}(-, X)$, the perception of X, this prompts us to refer to $\mathrm{QCoh}(\mathbf{X})$ as the manifestation of $\mathbf{X}$.

A sheaf of $\mathbb{E}_\infty$-rings $\mathcal{F}$ on an ∞-topos $\mathcal{X}$ is *strictly Henselian*[8] if $\pi_0\mathcal{F}$ is strictly Henselian. Here $\pi_0\mathcal{F}$ is regarded as a commutative ring object of the underlying topos $\mathcal{X}^\heartsuit = \tau_{\leq 0}\mathcal{X}$[9]. For a topos $\mathcal{X}$ and a commutative ring object $\mathcal{F}$ of $\mathcal{X}$, one says $\mathcal{F}$ is *strictly Henselian*[10] if for any commutative ring R and any finite collection of étale maps $R \to R_\alpha$ that induces a faithfully flat map $R \to \prod_\alpha R_\alpha$, the induced map $\coprod_\alpha \mathrm{Sol}_{R_\alpha}(\mathcal{F}) \to \mathrm{Sol}_R(\mathcal{F})$ is an effective epi. If one denotes by $\mathrm{CAlg}^\heartsuit$ the category of commutative rings, then $\mathrm{Sol}_R(\mathcal{F})$ denotes[11] an object of $\mathcal{X}$ characterized by the universal property

$$\mathrm{Hom}_{\mathcal{X}}(U, \mathrm{Sol}_R(\mathcal{F})) \cong \mathrm{Hom}_{\mathrm{CAlg}^\heartsuit}(R, \mathrm{Hom}_{\mathcal{X}}(U, \mathcal{F}))$$

for any object U of $\mathcal{X}$. The solution sheaves $\mathrm{Sol}_R(\mathcal{F})$ as objects of a topos $\mathcal{X}$ allow us to couple a commutative ring R with the commutative ring object $\mathcal{F}$ of $\mathcal{X}$ in such a fashion that those solution sheaves are well behaved with respect to the étale topology on the category of commutative rings. This allows us in particular to prove that the étale spectrum functor Spét on commutative rings is right adjoint to the global sections functor.[12] One can give a direct characterization of strictly Henselian sheaves on an ∞-topos along the same lines: consider a ringed ∞-topos $(\mathcal{X}, \mathcal{O}_{\mathcal{X}})$, and fix some $\mathbb{E}_\infty$-ring R. Then by Notation 1.4.3.1 of [54] the functor $\mathrm{Map}_{\mathrm{CAlg}}(R, \mathcal{O}_{\mathcal{X}}(-)) \in \mathrm{Fun}(\mathcal{X}, \mathcal{S})$ is a sheaf, which one can represent by $\mathrm{Map}_{\mathcal{X}}(-, \mathrm{Sol}_R(\mathcal{O}_{\mathcal{X}}))$. Lemma 1.4.3.9 of [54] then states that $\mathcal{O}_{\mathcal{X}}$ is strictly Henselian iff for all $A \in \mathrm{CAlg}$ and every faithfully flat étale morphism $A \to \prod_{1 \leq i \leq n} A_i$ in CAlg, the induced map $\coprod_i \mathrm{Sol}_{A_i}(\mathcal{O}_{\mathcal{X}}) \to \mathrm{Sol}_A(\mathcal{O}_{\mathcal{X}})$ is an effective epi in $\mathcal{X}$. Here we use Definition B.6.1.1 of [54], namely that a morphism $A \to B$ in CAlg is said to be *faithfully flat* if $\pi_0 A \to \pi_0 B$ is faithfully flat in the classical sense, and one has an induced isomorphism $\pi_0 B \otimes_{\pi_0 A} \pi_n A \to \pi_n B$ for all n.

15.3 On being locally almost of finite presentation

One says that $X \in \mathrm{SpDM}$ is *locally almost of finite presentation over* R[13] if for any CD in SpDM of the form (it being implied A and B are $\mathbb{E}_\infty$-rings)

8 Def. 1.4.2.1 [54].

9 Not. 1.3.2.3 [54].

10 Def. 1.2.2.5 [54].

11 Not. 1.2.2.1 [54].

12 Prop. 12.3.4 and Cor. 1.2.3.5 [54].

13 Def. 4.2.0.1 [54].

$$\begin{array}{ccc} \mathrm{Sp\acute{e}t}B & \longrightarrow & X \\ \downarrow & & \downarrow \\ \mathrm{Sp\acute{e}t}A & \longrightarrow & \mathrm{Sp\acute{e}t}R \end{array}$$

with étale horizontal maps, B is *almost of finite presentation* over A,[14] that is, B is an *almost compact object* of CAlg_A, meaning that $\tau_{\leq n}B$ is a compact object of $\tau_{\leq n}\mathrm{CAlg}_A$ for all $n \geq 0$.

Now if $f : \mathbf{X} \to \mathbf{Y}$ is a natural transformation in $\mathrm{Fun}(\mathrm{CAlg}^{\mathrm{cn}}, \mathcal{S})$, then one says it is *locally almost of finite presentation*[15] if for all $m \geq 0$ and any filtered diagram $\{A_\alpha\}$ of m-truncated objects of $\mathrm{CAlg}^{\mathrm{cn}}$ with colimit A, the canonical map

$$\varinjlim \mathbf{X}(A_\alpha) \to \mathbf{X}(A) \times_{\mathbf{Y}(A)} \varinjlim \mathbf{Y}(A_\alpha)$$

is a homotopy equivalence. If we assume this to hold for $\mathbf{X} \to \operatorname{Spec} R$, then by Proposition 17.4.3.1 of [54], X will be locally almost of finite presentation over R.

Here we are using the notion of *étale map in SpDM*.[16] A morphism of spectral DM stacks is in particular a morphism of spectrally ringed ∞-topos. A morphism $f : (\mathcal{X}, \mathcal{O}_{\mathcal{X}}) \to (\mathcal{Y}, \mathcal{O}_{\mathcal{Y}})$ between spectrally ringed ∞-topos[17] is given by a pair (f_*, ψ), where $f_* : \mathcal{X} \to \mathcal{Y}$ a geometric morphism, and $\psi : f^*\mathcal{O}_{\mathcal{Y}} \to \mathcal{O}_{\mathcal{X}}$ is the induced map on sheaves in $\mathrm{Sh}_{\mathrm{CAlg}}(\mathcal{X})$. One says f is étale if $f_* \rightsquigarrow (\mathcal{X} \simeq \mathcal{Y}_{/U})$ for some $U \in \mathrm{Ob}(\mathcal{Y})$ and ψ is an equivalence. This may seem to be an unusual way of defining étaleness. One may imagine that a morphism $X \to Y$ of spectral DM stacks is étale if it is locally étale.

One has the equivalence of those two characterizations by virtue of Corollary 1.4.10.3 of [54], which states that if $f : X \to Y$ is a morphism of nonconnective spectral DM stacks, then f is étale iff for every CD of the form

$$\begin{array}{ccc} \mathrm{Sp\acute{e}t}B & \longrightarrow & X \\ \downarrow & & \downarrow f \\ \mathrm{Sp\acute{e}t}A & \longrightarrow & Y \end{array}$$

where the horizontal maps are étale, the map $A \to B$ of $\mathbb{E}_\infty$-rings is étale. This also means that our definition of étale map is what it should be,[18] namely that $\mathrm{Sp\acute{e}t}B \to \mathrm{Sp\acute{e}t}A$ is étale iff the morphism $A \to B$ of $\mathbb{E}_\infty$-rings is étale.

---

14 Def. 7.2.4.26 [53].

15 Def. 17.4.1.1 [54].

16 Def. 1.4.0.1 [54].

17 Constr. 1.4.1.3 [54].

18 Thm. 1.4.10.2 [54].

15.4 Properties of functors X : $\mathrm{CAlg}^{\mathrm{cn}} \to \mathcal{S}$

15.4.1 Nilcompleteness

A functor $\mathbf{X} : \mathrm{CAlg}^{\mathrm{cn}} \to \mathcal{S}$ is said to be *nilcomplete*[19] if for all $R \in \mathrm{CAlg}^{\mathrm{cn}}$, we have a homotopy equivalence $\mathbf{X}(R) \xrightarrow{\simeq} \lim \mathbf{X}(\tau_{\leq n}R)$. In other terms, such a functor is defined by its restriction to truncated objects.

15.4.2 On being infinitesimally cohesive

Let A_1, A_2, and A_0 be objects of $\mathrm{CAlg}^{\mathrm{cn}}$ such that the maps $\pi_0 A_1 \twoheadrightarrow \pi_0 A_0$ and $\pi_0 A_2 \twoheadrightarrow \pi_0 A_0$ have nilpotent kernels in $\pi_0 A_1$ and $\pi_0 A_2$, respectively. Then a functor $\mathbf{X} : \mathrm{CAlg}^{\mathrm{cn}} \to \mathcal{S}$ is said to be *infinitesimally cohesive*[20] if it maps any pullback CD in $\mathrm{CAlg}^{\mathrm{cn}}$ of the form

$$\begin{array}{ccc} A_{12} & \longrightarrow & A_2 \\ \downarrow & & \downarrow \\ A_1 & \longrightarrow & A_0 \end{array}$$

to a pullback square in $\mathcal{S}$

$$\begin{array}{ccc} \mathbf{X}(A_{12}) & \longrightarrow & \mathbf{X}(A_2) \\ \downarrow & & \downarrow \\ \mathbf{X}(A_1) & \longrightarrow & \mathbf{X}(A_0). \end{array}$$

Being infinitesimally cohesive is a slight generalization of the definition of cohesiveness,[21] introduced for the sake of deformation theory. Indeed, as pointed out in Remark 17.3.1.8 of [54], if $\mathbf{X} : \mathrm{CAlg}^{\mathrm{cn}} \to \mathcal{S}$ is infinitesimally cohesive and has a cotangent complex $L_{\mathbf{X}}$, then the latter is intricately linked to the deformation theory of $\mathbf{X}$. With regards to being cohesive, this provides a functorial compatibility for gluing along closed immersion, as illustrated in Remark 17.3.1.4 of [54].

15.4.3 On being integrable

Let R be a local Noetherian $\mathbb{E}_\infty$-ring. Suppose R is complete with respect to its maximum ideal $\mathfrak{m} \subseteq \pi_0 R$. Let $\mathbf{X} : \mathrm{CAlg}^{\mathrm{cn}} \to \mathcal{S}$ be a functor. If

19 Def. 17.3.2.1 [54].

20 Def. 17.3.1.5 [54].

21 Def. 17.3.1.1 [54].

$$(\mathrm{Spf}R \hookrightarrow \mathrm{Spec}\,R) \rightsquigarrow (\mathrm{Map}_{\mathcal{C}}(\mathrm{Spec}\,R, \mathbf{X}) \xrightarrow{\simeq} \mathrm{Map}_{\mathcal{C}}(\mathrm{Spf}R, \mathbf{X}))$$

with $\mathcal{C} = \mathrm{Fun}(\mathrm{CAlg}^{\mathrm{cn}}, \mathcal{S})$, where on the right-hand side, we have a homotopy equivalence, then one says $\mathbf{X}$ is *integrable*.[22] This condition is a generalization of a result that holds for spectral DM stacks and has the advantage of being able to be tested on discrete rings, as illustrated by Proposition 17.3.5.1 of [54]. In this definition, recall that $\mathrm{Spec}\,R = \mathrm{Hom}_{\infty\mathcal{T}\mathrm{op}^{\mathrm{sHen}}_{\mathrm{CAlg}}}(\mathrm{Sp\acute{e}t}(-), \mathrm{Sp\acute{e}t}R)$, where $\mathrm{Sp\acute{e}t}R = (\mathrm{Sh}^{\text{ét}}_R, \mathcal{O})$, and $\mathrm{Spf}R$ is the *formal spectrum*,[23] which is constructed as follows: for an adic $\mathbb{E}_\infty$-ring R, a finitely generated ideal $I \subseteq \pi_0 R$ of definition, and the I-completion functor $R \mapsto R^\wedge_I$, denote by $\mathcal{O}_{\mathrm{Spf}R}$ the following composition:

$$\mathrm{CAlg}^{\text{ét}}_R \xrightarrow{\mathcal{O}_{\mathrm{Sp\acute{e}t}R}} \mathrm{CAlg}_R \xrightarrow{\wedge} \mathrm{CAlg}_R.$$

Then one defines $\mathrm{Spf}R = (\mathrm{Sh}^{\mathrm{ad}}_R, \mathcal{O}_{\mathrm{Spf}R})$. Here $\mathrm{Sh}^{\mathrm{ad}}_R$ is the subtopos of $\mathrm{Sh}^{\text{ét}}_R$ corresponding to the vanishing locus $X \subseteq |\mathrm{Spec}\,R|$ of I.[24] Going back to the definition of integrability, the inclusion $\mathrm{Spf}R \hookrightarrow \mathrm{Spec}\,R$ should be understood as $\mathrm{Hom}(\mathrm{Sp\acute{e}t}(-), \mathrm{Spf}R) \hookrightarrow \mathrm{Hom}(\mathrm{Sp\acute{e}t}(-), \mathrm{Sp\acute{e}t}R) = \mathrm{Spec}\,R$.

For a commutative ring R and an ideal $I \subseteq R$, $V_I = \{\mathfrak{p} \in |\mathrm{Spec}\,R| \,:\, I \subseteq \mathfrak{p}\}$ is referred to as the *vanishing locus of* I.[25]

For completeness, recall that an *adic* $\mathbb{E}_\infty$*-ring*[26] is an object $R \in \mathrm{CAlg}^{\mathrm{cn}}$ with a topology on $\pi_0 R$ that makes it an adic ring.

Let A be a commutative ring. Every ideal $I \subseteq A$ determines a topology on A with a basis of open sets given by $x + I^n \subseteq A$ for $x \in A$ and $n \geq 0$. One refers to this topology as the *I-adic topology*.[27] Now if A is a topological commutative ring, then one says an ideal $I \subseteq A$ is an *ideal of definition* if the I-adic topology coincides with the topology on A. An *adic ring* is a topological commutative ring with a finitely generated ideal of definition.

Regarding the completion functor, if R is an $\mathbb{E}_2$-ring, $I \subseteq \pi_0 R$ is a finitely generated ideal, and $\mathcal{C}$ is a stable R-linear ∞-category, the inclusion $\mathcal{C}^{\mathrm{Cpl}(I)} \hookrightarrow \mathcal{C}$ admits a L adjoint, which one refers to as the *I-completion functor*[28] and denotes by $C \mapsto C^\wedge_I$. The ∞-category $\mathcal{C}^{\mathrm{Cpl}(I)}$ is defined as follows: if R is an $\mathbb{E}_2$-ring, $\mathcal{C}$ is a stable R-linear ∞-category, and $I \subseteq \pi_0 R$ is an ideal, an object $C \in \mathcal{C}$ is said to be *I-complete*[29] if for

---

22 Def. 17.3.4.1 [54].

23 Constr. 8.1.1.10 [54].

24 Not. 8.1.1.8 [54].

25 Ex. 1.1.1.2 [54].

26 Def. 8.1.1.5 [54].

27 Def. 8.1.1.1 [54].

28 Def. 7.3.1.5 [54].

29 Def. 7.3.1.1. [54].

any I-local object $D \in \mathcal{C}$, $\mathrm{Map}_{\mathcal{C}}(D, C)$ is contractible. Then $\mathcal{C}^{\mathrm{Cpl}(I)}$ denotes the full subcategory of $\mathcal{C}$ spanned by the I-complete objects. In the same setting as above, except that I is not necessarily finitely generated, $C \in \mathrm{Ob}\,\mathcal{C}$ is said to be *I-local*[30] if for all I-nilpotent $D \in \mathcal{C}$, $\mathrm{Map}_{\mathcal{C}}(D, C)$ is contractible.

Finally, for $R \in \mathrm{CAlg}^{\mathrm{cn}}$ with an ideal $I \subseteq \pi_0 R$ and a prestable R-linear ∞-category $\mathcal{C}$, $C \in \mathrm{Ob}\,\mathcal{C}$ is said to be *I-nilpotent*[31] if for all $x \in I$, $R[x^{-1}] \otimes_R C$ vanishes.[32]

For a connective $\mathbb{E}_2$-ring R, an object $\mathcal{C} \in \mathrm{LinCat}_R^{\mathrm{add}}$ is a *stable R-linear ∞-category*[33] if $\mathcal{C}$ is stable as an ∞-category, where for a connective $\mathbb{E}_2$-ring R and the ∞-category $\mathcal{P}r^{\mathrm{L}}$ of presentable ∞-categories, one denotes

$$\mathrm{LinCat}_R^{\mathrm{add}} = \mathrm{LMod}_{\mathrm{LMod}_R^{\mathrm{cn}}}(\mathcal{P}r^{\mathrm{L}})$$

and its objects are referred to as *additive R-linear ∞-categories.*[34]

15.4.4 Cotangent complex

We now tackle $\mathbf{X}$ having a cotangent complex. The full definition is intricate, and rather than being repetitive, we refer the reader to [54] for a full coverage. We will limit ourselves to providing the great lines only, since we do not need a working definition, rather we just want to briefly expose the concept of tangent complex to put things in perspective. Let $\mathbf{X} : \mathrm{CAlg}^{\mathrm{cn}} \to \mathcal{S}$ be a functor, classifying a L fibration $\overline{\mathrm{CAlg}^{\mathrm{cn}}} \to \mathrm{CAlg}^{\mathrm{cn}}$. Let $\underline{\mathrm{Mod}} = \mathrm{Mod}(\mathrm{Sp})$ be the ∞-category of pairs (A, M) for $A \in \mathrm{CAlg}$ and $M \in \mathrm{Mod}_A$. Denote $\overline{\mathrm{CAlg}^{\mathrm{cn}}} \times_{\mathrm{CAlg}} \mathrm{Mod}$ by $\mathrm{Mod}^{\mathbf{X}}$, with object triples (A, M, η) with A connective and $\eta \in \mathbf{X}(A)$. The full subcategory thereof for which the objects are such that M is connective is denoted by $\mathrm{Mod}^{\mathbf{X}}_{\mathrm{cn}}$.[35] Now for two functors $\mathbf{X}, \mathbf{Y} : \mathrm{CAlg}^{\mathrm{cn}} \to \mathcal{S}$, a natural transformation $\alpha : \mathbf{X} \longrightarrow \mathbf{Y}$, and $\Psi : \mathrm{Mod}^{\mathbf{X}}_{\mathrm{cn}} \to \mathrm{Mor}\,\mathcal{S}$ defined by $\Psi(A, M, \eta) = (\mathbf{X}(A \oplus M) \to \mathbf{X}(A) \times_{\mathbf{Y}(A)} \mathbf{Y}(A \oplus M))$, let $F = \mathrm{fib}_{\bar{\eta}}(\Psi) : \mathrm{Mod}^{\mathbf{X}}_{\mathrm{cn}} \to \mathcal{S}$, and let $\bar{\eta} \in \mathbf{X}(A) \times_{\mathbf{Y}(A)} \mathbf{Y}(A \oplus M)$ be induced by $\eta \in \mathbf{X}(A)$. According to Proposition 17.2.3.2 of [54], one has a fully faithful embedding $\iota : \mathrm{QCoh}(\mathbf{X})^{\mathrm{acn}} \hookrightarrow \mathrm{Fun}(\mathrm{Mod}^{\mathbf{X}}_{\mathrm{cn}}, \mathcal{S})^{\mathrm{op}}$. Then morally, the *cotangent complex*[36] of α is defined by $\mathrm{L}_{\mathbf{X}/\mathbf{Y}} = \iota^{-1} F$. If $\mathbf{Y} = *$ is the final object of $\mathrm{Fun}(\mathrm{CAlg}^{\mathrm{cn}}, \mathcal{S})$, then one says $\mathbf{X}$ admits a cotangent complex $\mathrm{L}_{\mathbf{X}} = \mathrm{L}_{\mathbf{X}/*}$.

30 Def. 7.2.4.1 [54].

31 Def. 7.1.1.6 [54].

32 Def. 7.1.1.1. [54].

33 Def. D.1.4.1 [54].

34 Def. D.1.2.1 [54].

35 Not. 17.2.4.1. [54].

36 Def. 17.2.4.2 [54].

The cotangent complex, or rather its general form, the relative cotangent complex, can be seen as the analogous construct in Spectral Algebraic Geometry of the module of relative Kähler differentials in classical Algebraic Geometry; for a morphism $X \to Y$ of spectral DM stacks, one can define its relative cotangent complex $L_{X/Y}$, which can be further generalized for a natural transformation between objects of $\mathrm{Fun}(\mathrm{CAlg}^{\mathrm{cn}}, \mathcal{S})$. This latter object is instrumental in showing that a functor $\mathrm{CAlg}^{\mathrm{cn}} \to \mathcal{S}$ is representable by a spectral DM stack.[37]

15.5 Quasi-coherent sheaves

One thing we need to clearly define is the notion of quasi-coherent sheaf on stacks and functors. We start with functors, following Construction 6.2.1.7 of [54]. Let $q : \mathcal{D} \to \mathcal{E}$ be a Cartesian fibration in Set_Δ. A functor $F : \mathcal{C} \to \mathcal{D}$ over $\mathcal{E}$ is said to be *q-Cartesian*[38] if it maps $\mathcal{C}_1$ into q-Cartesian edges of $\mathcal{D}$. Let $\mathrm{Fun}^{\mathrm{cart}}_{\mathcal{E}}(\mathcal{C}, \mathcal{D})$ be the full subcategory of $\mathrm{Fun}_{\mathcal{E}}(\mathcal{C}, \mathcal{D})$ spanned by such maps. Now let $e : \mathcal{C} \to \mathcal{E}$ be a R fibration, object of $(\mathrm{Set}^{\mathrm{Rfib}}_\Delta)_{/\mathcal{E}}$, the simplicial category of right fibrations over $\mathcal{E}$. One has a functor

$$\begin{aligned} \mathrm{Fun} : ((\mathrm{Set}^{\mathrm{Rfib}}_\Delta)_{/\mathcal{E}})^{\mathrm{op}} &\to \mathrm{Cat}^\Delta_\infty, \\ \mathcal{C} &\mapsto \mathrm{Fun}^{\mathrm{cart}}_{\mathcal{E}}(\mathcal{C}, \mathcal{D}). \end{aligned}$$

Here $\mathrm{Cat}^\Delta_\infty$ denotes the simplicial category of small ∞-categories with morphisms space between $\mathcal{C}, \mathcal{D} \in \mathrm{Cat}_\infty$ the largest Kan complex in $\mathrm{Fun}(\mathcal{C}, \mathcal{D})$, and hence it is a simplicial category. One has $\mathrm{N}(\mathrm{Cat}^\Delta_\infty) = \mathrm{Cat}_\infty$. But it is also true that $\mathrm{N}((\mathrm{Set}^{\mathrm{Rfib}}_\Delta)_{/\mathcal{E}}) \simeq \mathrm{Fun}(\mathcal{E}^{\mathrm{op}}, \mathcal{S})$. Thus by taking the simplicial nerve of the functor Fun one obtains a functor

$$\Phi[q] : \mathrm{Fun}(\mathcal{E}^{\mathrm{op}}, \mathcal{S})^{\mathrm{op}} \to \mathrm{Cat}_\infty.$$

In practice, for a functor $\mathbf{X} : \mathcal{E}^{\mathrm{op}} \to \mathcal{S}$ classifying a R fibration $\mathcal{C} \to \mathcal{E}$, $\Phi[q](\mathbf{X}) = \mathrm{Fun}^{\mathrm{cart}}_{\mathcal{E}}(\mathcal{C}, \mathcal{D})$.[39] Dually, if $q : \mathcal{D} \to \mathcal{E}$ is a coCartesian fibration, then considering left fibrations over $\mathcal{E}$ and doing the same construction, one gets a functor $\Phi'[q] : \mathrm{Fun}(\mathcal{E}, \hat{\mathcal{S}})^{\mathrm{op}} \to \widehat{\mathrm{Cat}_\infty}$. We now apply this formalism to

$$q = \pi_1 : \mathcal{D} = \mathrm{CAlg}^{\mathrm{cn}} \times_{\mathrm{CAlg}} \mathrm{Mod} \to \mathrm{CAlg}^{\mathrm{cn}} = \mathcal{E}$$

a coCartesian fibration. Let $\mathbf{X} : \mathrm{CAlg}^{\mathrm{cn}} \to \hat{\mathcal{S}}$ be a functor classifying a L fibration $\mathcal{C} \to \mathrm{CAlg}^{\mathrm{cn}}$. Then one has

37 Thm. 18.1.0.2 [54].

38 Def. 6.2.1.1. [54].

39 Rmrk 6.2.1.8 [54].

$$\mathbf{QCoh} \equiv \Phi'[q] : \mathrm{Fun}(\mathrm{CAlg}^{\mathrm{cn}}, \hat{\mathcal{S}})^{\mathrm{op}} \to \widehat{\mathrm{Cat}_{\infty}},$$
$$\mathbf{X} \mapsto \mathrm{Fun}^{\mathrm{cocart}}_{\mathrm{CAlg}}(\mathcal{C}, \mathrm{CAlg}^{\mathrm{cn}} \times_{\mathrm{CAlg}} \mathrm{Mod})$$

which defines the *∞-category of quasi-coherent sheaves on* $\mathbf{X}$. In practice though, $\mathcal{C}$ is denoted $\overline{\mathrm{CAlg}^{\mathrm{cn}}}$, and $q : \overline{\mathrm{CAlg}^{\mathrm{cn}}} \times_{\mathrm{CAlg}} \mathrm{Mod} \to \overline{\mathrm{CAlg}^{\mathrm{cn}}} \to \mathrm{CAlg}^{\mathrm{cn}}$, since we consider functors over $\overline{\mathrm{CAlg}^{\mathrm{cn}}}$. We recognize the fiber product as $\mathrm{Mod}^{\mathbf{X}}$. Thus one has $\mathbf{QCoh}(\mathbf{X}) = \mathrm{Fun}^{\mathrm{cocart}}_{\mathrm{CAlg}}(\overline{\mathrm{CAlg}^{\mathrm{cn}}}, \mathrm{Mod}^{\mathbf{X}})$.

Quasi-coherent sheaves on spectral DM stacks are defined differently. Recall that Mod is the ∞-category whose objects are pairs (A, M) with $A \in \mathrm{CAlg}$ and $M \in \mathrm{Mod}_A$. We can generalize this as follows: one can consider triples $(\mathcal{X}, \mathcal{O}, \mathcal{F})$, where $\mathcal{X}$ is an ∞-topos, $\mathcal{O}$ is a sheaf of $\mathbb{E}_{\infty}$-rings on $\mathcal{X}$, and $\mathcal{F}$ is a sheaf of $\mathcal{O}$-module spectra on $\mathcal{X}$. Let $\infty\mathcal{T}\mathrm{op}_{\mathrm{Mod}}$ be the ∞-category whose objects are such triples, with morphism triples consisting of a geometric morphism between ∞-topoi and the two corresponding induced maps on sheaves. One can also speak of $(\mathcal{O}, \mathcal{F})$ as being a Mod-valued sheaf on $\mathcal{X}$. Let $\infty\mathcal{T}\mathrm{op}^{\mathrm{sHen}}_{\mathrm{Mod}}$ be the subcategory of $\infty\mathcal{T}\mathrm{op}_{\mathrm{Mod}}$ whose objects are such that $\mathcal{O}$ is strictly Henselian. The global sections functor $\Gamma : \infty\mathcal{T}\mathrm{op}^{\mathrm{sHen}}_{\mathrm{Mod}} \to \mathrm{Mod}^{\mathrm{op}}$ has a right adjoint, which one denotes by $\mathrm{Spét}_{\mathrm{Mod}}$.[40] Now for a (nonconnective) spectral DM stack $X = (\mathcal{X}, \mathcal{O}_{\mathcal{X}})$, if $\mathcal{F}$ is a sheaf of $\mathcal{O}_{\mathcal{X}}$-modules on $\mathcal{X}$, then one can regard $(\mathcal{X}, \mathcal{O}_{\mathcal{X}}, \mathcal{F})$ as an object of $\infty\mathcal{T}\mathrm{op}_{\mathrm{Mod}}$. One says $\mathcal{F}$ is *quasi-coherent*[41] if there exists a collection U_α of objects of $\mathcal{X}$ that cover it, such that for all α, there exist $\mathrm{A}_\alpha \in \mathrm{CAlg}$ and $M_\alpha \in \mathrm{Mod}_{\mathrm{A}_\alpha}$ with equivalences

$$(\mathcal{X}_{/U_\alpha}, \mathcal{O}|_{U_\alpha}, \mathcal{F}|_{U_\alpha}) \simeq \mathrm{Spét}_{\mathrm{Mod}}(\mathrm{A}_\alpha, M_\alpha).$$

One denotes by $\mathrm{QCoh}(X)$ the full subcategory of $\mathrm{Mod}_{\mathcal{O}_{\mathcal{X}}}$ spanned by such objects $\mathcal{F}$. We also have another characterization of quasi-coherent sheaves on spectral DM stacks: by virtue of Proposition 2.2.4.3 of [54], if $(\mathcal{X}, \mathcal{O})$ is a nonconnective spectral DM stack and $\mathcal{F}$ is a sheaf of $\mathcal{O}$-modules on $\mathcal{X}$, then $\mathcal{F}$ is a quasi-coherent sheaf iff for $f : U \to V$ in $\mathrm{Mor}\,\mathcal{X}$ such that $(\mathcal{X}_{/U}, \mathcal{O}|_U)$ and $(\mathcal{X}_{/V}, \mathcal{O}|_V)$ are affine, the map $\mathcal{F}(V) \otimes_{\mathcal{O}(V)} \mathcal{O}(U) \to \mathcal{F}(U)$ is an equivalence.

40 Cor. 2.2.1.5. [54].

41 Def. 2.2.2.1. [54].

16 Representation theory of formal moduli problems

In this section, we introduce formal moduli problems and one fundamental result from [54] for our purposes: Theorem 13.4.0.1, which states that we have a fully faithful embedding $\mathbf{QCoh}_{\mathrm{art}}(\mathbf{X}) \hookrightarrow \mathrm{Rep}_{\mathfrak{g}_*}$ for a dg Lie algebra $\mathfrak{g}_*$ over a field k of characteristic zero and the formal moduli problem $\mathbf{X} = \Psi(\mathfrak{g}_*)$ associated with $\mathfrak{g}_*$ in a sense to be specified. This result essentially says that the manifestation of $\mathbf{X}$ has a representation-theoretic interpretation.

We first define formal moduli problems, then quasi-coherent sheaves defined on them, and finally the morphism Ψ in the above statement.

16.1 Formal moduli problems

Let k be a field of characteristic zero, and let $\mathbf{X} : \mathrm{CAlg}_k^{\mathrm{art}} \to \mathcal{S}$ be a functor. It is said to be a *formal moduli problem*[1] if $\mathbf{X}(k)$ is contractible and for objects R_1, R_2, R_0 of $\mathrm{CAlg}_k^{\mathrm{art}}$ such that $\pi_0 R_1 \twoheadrightarrow \pi_0 R_0$ and $\pi_0 R_2 \twoheadrightarrow \pi_0 R_0$, $\mathbf{X}$ maps any pullback

$$\begin{array}{ccc} R_{12} & \longrightarrow & R_2 \\ \downarrow & & \downarrow \\ R_1 & \longrightarrow & R_0 \end{array}$$

to a pullback

$$\begin{array}{ccc} \mathbf{X}(R_{12}) & \longrightarrow & \mathbf{X}(R_2) \\ \downarrow & & \downarrow \\ \mathbf{X}(R_1) & \longrightarrow & \mathbf{X}(R_0). \end{array}$$

One denotes by Moduli_k the full subcategory of $\mathrm{Fun}(\mathrm{CAlg}_k^{\mathrm{art}}, \mathcal{S})$ spanned by such functors.

16.2 Quasi-coherent sheaves on formal moduli problems

Quasi-coherent sheaves on formal moduli problems are valued in commutative algebra objects of $\widehat{\mathrm{Cat}_\infty}$, and this can be seen from the following construction.

Consider the coCartesian fibration $q : \mathrm{Mod}(\mathrm{Mod}_k)^{\otimes} \to \mathrm{CAlg}_k \times \mathcal{F}\mathrm{in}_*$. There is a map $\xi : \mathrm{CAlg}_k \to \mathrm{CAlg}(\widehat{\mathrm{Cat}_\infty})$ that classifies q. Its restriction to artinian algebras admits an essentially unique factorization

1 Ch. 13 [54].

https://doi.org/10.1515/9783111334073-016

$$\mathrm{CAlg}_k^{\mathrm{art}} \to \mathrm{Fun}(\mathrm{CAlg}_k^{\mathrm{art}}, \mathcal{S}) \to \mathrm{CAlg}(\widehat{\mathrm{Cat}_\infty})$$

where the second map preserves small limits and is denoted by $\mathbf{QCoh}_{\mathrm{art}}$. For $\mathbf{X} \in \mathrm{Moduli}_k$, $\mathbf{QCoh}_{\mathrm{art}}(\mathbf{X})$ is called the *∞-category of quasi-coherent sheaves on* $\mathbf{X}$.[2] In this definition, if $\mathcal{C}$ is a symmetric monoidal ∞-category, then $\mathrm{Mod}(\mathcal{C})^{\otimes}$ has objects of the form $(A, M_1, \ldots, M_n)$, where $A \in \mathrm{CAlg}(\mathcal{C})$, and $M_i \in \mathrm{Mod}_A$, $1 \le i \le n$. $\mathcal{F}\mathrm{in}_*$ is Segal's category of finite pointed sets with objects of the form $\langle n \rangle$, $n \ge 0$, and morphisms are maps $\langle m \rangle \to \langle n \rangle$ that preserve the fixed point $*$.

Now one can derive a formula for $\mathbf{QCoh}_{\mathrm{art}}(\mathbf{X})$ following the method used to find one for $\mathbf{QCoh}(\mathbf{X})$, $\mathbf{X} : \mathrm{CAlg}^{\mathrm{cn}} \to \mathcal{S}$. Consider the projection $q : \overline{\mathrm{CAlg}_k^{\mathrm{art}}} \times_{\mathrm{CAlg}} \mathrm{Mod}(\mathrm{Mod}_k) \to \overline{\mathrm{CAlg}_k^{\mathrm{art}}}$. To understand the objects involved, note that $\mathrm{Mod}(\mathrm{Mod}_k) = (\mathrm{Mod}(\mathrm{Mod}_k)^{\otimes})_{<1>}$. It follows that

$$\mathrm{Mod}(\mathrm{Mod}_k) = \{(A, M) \in \mathrm{CAlg}(\mathrm{Mod}_k) \times_{\mathrm{CAlg}} \mathrm{Mod}\}.$$

Then observe that

$$\mathrm{CAlg}(\mathrm{Mod}_k) = \mathrm{CAlg}(\mathrm{Mod}_k(\mathrm{Sp})) = \mathrm{CAlg}_k,^{3}$$

and hence $\mathrm{Mod}(\mathrm{Mod}_k) = \mathrm{CAlg}_k \times_{\mathrm{CAlg}} \mathrm{Mod}$, so that one can write $\overline{\mathrm{CAlg}_k^{\mathrm{art}}} \times_{\mathrm{CAlg}} \mathrm{Mod}(\mathrm{Mod}_k) = \overline{\mathrm{CAlg}_k^{\mathrm{art}}} \times_{\mathrm{CAlg}} \mathrm{Mod}$, which one denotes by $\mathrm{Mod}_{\mathrm{art}}^{\mathbf{X}}$. Thus a more precise formulation of q is $q : \mathrm{Mod}_{\mathrm{art}}^{\mathbf{X}} \to \overline{\mathrm{CAlg}_k^{\mathrm{art}}}$. Let $\mathbf{X} : \mathrm{CAlg}_k^{\mathrm{art}} \to \mathcal{S}$ be a formal moduli problem classifying a L fibration $\overline{\mathrm{CAlg}_k^{\mathrm{art}}} \to \mathrm{CAlg}_k^{\mathrm{art}}$. Then one can show that $\mathbf{QCoh}_{\mathrm{art}}(\mathbf{X}) = \mathrm{Fun}^{\mathrm{cocart}}_{\mathrm{CAlg}_k^{\mathrm{art}}}(\overline{\mathrm{CAlg}_k^{\mathrm{art}}}, \mathrm{Mod}_{\mathrm{art}}^{\mathbf{X}})$.

Finally, we define the morphism Ψ used in the statement of Theorem 13.4.0.1 of [54]. This morphism is in fact introduced in Theorem 13.0.0.2 of the same reference. $\Psi : \mathrm{Lie}_k \to \mathrm{Moduli}_k$ is an equivalence of ∞-categories. One has $\Psi(\mathfrak{g}_*) = \mathrm{Map}_{\mathrm{Lie}_k}(\mathcal{D}(-), \mathfrak{g}_*)$, where $\mathcal{D} : (\mathrm{CAlg}_k^{\mathrm{aug}})^{\mathrm{op}} \to \mathrm{Lie}_k$ is the Koszul duality functor, right adjoint to the cohomological Chevalley–Eilenberg complex functor $C^* : \mathrm{Lie}_k \to (\mathrm{CAlg}_k^{\mathrm{aug}})^{\mathrm{op}}$. Originally, the functor C^* is valued in commutative dg algebras, which form a subcategory $\mathrm{CAlg}_k^{\mathrm{dg}} \subset \mathrm{Vect}_k^{\mathrm{dg}}$ of dg vector spaces over k, whose underlying ∞-category is the ∞-category Mod_k by virtue of Proposition 7.1.1.15 of [53]. It follows that at the level of ∞-categories, C^* is valued in $\mathrm{CAlg}(\mathrm{Mod}_k) = \mathrm{CAlg}_k$ (see Notation 13.2.5.4 of [54]). The interested reader will find ample details in Chapter 13 of [54].

Let k be a field of characteristic zero in what follows. A *dg Lie algebra over* k[4] is a chain complex $(\mathfrak{g}_*, d)$ in Vect_k equipped with a Lie bracket $[\cdot, \cdot] : \mathfrak{g}_p \otimes \mathfrak{g}_q \to \mathfrak{g}_{p+q}$ satisfying

2 Constr. 13.4.6.1 [54].

3 Var. 7.1.3.8 [54].

4 Def. 13.1.2.1 [54].

$$[x,y] + (-1)^{pq}[y,x] = 0$$

for $x \in \mathfrak{g}_p$ and $y \in \mathfrak{g}_q$ and

$$(-1)^{pr}[x,[y,z]] + (-1)^{pq}[y,[z,x]] + (-1)^{qr}[z,[x,y]] = 0$$

for $x \in \mathfrak{g}_p, y \in \mathfrak{g}_q$, and $z \in \mathfrak{g}_r$.

We also need compatibility with the differential:

$$d[x,y] = [dx,y] + (-1)^p[x,dy]$$

for $x \in \mathfrak{g}_p$ and $y \in \mathfrak{g}_q$.

A morphism $(\mathfrak{g}_*, d) \to (\mathfrak{g}'_*, d')$ of dg Lie algebras over k corresponds to a map of chain complexes $F : (\mathfrak{g}_*, d) \to (\mathfrak{g}'_*, d')$ such that $F[x,y] = [Fx, Fy]$. One denotes the category of dg Lie algebras over k by $\mathrm{Lie}_k^{\mathrm{dg}}$.

Now there exist an ∞-category Lie_k and a functor $\pi : \mathrm{Lie}_k^{\mathrm{dg}} \to \mathrm{Lie}_k$ such that for all $\mathcal{C} \in \mathrm{Cat}_\infty$, π induces an equivalence of $\mathrm{Fun}(\mathrm{Lie}_k, \mathcal{C})$ with the full subcategory of $\mathrm{Fun}(\mathrm{Lie}_k^{\mathrm{dg}}, \mathcal{C})$ spanned by those functors that map quasi-isomorphisms in $\mathrm{Lie}_k^{\mathrm{dg}}$ to equivalences in $\mathcal{C}$. One refers to Lie_k as the *∞-category of dg Lie algebras over k*.[5]

For $\mathfrak{g}_* \in \mathrm{Lie}_k^{\mathrm{dg}}$, one defines a *representation of* $\mathfrak{g}_*$[6] to be given by a dg vector space V_*, along with a map $\mathfrak{g}_* \otimes V_* \to V_*$ such that $[x,y]v = x(yv) + (-1)^{pq}y(xv)$ for $x \in \mathfrak{g}_p$, $y \in \mathfrak{g}_q$, $v \in V_n$. Representations of $\mathfrak{g}_*$ form a category that one denotes by $\mathrm{Rep}_{\mathfrak{g}_*}^{\mathrm{dg}}$. By Proposition 13.4.1.3 of [54] there is a model structure on this category. If one denotes by $W_{\mathfrak{g}_*}$ its set of equivalences, then one defines $\mathrm{Rep}_{\mathfrak{g}_*} = \mathrm{Rep}_{\mathfrak{g}_*}^{\mathrm{dg}}[W_{\mathfrak{g}_*}^{-1}]$.

5 Def. 13.1.4.1 [54].

6 Def. 13.4.1.1 [54].

17 Relations between quasi-coherent sheaves

For $\mathbf{X} \in \mathrm{Fun}(\mathrm{CAlg}^{\mathrm{cn}}, \mathcal{S})$, let $\iota^*\mathbf{X}$ be the pullback of $\mathbf{X}$ along the fully faithful embedding $\iota : \mathrm{CAlg}_k^{\mathrm{art}} \hookrightarrow \mathrm{CAlg}^{\mathrm{cn}}$. One says $\mathbf{X} : \mathrm{CAlg}^{\mathrm{cn}} \to \mathcal{S}$ is *artinian* if $\iota^*\mathbf{X} \in \mathrm{Moduli}_k$. On the one hand, one has $\mathbf{QCoh}(\mathbf{X}) = \mathrm{Fun}^{\mathrm{cocart}}_{\mathrm{CAlg}^{\mathrm{cn}}}(\overline{\mathrm{CAlg}^{\mathrm{cn}}}, \mathrm{Mod}^{\mathbf{X}})$, and on the other, one has

$$\mathbf{QCoh}_{\mathrm{art}}(\iota^*\mathbf{X}) = \mathrm{Fun}^{\mathrm{cocart}}_{\mathrm{CAlg}_k^{\mathrm{art}}}(\overline{\mathrm{CAlg}_k^{\mathrm{art}}}, \mathrm{Mod}_{\mathrm{art}}^{\iota^*\mathbf{X}}).$$

Consider the following CD:

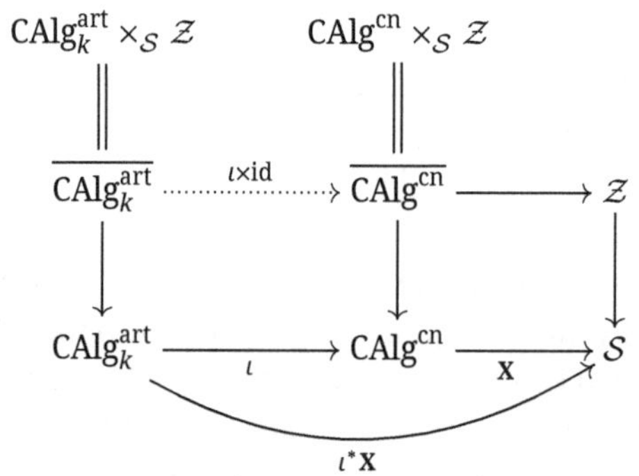

where $\mathcal{Z} \to \mathcal{S}$ is the universal L fibration (see [52]). The outside square and the rightmost square are Cartesian, and hence so is the leftmost square.

This means that

$$\overline{\mathrm{CAlg}_k^{\mathrm{art}}} = \mathrm{CAlg}_k^{\mathrm{art}} \times_{\mathrm{CAlg}^{\mathrm{cn}}} \overline{\mathrm{CAlg}^{\mathrm{cn}}}.$$

It follows that

$$\begin{aligned}\mathrm{Mod}_{\mathrm{art}}^{\iota^*\mathbf{X}} &= \overline{\mathrm{CAlg}_k^{\mathrm{art}}} \times_{\mathrm{CAlg}^{\mathrm{cn}}} \mathrm{Mod} \\ &= (\mathrm{CAlg}_k^{\mathrm{art}} \times_{\mathrm{CAlg}^{\mathrm{cn}}} \overline{\mathrm{CAlg}^{\mathrm{cn}}}) \times_{\mathrm{CAlg}^{\mathrm{cn}}} \mathrm{Mod} \\ &\simeq \mathrm{CAlg}_k^{\mathrm{art}} \times_{\mathrm{CAlg}^{\mathrm{cn}}} (\overline{\mathrm{CAlg}^{\mathrm{cn}}} \times_{\mathrm{CAlg}^{\mathrm{cn}}} \mathrm{Mod}) \\ &= \mathrm{CAlg}_k^{\mathrm{art}} \times_{\mathrm{CAlg}^{\mathrm{cn}}} \mathrm{Mod}^{\mathbf{X}}.\end{aligned}$$

Consider the following CD:

https://doi.org/10.1515/9783111334073-017

$$\begin{array}{ccccc}
\overline{\mathrm{CAlg}_k^{\mathrm{art}}} & & \mathrm{Mod}_{\mathrm{art}}^{\iota^*\mathbf{X}} & & \\
\| & & \| & & \\
\mathrm{CAlg}_k^{\mathrm{art}} \times_{\mathrm{CAlg}^{\mathrm{cn}}} \overline{\mathrm{CAlg}^{\mathrm{cn}}} & \dashrightarrow & \mathrm{CAlg}_k^{\mathrm{art}} \times_{\mathrm{CAlg}^{\mathrm{cn}}} \mathrm{Mod}^{\mathbf{X}} & \longrightarrow & \mathrm{CAlg}_k^{\mathrm{art}} \\
\downarrow{\scriptstyle \pi_2} & & {\scriptstyle \pi_2}\downarrow & & {\scriptstyle \iota}\downarrow \\
\overline{\mathrm{CAlg}^{\mathrm{cn}}} & \longrightarrow & \mathrm{Mod}^{\mathbf{X}} & \longrightarrow & \mathrm{CAlg}^{\mathrm{cn}}
\end{array}$$

from which we see that $\mathrm{Fun}(\overline{\mathrm{CAlg}_k^{\mathrm{art}}}, \mathrm{Mod}_{\mathrm{art}}^{\iota^*\mathbf{X}})$ is functorially obtained from $\mathrm{Fun}(\overline{\mathrm{CAlg}^{\mathrm{cn}}}, \mathrm{Mod}^{\mathbf{X}})$ by base change, which also preserves coCartesian maps. Indeed, recall that if $q : \mathrm{Mod}^{\mathbf{X}} \to \mathrm{CAlg}^{\mathrm{cn}}$, then a functor $F : \overline{\mathrm{CAlg}^{\mathrm{cn}}} \to \mathrm{Mod}^{\mathbf{X}}$ is q-coCartesian if it maps every edge of $\overline{\mathrm{CAlg}^{\mathrm{cn}}}$ to a q-coCartesian edge of $\mathrm{Mod}^{\mathbf{X}}$. Some of those same edges of $\overline{\mathrm{CAlg}^{\mathrm{cn}}}$ are also constituents of edges of $\overline{\mathrm{CAlg}_k^{\mathrm{art}}} = \mathrm{CAlg}_k^{\mathrm{art}} \times_{\mathrm{CAlg}^{\mathrm{cn}}} \overline{\mathrm{CAlg}^{\mathrm{cn}}}$ that map under $\mathrm{CAlg}_k^{\mathrm{art}} \times_{\mathrm{CAlg}^{\mathrm{cn}}} F$ to $\mathrm{CAlg}_k^{\mathrm{art}} \times_{\mathrm{CAlg}^{\mathrm{cn}}}$ q-coCartesian edges of $\mathrm{CAlg}_k^{\mathrm{art}} \times_{\mathrm{CAlg}^{\mathrm{cn}}} \mathrm{Mod}^{\mathbf{X}} = \mathrm{Mod}_{\mathrm{art}}^{\iota^*\mathbf{X}}$. Thus we have shown that $\mathbf{QCoh}_{\mathrm{art}}(\iota^*\mathbf{X}) = \mathrm{Fun}_{\mathrm{CAlg}_k^{\mathrm{art}}}^{\mathrm{cocart}}(\overline{\mathrm{CAlg}_k^{\mathrm{art}}}, \mathrm{Mod}_{\mathrm{art}}^{\iota^*\mathbf{X}})$ is functorially obtained from $\mathbf{QCoh}(\mathbf{X}) = \mathrm{Fun}_{\mathrm{CAlg}^{\mathrm{cn}}}^{\mathrm{cocart}}(\overline{\mathrm{CAlg}^{\mathrm{cn}}}, \mathrm{Mod}^{\mathbf{X}})$ by base change.

There are a few technical, peripheral results pertaining to quasi-coherent sheaves we go over presently. The first result is the following:

$$(\mathbf{QCoh}(\mathbf{X})|_{\mathrm{CAlg}_k^{\mathrm{art}}})^{\mathrm{cn}} \supseteq \mathbf{QCoh}(\mathbf{X})^{\mathrm{cn}}|_{\mathrm{CAlg}_k^{\mathrm{art}}}$$

where we have replaced, for notation purposes, the functorial fiber product with $\mathrm{CAlg}_k^{\mathrm{art}}$ in $\mathrm{Fun}(\overline{\mathrm{CAlg}^{\mathrm{cn}}}, \mathrm{Mod}^{\mathbf{X}})$ with a restriction. Recall from Remark 6.2.2.7 of [54] that for a functor $\mathbf{X} : \mathrm{CAlg}^{\mathrm{cn}} \to \mathcal{S}$, for $A \in \mathrm{CAlg}^{\mathrm{cn}}$ and $\eta \in \mathbf{X}(A)$, both of which are encapsulated in a lift $\tilde{A} \in \overline{\mathrm{CAlg}^{\mathrm{cn}}}$, where $\overline{\mathrm{CAlg}^{\mathrm{cn}}} \to \mathrm{CAlg}^{\mathrm{cn}}$ is a L fibration classified by $\mathbf{X}$, we have that $\mathcal{F} \in \mathbf{QCoh}(\mathbf{X})$ can be seen as a map that with (A, η) associates $\mathcal{F}(\tilde{A}) \in \mathrm{Mod}_A$, and we will use the notation of [54]: $\mathcal{F}(\tilde{A}) = (A, \mathcal{F}(\eta))$. Now by Definition 6.2.5.3 of [54], if P is a characteristic of pairs (A, M), $A \in \mathrm{CAlg}^{\mathrm{cn}}$, $M \in \mathrm{Mod}_A$, stable under base change, then $\mathcal{F} \in \mathbf{QCoh}(\mathbf{X})$ is said to have the property P if for all $A \in \mathrm{CAlg}^{\mathrm{cn}}$ and $\eta \in \mathbf{X}(A)$, $(A, \mathcal{F}(\eta))$ has the property P. By Proposition 6.2.5.2. of [54], if $P = \mathrm{cn}$ is the property of being connective, defined by asking that in a pair (A, M), M is connective seen as a spectrum object, then cn is stable under base change, which allows us to speak of connective quasi-coherent sheaves. To say that $\mathcal{F} \in \mathbf{QCoh}(\mathbf{X})$ is *connective* then means that for all $A \in \mathrm{CAlg}^{\mathrm{cn}}$, $\mathcal{F}(\eta) \in \mathrm{Mod}_A$ is connective, which is true in particular if $A \in \mathrm{CAlg}_k^{\mathrm{art}}$, that is, $\mathbf{QCoh}(\mathbf{X})^{\mathrm{cn}}|_{\mathrm{CAlg}_k^{\mathrm{art}}} \subseteq \mathbf{QCoh}_{\mathrm{art}}(\iota^*\mathbf{X})^{\mathrm{cn}}$. One may very well conceive that $\mathcal{F} \in \mathbf{QCoh}_{\mathrm{art}}(\iota^*\mathbf{X})^{\mathrm{cn}}$ comes from some object of $\mathbf{QCoh}(\mathbf{X})$ that was not necessarily connective on non-artinian objects of $\mathrm{CAlg}^{\mathrm{cn}}$, which means that we just have an inclusion. By denoting $\mathbf{QCoh}_{\mathrm{art}}(\iota^*\mathbf{X})$ by $\mathbf{QCoh}(X)|_{\mathrm{CAlg}_k^{\mathrm{art}}}$ we have the desired result.

We will see later that the affine perception of a spectral DM stack X is related to its manifestation by $\mathrm{Aff}_X^{\mathrm{op}} \simeq \mathrm{CAlg}(\mathrm{QCoh}(X)^{\mathrm{cn}})$. If we want to connect this with the representation theory aspect of the problem, then one may want to relate $\mathrm{CAlg}(\mathrm{QCoh}(X)^{\mathrm{tcn}}) \simeq \mathrm{CAlg}(\mathbf{QCoh}(\mathbf{X})^{\mathrm{cn}})$ with $\mathrm{CAlg}(\mathbf{QCoh}_{\mathrm{art}}(\iota^*\mathbf{X})^{\mathrm{cn}})$. From the work just done on connective sheaves it is clear that it suffices to establish a relation between $\mathrm{CAlg}(\mathbf{QCoh}(\mathbf{X}))$ and $\mathrm{CAlg}(\mathbf{QCoh}_{\mathrm{art}}(\iota^*\mathbf{X}))$, if any.

Recall from Definition 2.1.2.7 of [53] that for $\mathcal{C}$ an ∞-operad, $\mathrm{CAlg}(\mathcal{C}) \subseteq \mathrm{Fun}(\mathrm{N}(\mathcal{F}\mathrm{in}_*), \mathcal{C})$ is spanned by the ∞-operad maps, i. e., those maps for which the following diagram is commutative:

$$\begin{array}{ccc} \mathrm{N}(\mathcal{F}\mathrm{in}_*) & \longrightarrow & \mathcal{C} \\ & \searrow\!\!\!\!= \quad \swarrow & \\ & \mathrm{N}(\mathcal{F}\mathrm{in}_*) & \end{array}$$

and preserves inert morphisms, where by *inert morphism*[1] in $\mathcal{C}^{\otimes}$ we mean the following: for an ∞-operad $p : \mathcal{C}^{\otimes} \to \mathrm{N}(\mathcal{F}\mathrm{in}_*)$, a morphism f in $\mathcal{C}^{\otimes}$ is inert if $p(f)$ is inert and f is p-coCartesian. By Definition 2.1.1.8 of [53], a morphism $\psi : \langle m\rangle \to \langle n\rangle$ in $\mathcal{F}\mathrm{in}_*$ is inert if for all $i \in \langle n\rangle - \{*\}$, $\psi^{-1}(i)$ is a singleton. We will split the definition of being inert for a morphism f in $\mathcal{C}^{\otimes}$ into f being *p-inert*, that is, $p(f)$ is inert, and f being p-coCartesian.

Since we are interested in $\mathrm{CAlg}(\mathbf{QCoh}_{\mathrm{art}}(\iota^*\mathbf{X}))$ in particular, by the above definitions we should understand inert morphisms in $\mathbf{QCoh}_{\mathrm{art}}(\iota^*\mathbf{X})^{\otimes}$, so we consider the ∞-operad $p : \mathbf{QCoh}_{\mathrm{art}}(\iota^*\mathbf{X})^{\otimes} \to \mathrm{N}(\mathcal{F}\mathrm{in}_*)$. Observe that a p-inert edge of $p : \mathbf{QCoh}(\mathbf{X})^{\otimes} \to N(\mathcal{F}\mathrm{in}_*)$ (using the same notation p for simplicity) will produce a p-inert edge of $\mathbf{QCoh}_{\mathrm{art}}(\iota^*\mathbf{X})^{\otimes}$ once restricted to artinian algebras, but precisely because of that restriction, there can be edges of $\mathbf{QCoh}(\mathbf{X})^{\otimes}$ that become p-inert after restriction, so we have the following inclusion:

$$p\text{-inert}(\mathbf{QCoh}(\mathbf{X})|_{\mathrm{CAlg}_k^{\mathrm{art}}}) \supseteq p\text{-inert}(\mathbf{QCoh}(\mathbf{X}))|_{\mathrm{CAlg}_k^{\mathrm{art}}}. \tag{17.1}$$

Regarding p-coCartesian maps, suppose $f : \mathcal{F} \to \mathcal{G}$ is p-coCartesian in $\mathbf{QCoh}(\mathbf{X})$. This means that for all $\mathcal{H} \in \mathbf{QCoh}(\mathbf{X})$, we have a homotopy pullback square

$$\begin{array}{ccc} \mathrm{Map}_{\mathbf{QCoh}(\mathbf{X})}(\mathcal{G},\mathcal{H}) & \longrightarrow & \mathrm{Map}_{\mathbf{QCoh}(\mathbf{X})}(\mathcal{F},\mathcal{H}) \\ \downarrow & & \downarrow \\ \mathrm{Map}_{\mathrm{N}(\mathcal{F}\mathrm{in}_*)}(p\mathcal{G},p\mathcal{H}) & \longrightarrow & \mathrm{Map}_{\mathrm{N}(\mathcal{F}\mathrm{in}_*)}(p\mathcal{F},p\mathcal{H}). \end{array}$$

Denote the bottom left mapping space by $\mathcal{C}$, the top right one by $\mathcal{E}$, and the bottom right one by $\mathcal{D}$. Then we have $\mathrm{Map}_{\mathbf{QCoh}(\mathbf{X})}(\mathcal{G},\mathcal{H}) \simeq \mathcal{C} \times_{\mathcal{D}} \mathcal{E}$. Since $\mathbf{QCoh}_{\mathrm{art}}(\iota^*\mathbf{X}) = \mathrm{CAlg}_k^{\mathrm{art}} \times_{\mathrm{CAlg}^{\mathrm{cn}}} \mathbf{QCoh}(\mathbf{X})$, if $\mathcal{F} \in \mathbf{QCoh}(\mathbf{X})$, we have $\mathrm{CAlg}_k^{\mathrm{art}} \times_{\mathrm{CAlg}^{\mathrm{cn}}} \mathcal{F} \in \mathbf{QCoh}_{\mathrm{art}}(\iota^*\mathbf{X})$. We will denote such restrictions by the same letter $\mathcal{F}$ for simplicity. Note that we have

1 Def. 2.1.2.3 [53].

$$\begin{aligned}
\mathrm{Map}_{\mathbf{QCoh}(\mathbf{X})|_{\mathrm{CAlg}_k^{\mathrm{art}}}}(\mathcal{G},\mathcal{H}) &= \mathrm{CAlg}_k^{\mathrm{art}} \times_{\mathrm{CAlg}^{\mathrm{cn}}} \mathrm{Map}_{\mathbf{QCoh}(\mathbf{X})}(\mathcal{G},\mathcal{H}) && (17.2)\\
&\simeq \mathrm{CAlg}_k^{\mathrm{art}} \times_{\mathrm{CAlg}^{\mathrm{cn}}} (\mathcal{C} \times_{\mathcal{D}} \mathcal{E}) && (17.3)\\
&\simeq \mathcal{C} \times_{\mathcal{D}} (\mathrm{CAlg}_k^{\mathrm{art}} \times_{\mathrm{CAlg}^{\mathrm{cn}}} \mathcal{E}) && (17.4)\\
&= \mathcal{C} \times_{\mathcal{D}} \mathrm{Map}_{\mathbf{QCoh}(\mathbf{X})|_{\mathrm{CAlg}_k^{\mathrm{art}}}}(\mathcal{F},\mathcal{H})
\end{aligned}$$

where $\mathcal{C}$ and $\mathcal{D}$ in (17.4) are using restricted functors $p\mathcal{F}$, $p\mathcal{G}$, and $p\mathcal{H}$.

Having the above equivalence means precisely that the following diagram is a homotopy pullback square:

$$\begin{array}{ccc}
\mathrm{Map}_{\mathbf{QCoh}(\mathbf{X})|_{\mathrm{CAlg}_k^{\mathrm{art}}}}(\mathcal{G},\mathcal{H}) & \longrightarrow & \mathrm{Map}_{\mathbf{QCoh}(\mathbf{X})|_{\mathrm{CAlg}_k^{\mathrm{art}}}}(\mathcal{F},\mathcal{H}) \\
\downarrow & & \downarrow \\
\mathrm{Map}_{\mathrm{N}(\mathcal{F}\mathrm{in}_*)}(p\mathcal{G},p\mathcal{H}) & \longrightarrow & \mathrm{Map}_{\mathrm{N}(\mathcal{F}\mathrm{in}_*)}(p\mathcal{F},p\mathcal{H})
\end{array}$$

and hence f is p-coCartesian viewed as a morphism of $\mathbf{QCoh}_{\mathrm{art}}(\iota^*\mathbf{X})$. Thus the restriction to $\mathrm{CAlg}_k^{\mathrm{art}}$ of a p-coCartesian edge of $\mathbf{QCoh}(\mathbf{X})$ is again a p-coCartesian edge of $\mathbf{QCoh}(\mathbf{X})|_{\mathrm{CAlg}_k^{\mathrm{art}}}$, but by (17.2) and (17.3) one can very well have a p-coCartesian edge of $\mathbf{QCoh}(\mathbf{X})|_{\mathrm{CAlg}_k^{\mathrm{art}}}$ being the restriction to artinian objects of a map in $\mathbf{QCoh}(\mathbf{X})$ that is not necessarily p-coCartesian, and thus we have (with obvious notations)

$$p\text{-coCart}(\mathbf{QCoh}(\mathbf{X})|_{\mathrm{CAlg}_k^{\mathrm{art}}}) \supseteq p\text{-coCart}(\mathbf{QCoh}(\mathbf{X})|_{\mathrm{CAlg}_k^{\mathrm{art}}}. \tag{17.5}$$

It follows from (17.1) and (17.5) that

$$\mathrm{CAlg}(\mathbf{QCoh}_{\mathrm{art}}(\iota^*\mathbf{X})) \supseteq \mathrm{CAlg}(\mathbf{QCoh}(\mathbf{X}))|_{\mathrm{CAlg}_k^{\mathrm{art}}}.$$

By formally the same reasoning we conclude:

$$\mathrm{CAlg}(\mathbf{QCoh}_{\mathrm{art}}(\iota^*\mathbf{X})^{\mathrm{cn}}) \supseteq \mathrm{CAlg}(\mathbf{QCoh}(\mathbf{X})^{\mathrm{cn}})|_{\mathrm{CAlg}_k^{\mathrm{art}}}.$$

Observe that this could also be obtained from the inclusion $\mathbf{QCoh}_{\mathrm{art}}(\iota^*\mathbf{X})^{\mathrm{cn}} \supseteq \mathbf{QCoh}(\mathbf{X})^{\mathrm{cn}}|_{\mathrm{CAlg}_k^{\mathrm{art}}}$ and the fact that $\mathrm{CAlg}(\mathcal{C}|_{\mathrm{CAlg}_k^{\mathrm{art}}}) \supseteq \mathrm{CAlg}(\mathcal{C})|_{\mathrm{CAlg}_k^{\mathrm{art}}}$ by following the arguments above.

Part IV: **Applications**

18 A dual representation in spectral algebraic geometry

18.1 Introduction

For a small category $\mathcal{C}$, the Grothendieck construction provides a 2-functor ne $\mathrm{Fun}(\mathcal{C}^{\mathrm{op}}, \mathrm{Cat}) \to \mathrm{Cat}_{/\mathcal{C}}$ that gives rise to an equivalence of 2-categories $\mathrm{Fun}(\mathcal{C}^{\mathrm{op}}, \mathrm{Cat}) \simeq \mathrm{Fib}(\mathcal{C})$, where $\mathrm{Fib}(\mathcal{C})$ denotes the 2-category of Grothendieck fibrations over $\mathcal{C}$. In [52], this was generalized to the higher categorical setting by considering simplicial functors on a simplicial category $\mathfrak{C}[S]$, $S \in \mathrm{Set}_\Delta$, giving rise to a Quillen equivalence $(\mathrm{Set}_\Delta)^{\mathfrak{C}[S]} \simeq (\mathrm{Set}_\Delta)_{/S}$. Presently, we focus on the heuristics of the Grothendieck construction in a higher categorical setting, namely the change of point of view from functors from a given ∞-category $\mathcal{C}$ to some ∞-category $\mathcal{D}$, provided by the functor category $\mathrm{Fun}(\mathcal{C}^{\mathrm{op}}, \mathcal{D})$, to functors over $\mathcal{C}$, provided by some slice category $\mathcal{E}_{/\mathcal{C}}$ for some ∞-category $\mathcal{E}$. To be more precise, we will focus on *foliations* of such equivalences, that is, we consider constructs $\mathbb{K}$ on a full subcategory $\mathcal{F}[\mathcal{C}]$ of $\mathrm{Fun}(\mathcal{C}^{\mathrm{op}}, \mathcal{D})$ such that we have an equivalence with a full subcategory $\mathcal{G}_{/\mathcal{C}}$ of $\mathcal{E}_{/\mathcal{C}}$ as displayed below:

$$\begin{array}{ccc} & & \mathcal{F}[\mathcal{C}] \subset \mathrm{Fun}(\mathcal{C}^{\mathrm{op}}, \mathcal{D}) \\ & \swarrow^{\mathbb{K}} & \\ \mathcal{E}_{/\mathcal{C}} \supset \mathcal{G}_{/\mathcal{C}} \xrightarrow{\simeq} \mathbb{K}\mathcal{F}[\mathcal{C}]. & & \end{array}$$

To put an emphasis on the fact that $\mathcal{F}[\mathcal{C}]$ can be interpreted as a manifestation of $\mathcal{C}$, we show that it is embedded in an ∞-category of representations $\mathrm{Rep}[\mathcal{F}[\mathcal{C}]]$ after base change:

$$\begin{array}{ccccc} & & \mathcal{F}[\mathcal{C}] & & \\ & \swarrow^{\mathbb{K}} & & \searrow^{\text{base change}} & \\ \mathcal{G}_{/\mathcal{C}} \xrightarrow{\simeq} & \mathbb{K}\mathcal{F}[\mathcal{C}] & & & \mathrm{Rep}[\mathcal{F}[\mathcal{C}]]. \end{array}$$

Our aim then is to show that for a certain class of geometric objects X, we have a connection, as specified below, between certain collections of maps into X, which provides the so-called *perception* of X and the representation theory of those objects X. More generally, if $\mathbf{X}$ satisfies the hypotheses of the spectral Artin representability theorem (Theorem 16.0.1 of [54]), it is representable by a spectral Deligne–Mumford stack X: $\mathbf{X} = \mathrm{Hom}(\mathrm{Spét}(-), X) = h_X$. Further, by Proposition 6.2.4.1 of [54] we have an equivalence of ∞-categories $\mathrm{QCoh}(h_X) \simeq \mathrm{QCoh}(X)$. Finally, by Proposition 2.5.1.2 of [54], for the full subcategory Aff_X of the slice category $\mathrm{SpDM}_{/X}$ of spectral Deligne–Mumford stacks over X spanned by affine maps, we have an equivalence of ∞-categories $\mathrm{Aff}_X^{\mathrm{op}} \simeq$

https://doi.org/10.1515/9783111334073-018

CAlg(QCoh$(X)^{\mathrm{cn}}$). Collecting things together, we arrive at $\mathrm{Aff}_X^{\mathrm{op}} \simeq \mathrm{CAlg}(\mathbf{QCoh}(\mathbf{X})^{\mathrm{cn}})$ (we use boldface notations for concepts pertaining to functors). We refer to Aff_X as the *affine perception* of X.

On the other hand, Theorem 13.4.0.1 of [54] states that for a field k of characteristic zero, a formal moduli problem $\mathbf{X} : \mathrm{CAlg}_k^{\mathrm{art}} \to \mathcal{S}$ over k, and its associated dg Lie algebra $\Psi^{-1}(\mathbf{X}) = \mathfrak{g}_*$ over k, we have a fully faithful monoidal embedding $\mathbf{QCoh}_{\mathrm{art}}(\mathbf{X}) \hookrightarrow \mathrm{Rep}_{\mathfrak{g}_*}$. Here $\Psi : \mathrm{Lie}_k \to \mathrm{Moduli}_k$ is an equivalence of ∞-categories provided by Theorem 13.0.0.2 of the same reference, and $\mathbf{QCoh}_{\mathrm{art}}$ is the quasi-coherent sheaf functor on formal moduli problems. In this context, we refer to the ∞-category of quasi-coherent sheaves on $\mathbf{X}$ as the *manifestation* of $\mathbf{X}$, not to be confused with the manifestation as introduced in [54], and $\mathrm{Rep}_{\mathfrak{g}_*}$ as the *representation* mentioned in the previous paragraph. The connection between perceptions and representations is provided by the obvious functor $\iota : \mathrm{CAlg}_k^{\mathrm{art}} \to \mathrm{CAlg}^{\mathrm{cn}}$. For the pullback of the artinian object $\mathbf{X} : \mathrm{CAlg}^{\mathrm{cn}} \to \mathcal{S}$ along ι, we have a fully faithful embedding as pointed out above:

$$\mathbf{QCoh}_{\mathrm{art}}(\iota^*\mathbf{X}) \hookrightarrow \mathrm{Rep}_{\Psi^{-1}(\iota^*\mathbf{X})}$$

where functors $\mathbf{X} : \mathrm{CAlg}^{\mathrm{cn}} \to \mathcal{S}$ such that $\iota^*\mathbf{X} \in \mathrm{Moduli}_k$ are said to be artinian. The connection between quasi-coherent sheaves on a functor $\mathbf{X}$ and those on $\iota^*\mathbf{X}$ is simply given by base change:

$$\mathbf{QCoh}_{\mathrm{art}}(\iota^*\mathbf{X}) = \mathrm{CAlg}_k^{\mathrm{art}} \times_{\mathrm{CAlg}^{\mathrm{cn}}} \mathbf{QCoh}(\mathbf{X}).$$

Collecting things together, we have our first result.

Theorem 18.1.1. *If a functor* $\mathbf{X} : \mathrm{CAlg}^{\mathrm{cn}} \to \mathcal{S}$ *satisfies the hypotheses of the spectral Artin representability theorem, so that* $\mathbf{X} = \mathrm{Hom}(\mathrm{Sp\acute{e}t}(-), X)$ *for some spectral Deligne–Mumford stack* X, *and if* $\mathbf{X}$ *is artinian, then from the manifestation* $\mathbf{QCoh}(\mathbf{X})$ *of* $\mathbf{X}$ *one can extract a perception as* $\mathrm{CAlg}(\mathbf{QCoh}(\mathbf{X})^{\mathrm{cn}}) \simeq \mathrm{Aff}_X^{\mathrm{op}}$, *and one also has a representation-theoretic presentation of* $\mathbf{QCoh}(\mathbf{X})$ *as a full subcategory of* $\mathrm{Rep}_{\Psi^{-1}(\iota^*\mathbf{X})}$ *after base change.*

For a stronger result, we use Theorem 9.3.0.3 of [54], Tannaka duality for geometric stacks: if $\mathbf{X}$ is a geometric stack, then $\mathbf{X}$ can be functorially recovered from $\mathbf{QCoh}(\mathbf{X})$. In this situation, if $\mathbf{X}$ is representable, $\mathbf{X} = \mathrm{Hom}(\mathrm{Sp\acute{e}t}(-), X)$, then $\mathbf{X}$ would correspond to a functorial local perception of X. It follows that we have the following result.

Theorem 18.1.2. *If* $\mathbf{X}$ *is an artinian geometric stack and satisfies the hypotheses of the spectral Artin representability theorem, then from* $\mathbf{QCoh}(\mathbf{X})$ *one can recover the local perception* $\mathbf{X}$ *of the spectral Deligne–Mumford stack it represents, and one can obtain a full subcategory of* $\mathrm{Rep}_{\Psi^{-1}(\iota^*\mathbf{X})}$ *after base change.*

In both instances the manifestation $\mathbf{QCoh}(\mathbf{X})$ of $\mathbf{X}$ is pivotal in obtaining the perception of the object X representing $\mathbf{X}$ on the one hand and a subcategory $\mathbf{QCoh}_{\mathrm{art}}(\iota^*\mathbf{X})$ with a representation theoretic flavor on the other, a dual representation of sorts. This provides a bridge between the perception of a geometric object X and its actions through its representation theory.

18.2 Dual representation

By dual representation we mean representing a geometric object in two ways. This will involve its manifestation (its ∞-category of quasi-coherent sheaves), which can be defined for spectral DM stacks, for objects of $\mathrm{Fun}(\mathrm{CAlg}^{\mathrm{cn}}, \mathcal{S})$, and for formal moduli problems. We discuss our first dual representation result at the level of spectral DM stacks and then at the level of functors $\mathrm{CAlg}^{\mathrm{cn}} \to \mathcal{S}$.

18.2.1 Dual representation for spectral DM stacks

For a spectral DM stack X, denote by Aff_X the full subcategory of SpDM spanned by affine morphisms $Y \to X$, where one says such a map is *affine* if for any map $\mathrm{Spét} R \to X$, $\mathrm{Spét} R \times_X Y$ is affine, that is, it is of the form $\mathrm{Spét} A$ for some $\mathbb{E}_\infty$-ring A. We call Aff_X the *affine perception of* X. We have the following theorem.

Theorem 18.2.1. *Let $X \in \mathrm{SpDM}$, and let $\mathbf{X}$ be the functor it represents, which we suppose is artinian. Then we have a dual representation of the manifestation $\mathbf{QCoh}(\mathbf{X})$ of $\mathbf{X}$:*

$$\begin{array}{ccc} & \mathrm{QCoh}(X) \simeq \mathbf{QCoh}(\mathbf{X}) & \\ \swarrow^{\mathrm{CAlg}(-)^{\mathrm{cn}}} & & \searrow^{\mathrm{CAlg}_k^{\mathrm{art}} \times_{\mathrm{CAlg}^{\mathrm{cn}}} (-)} \\ \mathrm{Aff}_X^{\mathrm{op}} & & \mathbf{QCoh}_{\mathrm{art}}(\iota^* \mathbf{X}) \hookrightarrow \mathrm{Rep}_{\Psi^{-1}(\iota^* \mathbf{X})}. \end{array}$$

Proof. By Proposition 2.5.1.2 of [54] we have an equivalence of ∞-categories $\mathrm{Aff}_X^{\mathrm{op}} \simeq \mathrm{CAlg}(\mathrm{QCoh}(X)^{\mathrm{cn}})$. From Proposition 6.2.4.1 of [54], for a spectral DM stack X, we have an equivalence of ∞-categories $\mathrm{QCoh}(X) \simeq \mathbf{QCoh}(\mathbf{X})$, with $\mathbf{X} = \mathrm{Hom}(\mathrm{Spét}(-), X)$. On the other hand, since $\mathbf{X}$ is artinian, $\iota^* \mathbf{X}$ is a formal moduli problem by definition, and Theorem 13.4.0.1 of [54] provides a fully faithful monoidal embedding $\mathbf{QCoh}_{\mathrm{art}}(\iota^* \mathbf{X}) \hookrightarrow \mathrm{Rep}_{\mathfrak{g}_*}$, where $\mathfrak{g}_* = \Psi^{-1}(\iota^* \mathbf{X})$, where Ψ^{-1} is the homotopy inverse of Ψ. Further, $\mathbf{QCoh}_{\mathrm{art}}(\iota^* \mathbf{X})$ is obtained from $\mathbf{QCoh}(\mathbf{X})$ by base change, which we can formally write as

$$\mathbf{QCoh}_{\mathrm{art}}(\iota^* \mathbf{X}) = \mathrm{CAlg}_k^{\mathrm{art}} \times_{\mathrm{CAlg}^{\mathrm{cn}}} \mathbf{QCoh}(\mathbf{X})$$

□

Rmrk 18.2.2. A more precise presentation of the left side of this graph is given by

$$\mathrm{QCoh}(X) \xrightarrow{\mathrm{CAlg}(-)^{\mathrm{cn}} \equiv \mathbb{K}} \mathrm{CAlg}(\mathrm{QCoh}(X)^{\mathrm{cn}}), \qquad \mathrm{Aff}_X^{\mathrm{op}} \xrightarrow{\simeq} \mathrm{CAlg}(\mathrm{QCoh}(X)^{\mathrm{cn}})$$

which is clear from the proof of Proposition 2.5.1.2 of [54], which shows that the above equivalence arises from the construction that with $f : Y \to X$ associates $f_*\mathcal{O}_Y \in \mathrm{CAlg}(\mathrm{QCoh}(X)^{\mathrm{cn}})$ where $f_* : \mathrm{Mod}_{\mathcal{O}_Y} \to \mathrm{Mod}_{\mathcal{O}_X}$ is in fact a functor $f_* : \mathrm{QCoh}(Y) \to \mathrm{QCoh}(X)$ by Proposition 2.5.1.1.

18.2.2 Dual representation for geometric stacks

In the previous subsection, we introduced the affine perception of a spectral DM stack, the ∞-category of all affine maps $Y \to X$. If $Y = \mathrm{Sp\acute{e}t}R$, then we would have a notion of local perception. Observe that if a functor $\mathbf{X} : \mathrm{CAlg}^{\mathrm{cn}} \to \mathcal{S}$ is represented by X, $\mathbf{X} = \mathrm{Hom}(\mathrm{Sp\acute{e}t}(-), X)$, then $\mathbf{X}$ itself provides the *local perception of* X. In this subsection, we are interested in recovering $\mathbf{X}$ from the quasi-coherent sheaves defined on it. This is possible of $\mathbf{X}$ if in addition to being artinian, it is also a *geometric stack*,[1] that is, it satisfies descent for the fpqc topology, its diagonal map is affine, and there is a faithfully flat map $\mathrm{Spec}\, R \to \mathbf{X}$ for some $R \in \mathrm{CAlg}^{\mathrm{cn}}$.

Theorem 18.2.3. *Let $X \in \mathrm{SpDM}$, representing an artinian, geometric stack $\mathbf{X}$. Then we have a dual representation of the manifestation $\mathrm{QCoh}(X)$ of X:*

$$\mathbf{X} = \mathrm{Hom}(\mathrm{Sp\acute{e}t}(-), X) \xleftarrow{\text{Tannaka Duality}} \mathrm{QCoh}(X) \simeq \mathbf{QCoh}(\mathbf{X}) \xrightarrow{\mathrm{CAlg}_k^{\mathrm{art}} \times_{\mathrm{CAlg}^{\mathrm{cn}}} (-)} \mathbf{QCoh}_{\mathrm{art}}(\iota^*\mathbf{X}) \hookrightarrow \mathbf{Rep}_{\Psi^{-1}(\iota^*\mathbf{X})}.$$

Proof. From Theorem 9.3.0.3 of [54], a geometric stack $\mathbf{X}$ can be functorially recovered from $\mathbf{QCoh}(\mathbf{X})$ by Tannaka duality. The restriction of the latter to artinian algebras is $\mathbf{QCoh}_{\mathrm{art}}(\iota^*\mathbf{X})$, full subcategory of $\mathrm{Rep}_{\mathfrak{g}_*}$ (where $\mathfrak{g}_* = \Psi^{-1}(\iota^*\mathbf{X})$) is the ∞-category of representations of the dg Lie algebra $\mathfrak{g}_*$ associated with $\iota^*\mathbf{X}$. □

1 Def. 9.3.0.1 [54].

Rmrk 18.2.4. To be precise, $\mathbf{X}$ can be functorially recovered from $\mathbf{QCoh}(\mathbf{X})$, which really means we have the following suggestive diagram:

$$\begin{array}{ccc}
 & & \mathbf{QCoh}(\mathbf{X}) \\
 & \swarrow^{\text{Tannaka duality}} & \Big\downarrow \text{functorial} \\
\mathbf{X} & & \mathrm{Fun}^{\otimes}(\mathbf{QCoh}(\mathbf{X}), \mathbf{QCoh}(-)) \\
\text{functorial} \Big\uparrow & & \Big\downarrow \mathbb{K} \\
\mathrm{Map}_{\mathrm{Fun}(\mathrm{CAlg}^{\mathrm{cn}}, \hat{\mathcal{S}})}(-, \mathbf{X}) & \overset{\simeq}{\Longrightarrow} & \mathrm{Fun}^{\otimes}_{\mathrm{TD}^*}(\mathbf{QCoh}(\mathbf{X}), \mathbf{QCoh}(-))
\end{array}$$

where the bottom arrow above is a natural equivalence, and we remind the reader from Theorem 9.3.0.3 of [54] that the Tannaka duality in the present context is the statement that the construction that with $f : \mathbf{Y} \to \mathbf{X}$ associates $f^* : \mathbf{QCoh}(\mathbf{X}) \to \mathbf{QCoh}(\mathbf{X})$ is a fully faithful embedding

$$\mathrm{Map}_{\mathrm{Fun}(\mathrm{CAlg}^{\mathrm{cn}}, \hat{\mathcal{S}})}(\mathbf{Y}, \mathbf{X}) \hookrightarrow \mathrm{Fun}^{\otimes}(\mathbf{QCoh}(\mathbf{X}), \mathbf{QCoh}(\mathbf{Y}))$$

whose essential image $\mathrm{Fun}^{\otimes}_{\mathrm{TD}^*}(\mathbf{QCoh}(\mathbf{X}), \mathbf{QCoh}(\mathbf{Y}))$ is spanned by those functors that preserve small colimits, connective objects, and flat objects. We denote by $\mathbb{K}$ the construct that produces $\mathrm{Fun}^{\otimes}_{\mathrm{TD}^*}(\mathbf{QCoh}(\mathbf{X}), \mathbf{QCoh}(-))$ from $\mathrm{Fun}^{\otimes}(\mathbf{QCoh}(\mathbf{X}), \mathbf{QCoh}(-))$.

18.2.3 Variants

We can obtain stronger results, albeit with stricter hypotheses, by asking that $\mathbf{X}$ be representable, which necessitates the spectral Artin representability theorem.

Theorem 18.2.3.1. *Let $\mathbf{X} : \mathrm{CAlg}^{\mathrm{cn}} \to \mathcal{S}$ be a functor that satisfies the hypotheses of the spectral Artin representability theorem for some $n \geq 0$ and hence is representable by a spectral DM n-stack X. Then supposing $\mathbf{X}$ is artinian, the manifestation* $\mathrm{QCoh}(X)$ *of X has a dual representation*

$$\begin{array}{ccc}
 & \mathrm{QCoh}(X) \simeq \mathbf{QCoh}(\mathbf{X}) & \\
 {}^{\mathrm{CAlg}(-)^{\mathrm{cn}}}\swarrow & & \searrow^{\mathrm{CAlg}_k^{\mathrm{art}} \times_{\mathrm{CAlg}^{\mathrm{cn}}} (-)} \\
\mathrm{Aff}_X^{\mathrm{op}} & & \mathbf{QCoh}_{\mathrm{art}}(\iota^* \mathbf{X}).
\end{array}$$

Theorem 18.2.3.2. *Let $\mathbf{X}$ be an artinian, geometric stack that satisfies the hypotheses of the spectral Artin representability theorem for some $n \geq 0$, and let X be the n-spectral DM stack representing it. Then the manifestation* $\mathrm{QCoh}(X)$ *of X has a dual representation*

$$\mathrm{QCoh}(X) \simeq \mathbf{QCoh}(\mathbf{X})$$

Tannaka Duality ↙ ↘ $\mathrm{CAlg}_k^{\mathrm{art}} \times_{\mathrm{CAlg}^{\mathrm{cn}}} (-)$

$$\mathbf{X} = \mathrm{Hom}(\mathrm{Sp\acute{e}t}(-), X) \qquad\qquad \mathbf{QCoh}_{\mathrm{art}}(\iota^* \mathbf{X}).$$

As a side note, if Tannaka for geometric stacks allows us to recover **X** from **QCoh**(**X**), note that this is very much static. One can get a similar reconstruction in a smooth setting by considering the cohomology of **X**.

To do this, we first consider X, a simply connected *rational space*,[2] a space X such that $\pi_n X$ is a vector space over $\mathbb{Q}$ for all $n \geq 2$, and we ask also that X be of *finite type*, that is, the vector spaces $\pi_n X$ are finite-dimensional over $\mathbb{Q}$ for $n \geq 2$. Generally, for a field k, denote by k_X the unit in the functor category Mod_k^X. Regarding X as a Kan complex, we have a projection map $p : X \to \Delta^0$. Denote $p_* k_X$ by $C^*(X; k)$, regarded as an object of CAlg_k.[3] The construction $X \mapsto C^*(X; k)$ gives us a functor $\mathcal{S} \to \mathrm{CAlg}_k^{\mathrm{op}}$, which has a right adjoint $R \mapsto \mathrm{Map}_{\mathrm{CAlg}_k}(R, k)$.[4] Then by Proposition 1.3.3 of [54], if X is a simply connected rational space of finite type, then the unit map $X \to \mathrm{Map}_{\mathrm{CAlg}_k}(C^*(X; \mathbb{Q}), \mathbb{Q})$ is a homotopy equivalence. Here $C^*(X; k)$ provides a smooth representation of X from which the latter can be recovered, in contrast to its manifestation, which only provides a discrete reconstruction. Now this is true for a space. To make this functorial, apply this pointwise to a functor $\mathbf{X} : \mathrm{CAlg}^{\mathrm{cn}} \to \mathcal{S}$.

We would like to close this topic with an interesting observation about the perception of an object or, more generally, the Yoneda embedding $\mathcal{C} \to \mathrm{Fun}(\mathcal{C}^{\mathrm{op}}, \mathcal{S})$ defined by $X \mapsto \mathrm{Hom}_{\mathcal{C}}(-, X) = h_X$. A category can be defined starting with its objects and then with the morphism spaces between them, or from arrows only. Starting with objects, given a category $\mathcal{C}$, one can write

$$\mathcal{C} = \cup_{X \in \mathcal{C}} \{\mathrm{Hom}_{\mathcal{C}}(Y, X) \; : \; Y \in \mathcal{C}\}$$

which can be functorially represented as $\cup_{X \in \mathcal{C}} h_X = h$. We can also start from morphisms:

$$\begin{aligned} \mathcal{C} &= \{\mathrm{Hom}_{\mathcal{C}}(X, Y) \; : \; X, Y \in \mathrm{Ob}\, \mathcal{C}\} \\ &= \cup_{X \in \mathcal{C}} \cup_{Y \in \mathcal{C}} h_X(Y). \end{aligned}$$

The latter can be functorially written as $\cup_{X \in \mathcal{C}} h_X = h$. Thus an optimal functorial representation of $\mathcal{C}$ ought to be provided by the Yoneda embedding, and at a point X, the perception h_X is a crystallization of $\mathcal{C}$ at that particular object; h itself can be regarded

---

2 Def. 1.0.1 [54].

3 Not. 1.1.3 [54].

4 Rmrk 1.3.2 [54].

as a global perception. Now one may consider the following situation: suppose we have a mathematical theory Γ being defined at some point X: $\Gamma = \Gamma[X]$, X being an object of a category $\mathcal{C}$. To make this covariant, or functorial, one may want to have that concept defined on $\mathcal{C}$ in a natural way, and thus we regard Γ as a functor $\Gamma : \mathcal{C} \to \mathcal{D}$ for some target category $\mathcal{D}$. Suppose however that we would like Γ to not simply be defined pointwise, but to be defined on relative objects, such as perceptions. One way to implement this is to simply use the Yoneda embedding h of $\mathcal{C}$. As an illustrative example, pick $\Gamma = \mathbf{QCoh}$, $\mathcal{C} = \mathrm{SpDM}$, and $\mathcal{D} = \mathrm{Cat}_\infty$, whence the need to define quasi-coherent sheaves as a functor on $\mathrm{Fun}(\mathrm{CAlg}^{\mathrm{cn}}, \mathcal{S}) \ni h(\mathrm{SpDM})$. To be more precise, we knew how to define quasi-coherent sheaves on spectral DM stacks, as a functor $\mathrm{QCoh} : \mathrm{SpDM} \to \mathrm{Cat}_\infty$. If we want that functor to be defined on relative objects, then we use the fully faithful embedding $h : \mathrm{SpDM} \to \mathrm{Fun}(\mathrm{CAlg}^{\mathrm{cn}}, \mathcal{S})$ given by $h_X = \mathrm{Map}_{\infty\mathcal{T}\mathrm{op}^{\mathrm{loc}}_{\mathrm{CAlg}}}(\mathrm{Sp\acute{e}t}(-), X)$, $X \in \mathrm{SpDM}$, giving the local perception of that object. Thus in this setting, we have $\mathbf{QCoh} : \mathrm{Fun}(\mathrm{CAlg}^{\mathrm{cn}}, \mathcal{S}) \to \mathrm{Cat}_\infty$, and $\mathbf{QCoh}(h_X)$ is now used instead of $\mathrm{QCoh}(X)$.

19 A Connection SAG-DAG

The aim of this short chapter is to make a connection, which is by no means canonical, between SAG and DAG, in the hope that this will put those theories in perspective relative to each other for the reader.

To make a connection between the notion of derived stack on sk-CAlg, as developed by Toen and Vezzosi in Homotopical Algebraic Geometry, and the objects studied by Lurie in Spectral Algebraic Geometry, we first briefly remind the reader of the construction of derived stacks by Toen and Vezzosi.

One starts with a model category M and considers functors from M into Set_Δ, giving rise to a functor category $\mathrm{Fun}(M^{\mathrm{op}}, \mathrm{Set}_\Delta) = \mathrm{sPr}(M)$ on which we put the projective model structure. Taking the left Bousfield localization with respect to equivalences in M, one gets a simplicial category of prestacks $M^\wedge = L_W \mathrm{sPr}(M)$. Now suppose M is endowed with a topology τ so that we have a model site (M, τ). Then we can further take the left Bousfield localization of $M^\wedge$ with respect to τ-hypercovers (or local equivalences), giving rise to a simplicial category of stacks on (M, τ): $M^{\sim,\tau} = L_{H_\tau} M^\wedge$, where H_τ is a class of homotopy τ-hypercovers (resp., $M^{\sim,\tau} = L_{\mathrm{loc.equiv}} M^\wedge$).

Next, we take $M = \mathcal{A}\mathrm{ff}_\mathcal{C} = \mathrm{Comm}(\mathcal{C})^{\mathrm{op}}$, where $\mathcal{C}$ is a base symmetric monoidal model category. We have chosen to work with commutative monoids instead of $\mathbb{E}_\infty$-algebras, something that would seem to be the correct homotopical generalization of the notion of commutative monoid to the setting in which we work for a general base symmetric monoidal model category.

For our purposes, we take $\mathcal{C}$ to be the category of simplicial modules over some base commutative ring k, which we denote by sk-Mod. Take the commutative monoid objects thereof: $\mathrm{Comm}(\mathcal{C})$, which in this situation is denoted sk-CAlg and is referred to as the category of commutative simplicial k-algebras. The opposite category $\mathrm{Comm}(\mathcal{C})^{\mathrm{op}}$ is denoted k-D$^-$Aff, which we endow with the étale topology. Thus we get a model category k-D$^-$Aff$^\wedge$ of prestacks. Note that Ho (k-D$^-$Aff$^\wedge$) is identified with the full subcategory of Ho (sPr(k-D$^-$Aff)) consisting of functors preserving equivalences, and its objects are prestacks. Then the model category of stacks k-D$^-$Aff$^{\sim,\text{ét}}$ has for objects functors $F : sk\text{-CAlg} \to \mathrm{Set}_\Delta$ that preserve equivalences and satisfy ét-hyperdescent.

We also need to connect simplicial commutative rings over k with connective $\mathbb{E}_\infty$-rings. If k is a commutative ring and if Poly_k denotes the full subcategory of CRing_k spanned by commutative k-algebras $k[x_1, \ldots, x_n]$ for $n \geq 0$, then one defines $\mathrm{SCR}_k := \mathcal{P}_\Sigma(\mathrm{N}(\mathrm{Poly}_k))$ to be the ∞-category of simplicial commutative k-algebras.[1] Here for a small ∞-category $\mathcal{C}$ that admits finite coproducts, $\mathcal{P}_\Sigma(\mathcal{C})$ denotes the full subcategory of $\mathrm{Pr}(\mathcal{C})$ spanned by functors that preserve finite coproducts.[2] From Proposition 4.1.11 of [57], for a commutative ring k, there exists a functor $\theta : \mathrm{SCR}_k \to \mathrm{CAlg}_k^{\mathrm{cn}}$ that preserves

1 Def. 4.1.1 [57].

2 Def. 5.5.8.8 [52].

https://doi.org/10.1515/9783111334073-019

small limits and colimits, is conservative, has a left and a right adjoint, and is an equivalence of ∞-categories if k is a $\mathbb{Q}$-algebra. By Remark 4.1.2 of the same reference, if A denotes the ordinary category of simplicial commutative k-algebras, that is, $A = sk\text{-CAlg} = \mathrm{Comm}(sk\text{-Mod})$, then we have a canonical equivalence of ∞-categories $\mathrm{N}(A^\circ) \to \mathrm{SCR}_k$, which induces an equivalence of simplicial categories $\mathfrak{C}[\mathrm{N}(A^\circ)] \to \mathfrak{C}[\mathrm{SCR}_k]$, with inverse $\mathfrak{C}[\mathrm{SCR}_k] \to \mathfrak{C}[\mathrm{N}(A^\circ)]$. Since $\mathfrak{C} \dashv \mathrm{N}$, we also have a functor $\mathfrak{C}[\mathrm{N}(A^\circ)] \to A^\circ$ and hence a composite functor $\mathfrak{C}[\mathrm{SCR}_k] \to \mathfrak{C}[\mathrm{N}(A^\circ)] \to A^\circ$. Now from [57], $A^\circ \hookrightarrow A$ is a full subcategory, so we have a composite functor $\mathfrak{C}[\mathrm{SCR}_k] \to A^\circ \hookrightarrow A$.

Now let M be a model category, and let $M^\wedge = L_W \mathrm{sPr}(M)$ be its model category of prestacks. If further (M, τ) is a model site, then one can define a notion of π_*-equivalence on $M^\wedge$, and the model structure on this category for which the equivalences are the π_*-equivalences is denoted $M^{\sim,\tau}$. Hence we have a composite left Bousfield localization functor $\mathrm{sPr}(M) \to M^{\sim,\tau}$. Now recall from [40] that a left Bousfield localization is in particular a left localization, i. e., a left Quillen functor, so it has a right adjoint $M^{\sim,\tau} \to \mathrm{sPr}(M)$. Apply this to the situation where $M = \mathcal{A}\mathit{ff}_C$, where $C = sk\text{-Mod}$, $\tau = \text{ét}$. Thus $M = \mathrm{Comm}(C)^{\mathrm{op}} = sk\text{-CAlg}^{\mathrm{op}} = \text{k-D}^-\text{Aff}$, and we have a functor $\mathrm{sPr}(\text{k-D}^-\text{Aff}) \to \text{k-D}^-\text{Aff}^{\sim,\text{ét}}$ with right adjoint

$$\text{k-D}^-\text{Aff}^{\sim,\text{ét}} \to \mathrm{sPr}(\text{k-D}^-\text{Aff}). \tag{19.1}$$

We write

$$\mathrm{sPr}(\text{k-D}^-\text{Aff}) = \mathrm{Set}_\Delta^{\text{k-D}^-\text{Aff}^{\mathrm{op}}} = \mathrm{Set}_\Delta^{sk\text{-CAlg}} = \mathrm{Set}_\Delta^{A}.$$

Since we have a functor $\mathfrak{C}[\mathrm{SCR}_k] \to A$ from the previous paragraph, this induces a functor

$$\mathrm{Set}_\Delta^{A} \to \mathrm{Set}_\Delta^{\mathfrak{C}[\mathrm{SCR}_k]}. \tag{19.2}$$

By unstraightening [52] we have a Quillen equivalence

$$\mathrm{Set}_\Delta^{\mathfrak{C}[\mathrm{SCR}_k]} \to (\mathrm{Set}_\Delta)_{/\mathrm{SCR}_k}. \tag{19.3}$$

The functor $\theta : \mathrm{SCR}_k \to \mathrm{CAlg}_k^{\mathrm{cn}}$ induces a functor

$$\theta_* : (\mathrm{Set}_\Delta)_{/\mathrm{SCR}_k} \to (\mathrm{Set}_\Delta)_{/\mathrm{CAlg}_k^{\mathrm{cn}}}. \tag{19.4}$$

Now by Example 1.1.2.6 of [52], for an ordinary category $\mathcal{C}$, its nerve $\mathrm{N}(\mathcal{C})$ is an ∞-category, which allows us to regard any category as an ∞-category. Thus we regard Set_Δ as $\mathrm{N}(\mathrm{Set}_\Delta)$, an ∞-category, and in particular a simplicial set. Further by Proposition 5.1.3.1 of [52], for $K \in \mathrm{Set}_\Delta$, we have a Yoneda embedding $j : K \to \mathrm{Pr}(K) = \mathrm{Fun}(K^{\mathrm{op}}, \mathcal{S})$, a fully faithful functor. Finally, recall the definition of an ∞-topos: it is an accessible left exact localization of a presheaf category $\mathrm{Pr}(\mathcal{C}) \to \mathcal{X}$, where $\mathcal{C}$ is a small ∞-category. In this manner, $\mathrm{Set}_\Delta = N(\mathrm{Set}_\Delta) \hookrightarrow \mathrm{Pr}(N(\mathrm{Set}_\Delta)) = \mathcal{X}$ is regarded as an ∞-topos, and we have a functor

$$(\mathrm{Set}_\Delta)_{/\mathrm{CAlg}_k^{\mathrm{cn}}} = (N(\mathrm{Set}_\Delta))_{/\mathrm{CAlg}_k^{\mathrm{cn}}} \xrightarrow{j} \mathrm{Pr}(N(\mathrm{Set}_\Delta))_{/\mathrm{CAlg}_k^{\mathrm{cn}}}. \tag{19.5}$$

Collecting things together, we have a functor resulting from the composition of (19.1), (19.2), (19.3), (19.4), and (19.5):

$$\text{k-D}^-\mathrm{Aff}^{\sim,\text{ét}} \to \mathrm{sPr}(\text{k-D}^-\mathrm{Aff}) = \mathrm{Set}_\Delta^A \to \mathcal{X}_{/\mathrm{CAlg}_k^{\mathrm{cn}}}$$

where $\mathcal{X} = \mathrm{Pr}(N(\mathrm{Set}_\Delta))$, which allows us to map derived stacks on sk-CAlg in the setting of Derived Algebraic Geometry in the sense of Toen and Vezzosi to objects of the slice category $\mathcal{X}_{/\mathrm{CAlg}_k^{\mathrm{cn}}}$.

Conversely, from [54], for a locally spectrally ringed ∞-topos $X = (\mathcal{X}, \mathcal{O}_\mathcal{X})$ and an $\mathbb{E}_\infty$-ring R, define

$$h_X : \mathrm{CAlg}^{\mathrm{cn}} \to \mathcal{S},$$
$$R \mapsto \mathrm{Map}_{\infty\mathcal{T}\mathrm{op}^{\mathrm{loc}}_{\mathrm{CAlg}}}(\mathrm{Sp\acute{e}t}R, X).$$

By Proposition 1.6.4.2, $X \mapsto h_X$ is a fully faithful embedding of SpDM into Fun($\mathrm{CAlg}^{\mathrm{cn}}$, $\mathcal{S}$). Now if X is a quasi-affine spectral DM stack, then h_X is a hypercomplete sheaf with respect to the flat topology on $\mathrm{CAlg}^{\mathrm{cn}}$. This allows us to go directly from spectral DM stacks of Lurie to functors from $\mathrm{CAlg}^{\mathrm{cn}}$ into $\mathcal{S}$, the setting in which the brave new algebraic geometry of Toen and Vezzosi [80] can be developed. Though this is not exactly recovering derived stacks as we understand them, this is nevertheless a quick connection between Spectral Algebraic Geometry and Brave New Algebraic Geometry, which is still part of Homotopical Algebraic Geometry.

20 Motifs in DAG

20.1 Introduction

It has been observed in the past [13, 12, 78] that one can formalize Algebraic Geometry by writing it in purely categorical terms by simply starting with some symmetric monoidal base category $\mathcal{C}$, by considering its category $\mathrm{Comm}(\mathcal{C})$ of commutative and unital monoids, and by regarding $\mathcal{A}\mathrm{ff}_{\mathcal{C}} = (\mathrm{Comm}(\mathcal{C}))^{\mathrm{op}}$ as the category of affine schemes over $\mathcal{C}$. On $\mathcal{A}\mathrm{ff}_{\mathcal{C}}$ one then puts some topology τ, be it the Zariski, étale, ffqc, or any topology that one so wishes to then develop a notion of stacks and higher stacks on $(\mathcal{A}\mathrm{ff}_{\mathcal{C}}, \tau)$. One would obtain in this manner what is referred to as Relative Algebraic Geometry, classical Algebraic Geometry corresponding to the case of having $\mathcal{C} = (\mathbb{Z}\text{-Mod}, \otimes)$. It is in an attempt to develop Relative Algebraic Geometry over symmetric monoidal ∞-categories that Toen observed that one could use model categories since they give rise to ∞-categories via the Dwyer–Kan simplicial localization technique. Thus one starts with a symmetric monoidal model category $(\mathcal{C}, \otimes)$ in the sense of [42], thereby introducing Homotopical Algebraic Geometry [75, 78, 76, 79] or Algebraic Geometry over model categories. In particular, if for some fixed commutative ring k, one considers $\mathcal{C} = sk\text{-Mod}$, the category of simplicial objects in k-Mod, then one obtains Derived Algebraic Geometry [79].

The need to introduce stacks, higher stacks, and derived stacks can be seen from a classification problem perspective. As recounted in [74], one can start with a contravariant functor F from a category $\mathcal{C}$ of geometric objects to Set, where for a geometric object X, $F(X)$ classifies families of objects parameterized by X, and F being valued in Set, this classification is really done up to equality. One may relax that condition and ask that classification be done up to isomorphism as well, whence the introduction of contravariant functors into Gpd, which corresponds to considering 1-stacks. Another motivation for making such a generalization as pointed out in [81] is that such Set-valued moduli functors F may not be representable and only admit a coarse moduli space, which is further not of the expected dimension, so following [49], a natural approach amounts to seeing such spaces as truncations of higher spaces, or derived spaces, smooth (as opposed to being singular), and of the expected dimension. This corresponds to seeking natural extensions of F to functors $F_1 : \mathcal{C}^{\mathrm{op}} \to \mathrm{Gpd}$ that make the following diagram commutative, where we take $\mathcal{C} = k\text{-CAlg} = \mathrm{Comm}(k\text{-Mod})$, k a commutative ring, since that will be our main point of interest:

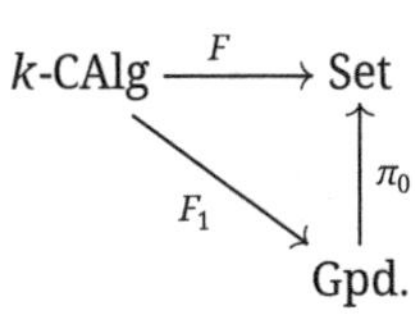

https://doi.org/10.1515/9783111334073-020

Next, one may want to further relax the classification scheme by also allowing classification up to equivalence. At an elementary level, this means having a Cat-valued functor, but morphisms in Cat would have to be inverted, and this is possible only if we have functors valued in ∞-categories, and hence we consider ∞-stacks, or stacks for short. This corresponds to looking at extensions F_∞ of F_1 and F that make the following diagram commutative:

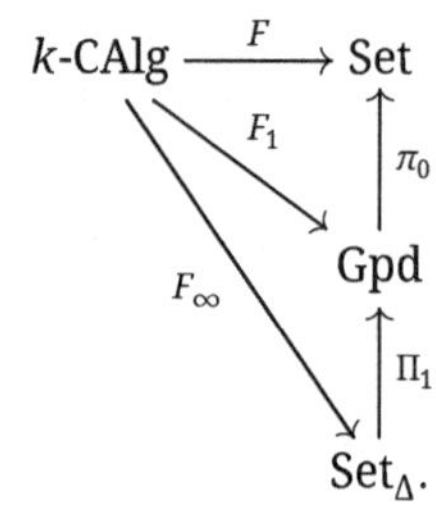

Finally, if one considers obstruction theory, then as pointed out in [74], Derived Algebraic Geometry is a natural formalism in which such a theory can be written out, and this corresponds to finding derived extensions $\mathbb{R}F$ of F, F_1, and F_∞ that make the following diagram commutative:

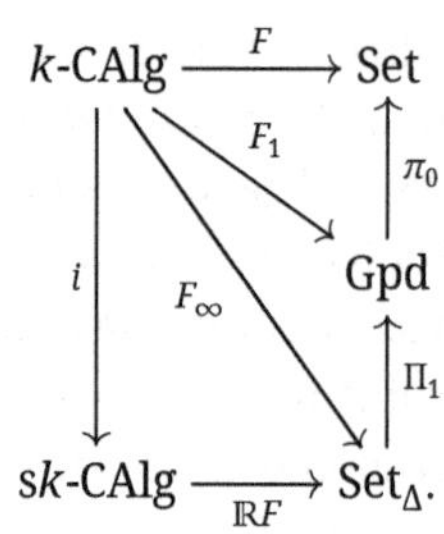

(20.1)

Now as discussed in [81], not all such extensions will work; there are constraints that have to be met in addition to having the above diagram commutative, for instance, having the right derived tangent stack, something that would be known at the onset. Moreover, there is no canonical choice of a derived extension. This is our main motivation for introducing a formalism where one would not have to worry about having to pick a derived extension. In addition, we would like to functorially construct a derived extension, as exposed in the section that follows.

20.2 Construction

We reproduce (20.1) above as it is presented in [74]:

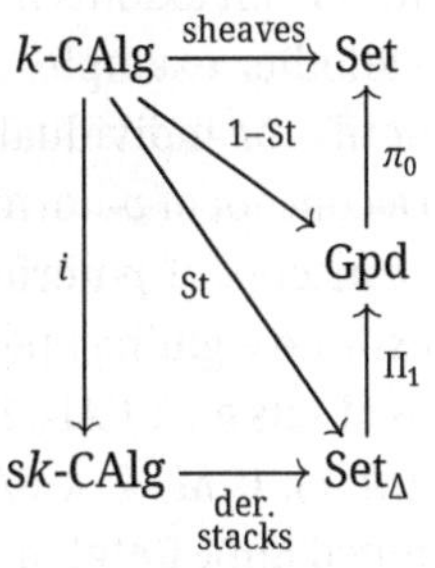

(20.2)

where for $X \in \mathrm{Set}_\Delta$, $\Pi_1(X) = G\gamma(X)$, $\gamma(X)$ the path category of X, and $G : \mathrm{Cat} \to \mathrm{Gpd}$, $A \mapsto GA = A[\Sigma^{-1}]$, $\Sigma = A_1$ [45, 28]. We have $\mathrm{Ob}\,(\Pi_1(X)) = X_0$, morphisms in $\Pi_1(X) = X_1$ and formal inverses. We also have the connected component functor $\pi_0 : \mathrm{Cat} \to \mathrm{Set}$ on categories [59].

This diagram is made commutative by selecting derived extensions to sk-CAlg of the notions of sheaves, 1-stacks, and stacks, which depends very much on the context, so one may inquire whether working with *motifs*, or variables, instead of specializations thereof, would provide something that is less than an ad hoc construction. This is motivated in particular by the fact that mentioning “sheaves” and “derived stacks” in the above diagram can be made precise using the formalism of motifs as defined presently. We briefly remind the reader of the definition of *motivic frame* as introduced in [24]: given a mathematical construct X, one can formalize its construction by using a motivic frame $x = \{x^{(n)}, x^{(n)} \xrightarrow{z^{(n)}} x^{(n+1)}\}_{n\geq 0}$ for its structural elements, with $z^{(n)}$ gluing maps and $x^{(n)}$ variables, or motifs, obtained from lower motifs by gluing, the idea being that X as a construct is regarded as being built from structures of a certain (higher) type, themselves being built from lower-type structural elements, and the manner in which this is performed can be abstracted with the use of motivic frames. Practically, lower-type objects are collected into classes, and the $x^{(i)}$ provide variables, or motifs, for objects of each such class. The gluing maps indicate in what manner the different motifs $x^{(i)}$ are put together to form subsequent motifs. The choice of the word motif is mainly one of semantics; “variable” is a rather crude way to refer to the $x^{(i)}$. A “motif”, or pattern, is meant to describe a unifying elementary model, and hence adopting such a concept in the present context captures the nature of the objects within each class in addition to being a simple variable. A motif depending on other motifs will simply be referred to as a higher motif.

For illustrative purposes, consider a certain space of matrices, consisting of a choice of a set of matrices and a choice of a norm. One can abstract this by considering normed vector spaces $(X, \|\cdot\|)$. In this picture, X is a motif for a set, and $\|\cdot\|$ is a motif for a norm, the combination of which is a motif for a normed vector space. From that perspective, both X and $\|\cdot\|$ are motifs of lowest degree 1, and $(X, \|\cdot\|)$ is a motif of degree 2, being a composite of motifs of lower degree, and the gluing consists in just collecting the two motifs of degree 1 together. This is a particularly simple example. Sometimes, collecting submotifs together is not sufficient, certain axioms have to be met, and those would be part of the gluing structure. As an another example, we have p-norms on matrices and p-norms on functions. Both are motifs for individual p-norms for specific values of p. However, they both correspond to a concept of p-norm, an overarching motif. From that perspective, collecting those two concepts of p-norms together to form the notion of p-norm, a higher motif, would constitute a gluing operation.

We let Sh be a motif for sheaves of sets on k-CAlg. Strictly speaking, we have an overarching motif Sh on a site motif (M, τ). If $M = k$-CAlg, then we specialize Sh to that particular site, so it should be denoted Sh$[k$-CAlg$]$, but for notation's sake, we will just denote it by Sh again, since those are the only sheaves we consider in this work. In this situation, Sh itself is a motif for sheaves on k-CAlg. Hence we are looking for a commutative diagram of motifs. Further we regard the move from sheaves to derived stacks as a functorial operation, thus we would like to replace $\pi_0 \circ \Pi_1$ on the right in (20.2) by an adjoint map Set $\to$ Set$_\Delta$ denoted j, which we would like to argue is formally identical to i on the left. To effectuate this, notice that on the one hand, the connected component functor can be defined from Set$_\Delta$ as the coequalizer [45]

$$X_1 \overset{d_1}{\underset{d_0}{\rightrightarrows}} X_0 \to \pi_0(X)$$

for $X \in \mathrm{Set}_\Delta$, whereas in the above diagram, for $X \in \mathrm{Set}_\Delta$, $\Pi_1(X) \in \mathrm{Gpd}$, from which we then apply the set of algebraic connected components functor as $\pi_0(N\Pi_1(X))$, N the nerve functor, so that $N\Pi_1(X) \in \mathrm{Set}_\Delta$. We typically abbreviate this composite by $\pi_0 \circ \Pi_1(X)$, taking the nerve functor as being understood. Observe then that we effectively start from an object $X \in \mathrm{Set}_\Delta$, consider its fundamental groupoid, an object of Gpd, and then use the nerve functor to obtain again an object of Set$_\Delta$, from which we finally use the connected component functor. Instead of going through the intermediate stage provided by groupoids, we consider an abbreviated, direct diagram

$$\begin{array}{ccc} k\text{-CAlg} & \xrightarrow{\text{Sh}} & \text{Set} \\ {\scriptstyle i}\downarrow & & \uparrow{\scriptstyle \pi_0} \\ sk\text{-CAlg} & \xrightarrow[\text{der. stacks}]{} & \text{Set}_\Delta . \end{array}$$

Recall that $\pi_0 \dashv \mathrm{cs}_*$, with cs_* the constant simplicial functor:

$$\begin{aligned} \mathrm{cs}_* : \mathrm{Set} &\to \mathrm{Set}_\Delta, \\ S &\mapsto \mathrm{cs}_*(S) \end{aligned}$$

where $\mathrm{cs}_*(S)_n = S$ for all $n \geq 0$, and all face and degeneracy maps are id_S. This provides us with the desired adjoint $j = \mathrm{cs}_*$, so that we are looking at

$$\begin{array}{ccc} k\text{-CAlg} & \xrightarrow{\mathrm{Sh}} & \mathrm{Set} \\ {\scriptstyle i}\downarrow & & \downarrow{\scriptstyle j} \\ sk\text{-CAlg} & \xrightarrow[\text{der. stacks}]{} & \mathrm{Set}_\Delta . \end{array}$$

On the other hand, by definition, $i = \mathrm{cs}_*$, where $i : k\text{-CAlg} \to sk\text{-CAlg}$ is the natural inclusion functor that sends a k-algebra A to the constant simplicial object A in sk-CAlg [74]. Let $s : A \to sA$ be the functor that maps a structured object A to its simplicial version, a motif for simplicialization. We have $s|_{k\text{-CAlg}}$ has i for specialization, and $s|_{\mathrm{Set}}$ has j for specialization; s is a notation we favor over cs_*.

Our aim is to show that there is a motif dSt for derived stacks, depending on the motifs s and Sh, hence a higher motif, and making the following diagram commutative:

$$\begin{array}{ccc} k\text{-CAlg} & \xrightarrow{\mathrm{Sh}} & \mathrm{Set} \\ {\scriptstyle s}\downarrow & & \downarrow{\scriptstyle s} \\ sk\text{-CAlg} & \xrightarrow[\mathrm{dSt}]{} & \mathrm{Set}_\Delta . \end{array} \tag{20.3}$$

The interest of having such a motif is that it would comprise all possible derived extensions given s and Sh. The aim of the present work is to prove that such a higher motif stands for a derived stack variable, i. e., that one can functorially construct derived stacks, and that such a construction is choice-free if written in the language of motifs.

Note that if one uses $s|_{k\text{-CAlg}}$ on the left and $s|_{\mathrm{Set}}$ on the right in the above diagram, both being motifs in their own right, in the language of motivic frames, one has a commutative diagram of motifs of lowest degree 1. Observing that both motifs are really the same, namely simplicializing motifs, if one denotes by s their overarching motif, then using s instead in the same commutative diagram elevates it to a diagram of motifs of degree 2. Here we use the convention that a motif of degree i can be seen as a motif of degree j for $j \geq i$ by taking all gluing maps beyond the ith level to be trivial. Thus though Sh and dSt are of degree 1, they are regarded as degree 2 motifs, so that our resulting diagram is indeed a commutative diagram of motifs of degree 2. For notation's sake however, we will just use s for either motif $s|_{k\text{-CAlg}}$ or $s|_{\mathrm{Set}}$. In other terms, we implicitly work with specializations for building up the desired commutative diagram.

We ask that s be a motif for a functorial simplicial frame, which in addition preserves finite limits. The important point being made here is that though one could just work with cs_*, one would like a functor that for all practical purposes is equivalent to it but is otherwise unspecified. Thus we work in greater generality. Functorial simplicial frames allow us to do that. To fix ideas, recall from [40] that for an object A of k-CAlg, we define a *simplicial frame* on A to be a simplicial object $\hat{A} \in (k\text{-CAlg})^{\Delta^{op}} = sk\text{-CAlg}$, together with an equivalence $\mathrm{cs}_*A \to \hat{A}$ in the Reedy model category structure on sk-CAlg such that the induced map $A \to \hat{A}_0$ is an isomorphism, and if A is fibrant in k-CAlg, so is $\hat{A}$ in sk-CAlg. We then define a *functorial simplicial frame* on k-CAlg to be given by a pair (G, j), where $G : k\text{-CAlg} \to sk\text{-CAlg}$ is a functor, $j : \mathrm{id} \Rightarrow G$ is such that $j_A : \mathrm{cs}_*A \to G(A)$ is a simplicial frame for any $A \in k$-CAlg. We abuse notation and refer to G from (G, j) as the functorial simplicial frame. Then s stands for a motif of functorial simplicial frame, G being a specialization thereof.

Given a sheaf motif Sh and a functorial simplicial frame motif s on k-CAlg, we have a corresponding higher motif $\mathrm{dSt}[\mathrm{Sh}, s] = \mathrm{dSt}$, which we aim to prove is a motif for derived stacks and so must satisfy the properties of stacks. Recall from [74] that in the definition of a derived stack, we have to use hypercovers, which necessitates the introduction of coaugmented cosimplicial objects $\mathbf{A} \to \mathbf{B}_*$ in sk-CAlg. Here we take $\mathbf{A} = s(A)$ for $A \in k$-CAlg, and $\mathbf{B}_*$ is a functor

$$\begin{aligned} \mathbf{B}_* : \Delta &\to sk\text{-CAlg}, \\ n &\mapsto \mathbf{B}_n = s(B_n). \end{aligned}$$

This is a fundamental assumption in our work. Since we generate derived stacks from sheaves by using the motif s for simplicialization, we regard simplicial commutative k-algebras $\mathbf{A}$ to come from some $A \in k$-CAlg, but we also regard the $\mathbf{B}_n$ above as being images of $B_n \in k$-CAlg. To fix notations, $\mathbf{B}_* \in \mathrm{cs}(sk\text{-CAlg})$, where cs stands for cosimplicial. Thus in this work, we will consider coaugmented cosimplicial objects of the form $sA \to sB_*$ for $A \in k$-CAlg and $B_* \in$ csk-CAlg. To see that this still produces a cosimplicial object sB_* of sk-CAlg, note that s is defined on k-CAlg. Thus sB_n makes sense for all $[n] \in \Delta$. Now functorially this makes sB_* into a cosimplicial object of sk-CAlg, since the functor $s : k\text{-CAlg} \to sk\text{-CAlg}$ preserves the cosimplicial identities.

The reason for introducing cosimplicial objects is elementary; we will argue the opposite of an algebra $A \in sk$-CAlg is an object $\mathrm{Spec}\, A \in dk$-Aff, and it is over such objects that we have hypercovers given by simplicial objects $\mathrm{Spec}\, B_* \to \mathrm{Spec}\, A$. To see that taking a cosimplicial coaugmentation produces just that, it suffices to start from $B_* : \Delta \to sk\text{-CAlg}$, so that $B_*^{op} = \mathrm{Spec}\, B_* : \Delta^{op} \to sk\text{-CAlg}^{op} = dk\text{-Aff}$ is indeed in sdk-Aff.

We are now ready to give the conditions dSt has to satisfy to be a derived stack. According to [74] applied to our setting, the following conditions must be met:

- For any equivalence $sA \to sB$ in sk-CAlg, $A, B \in k$-CAlg, the induced morphism $\mathrm{dSt}(sA) \to \mathrm{dSt}(sB)$ is an equivalence in Set_Δ.
- For any coaugmented cosimplicial object $sA \to sB_*$, corresponding to an ét-hypercovering in $dk\text{-Aff} = sk\text{-CAlg}^{op}$, the induced morphism $\mathrm{dSt}(sA) \to \mathrm{holim}_{n \in \Delta} \mathrm{dSt}(sB_n)$ is an equivalence in Set_Δ.
- For any finite family $\{sA_i\}$ in sk-CAlg, the natural morphism $\mathrm{dSt}(\prod sA_i) \to \prod \mathrm{dSt}(sA_i)$ is an equivalence in Set_Δ.

Since the construction of derived stacks is intimately linked to that of sheaves of sets on k-CAlg in this work, we also give their definition as given in [79]. We regard sets as constant simplicial sets. A functor $F : k\text{-CAlg} \to \text{Set}$ is a sheaf if

- For any equivalence $A \to B$ in k-CAlg, the induced morphism $F(A) \to F(B)$ is an equivalence in Set_Δ; and
- For any finite family $\{A_i\}_{i\in I}$ in k-CAlg, the natural morphism

  $$F(\prod A_i) \to \prod_{i\in I} F(A_i)$$

  is an equivalence in Set_Δ.
- For any cosimplicial object $A \to B_*$ in k-CAlg corresponding to an ét-hypercover $\operatorname{Spec} B_* \to \operatorname{Spec} A$ in k-Aff, the induced morphism $F(A) \to \operatorname{holim}_{n\in\Delta} F(B_n)$ is an equivalence in Set_Δ.

20.3 Statement of the theorem and proof

Theorem 20.3.1. *A higher motif* dSt *as defined by the commutative diagram* (20.3) *is a motif for derived stacks.*

There are three points to be checked, and each will be the subject of a subsection.

20.3.1 dSt preserves equivalences

Suppose $sA \to sB$ is an equivalence in sk-CAlg for $A, B \in k$-CAlg. If we denote equivalences by $\sim$, owing to the fact that s is a functorial simplicial frame, we have

$$\begin{array}{ccc} \mathrm{cs}_*(A) & \xrightarrow{\sim} & sA \\ \downarrow & & \downarrow{\scriptstyle\sim} \\ \mathrm{cs}_*(B) & \xrightarrow[\sim]{} & sB. \end{array}$$

$\mathrm{cs}_*(A) \to \mathrm{cs}_*(B)$ is an equivalence by using the 2–3 property twice in the above diagram, so componentwise this gives an equivalence $A \to B$ in k-CAlg. Now dSt is built from a motif of sheaves Sh, which is a sheaf, and the sheaf condition $A \xrightarrow{\sim} B$ implies $\mathrm{Sh}(A) \xrightarrow{\sim} \mathrm{Sh}(B)$ in Set. Since each of $\mathrm{Sh}(A)$ and $\mathrm{Sh}(B)$ is a set, regarded as a constant simplicial set in Set_Δ, this reads $\mathrm{cs}_*\mathrm{Sh}(A) \xrightarrow{\sim} \mathrm{cs}_*\mathrm{Sh}(B)$. Since s is a functorial simplicial frame motif, we have a commutative diagram

$$\begin{array}{ccc} \mathrm{cs}_* \mathrm{Sh}(A) & \xrightarrow{\sim} & s(\mathrm{Sh}(A)) \\ \sim \downarrow & & \downarrow p \\ \mathrm{cs}_* \mathrm{Sh}(B) & \xrightarrow[\sim]{} & s(\mathrm{Sh}(B)) \end{array}$$

and by using the 2–3 property twice in this diagram we have that p is a weak equivalence. Finally, by the commutativity of (20.3), $\mathrm{dSt}(sA) = s(\mathrm{Sh}(A)) \xrightarrow{\sim} s(\mathrm{Sh}(B)) = \mathrm{dSt}(sB)$, so dSt does preserve equivalences.

20.3.2 dSt satisfies hyperdescent

It is important to emphasize that dSt is a motif for stacks generated by s, so it suffices to check the hypercover assumption on objects obtained from k-CAlg using that particular functor s. Thus we consider a coaugmented cosimplicial object $sA \to sB_*$ in sk-CAlg corresponding to an ét-hypercover $\mathrm{Spec}\,(sB_*) \to \mathrm{Spec}\,(sA)$ in $\mathrm{d}k\text{-Aff} = sk\text{-CAlg}^{\mathrm{op}}$. We have to show $\mathrm{dSt}(sA) \xrightarrow{\sim} \mathrm{holim}_{n\in\Delta} \mathrm{dSt}(sB_n)$ in Set_Δ. Regarding hypercovers, recall from [78] that the étale topology on dk-Aff induces étale hypercoverings: if Spec sA is an object of the site (dk-Aff, ét), then a homotopy ét-hypercover of Spec sA is given by some Spec sB_* in Ho (sd$k$-Aff) along with a morphism Spec $sB_* \to$ Spec sA in Ho (sd$k$-Aff) such that for all $n \geq 0$,

$$\mathrm{Spec}\, sB_*^{\mathbb{R}\Delta^n} \to \mathrm{Spec}\, sB_*^{\mathbb{R}\partial\Delta^n} \times^h_{\mathrm{Spec}\, sA^{\mathbb{R}\partial\Delta^n}} \mathrm{Spec}\, sA^{\mathbb{R}\Delta^n}$$

is an ét-covering in Ho (dk-Aff). To understand what this means concretely, by [74] a finite family $\{f_i : sA \to sB_i\}_{i\in I}$ in $sk\text{-CAlg} = \mathrm{d}k\text{-Aff}^{\mathrm{op}}$ is an ét-covering if and only if there is a finite subset J of I such that

$$\coprod_{j\in J} \mathrm{Spec}\, \pi_0(sB_j) \to \mathrm{Spec}\, \pi_0(sA)$$

is étale and surjective, and for all $i \in I$,

$$\pi_*(sA) \otimes_{\pi_0(sA)} \pi_0(sB_i) \to \pi_*(sB_i)$$

is an isomorphism.

We first have to show that given the conditions on $sA \to sB_*$ above, $A \to B_*$ is a coaugmented cosimplicial object in k-CAlg corresponding to an ét-hypercovering $\mathrm{Spec}\, B_* \to \mathrm{Spec}\, A$ in k-Aff. The idea is that we are projecting down to the level of sheaves the descent condition, and we eventually use the functoriality of s to conclude that the motif dSt satisfies hyperdescent as well.

20.3.2.1 Spec B_* $\to$ Spec A is an ét-hypercovering in k-Aff

The aim of this subsection is to show that for all $n \geq 0$,

$$\operatorname{Spec} B_*^{\mathbb{R}\Delta^n} \to \operatorname{Spec} B_*^{\mathbb{R}\partial\Delta^n} \times^h_{\operatorname{Spec} A^{\mathbb{R}\partial\Delta^n}} \operatorname{Spec} A^{\mathbb{R}\Delta^n} \tag{20.4}$$

is an ét-covering in Ho (k-Aff).

To achieve this, we will start from the definition of an ét-hypercovering in dk-Aff and use the fact that s is a motif for a simplicial frame functor, which will allow us to equivalently work with expressions using the constant simplicial functor in lieu of s, and in particular this will provide us with

$$(\mathrm{cs}_*(B_*))^{\mathrm{op},\,\mathbb{R}\Delta^n} \to (\mathrm{cs}_*(B_*))^{\mathrm{op},\mathbb{R}\partial\Delta^n} \times^h_{(\mathrm{cs}_*A)^{\mathrm{op},\mathbb{R}\partial\Delta^n}} (\mathrm{cs}_*A)^{\mathrm{op},\,\mathbb{R}\Delta^n}$$

an ét-covering in Ho (dk-Aff), at which point we will then project down levelwise to the desired statement (20.4). For simplicity, we first rewrite

$$\operatorname{Spec} sB_*^{\mathbb{R}\Delta^n} \to \operatorname{Spec} sB_*^{\mathbb{R}\partial\Delta^n} \times^h_{\operatorname{Spec} sA^{\mathbb{R}\partial\Delta^n}} \operatorname{Spec} sA^{\mathbb{R}\Delta^n}$$

as

$$(sB_*)^{\mathrm{op},\,\mathbb{R}\Delta^n} \to (sB_*)^{\mathrm{op},\mathbb{R}\partial\Delta^n} \times^h_{(sA)^{\mathrm{op},\mathbb{R}\partial\Delta^n}} (sA)^{\mathrm{op},\,\mathbb{R}\Delta^n}. \tag{20.5}$$

From [76] sd$k$-Aff being a simplicial model category, it is tensored and cotensored over Set_Δ, and hence for $F_* \in$ sd$k$-Aff, since $\Delta^n \in \mathrm{Set}_\Delta$, we have a well-defined product $\underline{\Delta}^n \otimes F_*$, and with $(sB_*)^{\mathrm{op}} \in$ sd$k$-Aff, by adjunction we have

$$\operatorname{Hom}(\underline{\Delta}^n \otimes F_*, (sB_*)^{\mathrm{op}}) \cong \operatorname{Hom}(F_*, (sB_*)^{\mathrm{op},\underline{\Delta}^n})$$

and

$$\operatorname{Hom}(\underline{\Delta}^n \otimes F_*, (\mathrm{cs}_*(B_*))^{\mathrm{op}}) \cong \operatorname{Hom}(F_*, (\mathrm{cs}_*(B_*))^{\mathrm{op},\underline{\Delta}^n}).$$

Since the exponential map is natural in both arguments, having a map $(sB_*)^{\mathrm{op}} \to (\mathrm{cs}_*(B_*))^{\mathrm{op}}$, we have maps

$$\begin{array}{ccc}
(sB_*)^{\mathrm{op},\underline{\Delta}^n} & \xrightarrow{\phi} & (\mathrm{cs}_*(B_*))^{\mathrm{op},\underline{\Delta}^n} \\
\downarrow{\scriptstyle \partial_0} & & \downarrow{\scriptstyle \partial_0} \\
(sB_*)^{\mathrm{op},\Delta^n} & \xrightarrow{\phi_0} & (\mathrm{cs}_*(B_*))^{\mathrm{op},\Delta^n} \\
\downarrow{\scriptstyle \gamma} & & \downarrow{\scriptstyle \gamma} \\
(sB_*)^{\mathrm{op},\,\mathbb{R}\Delta^n} & \xrightarrow[\gamma\phi_0]{} & (\mathrm{cs}_*(B_*))^{\mathrm{op},\,\mathbb{R}\Delta^n}
\end{array} \tag{20.6}$$

where ∂_0 is the degree zero map with the map ϕ_0 on degree zero elements induced by ϕ, and γ is the canonical functor from a given category to its homotopy category. In the same manner, we have maps

$$(sB_*)^{\mathrm{op},\mathbb{R}\partial\Delta^n} \to (\mathrm{cs}_*(B_*))^{\mathrm{op},\mathbb{R}\partial\Delta^n} \tag{20.7}$$

$$(sA)^{\mathrm{op},\,\mathbb{R}\Delta^n} \to (\mathrm{cs}_*A)^{\mathrm{op},\,\mathbb{R}\Delta^n} \tag{20.8}$$

and

$$(sA)^{\mathrm{op},\mathbb{R}\partial\Delta^n} \to (\mathrm{cs}_*A)^{\mathrm{op},\mathbb{R}\partial\Delta^n} \tag{20.9}$$

which induce a map of homotopy fiber products

$$\begin{array}{c} (sB_*)^{\mathrm{op},\mathbb{R}\partial\Delta^n} \times^h_{(sA)^{\mathrm{op},\mathbb{R}\partial\Delta^n}} (sA)^{\mathrm{op},\,\mathbb{R}\Delta^n} \\ \downarrow \\ (\mathrm{cs}_*(B_*))^{\mathrm{op},\mathbb{R}\partial\Delta^n} \times^h_{(\mathrm{cs}_*A)^{\mathrm{op},\mathbb{R}\partial\Delta^n}} (\mathrm{cs}_*A)^{\mathrm{op},\,\mathbb{R}\Delta^n} \end{array} \tag{20.10}$$

which, when combined with the bottom horizontal map of (20.6) and (20.5), yields

$$\begin{array}{ccc} (sB_*)^{\mathrm{op},\,\mathbb{R}\Delta^n} & \longrightarrow & (sB_*)^{\mathrm{op},\mathbb{R}\partial\Delta^n} \times^h_{(sA)^{\mathrm{op},\mathbb{R}\partial\Delta^n}} (sA)^{\mathrm{op},\,\mathbb{R}\Delta^n} \\ \downarrow & & \downarrow \\ (\mathrm{cs}_*(B_*))^{\mathrm{op},\,\mathbb{R}\Delta^n} & & (\mathrm{cs}_*(B_*))^{\mathrm{op},\mathbb{R}\partial\Delta^n} \times^h_{(\mathrm{cs}_*A)^{\mathrm{op},\mathbb{R}\partial\Delta^n}} (\mathrm{cs}_*A)^{\mathrm{op},\,\mathbb{R}\Delta^n}. \end{array} \tag{20.11}$$

The top horizontal map is an ét-covering in Ho (dk-Aff) by the definition of ét-hyper-covering. We would like to fill the bottom map and show that it is an ét-covering as well. To do this, we will show that both vertical maps are equivalences, which will give us a bottom horizontal map since we work in Ho (dk-Aff). This will also tell us that this map gives us an ét-covering. Firstly, to show the vertical maps above are weak equivalences, we will need all maps in (20.6), (20.7), (20.8), and (20.9) to be equivalences, so we prove the more general fact.

Lemma 20.3.2.1.1. *For $K = \Delta^n$ or $K = \partial\Delta^n$ and $A \in$ csk-CAlg, the map $(sA)^{\mathrm{op},\mathbb{R}K} \to (\mathrm{cs}_*A)^{\mathrm{op},\mathbb{R}K}$ is a weak equivalence.*

Proof. We use the fact that $X_*^{\mathbb{R}K} = (X_*^{\mathbb{R}\underline{K}})_0$, as shown in [79] by first taking a fibrant replacement of X_*, followed by taking the exponential by $\underline{K}$, and then taking the degree zero component. We will use the following fact from [40]: sd$k$-Aff being a simplicial model category, for any fibrant objects X and Y in sd$k$-Aff, $g : X \to Y$ is an equivalence if and only if for any cofibrant Z in sd$k$-Aff, we have $\mathrm{Hom}(Z,X) \simeq \mathrm{Hom}(Z,Y)$ in Set_Δ. Thus we fix some cofibrant object F_* in sd$k$-Aff and consider

$$\begin{aligned}\operatorname{Hom}(F_*, (sA)^{\mathrm{op},\mathbb{R}\underline{K}}) &= \operatorname{Hom}(QF_*, (sA)^{\mathrm{op},\mathbb{R}\underline{K}})\\ &\simeq \operatorname{Hom}(\underline{K}\otimes QF_*, R((sA)^{\mathrm{op}}))\\ &= \operatorname{Hom}(Q(\underline{K}\otimes F_*), R((sA)^{\mathrm{op}}))\end{aligned} \tag{20.12}$$

Now again we can invoke that same result from [40] for $R((sA)^{\mathrm{op}}) \to R((\mathrm{cs}_*A)^{\mathrm{op}})$ since in the diagram below, vertical maps are trivial cofibrations, the top horizontal map is a weak equivalence by definition of s, so by the 2–3 property applied twice, the bottom horizontal map is a weak equivalence as well:

$$\begin{array}{ccc} (sA)^{\mathrm{op}} & \longrightarrow & (\mathrm{cs}_*A)^{\mathrm{op}} \\ \downarrow & & \downarrow \\ R((sA)^{\mathrm{op}}) & \longrightarrow & R((\mathrm{cs}_*A)^{\mathrm{op}}). \end{array}$$

Being an equivalence from (20.12) we have

$$\begin{aligned}\operatorname{Hom}(F_*, (sA)^{\mathrm{op},\mathbb{R}\underline{K}}) &\simeq \operatorname{Hom}(Q(\underline{K}\otimes F_*), R((sA)^{\mathrm{op}}))\\ &\simeq \operatorname{Hom}(Q(\underline{K}\otimes F_*), R((\mathrm{cs}_*A)^{\mathrm{op}}))\\ &\simeq \operatorname{Hom}(F_*, (\mathrm{cs}_*A)^{\mathrm{op},\mathbb{R}\underline{K}})\end{aligned}$$

and hence $(sA)^{\mathrm{op},\mathbb{R}\underline{K}} \to (\mathrm{cs}_*A)^{\mathrm{op},\mathbb{R}\underline{K}}$ is a weak equivalence, more precisely, a Reedy equivalence in the Reedy model structure for sdk-Aff. Those are levelwise equivalences, so its degree zero component $(sA)^{\mathrm{op},\mathbb{R}K} \to (\mathrm{cs}_*A)^{\mathrm{op},\mathbb{R}K}$ is an equivalence as well. □

Corollary 20.3.2.1.2. *The maps* (20.6), (20.7), (20.8), *and* (20.9) *are all weak equivalences.*

Proof. For the first two equations, it is immediate by just taking $A = B_*$ in the previous lemma. For the other two equations, $A \in k$-CAlg gives $sA \in sk$-CAlg, and hence $(sA)^{\mathrm{op}} \in dk$-Aff, regarded as a constant simplicial object $\mathrm{cs}_*((sA)^{\mathrm{op}}) \in$ sdk-Aff. Since $(sA)^{\mathrm{op}} \xrightarrow{\sim} (\mathrm{cs}_*A)^{\mathrm{op}}$, we also have $\mathrm{cs}_*((sA)^{\mathrm{op}}) \xrightarrow{\sim} \mathrm{cs}_*((\mathrm{cs}_*A)^{\mathrm{op}})$ since equivalences in the Reedy model structure of sdk-Aff are levelwise equivalences. Now for $K = \Delta^n$ or $K = \partial\Delta^n$, following the same reasoning as in the proof of the previous lemma, we find that $(\mathrm{cs}_*((sA)^{\mathrm{op}}))^{\mathbb{R}K} \to (\mathrm{cs}_*((\mathrm{cs}_*A)^{\mathrm{op}}))^{\mathbb{R}K}$ is a weak equivalence, and in a simplified notation, this reads $(sA)^{\mathrm{op},\mathbb{R}K} \to (\mathrm{cs}_*A)^{\mathrm{op},\mathbb{R}K}$ is an equivalence. □

As a consequence of having those equivalences, we prove that the fiber products in (20.10) are equivalent:

Lemma 20.3.2.1.3. *The map of homotopy fiber products*

$$\begin{array}{c} (sB_*)^{\mathrm{op},\mathbb{R}\partial\Delta^n} \times^h_{(sA)^{\mathrm{op},\mathbb{R}\partial\Delta^n}} (sA)^{\mathrm{op},\,\mathbb{R}\Delta^n} \\ \downarrow \\ (\mathrm{cs}_*(B_*))^{\mathrm{op},\mathbb{R}\partial\Delta^n} \times^h_{(\mathrm{cs}_*A)^{\mathrm{op},\mathbb{R}\partial\Delta^n}} (\mathrm{cs}_*A)^{\mathrm{op},\,\mathbb{R}\Delta^n} \end{array} \tag{20.13}$$

is a weak equivalence.

Proof. We are looking at the following diagram:

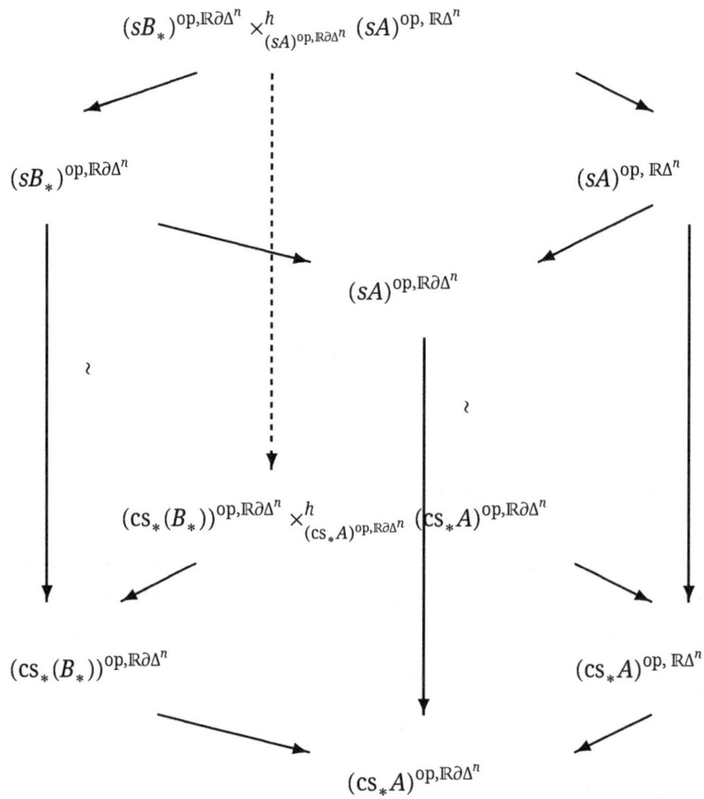

To prove that the vertical map in the back is an equivalence, we will use Proposition 15.10.10 from [40]: in a model category in which we have a diagram such as

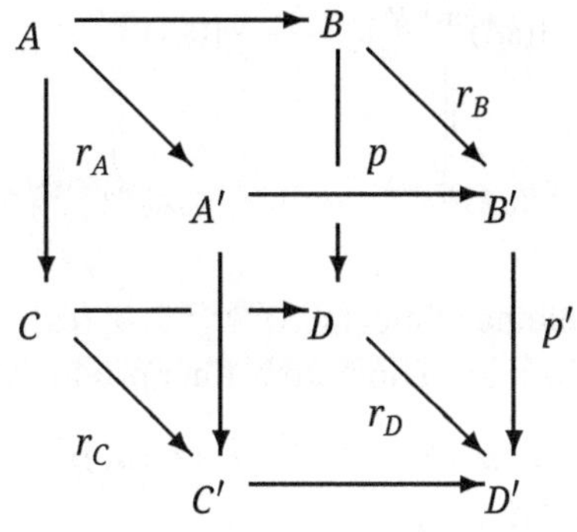

in which all objects are fibrant, the squares in the front and in the back are pullbacks, p and p' are fibrations, and r_B, r_C, and r_D are equivalences, then so is r_A. We apply this to our setting, where our initial model category is sd$k$-Aff, in which we took fibrant replacements of sA, sB_*, cs_*A, and cs_*B_* already. By Proposition 9.3.9 of [40] since sd$k$-Aff is a simplicial model category, the exponentials of such objects are fibrant as well, and so are their zero components by Proposition 15.3.11 of [40], so all objects are fibrant as needed. The pullback squares are given by the homotopy fiber products of the previous diagram, the maps r_B, r_C, and r_D are the vertical equivalences in that diagram, so r_A corresponds to the map of homotopy fiber products. We take for p and p' the following maps:

$$(sA)^{\mathrm{op},\,\mathbb{R}\Delta^n} \to (sA)^{\mathrm{op},\mathbb{R}\partial\Delta^n}$$

and

$$(\mathrm{cs}_*A)^{\mathrm{op},\,\mathbb{R}\Delta^n} \to (\mathrm{cs}_*A)^{\mathrm{op},\mathbb{R}\partial\Delta^n}.$$

We will use the following fact 9.3.9 2b) from [40]: if X is fibrant in a simplicial model category and $j : K \to L$ is an inclusion in Set_Δ, then $X^L \to X^K$ is a fibration. Applying this to the first map above for example, one gets that

$$\begin{array}{ccc} (R(sA)^{\mathrm{op}})^{\underline{\Delta}^n} & \longrightarrow & (R(sA)^{\mathrm{op}})^{\underline{\partial\Delta}^n} \\ \Vert & & \Vert \\ (sA)^{\mathrm{op},\mathbb{R}\underline{\Delta}^n} & & (sA)^{\mathrm{op},\mathbb{R}\underline{\partial\Delta}^n} \end{array}$$

is a fibration, more precisely, a Reedy fibration, and hence by Proposition 15.3.11 of [40] the degree zero component is a fibration as well:

$$\begin{array}{ccc}
[(sA)^{\mathrm{op},\mathbb{R}\underline{\Delta}^n}]_0 & \longrightarrow & [(sA)^{\mathrm{op},\mathbb{R}\partial\underline{\Delta}^n}]_0 \\
\| & & \| \\
(sA)^{\mathrm{op},\,\mathbb{R}\Delta^n} & & (sA)^{\mathrm{op},\mathbb{R}\partial\Delta^n}.
\end{array}$$

One shows in the same manner that $(\mathrm{cs}_*A)^{\mathrm{op},\,\mathbb{R}\Delta^n} \to (\mathrm{cs}_*A)^{\mathrm{op},\mathbb{R}\partial\Delta^n}$ is a fibration. This completes the proof that the map of homotopy fiber products is a weak equivalence. □

Lemma 20.3.2.1.4. *The map* $(\mathrm{cs}_*(B_*))^{\mathrm{op},\,\mathbb{R}\Delta^n} \to (\mathrm{cs}_*(B_*))^{\mathrm{op},\mathbb{R}\partial\Delta^n} \times^h_{(\mathrm{cs}_*A)^{\mathrm{op},\mathbb{R}\partial\Delta^n}} (\mathrm{cs}_*A)^{\mathrm{op},\,\mathbb{R}\Delta^n}$ *is an ét-covering in* Ho (dk-Aff).

Proof. The previous lemma shows that the right vertical map in the diagram

$$\begin{array}{ccc}
(sB_*)^{\mathrm{op},\,\mathbb{R}\Delta^n} & \longrightarrow & (sB_*)^{\mathrm{op},\mathbb{R}\partial\Delta^n} \times^h_{(sA)^{\mathrm{op},\mathbb{R}\partial\Delta^n}} (sA)^{\mathrm{op},\,\mathbb{R}\Delta^n} \\
\downarrow & & \downarrow \\
(\mathrm{cs}_*(B_*))^{\mathrm{op},\,\mathbb{R}\Delta^n} & \dashrightarrow & (\mathrm{cs}_*(B_*))^{\mathrm{op},\mathbb{R}\partial\Delta^n} \times^h_{(\mathrm{cs}_*A)^{\mathrm{op},\mathbb{R}\partial\Delta^n}} (\mathrm{cs}_*A)^{\mathrm{op},\,\mathbb{R}\Delta^n}
\end{array}$$

is an equivalence. The vertical map on the left has already been shown to be an equivalence as well. Hence in Ho (dk-Aff), we have a bottom horizontal map as shown in the diagram above. We show this is an ét-covering. From [79] the definition of coverings uses the higher homotopy groups π_i for $i \geq 0$. Recall that for a simplicial model category $\mathcal{C}$ and $A \in \mathcal{C}$, $|A| = \mathrm{Map}_{\mathcal{C}}(1, A) \in \mathrm{Ho}\,(\mathrm{Set}_\Delta)$. If $\mathcal{C}$ is pointed, then $|A|$ has a natural basepoint, and one can define $\pi_i(A) = \pi_i(|A|, *)$. Here $\mathrm{Map}_{\mathcal{C}}$ is the simplicial hom in the simplicial model category $\mathcal{C}$, which for us we take to be sk-CAlg. By Corollary 9.3.3 of [40] since 1 is cofibrant, if $A \xrightarrow{\sim} B$ is an equivalence of fibrant objects, then $\mathrm{Map}_{\mathcal{C}}(1, A) \to \mathrm{Map}_{\mathcal{C}}(1, B)$ is an equivalence in Set_Δ, i. e., $|A| \xrightarrow{\sim} |B|$, and by the definition of higher homotopy groups on sk-CAlg we conclude that being an ét-covering is an invariant on weak equivalence classes, whence the result. □

We finally prove the following:

Lemma 20.3.2.1.5. *The map* $B_*^{\mathrm{op},\,\mathbb{R}\Delta^n} \to B_*^{\mathrm{op},\mathbb{R}\partial\Delta^n} \times^h_{A^{\mathrm{op},\mathbb{R}\partial\Delta^n}} A^{\mathrm{op},\,\mathbb{R}\Delta^n}$ *gives an ét-covering in* Ho (k-Aff).

We start from

$$\begin{array}{ccc}
(\mathrm{cs}_*(B_*))^{\mathrm{op},\,\mathbb{R}\Delta^n} & \longrightarrow & (\mathrm{cs}_*(B_*))^{\mathrm{op},\mathbb{R}\partial\Delta^n} \times^h_{(\mathrm{cs}_*A)^{\mathrm{op},\mathbb{R}\partial\Delta^n}} (\mathrm{cs}_*A)^{\mathrm{op},\,\mathbb{R}\Delta^n} \\
\downarrow & & \downarrow \\
B_*^{\mathrm{op},\,\mathbb{R}\Delta^n} & & B_*^{\mathrm{op},\mathbb{R}\partial\Delta^n} \times^h_{A^{\mathrm{op},\mathbb{R}\partial\Delta^n}} A^{\mathrm{op},\,\mathbb{R}\Delta^n}.
\end{array}$$

We will show there is a bottom horizontal map making this diagram commutative, which in addition provides an ét-covering in Ho (k-Aff). First, we show that $(\mathrm{cs}_*(B_*))^{\mathrm{op},\mathbb{R}K} \cong \mathrm{cc}_*((B_*)^{\mathrm{op},\mathbb{R}K})$ in Ho (sd$k$-Aff) for $K = \Delta^n$ or $K = \partial\Delta^n$, where cc_* denotes the constant cosimplicial functor. Note that this isomorphism still holds if we considered A instead of B_*. This allows us to rewrite the top horizontal map above using the constant cosimplicial functor cc_*. Then we will argue that

$$\mathrm{cc}_*(B_*^{\mathrm{op},\mathbb{R}\partial\Delta^n}) \times^h_{\mathrm{cc}_*(A^{\mathrm{op},\mathbb{R}\partial\Delta^n})} \mathrm{cc}_*(A^{\mathrm{op},\,\mathbb{R}\Delta^n}) \cong \mathrm{cc}_*(B_*^{\mathrm{op},\mathbb{R}\partial\Delta^n} \times^h_{A^{\mathrm{op},\mathbb{R}\partial\Delta^n}} A^{\mathrm{op},\,\mathbb{R}\Delta^n})$$

and finally using the constant nature of the constant cosimplicial functor cc_*, we will show we have an ét-covering in Ho (k-Aff) as desired. First, we need the following:

Lemma 20.3.2.1.6. $(\mathrm{cs}_*(B_*))^{\mathrm{op},\mathbb{R}K} \cong \mathrm{cc}_*(B_*^{\mathrm{op},\mathbb{R}K})$ *in* Ho (sd$k$-Aff) *for* $K = \Delta^n$ *or* $K = \partial\Delta^n$.

Proof. First, observe that

$$\begin{aligned}(\mathrm{cs}_*(B_*))^{\mathrm{op},\mathbb{R}K} &= \{(R([\mathrm{cs}_*(B_*)]^{\mathrm{op}}))^{\underline{K}}\}_0 \\ &= \{(R(\mathrm{cc}_*(B_*^{\mathrm{op}})))^{\underline{K}}\}_0 \\ &\cong ((\mathrm{cc}_*(RB_*^{\mathrm{op}}))^{\underline{K}})_0 \end{aligned} \tag{20.14}$$

where the isomorphism in Ho (sd$k$-Aff) is derived as follows: we have a trivial cofibration $B_*^{\mathrm{op}} \to RB_*^{\mathrm{op}}$. This equivalence is preserved by the constant cosimplicial functor $\mathrm{cc}_*(B_*^{\mathrm{op}}) \xrightarrow{\sim} \mathrm{cc}_*(RB_*^{\mathrm{op}})$. On the other hand, we also have a trivial cofibration in sd$k$-Aff: $\mathrm{cc}_*(B_*^{\mathrm{op}}) \xrightarrow{\sim} R\mathrm{cc}_*(B_*^{\mathrm{op}})$. It follows that we have an isomorphism $R\mathrm{cc}_*(B_*^{\mathrm{op}}) \cong \mathrm{cc}_*(RB_*^{\mathrm{op}})$ in Ho (sd$k$-Aff), which gives us the isomorphism above. On the other hand, we have

$$\begin{aligned}\mathrm{cc}_*(B_*^{\mathrm{op},\mathbb{R}K}) &= \mathrm{cc}_*((RB_*^{\mathrm{op}})^K) \\ &= \mathrm{cc}_*(((RB_*^{\mathrm{op}})^{\underline{K}})_0)\end{aligned}$$

where the zeroth component is taken in the simplicial model structure of sk-Aff. To emphasize this, we can write this last object as $\mathrm{cc}_*(((RB_*^{\mathrm{op}})^{\underline{K}})_{0,-})$ for later purposes. Since this zeroth component is taken relative to B_*, we can write this as $(\mathrm{cc}_*((RB_*^{\mathrm{op}})^{\underline{K}}))_{0,-}$. It is the same zeroth component taken in (20.14). It follows that we have the desired result if we can show that

$$(\mathrm{cc}_*(RB_*^{\mathrm{op}}))^{\underline{K}} \cong \mathrm{cc}_*((RB_*^{\mathrm{op}})^{\underline{K}})$$

in Ho (sd$k$-Aff), where the exponentiation on the left is taken in the simplicial model structure of sd$k$-Aff and the one on the right with respect to the simplicial model structure of sk-Aff. We will show we have an equivalence $(\mathrm{cc}_*(RB_*^{\mathrm{op}}))^{\underline{K}} \simeq \mathrm{cc}_*((RB_*^{\mathrm{op}})^{\underline{K}})$ in sd$k$-Aff. For F_* cofibrant in sd$k$-Aff, we have

$$\mathrm{Hom}_{\mathrm{sd}k\text{-}\mathrm{Aff}}(F_*, (\mathrm{cc}_*(RB_*^{\mathrm{op}}))^{\underline{K}}) \simeq \mathrm{Hom}(F_{*,*} \underset{\mathrm{sd}k\text{-}\mathrm{Aff}}{\otimes} \underline{K}, \mathrm{cc}_*(RB_*^{\mathrm{op}}))$$

where the first index for F corresponds to the simplicial structure of sd$k$-Aff, and the second index corresponds to the cosimplicial structure of dk-Aff $=$ (sk-CAlg)^{op} $=$ cs(k-CAlg)^{op} $=$ cs$k$-Aff. Now observe that

$$F_{*,*} \underset{\mathrm{sd}k\text{-Aff}}{\otimes} \underline{K} = \coprod_{\sigma_n \in K_n} F_{n,*} = \coprod_p \Big(\coprod_{\sigma_n} F_{n,p} \Big) = \coprod_p F_{*,p} \underset{\mathrm{s}k\text{-Aff}}{\otimes} \underline{K}.$$

It follows that

$$\begin{aligned}
&\mathrm{Hom}_{\mathrm{sd}k\text{-Aff}}(F_{*,*} \otimes \underline{K}, \mathrm{cc}_*(RB_*^{\mathrm{op}})) \\
&\quad = \mathrm{Hom}\Big(\coprod_p F_{*,p} \underset{\mathrm{s}k\text{-Aff}}{\otimes} \underline{K}, \coprod_p \mathrm{cc}_p(RB_*^{\mathrm{op}})\Big) \\
&\quad = \coprod_p \mathrm{Hom}(F_{*,p} \otimes \underline{K}, RB_*^{\mathrm{op}}) \\
&\quad \simeq \coprod_p \mathrm{Hom}_{\mathrm{s}k\text{-Aff}}(F_{*,p}, (RB_*^{\mathrm{op}})^{\underline{K}}) \\
&\quad = \coprod_p \mathrm{Hom}(F_{*,p}, \mathrm{cc}_p(RB_*^{\mathrm{op},\underline{K}})) \\
&\quad = \mathrm{Hom}_{\mathrm{sd}k\text{-Aff}}(F_{*}, \mathrm{cc}_*(RB_*^{\mathrm{op},\underline{K}})).
\end{aligned}$$

□

By construction we formally have the same result if we use A instead of B_*.

We now show the following:

Lemma 20.3.2.1.7.

$$\mathrm{cc}_*(B_*^{\mathrm{op},\mathbb{R}\partial\Delta^n}) \times^h_{\mathrm{cc}_*(A^{\mathrm{op},\mathbb{R}\partial\Delta^n})} \mathrm{cc}_*(A^{\mathrm{op},\,\mathbb{R}\Delta^n}) \cong \mathrm{cc}_*(B_*^{\mathrm{op},\mathbb{R}\partial\Delta^n} \times^h_{A^{\mathrm{op},\mathbb{R}\partial\Delta^n}} A^{\mathrm{op},\,\mathbb{R}\Delta^n})$$

or in other terms, the functor cc$_*$ *preserves homotopy fiber products.*

Proof. We first focus on the homotopy fiber product on the left above. Following [40], it is defined by using a functorial factorization which we will denote by E: $A \times^h_C B = E(A) \times_C E(B)$. By the definition of the constant cosimplicial functor cc$_*$, a map $\mathrm{cc}_*(A) \to \mathrm{cc}_*(B)$ is given by a map $A \to B$, which is the same in all degrees hence can be denoted by $\mathrm{cc}_*(A \to B)$. By the definition of the functorial factorization, $E(\mathrm{cc}_* X) \cong \mathrm{cc}_*(EX)$ in Ho (cs$\mathcal{C}$) for $X \in \mathcal{C}$, which can be shown in the same fashion that we have shown $\mathrm{cc}_*(RB_*^{\mathrm{op}}) \cong R\mathrm{cc}_*(B_*^{\mathrm{op}})$, and hence we have a map $\mathrm{cc}_* A \leftarrow E(\mathrm{cc}_* A) \cong \mathrm{cc}_*(EA)$, which can be denoted $\mathrm{cc}_*(A \leftarrow E(A))$. It follows that

$$\begin{aligned}
&\mathrm{cc}_*(B_*^{\mathrm{op},\mathbb{R}\partial\Delta^n}) \times^h_{\mathrm{cc}_*(A^{\mathrm{op},\mathbb{R}\partial\Delta^n})} \mathrm{cc}_*(A^{\mathrm{op},\,\mathbb{R}\Delta^n}) \\
&= E(\mathrm{cc}_*(B_*^{\mathrm{op},\mathbb{R}\partial\Delta^n})) \times_{\mathrm{cc}_*(A^{\mathrm{op},\mathbb{R}\partial\Delta^n})} E(\mathrm{cc}_*(A^{\mathrm{op},\,\mathbb{R}\Delta^n})) \\
&= \lim \Big(E(\mathrm{cc}_*(B_*^{\mathrm{op},\mathbb{R}\partial\Delta^n})) \to \mathrm{cc}_*(A^{\mathrm{op},\mathbb{R}\partial\Delta^n}) \leftarrow E(\mathrm{cc}_*(A^{\mathrm{op},\,\mathbb{R}\Delta^n})) \Big) \\
&\cong \lim \Big(\mathrm{cc}_* E(B_*^{\mathrm{op},\mathbb{R}\partial\Delta^n}) \to \mathrm{cc}_*(A^{\mathrm{op},\mathbb{R}\partial\Delta^n}) \leftarrow \mathrm{cc}_* E(A^{\mathrm{op},\,\mathbb{R}\Delta^n}) \Big) \\
&= \lim \mathrm{cc}_*(E(B_*^{\mathrm{op},\mathbb{R}\partial\Delta^n}) \to A^{\mathrm{op},\mathbb{R}\partial\Delta^n} \leftarrow E(A^{\mathrm{op},\,\mathbb{R}\Delta^n}))
\end{aligned}$$

$$= \mathrm{cc}_*\big(\lim E(B_*^{\mathrm{op},\mathbb{R}\partial\Delta^n}) \to A^{\mathrm{op},\mathbb{R}\partial\Delta^n} \leftarrow E(A^{\mathrm{op},\,\mathbb{R}\Delta^n})\big)$$

since cc_* preserves limits as a right adjoint, and this equals $\mathrm{cc}_*(B_*^{\mathrm{op},\mathbb{R}\partial\Delta^n} \times^h_{A^{\mathrm{op},\mathbb{R}\partial\Delta^n}} A^{\mathrm{op},\,\mathbb{R}\Delta^n})$, as claimed. □

We can now complete the proof of Lemma 20.3.2.1.5: with the previous results the ét-covering in Ho (dk-Aff) given by Lemma 20.3.2.1.4 now reads

$$\mathrm{cc}_*(B_*^{\mathrm{op},\,\mathbb{R}\Delta^n}) \to \mathrm{cc}_*(B_*^{\mathrm{op},\mathbb{R}\partial\Delta^n} \times^h_{A^{\mathrm{op},\mathbb{R}\partial\Delta^n}} A^{\mathrm{op},\,\mathbb{R}\Delta^n}) \tag{20.15}$$

and again by the constant nature of cc_* this map being an ét-covering implies that

$$B_*^{\mathrm{op},\,\mathbb{R}\Delta^n} \to B_*^{\mathrm{op},\mathbb{R}\partial\Delta^n} \times^h_{A^{\mathrm{op},\mathbb{R}\partial\Delta^n}} A^{\mathrm{op},\,\mathbb{R}\Delta^n} \tag{20.16}$$

is an ét-covering in Ho (k-Aff), i. e., $A \to B_*$ corresponds to a homotopy ét-hypercovering in k-Aff.

This is not immediate, though fairly clear, so we flesh it out. To say that (20.15) is an étale covering means it is of the form $\{\mathrm{Spec}\, A_i \to \mathrm{Spec}\, A\}_{i\in I}$, a family for which there is a finite $J \subset I$ such that for all i, $\pi_* A \otimes_{\pi_0 A} \pi_0 A_i \to \pi_* A_i$ is an isomorphism, and $\coprod_{j\in J} \mathrm{Spec}\, \pi_0 A_i \to \mathrm{Spec}\, \pi_0 A$ is étale and surjective. It turns out here (20.15) is a one element family. To say (20.16) is an ét-covering in Ho (k-Aff) however means it is of the form $\{\mathrm{Spec}\, \underline{A}_i \to \mathrm{Spec}\, \underline{A}\}_{i\in I}$, and we have again a finite subset $J \subset I$ such that $\coprod_{j\in J} \mathrm{Spec}\, \underline{A}_j \to \mathrm{Spec}\, \underline{A}$ is faithfully flat and étale. We now show this.

We first write (20.15) in the form $\mathrm{Spec}\, \mathbf{B} \to \mathrm{Spec}\, \mathbf{A}$, where $\mathbf{A}$ and $\mathbf{B}$ are algebras. This means

$$\begin{aligned}\mathbf{B} &= \big(\mathrm{cc}_*(B_*^{\mathrm{op},\,\mathbb{R}\Delta^n})\big)^{\mathrm{op}}\\ &= \mathrm{cs}_*(B_*^{\mathrm{op},\,\mathbb{R}\Delta^n})^{\mathrm{op}}\end{aligned}$$

and

$$\begin{aligned}\mathbf{A} &= \big(\mathrm{cc}_*(B_*^{\mathrm{op},\mathbb{R}\partial\Delta^n} \times^h_{A^{\mathrm{op},\mathbb{R}\partial\Delta^n}} A^{\mathrm{op},\,\mathbb{R}\Delta^n})\big)^{\mathrm{op}}\\ &= \mathrm{cs}_*(B_*^{\mathrm{op},\mathbb{R}\partial\Delta^n} \times^h_{A^{\mathrm{op},\mathbb{R}\partial\Delta^n}} A^{\mathrm{op},\,\mathbb{R}\Delta^n})^{\mathrm{op}}.\end{aligned}$$

First, $\mathrm{Spec}\, \mathbf{B} \to \mathrm{Spec}\, \mathbf{A}$ being an ét-covering means $\pi_*\mathbf{A} \otimes_{\pi_0\mathbf{A}} \pi_*\mathbf{B} \to \pi_*\mathbf{B}$ is an isomorphism and $\mathrm{Spec}\, \pi_0\mathbf{B} \to \mathrm{Spec}\, \pi_0\mathbf{A}$ is étale and surjective. This means $\mathbf{A} \to \mathbf{B}$ is strongly étale by Definition 1.2.6.1 of [79], which in turns means $\mathbf{A} \to \mathbf{B}$ is étale by Theorem 2.2.2.6 of [79], which means flat and $\pi_0\mathbf{A} \to \pi_0\mathbf{B}$ étale by Corollary 2.2.2.11 of [79]. Being flat means $\mathbf{B} \otimes^{\mathbb{L}}_{\mathbf{A}} -$ is exact. However. both $\mathbf{A}$ and $\mathbf{B}$ are constant simplicial objects, so this means $\underline{\mathbf{B}} \otimes^{\mathbb{L}}_{\underline{\mathbf{A}}} -$ is exact (where $\mathbf{A} \in sC$ is generated by $\underline{\mathbf{A}} \in C$), that is, $\underline{\mathbf{A}} \to \underline{\mathbf{B}}$ is flat, and thus $\mathrm{Spec}\, \underline{\mathbf{B}} \to \mathrm{Spec}\, \underline{\mathbf{A}}$ is flat, where we used the notations

$$\underline{\mathbf{B}} = (B_*^{\mathrm{op},\,\mathbb{R}\Delta^n})^{\mathrm{op}},$$

$$\underline{\mathbf{A}} = (B_*^{\mathrm{op},\mathbb{R}\partial\Delta^n} \times^h_{A^{\mathrm{op},\mathbb{R}\partial\Delta^n}} A^{\mathrm{op},\,\mathbb{R}\Delta^n})^{\mathrm{op}}$$

that is, $B_*^{\mathrm{op},\,\mathbb{R}\Delta^n} \to B_*^{\mathrm{op},\mathbb{R}\partial\Delta^n} \times^h_{A^{\mathrm{op},\mathbb{R}\partial\Delta^n}} A^{\mathrm{op},\,\mathbb{R}\Delta^n}$ is flat. To show that Spec $\underline{\mathbf{B}} \to$ Spec $\underline{\mathbf{A}}$ is étale, that is, $B_*^{\mathrm{op},\,\mathbb{R}\Delta^n} \to B_*^{\mathrm{op},\mathbb{R}\partial\Delta^n} \times^h_{A^{\mathrm{op},\mathbb{R}\partial\Delta^n}} A^{\mathrm{op},\,\mathbb{R}\Delta^n}$ is étale, it suffices to observe Spec $\pi_0\mathbf{B} \to$ Spec $\pi_0\mathbf{A}$ is étale. This reads

$$\mathrm{Spec}\,\pi_0\mathrm{cs}_*(B_*^{\mathrm{op},\,\mathbb{R}\Delta^n})^{\mathrm{op}} \to \mathrm{Spec}\,\pi_0\mathrm{cs}_*(B_*^{\mathrm{op},\mathbb{R}\partial\Delta^n} \times^h_{A^{\mathrm{op},\mathbb{R}\partial\Delta^n}} A^{\mathrm{op},\,\mathbb{R}\Delta^n})^{\mathrm{op}}.$$

Now $\pi_0 \dashv \mathrm{cs}_*$, so this can be rewritten as

$$\mathrm{Spec}\,(B_*^{\mathrm{op},\,\mathbb{R}\Delta^n})^{\mathrm{op}} \to \mathrm{Spec}\,(B_*^{\mathrm{op},\mathbb{R}\partial\Delta^n} \times^h_{A^{\mathrm{op},\mathbb{R}\partial\Delta^n}} A^{\mathrm{op},\,\mathbb{R}\Delta^n})^{\mathrm{op}}$$

which simplifies to that

$$B_*^{\mathrm{op},\,\mathbb{R}\Delta^n} \to B_*^{\mathrm{op},\mathbb{R}\partial\Delta^n} \times^h_{A^{\mathrm{op},\mathbb{R}\partial\Delta^n}} A^{\mathrm{op},\,\mathbb{R}\Delta^n}$$

étale and flat so far. For the surjectivity, Spec $\pi_0\mathbf{B} \to$ Spec $\pi_0\mathbf{A}$ is surjective, and doing the same manipulation as above, this reads $B_*^{\mathrm{op},\,\mathbb{R}\Delta^n} \to B_*^{\mathrm{op},\mathbb{R}\partial\Delta^n} \times^h_{A^{\mathrm{op},\mathbb{R}\partial\Delta^n}} A^{\mathrm{op},\,\mathbb{R}\Delta^n}$ is surjective and so faithfully flat.

20.3.2.2 From Sh being a sheaf to dSt being a derived stack

Now that $A \to B_*$ corresponds to an ét-hypercovering in k-Aff, Sh being a sheaf, we obtain an equivalence in Set_Δ: $\mathrm{Sh}(A) \xrightarrow{\sim} \mathrm{holim}_{n\in\Delta}\,\mathrm{Sh}(B_n)$. Now by the definition of s we obtain the following commutative diagram:

$$\begin{array}{ccc}
\mathrm{cs}_*\,\mathrm{Sh}(A) & \xrightarrow{\sim} & \mathrm{cs}_*\,\mathrm{holim}_{n\in\Delta}\,\mathrm{Sh}(B_n) \\
\downarrow{\scriptstyle\sim} & & \downarrow{\scriptstyle\sim} \\
s(\mathrm{Sh}(A)) & \xrightarrow[\sim]{} & s(\mathrm{holim}_{n\in\Delta}\,\mathrm{Sh}(B_n)) \\
\| & & \| \\
\mathrm{dSt}(sA) & & \mathrm{holim}_{n\in\Delta}\,s(\mathrm{Sh}(B_n)) \\
 & & \| \\
 & & \mathrm{holim}\;\mathrm{dSt}(sB_n)
\end{array} \tag{20.17}$$

where $\mathrm{dSt}(sA) = s\mathrm{Sh}(A)$ by the commutativity of (20.3), $s(\mathrm{holim}\;\mathrm{Sh}(B_n)) = \mathrm{holim}\;s\mathrm{Sh}(B_n)$, since as shown in [40], homotopy limits are equalizers and hence limits, and s preserves finite limits by definition, and finally this last object is equal to $\mathrm{holim}\;\mathrm{dSt}(sB_n)$ again by the commutativity of (20.3). Further, Reedy equivalences being defined levelwise, $\mathrm{Sh}(A) \xrightarrow{\sim} \mathrm{holim}_{n\in\Delta}\,\mathrm{Sh}(B_n)$ implies $\mathrm{cs}_*\,\mathrm{Sh}(A) \xrightarrow{\sim} \mathrm{cs}_*\,\mathrm{holim}_{n\in\Delta}\,\mathrm{Sh}(B_n)$, and hence $s\mathrm{Sh}(A) \to s(\mathrm{holim}\;\mathrm{Sh}(B_n))$ is an equivalence by the 2–3 property, and it follows that $\mathrm{dSt}(sA) \xrightarrow{\sim} \mathrm{holim}\;\mathrm{dSt}(sB_n)$, that is, dSt satisfies descent.

20.3.3 dSt preserves finite products

We consider a finite family $\{sA_i\}$ in sk-CAlg. We need to show that the natural morphism

$$\mathrm{dSt}\Big(\prod sA_i\Big) \to \prod \mathrm{dSt}(sA_i)$$

is an equivalence in Set_Δ. We have $\mathrm{dSt}(\prod sA_i) = \mathrm{dSt}(s\prod A_i)$ since s preserves finite products. This latter object is equal to $s\mathrm{Sh}(\prod A_i)$ by the commutativity of (20.3). Since Sh is a sheaf, we have $\mathrm{Sh}(\prod A_i) \xrightarrow{\sim} \prod \mathrm{Sh}(A_i)$ in Set_Δ. Keeping in mind that we consider sets as constant simplicial sets in Set_Δ, this reads $\mathrm{cs}_*\mathrm{Sh}(\prod^h A_i) \xrightarrow{\sim} \mathrm{cs}_* \prod \mathrm{Sh}(A_i)$, so we have the following commutative diagram by virtue of the existence of the natural transformation j:

$$\begin{array}{ccc}
\mathrm{cs}_*\mathrm{Sh}(\prod A_i) & \xrightarrow{\sim} & \mathrm{cs}_* \prod \mathrm{Sh}(A_i) \\
\big\downarrow{\scriptstyle\sim} & & \big\downarrow{\scriptstyle\sim} \\
s\mathrm{Sh}(\prod A_i) & \xrightarrow[\sim]{} & s\prod \mathrm{Sh}(A_i) \\
\| & & \| \\
\mathrm{dSt}(s\prod A_i) & & \prod s(\mathrm{Sh}(A_i)) \\
\| & & \| \\
\mathrm{dSt}(\prod sA_i) & & \prod \mathrm{dSt}(sA_i)
\end{array}$$

where we used the 2–3 property to have an equivalence $s(\mathrm{Sh}(\prod A_i)) \to s\prod \mathrm{Sh}(A_i)$, and hence $\mathrm{dSt}(\prod sA_i) \xrightarrow{\sim} \prod \mathrm{dSt}(sA_i)$. This completes the proof that dSt is a derived stack, the claim of Theorem 20.3.1.

21 Introduction to a category of derived motivic spectra

21.1 Introduction

Grothendieck observed that in algebraic geometry one has a plethora of seemingly different cohomology theories that sometimes yield the same groups, which led him to think that perhaps there ought to be a common underlying theory to all these cohomology theories, something he dubbed motives [34], an object intermediate between a space and a cohomology. Formalizing this idea, one can construct derived categories of sheaves in an attempt to construct categories of motives [9], which if well-defined enough have objects within them that represent cohomologies, and if in addition those categories are endowed with a six-functor formalism, which is in a sense a shadow of cohomological properties, then one should regard those categories as categories of motives. Already at this point, one has shifted focus from cohomologies to derived categories of motives that predate cohomologies, something we will exploit to motivate our approach. Now Khan used the fact that many problems in classical algebraic geometry can be simplified in derived algebraic geometry (most notably for us intersection theory, which involves algebraic cycles, themselves needed to define Weil cohomologies, which are intimately tied to a notion of motives), to warrant developing a theory of motives in that context, which he did in [48]. We will follow the same point of view and use Segal categories for that purpose.

Most importantly, what we do presently is expose an alternative way of thinking about motives and how they come about. If classically motives hX associated with a space X appear as in the diagram

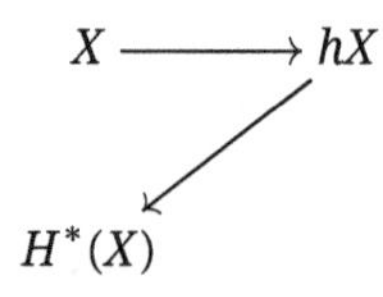

where $H^*(X)$ denotes a Weil cohomology, we suggest what initially seems to be a functorial formalism as in

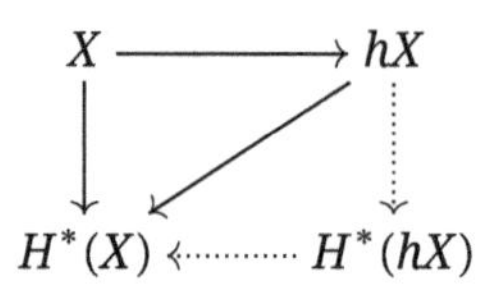

where $H^*(hX)$ would capture the cohomological information from a motive. Since, as we argued above, our attention can shift from cohomologies to studying derived categories, we seek diagrams of the following form instead:

https://doi.org/10.1515/9783111334073-021

$$
\begin{tikzcd}
X \arrow[r] \arrow[d] & hX \arrow[dl] \arrow[d, dotted] \\
\mathrm{D}(X) & \mathrm{D}(hX). \arrow[l, dotted]
\end{tikzcd}
\tag{21.1}
$$

We do have something formally similar to this picture where derived categories of sheaves are being used, in which setting the bottom map above can be regarded as a specialization functor, thereby prompting us to view the top map as a generalization. Indeed, from [9], for instance, we have

$$
\begin{tikzcd}
\mathrm{Sch}^{\mathrm{ft,sm}}_{/X} \arrow[r] \arrow[d] & \mathrm{Sch}^{\mathrm{ft}}_{/X} \arrow[d] \\
\mathrm{D}_{\mathbb{A}^1}\mathrm{Sh}_{\text{ét}}(\mathrm{Sch}^{\mathrm{ft,sm}}_{/X}, \mathbb{Q}) & \mathrm{D}_{\mathbb{A}^1}\mathrm{Sh}_t(\mathrm{Sch}^{\mathrm{ft}}_{/X}, \mathbb{Q}) \arrow[l]
\end{tikzcd}
$$

where $t = h$ or qfh. Coming back to (21.1), we regard the derived categories of sheaves therein as cohomological manifestations of schemes X and corresponding motives hX. To be fair, those are functor categories having slice categories $\mathcal{C}_{/X}$ and $\mathcal{C}'_{/hX}$ for base categories. This prompts us to regard the philosophy of motives as a multilayered process, the first step consisting in finding a large ambient category $\mathcal{C}'$ an object X of $\mathcal{C}$ can be embedded into and putting that object in relation with other objects of that category by means of a categorical perception such as a slice category $\mathcal{C}'_{/hX}$, the idea being that the object X would be fully displayed in this setting. The second step consists in turning slice categories into sites. We regard putting a Grothendieck topology on categories as providing a *filter* on such categories, the idea being that appropriate filters would put objects of such categories in a favorable light. The next step consists in studying manifestations of such sites. The proper way to do this coherently of course is to use sheaves, and hence the use of derived categories of sheaves. Finally, to highlight a particular characteristic of a geometric object X, one then has to select an appropriate modification of such derived categories to highlight such a characteristic, something we regard as a manifestation thereof. We refer to this last step as a *reduction*.

Thus we are considering multilayered diagrams of categories, each step of which being regarded as a filtering process, an example being the following diagram:

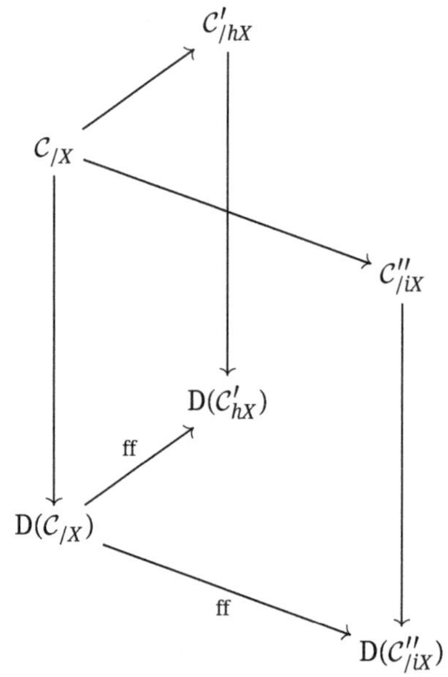

each vertical map connecting one layer with the next. Such diagrams are functorial in X.

For constructing such multilayered diagrams, in their first stages, we regard embeddings as a fundamental initial step towards highlighting the various facets of a geometric object X. Given an embedding $h : \mathcal{C} \hookrightarrow \mathcal{C}'$ with $X \in \mathcal{C}$ and $hX \in \mathcal{C}'$, the categorical perceptions figuring in the induced map $\mathcal{C}_{/X} \hookrightarrow \mathcal{C}'_{/hX}$ allow one to see how much of the original object X is perceived at each stage. In the language of derived categories of (21.1), if for any other embedding $i : \mathcal{C}' \hookrightarrow \mathcal{C}''$ no information is gained by studying the embedding of hX into $\mathcal{C}''$, that is, $\mathrm{D}'(hX) \cong \mathrm{D}''(ihX)$, where $\mathrm{D}'(hX)$ is a derived category of sheaves on $\mathcal{C}'_{/hX}$, $\mathrm{D}''(ihX)$ one on $\mathcal{C}''_{/ihX}$, and this for any D'', then we say that $\mathrm{D}'(hX)$ fully exhibits the nature of X via hX or that hX is the *complete* object associated with X. To reach this point, one already has to use the appropriate filters and manifestations on such categories $\mathcal{C}'$ and $\mathcal{C}''$, each such modification adding a corresponding layer to our construction. If one wants to highlight a particular characteristic T of a geometric object X, if $\mathfrak{R}_T$ designates its corresponding reduction on derived categories of sheaves for instance, and if after reduction, we have a similar statement as in $\mathfrak{R}'_T\mathrm{D}'(hX) \cong \mathfrak{R}''_T\mathrm{D}''(ihX)$ for any $\mathfrak{R}''_T$, then one says $\mathfrak{R}'_T\mathrm{D}'(hX)$ exhibits the T-character of X via hX, and it captures its T-ness. By convention, if $T = \emptyset$, $\mathfrak{R}_\emptyset = \mathrm{id}$, then the object X has been fully displayed in hX, which consequently becomes the complete object associated with X. If one considers those reductions $\mathfrak{R}_{\mathrm{mot}}$ that allow us to obtain motivic categories, namely those that turn derived categories of sheaves of the form $\mathrm{D}_{\mathbb{A}^1}\mathrm{Sh}_\tau(\mathcal{C}_{/X}, \mathbb{Q})$ as in (21.1), for example, into categories for which a six-functor formalism holds, then from our perspective the resulting categories $\mathfrak{R}_{\mathrm{mot}}\mathrm{D}_{\mathbb{A}^1}\mathrm{Sh}_\tau(\mathcal{C}_{/X}, \mathbb{Q})$ exhibit the motivic character of X. This is a formal heuristic approach to the philosophy of motives that we find helpful to put things into perspective. Nevertheless, for constructing a Segal category of derived motivic spectra, which is the main goal of this chapter, we will base our work on the notion of motives as they are usually understood. Our heuristics on motives will be used to smoothly introduce the Segal category derived motivic spectra are issued from.

We start with a scheme $X \in \mathrm{Sch}_k$, which we first embed in derived schemes, $iX \in \mathrm{dSch}_k$, and then derived stacks using the model Yoneda embedding, $F = h_{iX} \in \mathrm{dSt}_k$ (see [79]). We then focus on the slice category $(\mathrm{dSt}_k)_{/F}$. If most derived categories of sheaves in the traditional study of motives are of the form $\mathrm{DSh}_\tau(\mathcal{C}_{/X}, \mathcal{A})$, then we argue that the correct generalization of such an object in our context is constructed from a Segal category $\mathbb{R}\underline{\mathrm{Hom}}((\mathrm{dSt}_k)^{\mathrm{op}}_{/F}, \mathrm{Top})$. We develop a formalism of derived motivic spectra of objects in this category, the equivalent in our setting of motivic spectra in [48]. We define derived motivic spectra to be spectra that are $\mathbb{A}^1$-invariant (or rather its generalization) and satisfy Nisnevich excision instead of Nisnevich descent. One thing we do not do however is develop a six-functor formalism, and we do not work with pointed spaces for simplicity.

About notations, a hom set is denoted with the letter Hom, $\underline{\mathrm{Hom}}$ is used for internal homs, Map for simplicial mapping spaces, and map for a homotopy function complex. Cat_Δ will denote the category of simplicial categories, and Set_Δ the category of simplicial sets. We refer to $\mathrm{Hom}(-, X)$ as a functorial perception of X, $\mathcal{C}_{/X}$ as a categorical perception, $\mathrm{Hom}(X, -)$ as a functorial manifestation of X, and $\mathcal{C}_{X/}$ as a categorical manifestation.

21.2 From embeddings to comparison theorems

21.2.1 Concrete realization vs. abstraction

Working with motives [34, 70, 3] means in particular working with Weil cohomologies [44], since those factor through motives, and Weil cohomologies necessitate that we consider algebraic cycles among other things. This forces the standard conjectures upon us, which impose very concrete demands on our cohomology theories, which consequently appear as realizations of the overarching theory of motives they are issued from [62]. It follows that developing a theory of motives with Weil cohomologies in mind provides a facet only of a theory of motives. Adopting such a point of view subsumes motives are fundamental objects that take preeminence over cohomology theories. Thus what we are aiming for is a precursor to the notion of classical motive from which cohomologies can be derived.

To illustrate the need to seek an abstract theory of motives, recall that the notion of motive hX associated with a geometric object X was introduced since various cohomologies $H^*_{(i)}(X)$ could produce the same results, thereby exhibiting those cohomologies as representatives of a same class, the motive hX. We have the following picture:

$$\begin{array}{ccc} X & \longrightarrow & hX \\ & \swarrow & \\ H_i^*(X). & & \end{array} \tag{21.2}$$

Given that a motive is an object intermediate between a space and a cohomology theory, suppose for a moment we had a way of dissociating the cohomological information from a motive by means of a functor H^* in such a manner that (21.2) could be completed into a square diagram as follows:

$$\begin{array}{ccc} X & \longrightarrow & hX \\ \downarrow & \swarrow & \vdots \\ H_i^*(X) & \longleftarrow & H^*(hX) \end{array}$$

where each diagonal map factors through specialization functors $H_i^*(X) \leftarrow H^*(hX)$, thereby presenting $X \to hX$ as a generalization. The fact that we have such a formalism is not as far-fetched as it seems since we will see later that cohomological spectra are objects of derived categories of sheaves (or stacks) on certain slice categories and that the above statement about cohomological objects should read $\mathrm{DSh}_\tau(\mathcal{C}_{/X}, \mathcal{A}) \leftarrow \mathrm{DSh}_{\tau'}(\mathcal{C}'_{/hX}, \mathcal{A})$, a precomparison functor of some sort, which follows from having a map $X \to hX$. Making such a statement already shows a clear departure from standard motivic homotopy theory whereby $\mathrm{DSh}_\tau(\mathcal{C}_{/X}, \mathcal{A})$ is a category of motives, whereas for us the motivic character of a geometric object $X \in \mathcal{C}$ is apparent in its embedding hX in another category $\mathcal{C}'$ as discussed in the introduction, the various derived categories on the corresponding slice categories being different ways to make manifest the motivic character of such geometric objects. The rest of this chapter has as one of its aims to present this alternate formalism that provides an impetus for a theory of motives, in the hope it will spur new research.

21.2.2 Embeddings

If a motive is constructed starting from a variety, or a scheme, then we argue that the "motivic heart" of a geometric object can be revealed once the said object is first embedded in a larger category, and it is put in contrast with other objects of this category. Thus as a first step into abstraction proper, we are looking at a series of embeddings

$$\mathrm{Var}_k \hookrightarrow \mathrm{Sch}_k \xrightarrow{i} \mathrm{dSch}_k \xrightarrow{h} \mathrm{dSt}_k \tag{21.3}$$

for a commutative ring k, from varieties to schemes, if one is starting from varieties, then derived schemes, followed by derived stacks. To be precise, if k is an algebraically closed field, then we have a fully faithful functor $\mathrm{Var}_k \hookrightarrow \mathrm{Sch}_k$. From [73] we have a fully faithful functor from the category of k-schemes Sch_k to the Segal category dSch_k of derived schemes over k, which itself is embedded in the Segal category of derived stacks dSt_k. To be more precise, we have a Yoneda embedding $h : \mathrm{dSch}_k \to \mathrm{dSt}_k$ that sends a derived scheme iX, $X \in \mathrm{Sch}_k$, to $h_{iX} = \mathrm{Map}_{\mathrm{dSch}_k}(-, iX)$.

Regarding the latter embedding, recall from [79] that if M is a model category, $\Gamma_* : M \to M^\Delta$ a cofibrant resolution functor on M, if we let $\underline{h}_x(y) = \mathrm{Hom}_M(\Gamma_* y, x) = \mathrm{Map}_M(y, x)$, the functor $\underline{h} : M \to M^\wedge$ (where $M^\wedge$ is the left Bousfield localization of $\mathrm{sPr}(M)$ with respect to weak equivalences in M) can be right derived into $\mathbb{R}\underline{h}$. Further we have $h_x \cong \mathbb{R}\underline{h}_x$ in $\mathrm{Ho}(M^\wedge)$. Now to argue that h_x is a stack, it suffices to show $\mathbb{R}\underline{h}_x$ is one as well. Recall that a model pre-topology τ on M is said to be subcanonical if $\mathbb{R}\underline{h}_x$ is a stack for all $x \in M$. Formally, this means $\mathbb{R}\underline{h} : \mathrm{Ho}(M) \to \mathrm{Ho}(M^\wedge)$ factors through the category of stacks $\mathrm{Ho}(M^{\sim,\tau})$. Since this is what we need, we require that τ be subcanonical. Specifically, here we work with $M = \mathrm{D}^- k\text{-Aff} = sk\text{-Alg}^{\mathrm{op}}$. Now that τ is subcanonical in general follows from Assumption 1.3.2.2 of [79], which holds in the particular case of having $\tau = $ ét on $\mathrm{D}^- k$-Aff. Thus $\mathbb{R}\underline{h} : \mathrm{Ho}(\mathrm{D}^- k\text{-Aff}) \to \mathrm{Ho}(\mathrm{D}^- k\text{-Aff}^{\sim,\text{ét.}})$ is fully faithful, which is isomorphic to h, and thus when we say $h : \mathrm{dSch}_k \to \mathrm{dSt}_k = L\mathrm{D}^- k\text{-Aff}^{\sim,\text{ét.}}$ after simplicial localization of model categories is fully faithful, we really mean this in a homotopical sense.

For completeness, recall how dSt_k is constructed; one starts, as recalled in [79] and [74], with a symmetric monoidal model category $\mathcal{C}$, and in derived algebraic geometry, we are really thinking of $\mathcal{C} = sk\text{-Mod}$. Then $\mathrm{Comm}(\mathcal{C}) = sk\text{-Alg}$. Taking the Dwyer–Kan simplicial localization of this [15] produces a simplicial category $L(sk\text{-Alg})$ and hence a Segal category [78]. Its opposite Segal category is denoted $\mathrm{d}k\text{-Aff} = L(sk\text{-Alg})^{\mathrm{op}}$. Denoting $L\mathrm{Set}_\Delta$ by Top, we consider the Segal category of prestacks $\widehat{\mathrm{d}k\text{-Aff}} = \mathbb{R}\underline{\mathrm{Hom}}(\mathrm{d}k\text{-Aff}^{\mathrm{op}}, \mathrm{Top})$, and $\mathrm{dSt}_k \subset \widehat{\mathrm{d}k\text{-Aff}}$ is the sub-Segal category of derived stacks for the étale topology, which can also be obtained as $\mathrm{dSt}_k = L(\mathrm{D}^- k\text{-Aff}^{\sim,\text{ét.}})$, where $\mathrm{D}^- k\text{-Aff}^{\sim,\text{ét.}}$ is the model category of derived stacks on sk-Alg, as used in the preceding paragraph.

The tower (21.3) provides a chain of embeddings (that can possibly be further extended), and we seek the most general category a geometric object X can be embedded into, the idea being that if it appears difficult to find an optimal category $\mathcal{C}'$ such that any other embedding $\mathcal{C}' \hookrightarrow \mathcal{C}''$ would not provide any gain in information about the object X as discussed in the introduction, assuming the T-ness of X is already present in $hX \in \mathcal{C}'$, any further embedding thereafter would preserve the T-character of X, thus working with the most general category one can think of should suffice. Presently, we will settle on dSt_k for defining motives, which is sufficient since the category of smooth schemes already provides a base for constructing categories of motives. Putting X in contrast with other objects of this larger category dSt_k really means looking at a functorial perception $\mathrm{Hom}(-, X)$ or at a categorical perception such as a slice category $(\mathrm{dSt}_k)_{/X}$.

To fix ideas, using (21.3) above, for $X \in \mathrm{Sch}_k$, an embedding of slice categories is of the form

$$(\mathrm{Sch}_k)_{/X} \hookrightarrow (\mathrm{dSch}_k)_{/iX} \hookrightarrow (\mathrm{dSt}_k)_{/h_{iX}} \tag{21.4}$$

The motivation for working with slice categories follows from the relative point of view philosophy of Grothendieck, according to which the various aspects of an object can be made manifest once it is put in relation with other objects of the category in which it finds itself, something we refer to as *perception*. In particular, (21.4) amounts to working with categorical perceptions. Next, one puts topologies on such slice categories, something we regard as filters. Already at this point, we are starting to have layered diagrams with slice categories as base layer, followed by a layer of filtered categories, or sites. Then one studies sheaves on such sites, since they provide coherent manifestations, and this produces an additional layer. Another way of looking at this is that categories of sheaves formally correspond to an extrinsic comparison of a given base category with various target categories, something we will discuss in the next section. Finally, one studies appropriate functors on the resulting topos to study certain aspects of X, something we referred to as reductions in the introduction. If one desires to focus on the motivic aspect of X, then one works with reductions producing cohomological functors, and by this we mean functors that satisfy a six-functor formalism, which in a sense extract from X its cohomological character. This is in essence the program we will follow.

21.2.3 Manifestations/extrinsic representations

Following Grothendieck's philosophy of working with relative objects, we have introduced categorical perceptions $(\mathrm{Sch}_k)_{/X}$ and $(\mathrm{dSt}_k)_{/F}$, with $F = h_{iX}$. Categorical perceptions and their relative embeddings are primordial in our philosophy insofar, as once appropriately chosen, they provide a foundation for determining at which point a given geometric object has its T-ness made manifest. As briefly mentioned at the end of the preceding section, the next step consists in putting a topology on those slice categories, a filtering process of some sort, and we assume that has been done at this point.

Still following Grothendieck's relative point of view philosophy, analyzing those sites should involve other categories. One way to implement this is by studying their extrinsic representations, and we can do this functorially using sheaves for instance. Those functorial manifestations when appropriately modified via reductions can exhibit one aspect or another of geometric objects, and the resulting reduced manifestations tell us whether we have reached an optimal ambient category $\mathcal{C}'$ in a tower of embeddings, or whether we have to push further in such a tower.

This is actually very general: to highlight the character of type T of a geometric object $X \in \mathcal{C}$, one seeks ambient categories $\mathcal{C}'$ with corresponding embeddings $i : \mathcal{C} \hookrightarrow \mathcal{C}'$ such that for appropriately chosen functors F_T, one says the T-character of X is made manifest in $iX \in \mathcal{C}'$ if for any other category $\mathcal{C}''$ and any other embedding $j : \mathcal{C}' \hookrightarrow \mathcal{C}''$, for any other T-functor G_T, $F_T(\mathcal{C}'_{/iX}) \cong G_T(\mathcal{C}''_{/jiX})$, at which point we would also say $F_T(\mathcal{C}'_{/iX})$ displays the type T of X. If T is a motive and F_T corresponds to a derived category of sheaves on $\mathcal{C}'_{/iX}$ satisfying a six-functor formalism, then in this setting, $F_T(\mathcal{C}'_{/iX})$ is a category of motives in the traditional sense, which in our formalism exhibits the motivic character of X, or in other words, it captures the motivic aspect of $\imath X$.

Now to mirror the fact that in [9] we work with sheaves over slice categories, which in themselves are manifestations of such base categories, we now seek in our present context some manifestations of $(\mathrm{dSt}_k)_{/F}$, and those will be objects of a Segal category $\mathbb{R}\underline{\mathrm{Hom}}((\mathrm{dSt}_k)^{\mathrm{op}}_{/F}, \mathrm{Top})$. This will be the analog for us in the Segal setting of the categories of sheaves used in [9]. Note that we will not put topologies on $(\mathrm{dSt}_k)^{\mathrm{op}}_{/F}$ since this would put a filter on it, and we desire to work in wide generality.

21.2.4 A study of cohomology – deconstruction

Recall that one can associate with compact manifolds various cohomology theories that yield the same groups, supposing they are based on the Eilenberg–Steenrod axioms. This presents those cohomology theories as realizations of those axioms, which therefore appear as more fundamental, more motivic in a sense. This also shows that we should distance ourselves from cohomology theories proper since they appear secondary, and aim instead for precursors to cohomology theories, chief among them being motives. To further illustrate the necessity of a change of focus from cohomology to more abstract theories, observe that there are situations where cohomology is instrumentalized. An example of such a situation is sheaf cohomology. As recounted in [63] for example, recall that if $\mathrm{Sh}_{\text{ét}}(X)$ denotes the category of étale sheaves on a scheme X, then the functor $\mathrm{Sh}_{\text{ét}}(X) \to \mathrm{Ab}$, $\mathcal{F} \mapsto \Gamma(X, \mathcal{F}) = \mathcal{F}(X)$ is in a sense a first-order manifestation of X into Ab, relative to $\mathcal{F}$ of course. Since the global sections functor $\Gamma(X, -)$ is left exact, one can find its right derived functor. To do this, since $\mathrm{Sh}_{\text{ét}}(X)$ has enough injectives, one can find an injective resolution $0 \to \mathcal{F} \to \mathcal{I}^0 \to \mathcal{I}^1 \to \cdots$ of $\mathcal{F}$. Because the complex $\Gamma(X, \mathcal{I}^0) \to \Gamma(X, \mathcal{I}^1) \to \cdots$ is in general not exact, one can find its cohomology, which is none other than the étale cohomology of X. Note here that studying the cohomology of $\Gamma(X, \mathcal{I}^\bullet)$ complements its nonexactness, it restores a semblance of coherence. What transpires is that cohomology is necessary to study the manifestation of $X \in \mathrm{Sch}_k$ into Ab via $\mathcal{F} \in \mathrm{Sh}_{\text{ét}}(X)$ for having a complete picture of X but does not take center stage otherwise. Thus if one desires to develop a theory of motives, regarded as fundamental objects, it seems appropriate not to use cohomology directly, which sometimes seems to be secondary, but to base a theory of motives on more abstract theories instead.

21.2.5 Shift from cohomology to studying derived categories

As we pointed out in the preceding subsection, cohomology from our perspective is useful to study geometric objects. However, to formalize a theory of cohomologies, one ought to aim for theories that predate cohomologies. Motives is one such theory. The way motives were originally introduced, they were presented as an object in between a space and a cohomological object, in that any good cohomology would factor as in

$$\begin{array}{ccc} X & \longrightarrow & hX \\ & \searrow & \downarrow \\ & & H^*(X). \end{array}$$

For motives hX to appear, it is necessary that they carry some cohomological information, and for this, it is typically required that motives be objects of a category satisfying a six-functor formalism. This is not something we focus on; we will not ask that our categories of coefficients, as they will later be called, satisfy a six-functor formalism. We will explain later why that is the case. This simplifies our task greatly, and developing a six-functor formalism is still possible using our categories of derived motivic spectra as a starting point.

Categories of coefficients are derived categories that predate cohomology theories. In [9] in particular, many of those in the context of motives are derived categories of sheaves, and some of them have cohomology spectra as objects that represent cohomology theories. This is our desired abstraction. In this manner, one can eschew cohomology theories in favor of studying such derived categories and their comparisons. This is typically what is done since those derived categories are categories of motives. In the present chapter, we broaden the picture a bit by seeing those derived categories as manifestations of geometric objects. As such, they are meant to exhibit the motivic character of such objects, as recounted in the introduction.

21.2.6 Categories of motives

We use the presentation given in [9] of the various categories of motives of interest to us. Central for us is the derived category of Beilinson motives $\mathrm{DM}_B(X)$, the Verdier quotient of $\mathrm{D}_{\mathbb{A}^1}(\mathrm{Sh}_{\mathrm{Nis}}(\mathrm{Sm}_{/X}, \mathbb{Q}))$ by a subcategory of acyclic objects. It is worth pointing out again that although this is generally accepted as providing a category of motives, we propose an alternative way of seeing the study of motives as equivalent to first placing a given geometric object X in the most general ambient category it can be seen as an object of, and seeing it relative to other objects in this category. Thus we are looking at slice categories $\mathcal{C}_{/X}$. Those can be analyzed by using what we referred to as filters, namely Grothendieck topologies that turn them into sites. At this point, one projects such sites into other categories using sheaves, thereby providing us with topos. The various facets, or characteristics of type T, of the geometric object X we started with can be studied using T-specific manifestations called reductions, which hone in on the T-character of geometric objects. From that perspective we can regard $\mathrm{DM}_B(X)$ as one reduction that provides a manifestation of the motivic character of $X \in \mathrm{Sm}$ by using Nisnevich sheaves with rational coefficients. Observe that this in no way changes all the work done on motives so far, but it just provides another approach to the philosophy of motives; if $\mathrm{DM}_B(X)$ is a category of motives in the traditional sense, then we merely suggest seeing it as exhibiting the motivic character of X.

To come back to $\mathrm{DM}_B(X)$, by Sm we mean the class of smooth morphisms of finite type in a category of schemes over a scheme S fixed at the onset. By $\mathrm{D}_{\mathbb{A}^1}(\mathcal{A})$ for some category $\mathcal{A}$ we mean $\mathrm{D}_{\mathbb{A}^1}(\mathcal{A}) = \mathrm{D}(\mathrm{Sp}(\mathcal{A}))[\mathcal{W}_{\Omega,\mathbb{A}^1}^{-1}]$, where $\mathrm{D}(\mathcal{C})$ is the usual derived category of some category $\mathcal{C}$, $\mathrm{Sp}(\mathcal{A})$ is a category of spectra in $\mathcal{A}$, and here the localization is with respect to a class $\mathcal{W}_{\Omega,\mathbb{A}^1} = \mathcal{W}_\Omega \cup \mathcal{W}_{\mathbb{A}^1}$, the collection of two families of morphisms of Tate spectra. The reader is invited to read [9] for more details; our very cursory presentation is just meant to present the main actors and put things in perspective, and we will not deal with details beyond this level of presentation in the rest of this work. For us, it is important to keep in mind that the derived category of Beilinson motives is a derived category of rational, Nisnevich sheaves on the smooth site of a scheme X.

We also have the category of Voevodsky's motives $\mathrm{DM}_{t,\mathbb{Q}}(X)$ where the topology $t = h$ or qfh, defined as the localizing subcategory of $\underline{\mathrm{DM}}_{t,\mathbb{Q}}(X)$ spanned by objects of the form $\Sigma^\infty \mathbb{Q}_X^t(S)(n)$ for $S \in \mathrm{Sm}_{/X}$, $n \leq 0$, where $\mathbb{Q}_X^t(S)$ is the free $\mathbb{Q}$-linear t-sheaf represented by S, $\underline{\mathrm{DM}}_{t,\mathbb{Q}}(X) = \mathrm{D}_{\mathbb{A}^1}(\mathrm{Sh}_t(\mathcal{S}_{/X}^{\mathrm{ft}}, \mathbb{Q}))$, $\mathcal{S}^{\mathrm{ft}}$ a class of morphism of finite type in a category $\mathcal{S}$ of Noetherian schemes over X satisfying certain conditions (see [9] for details).

Another category of motives is $\mathrm{DM}_{\mathbb{Q}}(X)$ defined as a derived category of sheaves with transfer $\mathrm{D}_{\mathbb{A}^1}(\mathrm{Sh}^{\mathrm{tr}}(X, \mathbb{Q})) = \mathrm{D}_{\mathbb{A}^1}(\mathrm{Sh}_{\mathrm{Nis}}(\mathrm{Sm}_{\mathbb{Q},X}^{\mathrm{cor}}))$.

The category of Morel motives $\mathrm{D}_{\mathbb{A}^1}(X, \mathbb{Q})_+$ is a subcategory of $\mathrm{D}_{\mathbb{A}^1}(X, \mathbb{Q}) = \mathrm{D}_{\mathbb{A}^1}(\mathrm{Sh}_{\mathrm{Nis}}(\mathrm{Sm}_{/X}, \mathbb{Q}))$ obtained by projection (see [9] for details).

21.2.7 Realization functors

We can start our discussion of the notion of realization functors by the following result presented in [9]: for a scheme $X \in \mathrm{Sch}_{/S}$ with structural morphism $f : X \to S$ and a given stable cohomology E with associated ring spectrum $\mathcal{E}$, we have a realization functor $\mathrm{DM}_B(X) \to \mathrm{Ho}\,(\mathcal{E}_X\text{-Mod})$, where $\mathcal{E}_X = \mathbb{L}f^*\mathcal{E}$, a map that with M associates $\mathcal{E}_X \otimes_X^{\mathbb{L}} M$.

In [9] an abstraction of this situation is given as follows. If $\mathcal{M}$ and $\mathcal{M}'$ are monoidal model categories satisfying a host of conditions as spelled out in [9], $\phi^* : \mathcal{M} \rightleftarrows \mathcal{M}' : \phi_*$ is a left Quillen adjunction, $\mathcal{E}'$ is a fibrant resolution of $\mathbb{1}$ in $\mathcal{M}'(\mathrm{Spec}\, k)$, and $\mathcal{E}$ is a cofibrant resolution of $\phi_*(\mathcal{E}')$ in $\mathcal{M}(\mathrm{Spec}\, k)$, then by Theorem 17.1.5 of [9] we have a realization functor $\mathrm{Ho}\,(\mathcal{M}) \to \mathrm{Ho}\,(\mathcal{M}') \cong \mathrm{Ho}\,(\mathcal{E}\text{-Mod})$, $M \mapsto \mathcal{E} \otimes^{\mathbb{L}} M$. Note that having $\mathrm{Ho}\,(\mathcal{M}') \cong \mathrm{Ho}\,(\mathcal{E}\text{-Mod})$ means the representation theory of $\mathcal{E}$ is given by $\mathcal{M}'$, and if $\mathcal{E}$ represents a cohomology theory, then $\mathcal{M}'$ appears as the ambient category within which the cohomology theory $\mathcal{E}$ is supposed to represent can be developed. For this to happen, $\mathcal{M}$ must be a reference category in a sense, and the fact that it provides "clean" representations of cohomologies such as Ho ($\mathcal{E}$-Mod) justifies that it be considered as being associated with a theory of motives.

Having presented this abstract set up, observe that if $H_{B,X}$ is the Beilinson motivic cohomology spectrum, then by Theorem 14.2.9 of [9] the functor $\mathrm{DM}_B(X) \to \mathrm{Ho}\,(H_{B,X}\text{-Mod})$ is an equivalence of triangulated monoidal categories, and thus we have a trivial realization functor illustrating the fact that $\mathrm{DM}_B(X)$ is an optimal category of motives. In this manner, realization functors have enabled us to justify that derived categories such as $\mathrm{DM}_B(X)$ effectively be categories of motives. This is nothing new of course, but in our formalism, this has the added benefit of showing that of the two cohomological manifestations $\mathrm{DM}_B(X)$ and Ho ($\mathcal{E}_X$-Mod), the former appears as an optimal functor of X that exhibits its motivic character.

21.2.8 Comparison functors

Comparison theorems allow us to link all the motivic categories of interest. One thing worth pointing out before doing so is that if we have such equivalences of categories, then this indicates that each of the categories involved are not absolute in a sense, but are merely facets of a yet to be determined ultimate category of motives, each of those categories figuring in those comparison theorems being in a sense representatives of the class corresponding to a universal motivic homotopy theory. This point of view justifies the necessity of seeking a higher theory to extract a category of motives all the categories below being representatives of. For the time being though, observe that all the categories of motives involved in the comparison theorems we will cover come from reductions of various derived categories of sheaves as we will see and exhibit the motivic character of X in our formalism.

A first comparison theorem is provided by Theorem 16.1.2 of [9], which states that $\mathrm{DM}_B(X) \cong \mathrm{DM}_{t,\mathbb{Q}}(X)$, where $t = h$ or $t = qfh$. One thing we will be careful to clearly enunciate is what property X is supposed to satisfy in each of the comparison theorems we will give, for a reason we will expound upon below. Here X is an excellent Noetherian scheme of finite dimension.

If now X is excellent and geometrically unibranch in addition to being Noetherian and of finite dimension, then we have an equivalence of triangulated monoidal categories $\mathrm{DM}_B(X) \xrightarrow{\simeq} \mathrm{DM}_{\mathbb{Q}}(X)$.

By Theorem 16.2.13 of [9], for any Noetherian scheme X of finite dimension, we have a canonical equivalence of triangulated monoidal categories $\mathrm{DM}_B(X) \simeq \mathrm{D}_{\mathbb{A}^1}(X,\mathbb{Q})_+$.

Finally, for any Noetherian scheme X of finite dimension, we also have a canonical equivalence of categories $\mathrm{DM}_{\mathbb{Q}}(X) \simeq \mathrm{D}_{\mathbb{A}^1,\text{ét}}(X,\mathbb{Q}) = \mathrm{D}_{\mathbb{A}^1}(\mathrm{Sh}_{\text{ét}}(\mathrm{Sm}_{/X},\mathbb{Q}))$.

It is important to note at this point that if we modify the characteristics of our base scheme X, then we are shifting our point of view, and consequently the motives, understood in the traditional sense, will be different as well. Thus for illustrative purposes, and to fix ideas, we work with one type of scheme only. Thus for all the above comparison theorems to be used in unison, we fix X to be an excellent, geometrically unibranch Noetherian scheme of finite dimension.

21.3 Abstraction of a category of motivic spectra

21.3.1 General form of reductions

All the triangulated monoidal motivic categories used in the comparison theorems above are constructed from derived categories of sheaves on the small site of some scheme X, and thus we will seek an abstract category of motives that retains the same form, namely a category of sheaves, or stacks, on some slice category.

To illustrate the fact that the motivic categories we are dealing with are essentially what we call *reductions* of derived categories of sheaves, we present them in an itemized fashion:

- $\mathrm{DM}_{\mathcal{B}}(X) = \mathrm{D}_{\mathbb{A}^1}(\mathrm{Sh}_{\mathrm{Nis}}(\mathrm{Sm}_{/X}, \mathbb{Q}))_{/\{\mathrm{H}_{\mathcal{B}}\text{-acyclic}\}}$
- $\mathrm{DM}_{t,\mathbb{Q}}(X) = \langle \Sigma^{\infty}\mathbb{Q}^t_X(S)(n)\rangle \subset \mathrm{D}_{\mathbb{A}^1}(\mathrm{Sh}_t(\mathcal{S}^{\mathrm{ft}}_{/X}, \mathbb{Q}))$
- $\mathrm{DM}_{\mathbb{Q}}(X) = \mathrm{D}_{\mathbb{A}^1}(\mathrm{Sh}_{\mathrm{Nis}}(\mathrm{Sm}^{\mathrm{cor}}_{\mathbb{Q},X}))$
- $\mathrm{D}_{\mathbb{A}^1}(X, \mathbb{Q})_+ \subset \mathrm{D}_{\mathbb{A}^1}(\mathrm{Sh}_{\mathrm{Nis}}(\mathrm{Sm}_{/X}, \mathbb{Q}))$
- $\mathrm{D}_{\mathbb{A}^1,\text{ét}}(X, \mathbb{Q}) = \mathrm{D}_{\mathbb{A}^1}(\mathrm{Sh}_{\text{ét}}(\mathrm{Sm}_{/X}, \mathbb{Q}))$

If the symbol $\mathfrak{R}$ designates reduction, then all the above categories can be written in the form $\mathfrak{R}\mathrm{D}_{\mathbb{A}^1}(\mathrm{Sh}_\tau(\mathcal{C}_{/X}, \mathbb{Q}))$, where $\mathfrak{R}$ above is the Verdier quotient, projection, subset, or identity. The categories $\mathcal{C}_{/X}$ are slice categories over a scheme X. The larger categories $\mathrm{D}_{\mathbb{A}^1}(\mathrm{Sh}_\tau(\mathcal{C}_{/X}, \mathbb{Q}))$ are ambient categories where motives in the traditional sense live.

21.3.2 Multilayered diagrams of categories

Now that we have introduced comparison theorems and reductions, we digress momentarily and present an example of a multilayered diagram. It initially provides slice categories in one layer, followed by sites, then derived categories of sheaves, and in the last layer the reductions of such derived categories to provide categories of motives. Here is one example, where we have used the fact that $\mathrm{DM}_{\mathcal{B}}(X) \simeq \mathrm{D}_{\mathbb{A}^1}\mathrm{Sh}_{\text{ét}}(\mathrm{Sm}_{/X}, \mathbb{Q})$:

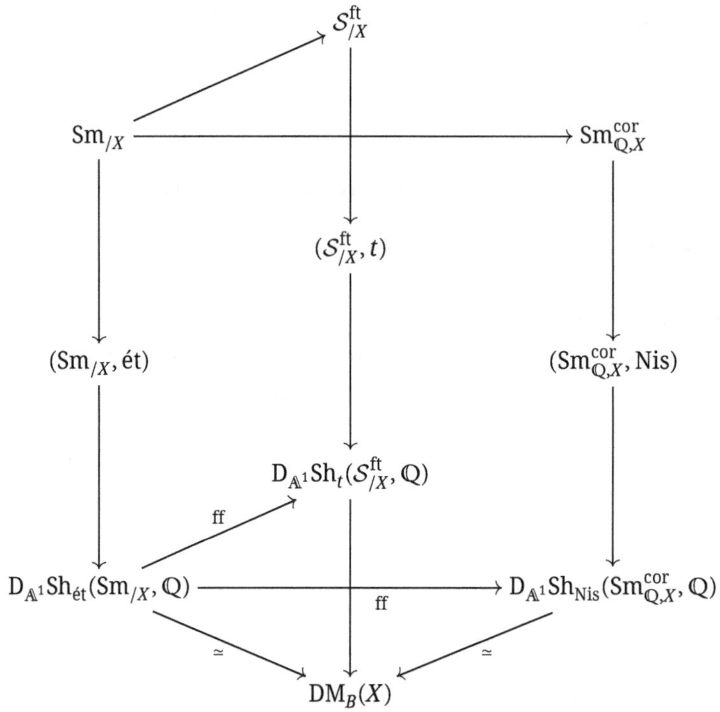

where the bottom diagonal maps are equivalences, and the last vertical map is a reduction consisting in taking a localizing subcategory, followed by an equivalence. The last layer shows that Beilinson motives highlight the motivic character of X. We will be working at the level of unrestricted unreduced derived categories of sheaves however, the next to last layer, which is not shown here.

At this point, it is fitting to expound on the notion of optimal manifestation of a geometric object. A statement about reductions of the form $F\mathrm{Sh}_{\tau'}(\mathcal{C}'_{/hX}, \Lambda') \cong G\mathrm{Sh}_{\tau''}(\mathcal{C}''_{/ihX}, \Lambda'')$ for any embedding $i : \mathcal{C}' \hookrightarrow \mathcal{C}''$, any topology τ'' on $\mathcal{C}''_{/ihX}$, any category Λ'', and any reduction G that produces a derived category of sheaves that satisfies a six-functor formalism means that $F\mathrm{Sh}_{\tau'}(\mathcal{C}'_{/hX}, \Lambda')$ is a category of motives in the traditional sense, which from our perspective is optimal in highlighting the motivic character of X through $hX \in \mathcal{C}'$ and corresponds to an optimal vertical thread in a multilayered diagram of categories such as the one above. Now we have an ordering on slice categories given by embeddings. Thus the first slice category $\mathcal{C}'$ for which we have isomorphisms of reductions is optimal in that regard. $\mathcal{C}'$ being fixed, we also have an ordering on topologies τ', a coarser topology being optimal if it leads to isomorphisms of reductions. However, $\mathcal{C}'$ and τ' being fixed, various categories Λ' and reductions F may be difficult to compare, so we see it is at the level of categories of sheaves and reductions thereof that we are starting to have classes of categories of motives in the absence of a clear identification of an optimal category of motives.

21.3.3 Abstraction of a category of coefficients

The derived categories of motives introduced in Section 21.3.1 are what Khan refers to as categories of coefficients. Those follow a six-functor formalism, which abstractly encodes what makes cohomology theories what they are. Hence the latter should be derivable from such categories of coefficients.

Since we are constructing categories that will ultimately be used to produce cohomology theories, we are interested in setting up a background from which one can develop a six-functor formalism. However, our focus is really on the functorial manifestations of motivic objects in our heuristic picture, not so much on cohomology per se. Because of this, we do not see an interest in developing a six-functor formalism for a category of coefficients, and we leave this task to the motivated reader.

For an ∞-category $\mathcal{C}$ with fiber products and the ∞-category $\mathrm{Cat}_\infty^{\mathrm{Pres}}$ of presentable ∞-categories, Khan defines a *category of coefficients* to be given by a symmetric monoidal functor $D : \mathcal{C}^{\mathrm{op}} \to \mathrm{Cat}_\infty^{\mathrm{Pres}}$. If $\mathcal{C}$ is a category of schemes and $X \in \mathcal{C}$, then $D(X)$ is a presentable ∞-category whose objects are referred to as sheaves on X. The reductions $\mathfrak{R}\mathrm{D}_{\mathbb{A}^1}(\mathrm{Sh}_\tau(\mathcal{C}_{/X}, \mathbb{Q}))$ we have considered before are of the form $D(X)$ and are regarded as categories of coefficients by abuse of language (albeit after passing to homotopy categories). We mention categories of coefficients presently since they provide a clear formalization of the notion of motivic category.

Since we have embeddings $\mathrm{Sch}_k \hookrightarrow \mathrm{dSch}_k \hookrightarrow \mathrm{dSt}_k$, moving a k-scheme X to a derived scheme iX and then to a derived stack $F = h_{iX}$, it is therefore natural to consider a category of coefficients to be based on a Segal category of the form $\mathbb{R}\underline{\mathrm{Hom}}((\mathrm{dSt}_k)^{\mathrm{op}}_{/F}, \mathrm{Top})$ with $\mathrm{Top} = L\mathrm{Set}_\Delta$, $\mathrm{dSt}_k = L(\mathrm{D}^- k\text{-}\mathrm{Aff}^{\sim, \text{ét.}})$, $\mathrm{D}^- k\text{-}\mathrm{Aff}^{\sim, \text{ét.}}$ the model category of stacks on sk-Alg, L stands for the simplicial localization of Dwyer–Kan, thereby producing Segal categories, dSt_k is the Segal category of derived stacks on sk-Alg, sub-Segal category of $\widehat{dk\text{-}\mathrm{Aff}} = \mathbb{R}\underline{\mathrm{Hom}}(dk\text{-}\mathrm{Aff}^{\mathrm{op}}, \mathrm{Top})$, where $dk\text{-}\mathrm{Aff} = L(sk\text{-}\mathrm{Alg})^{\mathrm{op}}$. The reason for having Top-valued functors will be given in Section 21.4.2.

Independently, Gallauer defines *coefficients systems* in [22], functors $F : (\mathrm{Sch}_k)^{\mathrm{op}}_{/X} \to \mathrm{Cat}_\infty^{\mathrm{st},\otimes}$ satisfying certain conditions, which are closely related to the categories of coefficients of Khan in the sense that presentable coefficient systems correspond to motivic categories of coefficients in the sense of Khan. We use Khan's coverage since for our purposes, it is easier to generalize to the Segal category setting. But coefficient systems bridge the gap between Khan's work and that of [9] in the sense that the category of homotopy functors $\mathrm{Ho}\,(F)$ for a coefficient system F yields in particular what is referred to as motivic triangulated categories in [9]. Another reason for alluding to Gallauer's work is that the universal coefficient system, which consequently would be the category of motives we are looking for, is none other than the stable motivic homotopy category. We use this fact as an imprint to seek an initial construction of such a category in our setting, before being made six-functor formalism ready, and we refer to it as the *category of derived motivic spectra*. We then turn to [48] to construct such a category by using the definition of the stable motivic homotopy category of Khan as motivation.

To be clear, Khan defines the stable motivic homotopy category $\mathcal{SH}(X)$ as the category of motivic $\mathbb{P}^1_X$-spectra, which is then enhanced for making it support a six-functor formalism. Since again this is not something we are aiming for, we stop at the category of motivic spectra, and thus we are really only looking at a category of derived motivic spectra in our setting.

To further highlight the central role played by the stable homotopy category, which we momentarily denote by SH as in [22], one can connect this category with any coefficient system C using an essentially unique functor $\rho^* : \mathrm{SH} \to C$ such that for $S \in (\mathrm{Sch}_k)_{/X}$, there is a right adjoint to $\rho^*(S)$: $\rho_*(S) : C(S) \to \mathrm{SH}(S)$ in such a manner that $\rho_*(S)\mathbb{1} := \mathcal{E} \in \mathrm{CAlg}(\mathrm{SH}(S))$ is a motivic spectrum that represents C-cohomology.

21.4 Construction of a category of derived motivic spectra

21.4.1 Khan's construction

To motivate our definition of a category of derived motivic spectra, we briefly recall Khan's construction from [48]. He defines the category of motivic spectra as $\mathcal{SH}(S) = \mathcal{SH}_{\mathbb{P}^1}(S)$, the category of motivic $\mathbb{P}^1_S$-spectra for a scheme S. Here $\mathbb{P}^1_S$ is shorthand for $M_S(\mathbb{P}^1_S)$ (see full definition in Section 21.4.3.3) pointed at ∞, and $\mathcal{SH}_{\mathcal{T}}(S)$ denotes the category of motivic $\mathcal{T}$-spectra, whose objects are $\mathcal{T}$-spectra over S that satisfy Nisnevich descent and $\mathbb{A}^1$-homotopy invariance, and this is defined levelwise. Recall that a $\mathcal{T}$-spectrum is given by a sequence $\{F_n\}_{n\geq 0}$ of fibered pointed spaces over S, along with isomorphisms $F_n \xrightarrow{\cong} \Omega_{\mathcal{T}}(F_{n+1})$ for all $n \geq 0$, compatible with morphisms of spectra. Here $\Omega_{\mathcal{T}}$ is the $\mathcal{T}$-loop functor $\underline{\mathrm{Hom}}_S(\mathcal{T}, -)$, where $\underline{\mathrm{Hom}}_S$ is the internal hom of the category of smooth fibred spaces over S, which are functors $F : \mathrm{Sm}^{\mathrm{op}}_{/S} \to \mathrm{Spc}$ valued in the ∞-category of spaces.

Satisfying Nisnevich descent for a fibered space F over S is equivalent to satisfying Nisnevich excision, which means that for any Cartesian square of smooth derived S-schemes of the form

$$\begin{array}{ccc} U \times_X V & \longrightarrow & V \\ \downarrow & & \downarrow{\scriptstyle p} \\ U & \underset{j}{\longrightarrow} & X \end{array}$$

with étale p, an open immersion j, and a complementary closed immersion $Z \hookrightarrow X$ such that $p^{-1}Z \xrightarrow{\cong} Z$, the following commutative square in Spc is Cartesian:

$$\begin{array}{ccc} \Gamma(X,F) & \longrightarrow & \Gamma(U,F) \\ \downarrow & & \downarrow \\ \Gamma(V,F) & \longrightarrow & \Gamma(U \times_X V, F). \end{array}$$

21.4.2 Generalization

Since we have an embedding $(\mathrm{Sch}_k)_{/X} \hookrightarrow (\mathrm{dSt}_k)_{/F}$, the role played for us by fibered spaces over a scheme S, that is, functors $F : \mathrm{Sm}^{\mathrm{op}}_{/S} \to \mathrm{Spc}$, will be that of functors $\mathcal{F} : (\mathrm{dSt}_k)^{\mathrm{op}}_{/F} \to \mathrm{Top}$. Since we work with Segal categories, we are therefore looking at the Segal category $\mathbb{R}\underline{\mathrm{Hom}}((\mathrm{dSt}_k)^{\mathrm{op}}_{/F}, \mathrm{Top})$ of all such functors, the categorical manifestation in $\mathrm{Top} = L\mathrm{Set}_\Delta$ of $(\mathrm{dSt}_k)_{/F}$. This is a Segal category from which the category of derived motivic spectra can be constructed. Using strictification [78], which states that $\mathbb{R}\underline{\mathrm{Hom}}((\mathrm{dSt}_k)^{\mathrm{op}}_{/F}, \mathrm{Top}) \cong L\mathrm{Fun}((\mathrm{dSt}_k)^{\mathrm{op}}_{/F}, \mathrm{Set}_\Delta)$, we model objects $\mathcal{F} \in \mathbb{R}\underline{\mathrm{Hom}}((\mathrm{dSt}_k)^{\mathrm{op}}_{/F}, \mathrm{Top})$ by simplicial functors $\mathcal{F} : (\mathrm{dSt}_k)^{\mathrm{op}}_{/F} \to \mathrm{Set}_\Delta$, and we will be working with those objects. Here L refers to Dwyer–Kan's simplicial localization functor [15], which produces a simplicial category we regard as a Segal category.

Recall how this is defined: if $B \in \mathrm{Cat}_\Delta$ and $V \subset B$ is a subcategory, then one defines $L(B,V) = \mathrm{diag} F_* B[F_* V^{-1}] = (F_* B_n[F_* V_n^{-1}])_n$, where the standard resolution $F_* C \in \mathrm{Cat}_\Delta$ for some $C \in \mathrm{Cat}$ is defined by $F_n C = F^{n+1} C$, where FC is the free category on C with a generator Fc for every nonidentity morphism c of C. This definition is important for us since it shows that objects of LB and B are the same and that V-morphisms are sequentially crunched but other morphisms are otherwise left alone. We will use this quite a bit in projecting down notions of morphisms at the level of $\mathrm{D}^- k\text{-}\mathrm{Aff}^{\sim,\text{ét.}}$ to the level of $\mathrm{dSt}_k = L\mathrm{D}^- k\text{-}\mathrm{Aff}^{\sim,\text{ét.}}$ by essentially saying that morphisms in dSt_k have a property P if their lifts in $\mathrm{D}^- k\text{-}\mathrm{Aff}^{\sim,\text{ét.}}$ do have this property. Here we use the shorthand LC for $L(C,W)$.

21.4.3 The category of derived motivic spectra

We define the category of derived motivic spectra to be the category of motivic $\mathbb{P}^1_X$-spectra in $\mathbb{R}\underline{\mathrm{Hom}}((\mathrm{dSt}_k)^{\mathrm{op}}_{/F}, \mathrm{Top})$ ($\mathbb{P}^1_X$ will be defined later in Subsection 21.4.3.3). Objects $\mathcal{F}$ therein are said to be motivic if they are $\mathbb{A}^1$-local (or rather a generalization thereof, as we will soon see) and are Nisnevich local in a sense defined below. The difference from the classical case is that we do not insist on our functors to satisfy Nisnevich descent, since for one thing, this is classically equivalent to satisfying Nisnevich excision, but also because homotopy functors in Goodwillie calculus [30, 31] already satisfy a form of excision, and objects $\mathcal{F} \in \mathbb{R}\underline{\mathrm{Hom}}((\mathrm{dSt}_k)^{\mathrm{op}}_{/F}, \mathrm{Top})$ being homotopy functors, it is therefore natural to ask that they satisfy a form of excision.

To define $\mathbb{P}^1_X$, one needs a notion of $\mathbb{A}^1$-localization and a notion of Nisnevich localization in our context, so we first define the generalization of $\mathbb{A}^1$ in our setting and the notion of Nisnevich square as well, along with the accompanying left Bousfield localizations.

21.4.3.1 $\mathbb{A}^1$-localization

- *Generalization of* $\mathbb{A}^1$. Recall from [48] that a space F over S is $\mathbb{A}^1$-homotopy invariant if for all smooth $X \in \mathrm{Sm}_{/S}$, the canonical morphism of spaces $\Gamma(X, F) \to \Gamma(X \times \mathbb{A}^1, F)$ is an isomorphism, where $\mathbb{A}^1 = \mathrm{Spec}\, \mathbb{Z}[t]$ is the usual affine line. Presently, we generalize this to $\mathbb{A}^1_k = \mathrm{Spec}\, k[x]$, which we regard as a constant simplicial object of $sk\text{-Alg}^{\mathrm{op}}$. Then we let $\underline{\mathbb{A}^1_k} = \mathbb{R}\underline{h}_{\mathbb{A}^1_k}$, $\mathbb{R}\underline{h} : \mathrm{Ho}\,(sk\text{-Alg}^{\mathrm{op}}) \to \mathrm{Ho}\,(\mathrm{D}^-k\text{-Aff}^{\sim,\text{ét.}})$ the Yoneda embedding, and $\mathrm{D}^-k\text{-Aff}^{\sim,\text{ét.}} \subset \mathrm{Fun}(sk\text{-Alg}, \mathrm{Set}_\Delta)$ the model category of stacks on the site $(sk\text{-Alg}, \text{ét})$ [79].
- $\mathbb{A}^1_k$*-homotopy invariance.* We say $\mathcal{F} : (\mathrm{dSt}_k)^{\mathrm{op}}_{/F} \to \mathrm{Top}$ is $\underline{\mathbb{A}^1_k}$*-homotopy invariant* if for all $X \in (\mathrm{dSt}_k)_{/F}$, its lift $\mathcal{F} : (\mathrm{dSt}_k)^{\mathrm{op}}_{/F} \to \mathrm{Set}_\Delta$ satisfies $\mathcal{F}(X) \simeq \mathcal{F}(X \times \underline{\mathbb{A}^1_k})$. We regard $\underline{\mathbb{A}^1_k}$ as an object of dSt_k by using the fact that $\mathrm{Ho}\,(\mathrm{D}^-k\text{-Aff}^{\sim,\text{ét.}})$ and $\mathrm{D}^-k\text{-Aff}^{\sim,\text{ét.}}$ have the same objects and the fact that $\mathrm{dSt}_k = L\mathrm{D}^-k\text{-Aff}^{\sim,\text{ét.}}$ and $\mathrm{D}^-k\text{-Aff}^{\sim,\text{ét.}}$ have the same objects as well.
- $\mathbb{A}^1_k$*-localization.* Going over the work done in [48], it is clear that one can recast it in an equivalent formalism using left Bousfield localizations after a bit of an enhancement. By [40], Set_Δ is a left proper cellular model category, and hence so is $\mathrm{Fun}((\mathrm{dSt}_k)^{\mathrm{op}}_{/F}, \mathrm{Set}_\Delta) := \mathcal{M}$, so $\mathcal{C}$ being a class of morphisms therein, one can define the left Bousfield localization of this functor category with respect to $\mathcal{C}$; it is a model structure on $\mathcal{M}$ for which the weak equivalences are the $\mathcal{C}$-local equivalences, cofibrations are those of $\mathcal{M}$, and fibrations are those maps that have the right lifting property with respect to $\mathcal{C}$-local equivalences that are also cofibrations. Also, because $\mathcal{M}$ is a model category, one can define a homotopy function complex map on it. Recall that if $\mathcal{M}$ is a model category and $\mathcal{C}$ is a class of morphisms in $\mathcal{M}$, then $W \in \mathrm{Ob}\,(\mathcal{M})$ is said to be $\mathcal{C}$-local if it is fibrant and for all $f : A \to B$ in $\mathcal{C}$, the induced map $\mathrm{map}(B, W) \to \mathrm{map}(A, W)$ is a weak equivalence in Set_Δ. Then a morphism $g : X \to Y$ in $\mathcal{M}$ is said to be a $\mathcal{C}$-local equivalence if for all $\mathcal{C}$-local object W of $\mathcal{M}$, the induced map $\mathrm{map}(Y, W) \to \mathrm{map}(X, W)$ is a weak equivalence. Finally, by Theorem 4.1.1 of [40] fibrant objects of the left Bousfield localization $L_{\mathcal{C}}\mathcal{M}$ of $\mathcal{M}$ with respect to $\mathcal{C}$ are $\mathcal{C}$-local objects, and objects of model categories have fibrant replacements. Thus $L_{\mathcal{C}}\mathcal{M}$ is effectively a model category of $\mathcal{C}$-local objects, and $\mathrm{Ho}\,(L_{\mathcal{C}}\mathcal{M})$ is consequently a category of $\mathcal{C}$-local objects.

It is easy to define $\underline{\mathbb{A}_k^1}$-local equivalences; if $\mathcal{C} = \{X \times \mathbb{A}_k^1 \to X\}$ is a class of morphisms in $(\mathrm{dSt}_k)_{/F}$, it gives rise to a class $h_{\mathcal{C}} = \{h_{X\times\underline{\mathbb{A}_k^1}} \to h_X\}$ in $\mathbb{R}\underline{\mathrm{Hom}}((\mathrm{dSt}_k)_{/F}^{\mathrm{op}}, \mathrm{Top})$. Indeed, $(\mathrm{dSt}_k)_{/F}$ being a simplicial model category, it has its own simplicial mapping space $\mathrm{Map}_F := \mathrm{Map}_{(\mathrm{dSt}_k)_{/F}}$. This allows us to define $h_X = \mathrm{Map}_F(-, X) : (\mathrm{dSt}_k)_{/F}^{\mathrm{op}} \to \mathrm{Set}_\Delta$. As a first approximation, an object $\mathcal{F}$ of $\mathbb{R}\underline{\mathrm{Hom}}((\mathrm{dSt}_k)_{/F}^{\mathrm{op}}, \mathrm{Top})$ is said to be $\underline{\mathbb{A}_k^1}$-local if it is $\underline{\mathbb{A}_k^1}$-homotopy invariant, that is, for all $X \in (\mathrm{dSt}_k)_{/F}$, $\mathcal{F}(X) \to \mathcal{F}(X \times \underline{\mathbb{A}_k^1})$ is a weak equivalence of simplicial sets. Using Yoneda and the simplicial mapping space of $\mathrm{Fun}((\mathrm{dSt}_k)_{/F}^{\mathrm{op}}, \mathrm{Set}_\Delta)$, this reads $\mathrm{Map}(h_X, \mathcal{F}) \xrightarrow{\simeq} \mathrm{Map}(h_{X\times\underline{\mathbb{A}_k^1}}, \mathcal{F})$, which we generalize to the statement $\mathrm{map}(h_X, \mathcal{F}) \xrightarrow{\simeq} \mathrm{map}(h_{X\times\underline{\mathbb{A}_k^1}}, \mathcal{F})$, where map is the homotopy function complex of $\mathrm{Fun}((\mathrm{dSt}_k)_{/F}^{\mathrm{op}}, \mathrm{Set}_\Delta)$. We take this as our definition of being $\underline{\mathbb{A}_k^1}$*-local.* Then a morphism $\mathcal{F} \to \mathcal{F}'$ is said to be an $\underline{\mathbb{A}_k^1}$*-local equivalence* if for all $\underline{\mathbb{A}_k^1}$-local objects $\mathcal{G}$, $\mathrm{map}(\mathcal{F}', \mathcal{G}) \to \mathrm{map}(\mathcal{F}, \mathcal{G})$ is a weak equivalence.

21.4.3.2 Nisnevich localization

- *Choice of Nisnevich excision over Nisnevich descent – Goodwillie calculus.* We generalize the work done by Hoyois [39] and Hall and Rydh [38] on algebraic stacks to the level of derived stacks. The problem we encounter is that for a presheaf $\mathcal{F}$ on stacks to satisfy Nisnevich descent, one naturally has to use Nisnevich coverings, which use the notion of residual gerbes that seem very difficult to generalize to functors on derived stacks. However, Hoyois shows that a presheaf $\mathcal{F}$ on stacks satisfies Nisnevich descent if and only if it satisfies Nisnevich excision. Thus having this equivalence at the level of presheaves on algebraic stacks, we focus on functors $\mathcal{F} : (\mathrm{dSt}_k)_{/F}^{\mathrm{op}} \to \mathrm{Top}$ that satisfy Nisnevich excision, in a sense to be specified.
  Presently, we make an allusion to Goodwillie calculus [30, 31]. Goodwillie considers homotopy functors, functors from spaces to spaces, or spectra, that preserve weak equivalences, and for us, this would correspond to working with prestacks. Such functors are said to be excisive if they map homotopy pushout squares to homotopy pullback squares, and this again is reminiscent of the Nisnevich excision condition. But excisive functors are really helpful in developing calculus on functors, a functorial differential approach to working with prestacks, which is pertinent for us for the sake of deformations. In particular, Goodwillie [30] shows that with any given homotopy functor F, one can associate what he calls an n-excisive functor P_nF together with a map $F \to P_nF$, a sort of nth Taylor polynomial for F, and those agree to order n. Further, we have a tower of such polynomial approximations with a map $F \to \lim P_nF$. If F is ρ-analytic, a condition based on excision, and if $X \to Y$ is connected in a certain sense, then the connectivity of the map $F(X) \to (P_nF)(X)$ tends to infinity with n. Thus excision is at the heart of Goodwillie calculus. Since we work with motivic objects that are inherently differential in character, what we have here is yet another reason for choosing excision as a requirement for our functors.

We will adopt the definition of *Nisnevich square* of stacks by adapting that provided in [39] for algebraic stacks, that is, we consider a Cartesian square of stacks of the form

$$\begin{array}{ccc} W & \longrightarrow & V \\ \downarrow & & \downarrow f \\ U & \xrightarrow[e]{} & X \end{array}$$

with étale f and an open immersion e with reduced complement Z such that $Z \times_X V \xrightarrow{\cong} Z$. It turns out that we will use Zariski open immersions for derived stacks, but otherwise all the other notions of morphisms will be preserved. For Hoyois, a presheaf of spaces, or spectra, $\mathcal{F}$ on the category of algebraic stacks satisfies Nisnevich excision if the induced square

$$\begin{array}{ccc} \mathcal{F}(X) & \xrightarrow{e^*} & \mathcal{F}(U) \\ \downarrow f^* & & \downarrow \\ \mathcal{F}(V) & \longrightarrow & \mathcal{F}(W) \end{array}$$

is homotopy Cartesian. We will use the same formalism for functors $\mathcal{F} : (\mathrm{dSt}_k)^{\mathrm{op}}_{/F} \to \mathrm{Top}$ by simply replacing open immersions by Zariski open immersions as defined in [79]. As usual, we will work with representatives $\mathcal{F} : (\mathrm{dSt}_k)^{\mathrm{op}}_{/F} \to \mathrm{Set}_\Delta$ of such functors by way of the strictification theorem $\mathbb{R}\underline{\mathrm{Hom}}((\mathrm{dSt}_k)^{\mathrm{op}}_{/F}, \mathrm{Top}) \cong L\mathrm{Fun}((\mathrm{dSt}_k)^{\mathrm{op}}_{/F}, \mathrm{Set}_\Delta)$.

- *Nisnevich squares.*
  - *Smooth morphisms.* Classically, one works with Nisnevich squares in $\mathrm{Sm}_{/S}$, which have $(\mathrm{dSt}_k)_{/F}$ as analog for us, so we seek to define a notion of smoothness in dSt_k. First, the notions of smoothness in [48] and [79] at the level of simplicial algebras coincide. Then one uses the model Yoneda embedding. By [79], for a class Q of morphisms in $(sk\text{-Alg})^{\mathrm{op}} = \mathrm{D}^-k\text{-Aff}$ stable by equivalences and homotopy pullbacks and compositions, the Yoneda embedding $\mathbb{R}\underline{h} : \mathrm{Ho}(\mathrm{D}^-k\text{-Aff}) \to \mathrm{Ho}(\mathrm{D}^-k\text{-Aff}^{\sim,\text{ét.}})$ allows one to extend the notion of Q-morphism to the level of representable stacks. $Q = \mathrm{Sm}$ is one such class by Definition 1.2.7.2 of [79] and by virtue of Propositions 1.2.3.3 and 1.2.7.3. Thus one can talk about smooth morphisms between representable stacks. Finally, one uses the fact that $\mathrm{dSt}_k = L(\mathrm{D}^-k\text{-Aff}^{\sim,\text{ét.}})$. Thus smooth morphisms are well-defined in $(\mathrm{dSt}_k)_{/F}$.
  - *Zariski open immersions.* In [79], it is argued that Zariski open immersions reduce to the classical notion of open immersion in the event that we put the trivial model structure on the category of commutative monoid objects in our base model categories. Thus Zariski open immersions appear to be the correct

generalization in homotopical algebra of the classical notion of open immersion, and this is the notion we will use. By Proposition 2.2.2.5 of [79] a morphism $f : A \to B$ in sk-Alg is a Zariski open immersion if the induced morphism of affine schemes $\operatorname{Spec} \pi_0 B \to \operatorname{Spec} \pi_0 A$ is a Zariski open immersion and $A \to B$ is strong, that is, $\pi_* A \otimes_{\pi_0 B} \pi_0 A \to \pi_* B$ is an isomorphism. Here we use the standard definitions from [79]: $|A| = \operatorname{Map}_{sk\text{-Alg}}(1, A) \in \operatorname{Ho}(\mathrm{Set}_\Delta)$, and $\pi_i A = \pi_i(|A|, *)$. To say that a morphism of affine schemes $\operatorname{Spec} B \to \operatorname{Spec} A$ is a Zariski open immersion means that $A \to B$ is a Zariski open immersion in k-Alg, meaning it is flat, of finite presentation, and is such that $B \cong B \otimes_A B$. Then we use the fact that we identify sk-Alg with the category of representable stacks on it to transfer that notion of Zariski open immersion at the level of sk-Alg to the level of stacks. To be precise, if Q is a class of morphisms in $sk\text{-Alg}^{\mathrm{op}}$ stable by equivalences and homotopy pullbacks and compositions, then the Yoneda embedding $\mathbb{R}\underline{h} : \operatorname{Ho}(sk\text{-Alg}^{\mathrm{op}}) \to \operatorname{Ho}(\mathrm{D}^- k\text{-Aff}^{\sim, \text{ét.}})$ allows one to lift the notion of morphism in Q at the level of sk-Alg to the level of morphism of representable stacks, a fact we have used before. This is true for Zariski open immersions, which by Definition 1.2.6.7 of [79] are finitely presented and are formal Zariski open immersions. Both classes are stable by equivalences, compositions, and homotopy pushouts by Propositions 1.2.3.3 and 1.2.6.3. Thus we can talk of Zariski open immersions between representable stacks. We use the notation $\mathbb{R}\underline{\operatorname{Spec}} A = \mathbb{R}\underline{h}_{\operatorname{Spec} A}$ for $A \in sk$-Alg.

Observe that Toen and Vezzosi define a morphism of D^--stacks to be a Zariski open immersion if it is a locally finitely presented and flat monomorphism. Since these notions do not allow us to find a transparent notion of reduced complement for defining a Nisnevich square of stacks, we use representable stacks instead for simplicity. If this seems to be a bit restrictive, then recall that motives can be defined starting from schemes, which through their functors of points correspond to representable stacks, thus working with representable stacks within $(\mathrm{dSt}_k)_{/F}$ is sufficient for our purposes.

– *Reduced complement*. For having a Nisnevich square, we need the notion of a reduced complement of a Zariski open immersion of representable stacks. Following [80], if $A \to B$ is a finitely presented morphism of commutative rings, then $\operatorname{Spec} B \to \operatorname{Spec} A$ is an open immersion if and only if the restriction functor $D^-(B) \to D^-(A)$ between derived categories is fully faithful. Now starting from $\mathbb{R}\underline{\operatorname{Spec}} B \to \mathbb{R}\underline{\operatorname{Spec}} A$ being a Zariski open immersion of representable stacks, means $\operatorname{Spec} B \to \operatorname{Spec} A$ is a Zariski open immersion, that is, $A \to B$ is a Zariski open immersion in sk-Alg, so it is strong, and $\operatorname{Spec} \pi_0 B \to \operatorname{Spec} \pi_0 A$ is a Zariski open immersion, that is, $\pi_0 A \to \pi_0 B$ is a Zariski open immersion, so finitely presented and a formal Zariski open immersion by Definition 1.2.6.7 of [79], which means flat and $f_* : \operatorname{Ho}(\pi_0 B\text{-Mod}) \to \operatorname{Ho}(\pi_0 A\text{-Mod})$ fully faithful, inducing a fully faithful map at the level of derived categories. It follows that $\operatorname{Spec} \pi_0 B \to \operatorname{Spec} \pi_0 A$ is an open immersion.

To recap, $\underline{\mathbb{R}\mathrm{Spec}\, B} \to \underline{\mathbb{R}\mathrm{Spec}\, A}$ Zariski open immersion means $\mathrm{Spec}\, \pi_0 B \to \mathrm{Spec}\, \pi_0 A$ is an open immersion, so $|\mathrm{Spec}\, \pi_0 B| \cong U^{\mathrm{open}} \subset \mathrm{Spec}\, \pi_0 A$, and it has a closed complement $Z = |\mathrm{Spec}\, \pi_0 A| - U$ on which we can put a reduced closed subscheme structure.

Now because Z is a closed subscheme of an affine scheme $\mathrm{Spec}\, \pi_0 A$, it is affine itself, $Z = \mathrm{Spec}\, C$, regard $C = \pi_0 C_\bullet$, $C_\bullet \in sk\text{-Alg}$ being C viewed as a constant simplicial object of k-Alg, with accompanying representable stack $\underline{\mathbb{R}\mathrm{Spec}\, C_\bullet}$, along with a morphism to $\underline{\mathbb{R}\mathrm{Spec}\, A}$, which we regard as the reduced complement to the Zariski open immersion $\underline{\mathbb{R}\mathrm{Spec}\, B} \to \underline{\mathbb{R}\mathrm{Spec}\, A}$.

- *Étale morphisms*. By Definition 1.2.6.7 of [79], étale morphisms are finitely presented and formally étale, and by Propositions 1.2.3.3 and 1.2.6.3, respectively, those are stable by equivalences, compositions, and homotopy pushouts, and thus we can lift the notion of étale morphism at the level of sk-Alg to the level of representable stacks in $\mathrm{D}^- k\text{-Aff}^{\sim, \text{ét.}}$ by using the usual argument. Finally, we invoke the fact that $\mathrm{dSt}_k = L\mathrm{D}^- k\text{-Aff}^{\sim, \text{ét.}}$ to argue that étale morphisms are well-defined in $(\mathrm{dSt}_k)_{/F}$.

- *Nisnevich localization*. In our present situation of working with functors $\mathcal{G} : (\mathrm{dSt}_k)^{\mathrm{op}}_{/F} \to \mathrm{Set}_\Delta$, in a first approximation, $\mathcal{G}$ is said to be *Nis-local* if it satisfies Nisnevich excision, and a morphism $\mathcal{F} \to \mathcal{F}'$ of objects of $\mathrm{Fun}((\mathrm{dSt}_k)^{\mathrm{op}}_{/F}, \mathrm{Set}_\Delta)$ is said to be a *Nis-local equivalence* if for all Nis-local $\mathcal{G}$ in this functor category, $\mathrm{map}(\mathcal{F}', \mathcal{G}) \xrightarrow{\simeq} \mathrm{map}(\mathcal{F}, \mathcal{G})$ in Set_Δ.

It is more difficult to define a left Bousfield localization with respect to Nisnevich squares than it was to define a left Bousfield localization with respect to maps of the form $X \times \underline{\mathbb{A}^1_k} \to X$. The difficulty resides in taking as our class of morphisms for left Bousfield localization purposes a class of squares. We show this can be done presently.

Recall that to satisfy Nisnevich excision means that for any Nisnevich square

$$\begin{array}{ccc} W & \longrightarrow & V \\ \downarrow & & \downarrow \\ U & \longrightarrow & X \end{array}$$

the square

$$\begin{array}{ccc} \mathcal{F}(X) & \longrightarrow & \mathcal{F}(U) \\ \downarrow & & \downarrow \\ \mathcal{F}(V) & \longrightarrow & \mathcal{F}(W) \end{array}$$

is homotopy Cartesian, which means $\mathcal{F}(X) \xrightarrow{\simeq} \mathcal{F}(V) \times^h_{\mathcal{F}(W)} \mathcal{F}(U)$. Now Theorem 18.1.10 of [40] states that if $\mathcal{M} \in \mathrm{Cat}_\Delta$, $\mathcal{C}$ is a small category, $\mathcal{X}$ is a $\mathcal{C}$-diagram in

$\mathcal{M}$, and Y is an object of $\mathcal{M}$, then we have an isomorphism $\mathrm{Map}(\mathrm{hocolim}_{\mathcal{C}}\mathcal{X}, Y) \cong \mathrm{holim}_{\mathcal{C}^{\mathrm{op}}}\mathrm{Map}(\mathcal{X}, Y)$, Map being the simplicial mapping space of $\mathcal{M}$. Applying this to $\mathcal{M} = \mathrm{Fun}((\mathrm{dSt}_k)_{/F}^{\mathrm{op}}, \mathrm{Set}_\Delta)$ with $\mathcal{C} = \{U \leftarrow W \rightarrow V\}$, $Y = \mathcal{F}$, and $h = \mathcal{X} : \mathcal{C} \rightarrow \mathcal{M}$, we have $\mathcal{F}(U) \times^h_{\mathcal{F}(W)} \mathcal{F}(V) \simeq \mathcal{F}(V \coprod^h_W U)$. Thus we see that $\mathcal{F}$ satisfying Nisnevich excision can be recast as saying $\mathcal{F}$ is local with respect to maps in $\mathcal{C} = \{V \coprod^h_W U \rightarrow X\}$ induced by Nisnevich squares. Thus satisfying Nisnevich excision is equivalent to being local with respect to Nisnevich squares, which we just refer to as being Nis-local. This is a first approximation. For the sake of using left Bousfield localizations, we generalize this as follows. Let $h_{\mathcal{C}} = \{h_{U \coprod^h_W V} \rightarrow h_X\}$ be the class of morphisms of $\mathrm{Fun}((\mathrm{dSt}_k)_{/F}^{\mathrm{op}}, \mathrm{Set}_\Delta)$ induced by the above class $\mathcal{C} = \{U \coprod^h_W V \rightarrow X\}$ in $(\mathrm{dSt}_k)_{/F}$, and replace the statement

$$\mathrm{Map}(h_X, \mathcal{F}) \simeq \mathcal{F}(X) \xrightarrow{\simeq} \mathcal{F}(U \coprod_W^h V) \simeq \mathrm{Map}(h_{V \coprod^h_W U}, \mathcal{F})$$

by $\mathrm{map}(h_X, \mathcal{F}) \xrightarrow{\simeq} \mathrm{map}(h_{V \coprod^h_W U}, \mathcal{F})$. We take this as our definition of being Nis-local. Then $\mathcal{F} \rightarrow \mathcal{F}'$ is a Nis-local equivalence if for any Nis-local object $\mathcal{G}$, we have that $\mathrm{map}(\mathcal{F}', \mathcal{G}) \rightarrow \mathrm{map}(\mathcal{F}, \mathcal{G})$ is a weak equivalence of simplicial sets.

21.4.3.3 Generalization of $\mathbb{P}^1_S$

Recall how $\mathbb{P}^1_S$ was constructed in [48]: $\mathbb{P}^1_S = (M_S(\mathbb{P}^1_S), \infty)$, where if $X \in \mathrm{Sm}_{/S}$, $M_S(X) = \mathrm{L}_{\mathrm{mot}}(h_S(X))$, where for a space F over S, $\mathrm{L}_{\mathrm{mot}}(F) = \lim(\mathrm{L}_{\mathbb{A}^1} \circ \mathrm{L}_{\mathrm{Nis}})^n(F)$ where $\mathrm{L}_{\mathbb{A}^1}$ is an accessible localization onto $\mathbb{A}^1$-homotopy invariant spaces, $\mathrm{L}_{\mathrm{Nis}}$ is the left exact localization onto Nisnevich local spaces. We also have $h_S(X) = \mathrm{Map}_S(-, X) : \mathrm{Sm}_{/S}^{\mathrm{op}} \rightarrow \mathrm{Spc}$. Going over the construction of $\mathbb{P}^1_S$ in [48], it becomes clear that one can generalize it to our setting where we work with derived schemes in dSch_k, and thus $\mathbb{P}^1_{iX} \in \mathrm{dSch}_k$ is well-defined.

We then use the fact that $\mathrm{dSch}_k \in \mathrm{Cat}_\Delta$, hence it has a simplicial mapping object, and hence so does $(\mathrm{dSch}_k)_{/iX}$. We denote its simplicial mapping space by Map_X. This allows us to see $h_{\mathbb{P}^1_{iX}} = \mathrm{Map}_X(-, \mathbb{P}^1_{iX}) : (\mathrm{dSch}_k)_{/iX}^{\mathrm{op}} \rightarrow \mathrm{Set}_\Delta$ as an object of $(\mathrm{dSt}_k)_{/F}$. Now we need to see this as a functor $(\mathrm{dSt}_k)_{/F}^{\mathrm{op}} \rightarrow \mathrm{Set}_\Delta$. This is achieved by taking its functor of points $\mathrm{Map}_F(-, h_{\mathbb{P}^1_{iX}}) = h_F \circ h_{\mathbb{P}^1_{iX}}$. We then take the motivic localization $\mathrm{L}_{\mathrm{mot}} = \mathrm{L}_{\underline{\mathbb{A}^1_k} \cup \mathrm{Nis}}$ of this object. Here $\mathrm{L}_{\underline{\mathbb{A}^1_k} \cup \mathrm{Nis}}$ is a left Bousfield localization of $\mathrm{Fun}((\mathrm{dSt}_k)_{/F}^{\mathrm{op}}, \mathrm{Set}_\Delta)$ with respect to the class $\{h_{X \times \underline{\mathbb{A}^1_k}} \rightarrow h_X\} \cup \{h_{U \coprod^h_W V} \rightarrow h_X\}$, the first subclass being informally referred to as $\underline{\mathbb{A}^1_k}$, and the second one by Nis, and again it is induced by Nisnevich squares $W \rightrightarrows U, V \rightrightarrows X$.

We define the category of derived motivic spaces by

$$\mathrm{Mot}(F) = \mathrm{Ho}\,(\mathrm{L}_{\underline{\mathbb{A}^1_k} \cup \mathrm{Nis}} \mathrm{Set}_\Delta^{(\mathrm{dSt}_k)_{/F}^{\mathrm{op}}}).$$

We regard $\mathrm{L}_{\mathrm{mot}}h_F \circ h_{\mathbb{P}^1_{iX}}$ as an object of this category, and we still denote it by $\mathbb{P}^1_X$ for simplicity.

21.4.3.4 Localizations

Presently we have the following picture:

$$\begin{array}{ccccc}
\mathrm{Set}_\Delta^{(\mathrm{dSt}_k)^{\mathrm{op}}_{/F}} & \xrightarrow{\mathrm{L}_{\underline{\mathbb{A}^1_k}\cup\mathrm{Nis}}} & \mathrm{L}_{\underline{\mathbb{A}^1_k}\cup\mathrm{Nis}}\mathrm{Set}_\Delta^{(\mathrm{dSt}_k)^{\mathrm{op}}_{/F}} & \xrightarrow{\mathrm{Ho}} & \mathrm{Mot}(F) \\
\uparrow{\scriptstyle L^{-1}} & & & & \downarrow{\scriptstyle L} \\
\mathbb{R}\underline{\mathrm{Hom}}((\mathrm{dSt}_k)^{\mathrm{op}}_{/F}, \mathrm{Top}) & & & & \mathrm{SeMot}(F)
\end{array}$$

where $\mathrm{SeMot}(F) = L\mathrm{Mot}(F)$ is the Segal category of derived motivic spaces on F, whose objects are also those of $\mathbb{R}\underline{\mathrm{Hom}}((\mathrm{dSt}_k)^{\mathrm{op}}_{/F}, \mathrm{Top})$. This holds by virtue of the fact that the equivalences in $\mathrm{Fun}((\mathrm{dSt}_k)^{\mathrm{op}}_{/F}, \mathrm{Set}_\Delta)$ used in the simplicial localization on the left in the above diagram are not involved in the left Bousfield localization $\mathrm{L}_{\underline{\mathbb{A}^1_k}\cup\mathrm{Nis}}$ and hence are preserved throughout and can be used later as in the right-hand side. This also means that objects of the Segal category $\mathrm{SeMot}(F)$ can by modeled in $\mathrm{Mot}(F)$ by functors $\mathcal{F} : (\mathrm{dSt}_k)^{\mathrm{op}}_{/F} \to \mathrm{Set}_\Delta$. We will use this fact in the next subsection.

21.4.3.5 $\mathcal{T}$-spectra

Recall from [48] that objects of the category $\mathcal{SH}_\mathcal{T}(S)$ of motivic $\mathcal{T}$-spectra are sequences $\{F_n\}_{n\geq 0}$ of fibered pointed spaces with structural isomorphism $F_n \to \Omega_\mathcal{T}(F_{n+1})$, where $\Omega_\mathcal{T} = \underline{\mathrm{Hom}}(\mathcal{T}, -)$, using the internal hom of the category of spaces over S. We mention in passing the fact that we will not work with pointed spaces, since we will not need it in our work, but the conscientious reader will have no difficulty making the proper changes for the pointed setting.

We generalize this in our context by defining the category of derived motivic $\mathcal{T}$-spectra, denoted $\mathrm{MotSp}_\mathcal{T}(F)$, $F = h_{iX} \in \mathrm{dSt}_k$ if $X \in \mathrm{Sch}_k$, with $\mathcal{T}$ an object of $\mathrm{SeMot}(F)$. The objects of $\mathrm{MotSp}_\mathcal{T}(F)$ are sequences $\{\mathcal{F}_n\}_{n\geq 0}$ of objects of $\mathrm{SeMot}(F)$ along with structural isomorphisms $\mathcal{F}_n \cong \Omega_\mathcal{T}(\mathcal{F}_{n+1})$, $\Omega_\mathcal{T} = \underline{\mathrm{Hom}}_F(\mathcal{T}, -)$, $\underline{\mathrm{Hom}}_F$ the internal Hom of $\mathrm{SeMot}(F)$, which we define as follows.

Set_Δ is a symmetric monoidal model category with an internal hom defined as follows. For $K, L \in \mathrm{Set}_\Delta$, one defines the function complex $\mathrm{Map}(K, L)$ in degree n by $\mathrm{Map}(K, L)_n = \mathrm{Hom}_{\mathrm{Set}_\Delta}(K \times \Delta_n, L)$. For three objects $\mathcal{F}$, $\mathcal{G}$, $\mathcal{H}$ of $\mathrm{SeMot}(F)$ modeled as functors $(\mathrm{dSt}_k)^{\mathrm{op}}_{/F} \to \mathrm{Set}_\Delta$, we have

$$\mathrm{Hom}_{\mathrm{Set}_\Delta}(\mathcal{F}(U) \times \mathcal{G}(U), \mathcal{H}(U)) \cong \mathrm{Hom}_{\mathrm{Set}_\Delta}(\mathcal{F}(U), \mathrm{Map}(\mathcal{G}(U), \mathcal{H}(U))$$

for $U \in (\mathrm{dSt}_k)_{/F}$. For a given functor $\mathcal{F} : \mathcal{C}^{\mathrm{op}} \to \mathcal{D}$, define $\mathcal{F}^{/\mathrm{op}} : \mathcal{C} \to \mathcal{D}$. Now define $\mathcal{F} \otimes \mathcal{G}$ and $\underline{\mathrm{Hom}}(\mathcal{G}, \mathcal{H})$ pointwise by $(\mathcal{F} \otimes \mathcal{G})(U) = \mathcal{F}(U) \times \mathcal{G}(U)$ and $\underline{\mathrm{Hom}}(\mathcal{G}, \mathcal{H})(U) =$

$\mathrm{Map}(\mathcal{G}^{/\mathrm{op}}(U), \mathcal{H}(U))$. In this manner, we have $\underline{\mathrm{Hom}}(\mathcal{G}, \mathcal{H})$ and $\mathcal{F} \otimes \mathcal{G}$ being functors $(\mathrm{dSt}_k)^{\mathrm{op}}_{/F} \to \mathrm{Set}_\Delta$. These being defined, we have

$$\mathrm{Hom}_{\mathrm{Mot}(F)}(\mathcal{F} \otimes \mathcal{G}, \mathcal{H}) \cong \mathrm{Hom}_{\mathrm{Mot}(F)}(\mathcal{F}, \underline{\mathrm{Hom}}(\mathcal{G}, \mathcal{H}))$$

which projects down to the same statement at the level of $\mathrm{SeMot}(F)$ after simplicial localization, if we look at individual elements of these hom sets, and we use the fact that objects of $\mathrm{Mot}(F)$ are also those of $\mathrm{SeMot}(F)$. Thus

$$\mathrm{Hom}_{\mathrm{SeMot}(F)}(\mathcal{F} \otimes \mathcal{G}, \mathcal{H}) \cong \mathrm{Hom}_{\mathrm{SeMot}(F)}(\mathcal{F}, \underline{\mathrm{Hom}}(\mathcal{G}, \mathcal{H})).$$

21.4.3.6 Construction of MotSp(*X*)

We define $\mathrm{MotSp}(X) = \mathrm{MotSp}_{\mathbb{P}^1_X}(F)$ to be the category of derived motivic spectra on $X \in \mathrm{Sch}_k$ with $F = h_{iX}$ whose objects are $\mathbb{P}^1_X$-spectra of functors $\mathcal{F} : (\mathrm{dSt}_k)^{\mathrm{op}}_{/F} \to \mathrm{Set}_\Delta$ – modeling objects of $\mathbb{R}\underline{\mathrm{Hom}}((\mathrm{dSt}_k)^{\mathrm{op}}_{/F}, \mathrm{Top})$ – that are $\underline{\mathbb{A}^1_k}$-local and Nis-local. $\mathrm{MotSp}(X)$ appears first as a delocalization, followed by a left Bousfield localization, upon which we take the corresponding homotopy category, followed by a simplicial localization, to finish with a spectralization as in the diagram

$$\begin{array}{ccccc}
\mathrm{Set}_\Delta^{(\mathrm{dSt}_k)^{\mathrm{op}}_{/F}} & \xrightarrow{\mathrm{L}_{\underline{\mathbb{A}^1_k} \cup \mathrm{Nis}}} & \mathrm{L}_{\underline{\mathbb{A}^1_k} \cup \mathrm{Nis}} \mathrm{Set}_\Delta^{(\mathrm{dSt}_k)^{\mathrm{op}}_{/F}} & \xrightarrow{\mathrm{Ho}} & \mathrm{Mot}(F) \\
\uparrow {\scriptstyle L^{-1}} & & & & \downarrow {\scriptstyle L} \\
\mathbb{R}\underline{\mathrm{Hom}}((\mathrm{dSt}_k)^{\mathrm{op}}_{/F}, \mathrm{Top}) & & & & \mathrm{SeMot}(F) \\
 & & & & \downarrow {\scriptstyle \mathrm{Sp}_{\mathbb{P}^1_X}} \\
 & & & & \mathrm{MotSp}(X).
\end{array}$$

Looking ahead, if one is interested in making contact with the theory of motives, then one has to find an appropriate topology on $(\mathrm{dSt}_k)^{\mathrm{op}}_{/F}$ to define stacks in $\mathrm{MotSp}(X)$, take the derived category thereof, take the analog in our situation of a $\mathrm{D}_{\mathbb{A}^1}$ localization, producing the equivalent for us of the derived categories of sheaves of [9], and possibly use an appropriate reduction to get a category equivalent to $\mathrm{DM}_B(X)$. Observe that this subsumes developing a six-functor formalism in the present setting, something we have not touched upon.

22 Goodwillie calculus and geometric stacks

22.1 Introduction

The present section provides an illustration of a (putative) higher categorical generalization of the bridge technique of [7] relating Morita equivalent theories $\mathbb{T}$ and $\mathbb{T}'$ via the equivalence of their respective classifying topos $\mathcal{E}_{\mathbb{T}}$ and $\mathcal{E}_{\mathbb{T}'}$ as in

$$\begin{array}{ccc} & \mathcal{E}_{\mathbb{T}} \simeq \mathcal{E}_{\mathbb{T}'} & \\ \nearrow & & \searrow \\ \mathbb{T} & & \mathbb{T}'. \end{array}$$

We regard theories as being inert objects in contrast to their classifying topoi, which we regard as dynamical representations thereof. Indeed, for a geometric theory $\mathbb{T}$, its classifying topos $\mathcal{E}_{\mathbb{T}}$ is $\mathrm{Sh}(\mathcal{C}_{\mathbb{T}}, J_{\mathbb{T}})$, the Grothendieck topos of sheaves on its syntactic category $\mathcal{C}_{\mathbb{T}}$. This provides a dynamic manifestation of $\mathbb{T}$ insofar as it is functorial. On can generalize this to the higher categorical setting by considering functors valued in Set_{Δ}, Sp, the category of spectra, or $\mathcal{S}$, the category of spaces. In particular, we consider the category of homotopy functors $\mathcal{F}(\mathcal{C}, \mathcal{D})$ from $\mathcal{C}$ to $\mathcal{D}$, both being categories of spaces with a notion of weak equivalence, and $\text{k-D}^{-}\mathrm{Aff}^{\sim, \text{ét.}}$, the category of D^{-}-stacks on sk-Alg for the étale topology, both of which are higher categories. Goodwillie calculus has homotopy functors $F : \mathcal{C} \to \mathcal{D}$ as objects of study, and such a “theory”, in the pedestrian sense, is built upon a theory that we denote by $\mathbb{G}$. In the same manner, derived stacks are functors $F : sk\text{-Alg} \to \mathrm{Set}_{\Delta}$ satisfying certain conditions, objects of $\text{k-D}^{-}\mathrm{Aff}^{\sim, \text{ét.}}$, whose formalism is developed following a theory that we will denote by $\mathbb{S}$. Whether it be $\mathbb{G}$ or $\mathbb{S}$, those are abstractions of the defining features of the respective higher categorical constructs they are meant to govern.

We argue different formalisms, such as Goodwillie calculus and the theory of n-geometric D^{-}-stacks, without necessarily being equivalent, at least sharing some attributes, which we refer to as aspects, collectively denoted by a, and we write

$$\mathcal{F}(\mathcal{C}, \mathcal{D}) \overset{a}{\simeq} \text{k-D}^{-}\mathrm{Aff}^{\sim, \text{ét.}}$$

to indicate that Goodwillie calculus in $\mathcal{F}(\mathcal{C}, \mathcal{D})$ and n-geometric D^{-}-stacks in $\text{k-D}^{-}\mathrm{Aff}^{\sim, \text{ét.}}$ share some common features, namely those in a. The object of the present work is to provide evidence for the existence of such features. Ultimately, our aim is to show that this leads to an underlying theory $\mathbb{T}_0$, from which both $\mathbb{G}$ and $\mathbb{S}$ originate. In the bridge technique formalism, this may be presented as follows:

https://doi.org/10.1515/9783111334073-022

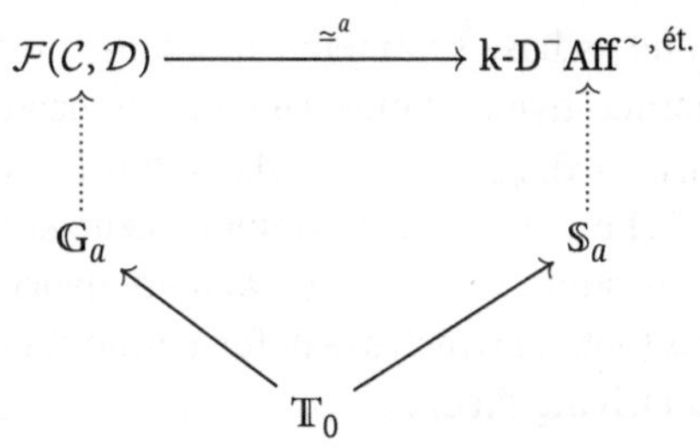

where $\mathbb{T}_a$ is that part of $\mathbb{T}$ necessary to develop those aspect a of a formalism $\mathcal{E}_{\mathbb{T}}$. In the above diagram, $\mathbb{G}_a$ and $\mathbb{S}_a$ are seen as expansions of $\mathbb{T}_0$. Observe that a theory $\mathbb{T}$ can be seen as an expansion of $\mathbb{T}_a$, and thus the bottom part of the above diagram can be completed as

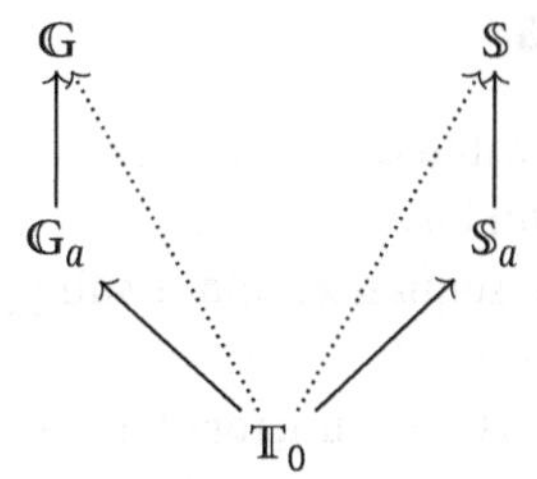

a maximal case corresponding to considering all aspects of both formalisms, in which case one recovers a higher categorical generalization of the bridge technique per se.

Though written with the aim to extract a theory $\mathbb{T}_0$ from $\mathbb{G}$ and $\mathbb{S}$, the latter theories are expansions of, we will not discuss theories themselves until the very end of this chapter. When we use the word “theory” throughout, we really mean formalism.

In a first time, we draw comparisons between certain results pertaining to the calculus of functors of Goodwillie [29, 30, 31] and some aspects of the theory of n-geometric D^--stacks [76, 79], rather than being a direct application of the former theory to the latter. Thus our aim is to show that those two different theories share some common features, something we would like to abstract further. We focus in particular on the Taylor towers for homotopy functors, on their built-in polynomials and derivatives thereof on the one hand, and on the convergence of towers built from truncations $\mathbb{R}F(A_{\leq n})$ for $A \in sk$-Alg and an n-geometric D^--stack F on the other. After having drawn some comparisons, we study a few parallel results, namely the similarities between the construction of polynomial approximations, pointwise equivalences, and a reconstruction theorem for towers. In the last part of the present work, we go a little deeper into each theory, which allows us to formally abstract their common features into an overarching theory.

Goodwillie observed that with a homotopy functor $F : \mathcal{C} \to \mathcal{D}$ between categories of spaces, that is, a functor that preserves equivalences, one can associate a Taylor tower $\cdots \to P_nF(X) \to \cdots \to P_1F(X) \to P_0F(X)$ for $X \in \mathcal{C}$ and that under a certain condition called analyticity, this Taylor tower converges to $F(X)$. In Derived Algebraic Geometry, one starts with sk-Mod as base symmetric monoidal model category, with sk-Alg as model category of commutative monoids. One can associate with any $A \in sk$-Alg a Postnikov tower $\cdots \to A_{\leq n} \to A_{\leq n-1} \to \cdots \to A_0$, and if F is an n-geometric D^--stack, then Theorem C.0.9 of [79] shows one has an isomorphism $\mathbb{R}F(A) \to \operatorname{holim}_n \mathbb{R}F(A_{\leq n})$, a fact that spurred the present study. It turns out that both theories, though very different, share many common traits (both of which are deformation-theoretic in nature), thereby pointing to a common underlying theory.

Goodwillie calculus was initiated in the seminal papers [29], [30], and [31] and pursued further in a host of later publications, some of which the reader can find in the reference section. In particular, we use [2]. For stacks, we essentially use the foundational papers [76] and [79].

22.2 Goodwillie calculus

We will start our study of Goodwillie calculus by one of the main results of [31], namely that certain types of functors can have approximations by polynomial-like approximations displayed in a Taylor-like sequence, and a subtype of such functors have those Taylor towers converging to them.

To be precise, Goodwillie considers functors $F : \mathcal{C} \to \mathcal{D}$ between categories of spaces (based or unbased), and sometimes $\mathcal{D}$ is even taken to be a category of spectra. In particular, one considers functors that preserve weak equivalences. Such functors are referred to as *homotopy functors*.

Having a Taylor expansion presupposes that a point be fixed, and the expansion is relative to some variable in a neighborhood of such a fixed point. This is implemented in the present situation by working with slice categories; if Y is a fixed space, then we look at $\mathcal{S}_{/Y}$, the category of based spaces over Y, or $\mathcal{U}_{/Y}$, the category of unbased spaces over Y, which we generically denote by $\mathcal{C}_Y$.

The notion of a neighborhood is then implemented by considering the connectedness of maps $X \to Y$ for objects $X \in \mathcal{C}_Y$.

One can associate with any homotopy functor $F : \mathcal{C}_Y \to \mathcal{D}$ a *Taylor tower* of homotopy functors P_nF:[1]

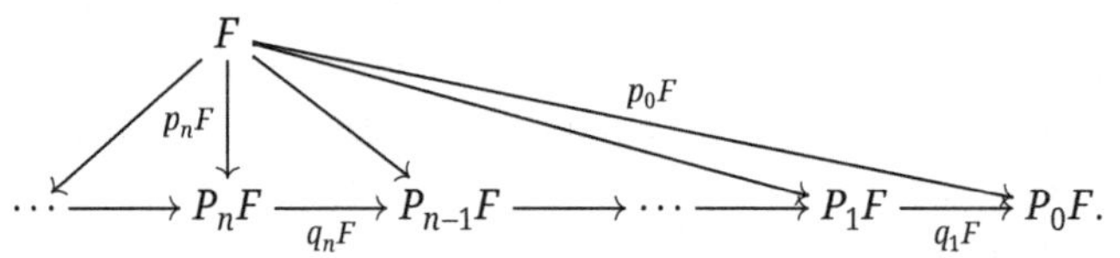

The different constituents in this tower are constructed as follows. Let $F : \mathcal{C}_Y \to \mathcal{D}$ be a homotopy functor, where $\mathcal{C}$ and $\mathcal{D}$ are the category of based or unbased spaces, and $\mathcal{D}$

1 Thm. 1.13 [31].

could also be the category Sp of spectra. For $X \in \mathcal{C}_Y$ and any space U, one defines the *fiberwise join over* Y [30] by

$$X *_Y U = \operatorname{hocolim}(X \leftarrow X \times U \rightarrow Y \times U).$$

Let $\underline{n+1} = \{1, \ldots, n\}$, and denote by $\mathcal{P}(\underline{n+1})$ the poset of subsets of $\underline{n+1}$. Let $\mathcal{P}_0(\underline{n+1}) \subset \mathcal{P}(\underline{n+1})$ be the poset of nonempty subsets of $\underline{n+1}$. Define

$$T_n F(X) = \operatorname*{holim}_{U \in \mathcal{P}_0(\underline{n+1})} F(X *_Y U).$$

This provides a homotopy functor $T_nF : \mathcal{C}_Y \to \mathcal{D}$, along with a natural transformation $F \xrightarrow{t_nF} T_nF$, which pointwise reads

$$F(X) = F(X *_Y \emptyset) \xrightarrow{t_nF(X)} \operatorname*{holim}_{U \in \mathcal{P}_0(\underline{n+1})} F(X *_Y U) = T_nF(X).$$

We denote by $P_nF(X)$ the sequential homotopy colimit of

$$F(X) \xrightarrow{t_nF(X)} T_nF(X) \xrightarrow{t_n(T_nF)(X)} T_n^2F(X) \rightarrow \cdots.$$

This provides yet another homotopy functor $P_nF : \mathcal{C}_Y \to \mathcal{D}$, along with a natural map $F \xrightarrow{p_nF} P_nF$. By construction we also have maps $q_{n,k} : T_n^kF \to T_{n-1}^kF$. To see that this is the case, observe that

$$T_n^kF(X) \cong \operatorname*{holim}_{(U_1,\ldots,U_k) \in \mathcal{P}_0(\underline{n+1})^k} F(X *_Y (U_1 * \cdots * U_k)).$$

Since we have an inclusion $\mathcal{P}_0(\underline{n})^k \hookrightarrow \mathcal{P}_0(\underline{n+1})^k$, it follows that we have an induced map $T_n^kF \xrightarrow{q_{n,k}} T_{n-1}^kF$ as desired and hence an induced map $q_nF : P_nF \to P_{n-1}F$ of horizontal homotopy colimits. Along with the maps $F \xrightarrow{p_nF} P_nF$, this provides us with the desired Taylor tower.

We now discuss one important characteristic of the functor P_nF. To do so, we have to introduce *cubical diagrams* [30], functors $\mathcal{X} : \mathcal{P}(S) \to \mathcal{C}$, a finite set S, the poset $\mathcal{P}(S)$ of all subsets of S, and $\mathcal{C}$ one of our categories of spaces. For instance, if $S = \{1, 2\}$, then $\mathcal{P}(S) = \{\emptyset, \{1\}, \{2\}, \{1, 2\}\}$, so that the diagram $\mathcal{X}$ looks like

$$\begin{array}{ccc}
\mathcal{X}(\emptyset) = \lim \mathcal{X} & \longrightarrow & \mathcal{X}(1) \\
\downarrow & & \downarrow \\
\mathcal{X}(1) & \longrightarrow & \mathcal{X}(1,2)
\end{array}$$

whence the name cubical. If S has cardinality n, then $\mathcal{X}$ is referred to as an n-cube. Denote by $\mathcal{P}_1(S)$ the set $\{T \mid T \subsetneq S\}$. We say $\mathcal{X}$ is *coCartesian*[2] if the map

$$\operatorname{hocolim}(\mathcal{X}|_{\mathcal{P}_1(S)}) \to \operatorname{colim}(\mathcal{X})$$

is a weak equivalence. We say $\mathcal{X}$ is *strongly coCartesian*[3] if each of its faces of dimension greater than 2 is coCartesian. In contrast to being coCartesian, a cubical diagram $\mathcal{X}$ is said to be *Cartesian*[4] if the map

$$\mathcal{X}(\emptyset) = \lim \mathcal{X} \to \operatorname{holim} \mathcal{X}|_{\mathcal{P}_0(S)}$$

is a weak equivalence. We can now characterize our functors P_nF: a homotopy functor $F : \mathcal{C} \to \mathcal{D}$ is said to be *n-excisive*[5] if for every strongly coCartesian $(n+1)$-cubical diagram $\mathcal{X}$, the composite cubical diagram $F \circ \mathcal{X} : \mathcal{P}(S) \to \mathcal{D}$ is Cartesian. By Theorem 1.8 of [31], P_nF is n-excisive. A weaker condition referred to as stable nth-order excision (see below) is satisfied in particular by analytic functors. By Proposition 1.5 of [31], F n-stably excisive agrees with P_nF (which is nth-order excisive) to order n in a certain sense. Essentially, the discrepancy between the two notions morally vanishes as n tends to ∞: if F is ρ-analytic (in particular, n-stably excisive) for any $X \in \mathcal{C}_Y$ such that $X \to Y$ is $(\rho+1)$-connected, i. e., X is close enough to Y, then $F(X) \simeq \operatorname{holim} P_nF(X)$.[6]

The interpretation of P_nF as a Taylor polynomial however is not immediate. To see this clearly, one has to consider the homotopy fiber of the map $P_nF(X) \to P_{n-1}F(X)$ for $X \in \mathcal{C}$, which is denoted $D_nF(X)$ in [31]. This should correspond to $\frac{f^{(n)}(a)}{n!}(x-a)^n$ in calculus. To see that we have something similar in form, it is convenient to work with spectra: $\mathcal{D} = \mathrm{Sp}$. In this case,

$$D_nF(X) = (C_n \wedge (X^{\wedge n}))_{h\Sigma_n}$$

where C_n is a spectrum with a Σ_n-action, and $h\Sigma_n$ stands for the homotopy orbit spectrum.

We now focus on the conditions under which we have the convergence of the Taylor tower of a homotopy functor F. As mentioned above, by Theorem 1.13 of [31], if F is ρ-analytic and $X \to Y$ is $(\rho+1)$-connected, then $F(X) \simeq \operatorname{holim} P_nF(X)$. Being analytic is a variant on the n-excisiveness condition. To introduce that concept, we need a few definitions. First, a map of spaces is k-connected if all its homotopy fibers are $(k-1)$-connected. Recall that above we defined a cubical diagram $\mathcal{X}$ to be Cartesian if the map

2 Def. 1.4 [30].

3 Def. 2.1 [30].

4 Def. 1.3 [30].

5 Def. 3.1 [30].

6 Thm. 1.13 [31].

$$\mathcal{X}(\emptyset) \to \operatorname{holim} \mathcal{X}|_{\mathcal{P}_0(S)}$$

is a weak homotopy equivalence. We now say $\mathcal{X}$ is *k-Cartesian*[7] if this map is k-connected. All of this being defined, we can define a weaker form of n-excisiveness: F as a functor is *stably n-excisive*[8] if it satisfies the following condition $E(n, c, K)$: for any strongly coCartesian $(n+1)$-cubical diagram $\mathcal{X} : \mathcal{P}(S) \to \mathcal{C}$ such that for all $s \in S$, $\mathcal{X}(\emptyset) \to \mathcal{X}(s)$ is k_s-connected for $k_s \geq K$, the diagram $F \circ \mathcal{X}$ is $(-c + \sum k_s)$-Cartesian. Finally, F is said to be *ρ-analytic*[9] if it satisfies $E(n, n\rho - q, \rho + 1)$ for some q and this for all $n \geq 1$.

22.3 Geometric stacks

22.3.1 General context

We briefly go over the construction of stacks. For a complete coverage, see [79] and [76]. Given a homotopical algebraic context $(\mathcal{C}, \mathcal{C}_0, \mathcal{A})$ (to be discussed below), let $\mathrm{Comm}(\mathcal{C})$ be the model category of commutative monoid objects of $\mathcal{C}$ with $\mathcal{A}\mathrm{ff}_{\mathcal{C}}$ as opposite category, which we refer to as the model category of affine schemes, or affine stacks. Let $\mathcal{A}\mathrm{ff}_{\mathcal{C}}^{\wedge} = L_W \mathrm{sPr}(\mathcal{A}\mathrm{ff}_{\mathcal{C}})$ be the model category of prestacks on $\mathcal{A}\mathrm{ff}_{\mathcal{C}}$, left Bousfield localization of the model category of simplicial functors on $\mathcal{A}\mathrm{ff}_{\mathcal{C}}$ with respect to weak equivalences in $\mathcal{A}\mathrm{ff}_{\mathcal{C}}$. After having fixed a model topology τ on this latter model category, one can then define the model category of stacks $\mathcal{A}\mathrm{ff}_{\mathcal{C}}^{\sim,\tau}$ on $\mathcal{A}\mathrm{ff}_{\mathcal{C}}$ as the left Bousfield localization of $\mathcal{A}\mathrm{ff}_{\mathcal{C}}^{\wedge}$ with respect to homotopy τ-hypercovers.

To define geometric stacks, one needs to introduce several assumptions, collected together into what is called a *Homotopical Algebraic context* (referred to as HA), and a *Homotopical Algebraic Geometry* context (referred to as HAG). For a full list of assumptions, the reader is referred to the main reference [79], where those are introduced. We will just present those that are most useful for our purposes. An HA context is a triple $(\mathcal{C}, \mathcal{C}_0, \mathcal{A})$ consisting of a symmetric monoidal model category $(\mathcal{C}, \otimes, 1)$. If we fix universes $\mathbb{U} \in \mathbb{V}$, then we suppose $\mathcal{C}$ is $\mathbb{V}$-small and $\mathbb{U}$-combinatorial. It is also assumed proper and pointed, and $\mathrm{Ho}(\mathcal{C})$ is an additive category. For $A \in \mathrm{Comm}(\mathcal{C})$, A-Mod is a $\mathbb{U}$-combinatorial model category, and it is proper. One can define a tensor product on A-Mod, for which it is a symmetric monoidal model category. $\mathcal{C}_0$ is a full subcategory of $\mathcal{C}$ that one needs in obstruction theory, and it is necessary to define various kinds of geometric stacks. $\mathcal{A}$ is a full subcategory of $\mathrm{Comm}(\mathcal{C})$ satisfying a technical condition relative to $\mathcal{C}_0$.

7 Def. 1.3 [30].

8 Def. 4.1 [30].

9 Def. 4.2 [30].

A HAG context consists of an HA context, together with a choice of a model topology τ on $\mathcal{A}\mathrm{ff}_{\mathcal{C}}$, and a class of morphisms P in $\mathcal{A}\mathrm{ff}_{\mathcal{C}}$, which is stable by equivalences. There are further assumptions on τ and P, so that $(\mathcal{C}, \mathcal{C}_0, \mathcal{A}, \tau, P)$ becomes a HAG context, some more technical than others. One can mention that covering families use morphisms in P, and morphisms in this class are stable by equivalences and homotopy pullbacks and compositions. The topology is necessary to define stacks, and the choice of P will dictate what kind of geometric stack we get, assuming that morphisms in P are compatible with the topology.

We suppose fixed a HAG contex $(\mathcal{C}, \mathcal{C}_0, \mathcal{A}, \tau, P)$. One can then define the notion of *n-geometric stack*[10] using induction: a stack F is *n-geometric* if the diagonal morphism $F \to F \times^h F$ is $(n-1)$-representable and if F admits an n-atlas. One defines an *n-atlas* for F to be a $\mathbb{U}$-small family of morphisms $\{U_i \to F\}_{i \in I}$ for which each U_i is representable, $\coprod U_i \to F$ is an epi, and each morphism in this family is in $(n-1)$-P. One says a morphism of stacks $F \to G$ is in n-P if it is n-representable, and for any morphism of stack $X \to G$, there is an n-atlas $\{U_i\}$ for $F \times_G^h X$ such that each composite $U_i \to X$ is in P. Finally, a morphism of stacks $F \to G$ is *n-representable* if for any morphism of stack $X \to G$, $F \times_G^h X$ is n-geometric. To initialize the induction, a representable stack is said to be (-1)-geometric, and a morphism of stacks $F \to G$ is (-1)-representable if for any morphism $X \to G$ from a representable stack X, $F \times_G^h X$ is representable as well, i. e., it is locally representable, and $F \to G$ is said to be in (-1)-P if in addition the morphism $F \times_G^h X \to X$ is in P, that is, $F \to G$ is locally in P. Observe that it follows from Proposition 1.3.3.3 of [79] that an $(n-1)$-geometric stack is n-geometric.

We are interested in particular in the setting where $\mathcal{C} = sk\text{-Mod} = \mathcal{C}_0$, $\mathcal{A} = sk\text{-Alg}$, $\tau = \text{ét}$, and P is the class of smooth morphisms which we define below. This provides us with a HAG context, which allows us to talk about *n-geometric D^--stacks*. We denote $\mathrm{Comm}(sk\text{-Mod})^{\mathrm{op}} = \text{k-}D^-\mathrm{Aff}$, with the homotopy category of D^--stacks denoted by $D^-\mathrm{St}(k) = \mathrm{Ho}\,(\text{k-}D^-\mathrm{Aff}^{\sim,\text{ét.}})$.

A morphism $f : A \to B$ of commutative algebras is said to be *smooth*[11] if it is finitely presented in sk-Alg = Comm(sk-Mod), and it is *formally smooth*,[12] that is, $\mathbb{L}_{B/A} \in B\text{-Mod}$ is projective, and there is a retraction of the morphism $\mathbb{L}_A \otimes_A^{\mathbb{L}} B \to \mathbb{L}_B$ in Ho (B-Mod). Recall that for $A \in sk\text{-Alg}$, to say that $B \in A\text{-Mod}$ is *projective*[13] means that it is a retract of $\coprod_S^{\mathbb{L}} A$ for some small $\mathbb{U}$-set S. The B-module $\mathbb{L}_{B/A}$ is the *cotangent complex of B over A*,[14] an object of Ho (B-Mod), constructed as

$$\mathbb{L}_{B/A} = \mathbb{L}Q\mathbb{R}I(B \otimes_A^{\mathbb{L}} B)$$

10 Def. 1.3.3.1 [79].

11 Def. 1.2.7.2 [79].

12 Def. 1.2.7.1 [79].

13 Def. 1.2.4.1 [79].

14 Def. 1.2.1.5 [79].

where $Q : B - sk\text{-Alg}_{\text{nu}} \to B\text{-Mod}$ is the morphism from nonunital B-algebras defined by the pushout

$$\begin{array}{ccc} C \otimes_B C & \xrightarrow{\mu} & C \\ \downarrow & & \downarrow \\ * & \longrightarrow & Q(C) \end{array}$$

and $I : B - sk\text{-Alg}_{/B} \to B - sk\text{-Alg}_{\text{nu}}$ takes a diagram $B \xrightarrow{f} C \xrightarrow{g} B$ to $\text{Ker}(g)$. Note that if $A = 1$, then we let $\mathbb{L}_{B/1} = \mathbb{L}_B$, which is referred to as the cotangent complex of B.[15]

To justify the existence of $\mathbb{L}_{B/A}$, we have to introduce another definition; if $A \to B$ is a morphism in sk-Alg, and $M \in B$-Mod, then one defines the *simplicial set of derived A-derivations*[16] $B \to M$ by

$$\mathbb{D}\text{er}_A(B, M) = \text{Map}_{A\text{-}sk\text{-Alg}_{/B}}(B, B \oplus M)$$

an object of $\text{Ho}(\text{Set}_\Delta)$, where $B \oplus M \in sk$-Alg has $A \coprod M$ as an underlying object with product as defined in [79]. Now we have the following result from [79], which guarantees the existence of a cotangent complex:[17] given a morphism $A \to B$ in sk-Alg, there are some $\mathbb{L}_{B/A} \in B$-Mod and $d \in \pi_0(\mathbb{D}\text{er}_A(B, \mathbb{L}_{B/A}))$ such that for all $M \in B$-Mod, the natural map $d^* : \text{Map}_{B\text{-Mod}}(\mathbb{L}_{B/A}, M) \to \mathbb{D}\text{er}_A(B, M)$ induced by d is an isomorphism in $\text{Ho}(\text{Set}_\Delta)$.

An alternate characterization of smooth morphisms in sk-Alg is provided by Theorem 2.2.2.6 of [79]: a morphism is smooth if and only if it is *strongly smooth*,[18] that is, it is *strong* (for all k, $\pi_k(A) \otimes_{\pi_0(A)} \pi_0(B) \to \pi_k(B)$ is an isomorphism), and $\text{Spec}\, \pi_0 B \to \text{Spec}\, \pi_0 A$ is smooth.

22.3.2 Representability theorem

Central to our transition from Goodwillie calculus to its essential realization within the realm of derived stacks is the following representability theorem, a pared down version of the original result of Lurie [51] adapted to the setting of geometric stacks by Toen and Vezzosi [79]: a D^--stack F is n-geometric if and only its truncation $t_0(F)$ is an Artin $(n+1)$-stack, F has an obstruction theory, and for all $A \in sk$-Alg, we have a natural isomorphism $\mathbb{R}F(A) \to \text{holim}_k\, \mathbb{R}F(A_{\leq k})$ in $\text{Ho}(\text{Set}_\Delta)$. We discuss these conditions in turn.

15 Def. 1.2.1.5 [79].

16 Def. 1.2.1.1 [79].

17 Prop. 1.2.1.2 [79].

18 Def. 2.2.2.3 [79].

22.3.2.1 Truncation statement

For the *truncation functor*,[19] we have a natural inclusion functor $i : k\text{-Aff} \hookrightarrow \text{k-D}^{-}\text{Aff}$, inducing a right Quillen functor $i^* : \text{k-D}^{-}\text{Aff}^{\wedge} \to k\text{-Aff}^{\wedge}$, with a right derived functor $\mathbb{R}i^* : \text{Ho}\,(\text{k-D}^{-}\text{Aff}^{\wedge}) \to \text{Ho}\,(k\text{-Aff}^{\wedge})$ that preserves the subcategory of stacks, and hence provides a right Quillen restriction functor $i^* : \text{k-D}^{-}\text{Aff}^{\sim,\text{ét.}} \to k\text{-Aff}^{\sim,\text{ét.}}$, whose right derived functor $t_0 = \mathbb{R}i^* : D^{-}\text{St}(k) \to \text{St}(k)$ is our desired truncation functor. Here $\text{St}(k)$ is the category of stacks obtained by starting with k-Mod instead of sk-Mod.

An *Artin n-stack*[20] is an object of $\text{St}(k)$ that is n-truncated and p-geometric for some integer p. Here the HAG context is based on $\mathcal{C} = k\text{-Mod} = \mathcal{C}_0$, $\mathcal{A} = k\text{-Alg}$, and $\tau = \text{ét}$, and P is the class of smooth morphisms, where we say $A \to B$ in k-Alg is *smooth* if $B \in A\text{-}k\text{-Alg}$ is finitely presented, $\Omega_{B/A}$ is projective in B-Mod, and $\pi_k(\mathbb{L}_{B/A}) = 0$ for $k \geq 1$. To say that $t_0(F)$ is an Artin $(n+1)$-stack is due to the fact that if F is an n-geometric stack in $\text{St}(k)$, then it is $(n+1)$-truncated.

22.3.2.2 Obstruction theory statement

The notion of obstruction theory is slightly more involved. For a pointed model category $\mathcal{C}$, we have a suspension functor $\Sigma : \text{Ho}\,(\mathcal{C}) \to \text{Ho}\,(\mathcal{C})$, $x \mapsto * \coprod_x^{\mathbb{L}} *$. Denote by $\mathcal{C}_1 = \Sigma\mathcal{C}_0$ the full subcategory of $\mathcal{C}$ spanned by objects equivalent to the suspension of some object of $\mathcal{C}_0$. The obstruction for n-geometric D^{-}-stacks is relative to $sk\text{-Mod}_1$.

A stack has an *obstruction theory*[21] relative to an HA context $(\mathcal{C}, \mathcal{C}_0, \mathcal{A})$ if it is infinitesimally Cartesian and if it has a global cotangent complex.

A stack F is said to be *infinitesimally Cartesian*[22] relative to a given HA context $(\mathcal{C}, \mathcal{C}_0, \mathcal{A})$, if for all $A \in \mathcal{A}$, $M \in A\text{-Mod}_1 = \Sigma A\text{-Mod}_0$, and $d \in \pi_0(\mathbb{D}\text{er}_A(A, M))$ with a corresponding morphism $A \to A \oplus M$, still denoted by d, the square

$$\begin{array}{ccc} \mathbb{R}F(A \oplus_d \Omega M) & \longrightarrow & \mathbb{R}F(A) \\ \downarrow & & \downarrow{\scriptstyle d} \\ \mathbb{R}F(A) & \xrightarrow[s]{} & \mathbb{R}F(A \oplus M) \end{array}$$

is homotopy Cartesian. Recall what this means: from [28] for instance, if $\mathcal{M}$ is a right proper model category and $A, B, C, D \in \text{Ob}\,\mathcal{M}$, then the square

$$\begin{array}{ccc} A & \longrightarrow & C \\ \downarrow & & \downarrow \\ B & \longrightarrow & D \end{array}$$

19 Def. 2.2.4.3 [79].

20 Def. 2.1.1.4 [79].

21 Def. 1.4.2.1. [79].

22 Def. 1.4.2.1 [79].

is said to be homotopy Cartesian, if for any factorization of $C \to D$ into a trivial cofibration $C \to E(C)$ followed by a fibration $E(C) \to D$, the induced map $A \xrightarrow{i_*} B \times_D E(C)$ is a weak equivalence. Note also that since we assume $\mathcal{C}$ to be pointed, it has a suspension functor as mentioned above, with the loop functor $\Omega : \mathrm{Ho}(\mathcal{C}) \to \mathrm{Ho}(\mathcal{C}), x \mapsto * \times_x^h *$ as right adjoint. Then $A \oplus_d \Omega M$ denotes the square zero extension of A by ΩM, defined by the homotopy Cartesian square

$$\begin{array}{ccc} A \oplus_d \Omega M & \longrightarrow & A \\ \downarrow & & \downarrow{\scriptstyle d} \\ A & \xrightarrow[s]{} & A \oplus M \end{array}$$

where $A \in \mathrm{Comm}(\mathcal{C})$, $M \in A\text{-Mod}$, $d : A \to A \oplus M$ is a derivation, and $s : A \to A \oplus M$ is the trivial derivation. In other terms, F infinitesimally Cartesian means that it preserves this homotopy square. To explain the notation, denote $A \oplus_d \Omega M$ by B for the moment. The morphism $B \to A$ has a fiber isomorphic to $\Omega M = * \times_M^h *$, with the zero multiplication, and that is why B is called a zero extension of A by ΩM. Recall the definition of such objects; for a commutative ring R, a square zero extension of R is given by a surjective morphism of commutative rings $\phi : R' \to R$ such that $(\mathrm{Ker}\phi)^2 = 0$, which is what we have here.

A stack F is said to have a *global cotangent complex*[23] relative to a given HA context $(\mathcal{C}, \mathcal{C}_0, \mathcal{A})$ if it has a local cotangent complex that behaves well under base change. More precisely, for all $A \in \mathcal{A}$ and $x : \mathbb{R}\underline{\mathrm{Spec}}\,A \to F$, F has a (local) cotangent complex $\mathbb{L}_{F,x}$ at x such that for any morphism $A \to B$ in $\mathcal{A}$ and any morphism

$$\begin{array}{ccc} \mathbb{R}\underline{\mathrm{Spec}}\,B & \xrightarrow{u} & \mathbb{R}\underline{\mathrm{Spec}}\,A \\ & {\scriptstyle y}\searrow \quad \swarrow{\scriptstyle x} & \\ & F & \end{array}$$

we have an induced isomorphism $u^* : \mathbb{L}_{F,x} \otimes_A^{\mathbb{L}} B \to \mathbb{L}_{F,y}$ in $\mathrm{Ho}(\mathrm{Sp}(B\text{-Mod}))$. In this definition, one uses the Yoneda embedding $\underline{h} : \mathcal{A}\mathrm{ff}_{\mathcal{C}} \to \mathcal{A}\mathrm{ff}_{\mathcal{C}}^{\wedge}$ with a total right derived functor $\mathbb{R}\underline{h} : \mathrm{Ho}(\mathcal{A}\mathrm{ff}_{\mathcal{C}}) \to \mathrm{Ho}(\mathcal{A}\mathrm{ff}_{\mathcal{C}}^{\wedge})$, which turns out to be a fully faithful functor into the category of stacks $\mathrm{St}(\mathcal{C}, \tau)$. This follows from our assumption on the model topology τ to have a HAG context, which implies in particular that it is subcanonical. It follows that we have $\mathbb{R}\underline{h} : \mathrm{Ho}(\mathcal{A}\mathrm{ff}_{\mathcal{C}}) \to \mathrm{St}(\mathcal{C}, \tau)$, and one denotes $\mathbb{R}\underline{h}_{\mathrm{Spec}\,A}$ by $\mathbb{R}\underline{\mathrm{Spec}}\,A$. Regarding the (local) cotangent complex, it is defined as follows. If F is a stack, $A \in \mathcal{A}$, and $x : \mathbb{R}\underline{\mathrm{Spec}}\,A \to F$, then we say F has a *cotangent complex*[24] at x if there exists a $(-n)$-connective stable A-module $\mathbb{L}_{F,x} \in \mathrm{Ho}(\mathrm{Sp}(A\text{-Mod}))$ and an isomorphism $\mathbb{D}\mathrm{er}_F(X, -) \cong$

23 Def. 1.4.1.7 [79].

24 Def. 1.4.1.5 [79].

$\mathbb{R}\underline{h}_s^{\mathbb{L}_{F,x}}$ in Ho $((A\text{-Mod}_0^{\text{op}})^\wedge)$. In such a definition, the derivations functor $\mathbb{D}\text{er}_F(X,-)$ is defined like so: given $x : X = \mathbb{R}\underline{\text{Spec}}\,A \to F$, denote $X[M] = \mathbb{R}\underline{\text{Spec}}\,(A \oplus M)$. One can then define the *simplicial set of derivations from F to M at the point x*[25] by

$$\mathbb{D}\text{er}_F(X,M) = \text{Map}_{X/\mathcal{A}\text{ff}_C^{\sim,\tau}}(X[M],F).$$

Regarding stable modules, recall that $\mathcal{C}$ is pointed, so we have suspension and loop functors, as mentioned above. Fix a cofibrant model S_C^1 for the object $\Sigma(1) \in \text{Ho}\,(\mathcal{C})$, and for $A \in \text{Comm}(\mathcal{C})$, define $S_A^1 = S_C^1 \otimes A \in A\text{-Mod}$. Then one defines the *model category of stable A-modules*[26] to be the model category of spectra Sp(A-Mod) with the connecting map $S_A^1 \otimes_A -$, whose objects will simply be referred to as stable modules. From there one defines a functor

$$\begin{aligned} \underline{h}_s^- : \text{Sp}(A\text{-Mod})^{\text{op}} &\to (A\text{-Mod}_0^{\text{op}})^\wedge, \\ M_* &\mapsto \underline{h}_s^{M_*} \end{aligned}$$

where

$$\begin{aligned} \underline{h}_s^{M_*} : A\text{-Mod}_0 &\to \text{Set}_\Delta, \\ N &\mapsto \text{Hom}(M_*, \Gamma_*(S_A(N))) \\ &= \text{Map}_{\text{Sp}(A\text{-Mod})}(M_*, S_A(N)) \end{aligned}$$

with Γ_* the simplicial resolution functor on Sp(A-Mod), and $S_A : A\text{-Mod} \to \text{Sp}(A\text{-Mod})$ has the nth space defined by $S_A(N)_n = (S_A^1)^{\otimes_A n} \otimes_A N$. By Proposition 1.2.11.3 of [79] we have a right derived functor $\mathbb{R}\underline{h}_s^- : \text{Ho}\,(\text{Sp}(A\text{-Mod}))^{\text{op}} \to \text{Ho}\,((A\text{-Mod}_0^{\text{op}})^\wedge)$. *Connectivity* is defined by induction: $M_* \in \text{Ho}\,(\text{Sp}(A\text{-Mod}))$ is said to be 0-connective if it is isomorphic to some object of $S_A(\text{Ho}\,(A\text{-Mod}_0))$. Then M_* is said to be $(-n)$-connective if it is isomorphic to $\Omega(M'_*)$ for some $-(n-1)$-connective A-module M'_*.

To summarize that part of having an obstruction that involves having a cotangent complex, for a stack to have a local cotangent complex means that functorially

$$\begin{array}{ccc} \mathbb{D}\text{er}_F(X,-) & \xrightarrow{\ \cong\ } & \mathbb{R}\underline{h}_s^{\mathbb{L}_{F,x}} \\ \| & & \| \\ \text{Map}_{x/\mathcal{A}\text{ff}_C^{\sim,\tau}}(\mathbb{R}\underline{\text{Spec}}\,(A \oplus -), F) & & \mathbb{R}\text{Hom}(\mathbb{L}_{F,x}, \Gamma_*(S_A(-))). \end{array}$$

To see that the cotangent complex does exhibit a notion of cotangency, we have to relate this notion to that of a *tangent complex* introduced in the same Definition 1.4.1.5 of [79]

25 Def. 1.4.1.4 [79].

26 Def. 1.2.11.1 [79].

that the cotangent complex was introduced. By definition the tangent complex of F at the same point x is defined by $\mathbb{T}_{F,x} = \mathbb{R}\underline{\mathrm{Hom}}(\mathbb{L}_{F,x}, A) \in \mathrm{Ho}\,(\mathrm{Sp}(A\text{-Mod}))$, where we used the fact that $\mathrm{Ho}\,(\mathrm{Sp}(A\text{-Mod}))$ is a closed symmetric monoidal category and hence has a stable A-module of morphisms $\mathbb{R}\underline{\mathrm{Hom}}(-,-)$. The tangent complex can be related to the *tangent stack* TF[27] of a stack F. The latter is defined as $TF = \mathbb{R}_\tau\underline{\mathcal{H}om}(\mathbb{D}_\epsilon, F)$, where $\mathbb{R}_\tau\underline{\mathcal{H}om}$ is the internal Hom of $\mathrm{Ho}\,(\mathcal{A}\mathrm{ff}_{\mathcal{C}}^{\sim,\tau})$, and $\mathbb{D}_\epsilon = \mathbb{R}\underline{\mathrm{Spec}}\,(1[\epsilon])$ is the infinitesimal disk, with $1 \in \mathrm{Comm}(\mathcal{C})$ having a trivial square zero extension $1 \oplus 1 = 1[\epsilon]$. In this manner, $TF = \mathbb{R}_\tau\underline{\mathcal{H}om}(\mathbb{R}\underline{\mathrm{Spec}}\,(1 \oplus 1), F)$. Note the similarity with $\mathbb{D}\mathrm{er}_F(X, M) = \mathrm{Map}(\mathbb{R}\underline{\mathrm{Spec}}\,(A \oplus M), F)$. Now by Proposition 1.4.1.6 of [79], for $F \in \mathrm{St}(\mathcal{C}, \tau)$, $A \in \mathcal{A}$, and $x : X = \mathbb{R}\underline{\mathrm{Spec}}\,A \to F$, supposing F has a cotangent complex $\mathbb{L}_{F,x}$ at x, it follows that we have natural isomorphisms

$$\mathbb{R}\underline{\mathrm{Hom}}_{\mathcal{A}\mathrm{ff}_{\mathcal{C}}^{\sim,\tau}}(X, TF) \cong \mathrm{Map}_{\mathrm{Sp}(A\text{-Mod})}(\mathbb{L}_{F,x}, A) \cong \mathrm{Map}_{\mathrm{Sp}(A\text{-Mod})}(A, \mathbb{T}_{F,x})$$

thereby showing that $\mathbb{T}_{F,x}$ is the analog among spectra of the tangent stack, which justifies calling $\mathbb{L}_{F,x}$ the cotangent complex of F at x.

To see that we have those isomorphisms, it suffices to write

$$\begin{aligned}
\mathbb{R}\underline{\mathrm{Hom}}(X, TF) &= \mathbb{R}\underline{\mathrm{Hom}}(\mathbb{R}\underline{\mathrm{Spec}}\,A, \mathbb{R}\underline{\mathrm{Hom}}(\mathbb{D}_\epsilon, F)) \\
&\cong \mathbb{R}\underline{\mathrm{Hom}}(\mathbb{R}\underline{\mathrm{Spec}}\,A \times^h \mathbb{D}_\epsilon, F) \\
&\cong \mathbb{R}\underline{\mathrm{Hom}}(\mathbb{R}\underline{\mathrm{Spec}}\,(A[\epsilon]), F) \\
&= \mathbb{R}\underline{\mathrm{Hom}}(\mathbb{R}\underline{\mathrm{Spec}}\,(A \oplus A), F) \\
&= \mathbb{R}\underline{\mathrm{Hom}}(X[A], F) \\
&= \mathbb{D}\mathrm{er}_F(X, A) \\
&\cong \mathbb{R}\underline{h}_s^{\mathbb{L}_{F,x}}(A) = \mathrm{Map}(\mathbb{L}_{F,x}, S_A(A)).
\end{aligned}$$

Since sk-Mod has a fully faithful suspension functor, by Lemma 1.2.11.2 of [79] the stabilization functor S_A is fully faithful as well, so we just write A for $S_A(A)$ in what follows:

$$\begin{aligned}
\mathbb{R}\underline{\mathrm{Hom}}(X, TF) &\cong \mathrm{Map}(\mathbb{L}_{F,x}, A) \\
&\cong \mathrm{Map}(\mathbb{L}_{F,x} \otimes_A^{\mathbb{L}} A, A) \\
&\cong \mathrm{Map}(A, \mathbb{R}\underline{\mathrm{Hom}}(\mathbb{L}_{F,x}, A)) \\
&= \mathrm{Map}(A, \mathbb{T}_{F,x}).
\end{aligned}$$

22.3.2.3 Convergence statement

The last claim of that representability theorem states that for any $A \in sk$-Alg, $\mathbb{R}F(A) \xrightarrow{\cong} \mathrm{holim}_k\, \mathbb{R}F(A_{\leq k})$. As pointed out in [79], any $A \in sk$-Alg has a functorial Postnikov tower of the form

27 Def. 1.4.1.2 [79].

$$A \to \cdots \to A_{\le n} \to A_{\le n-1} \to \cdots \to A_{\le 0} = \pi_0(A)$$

where $A_{\le k}$ is k-truncated, the morphisms $A \to A_{\le k}$ induce isomorphisms on fundamental groups in degree $l \le k$, and most importantly for us, this Postnikov tower is unique up to equivalence. In other terms, one can define the *k-truncation* of A to be given by $A_{\le k}$ in its Postnikov tower. In this manner, one can write

$$F(A_{\le k}) = F(\mathrm{tr}_k(A)) = F \circ \mathrm{tr}_k(A) := F_k(A)$$

where tr_k is the degree k truncation functor, which exists by virtue of the above discussion, and F_k is defined as above. We call it the degree k approximation to the functor F.

22.4 The bridge

This section aims to show that ρ-analyticity for Goodwillie translates into a stack being n-geometric, plus some added conditions.

22.4.1 Similitudes in towers

$F \in \text{k-D}^{-}\text{Aff}^{\sim,\text{ét.}}$ maps the Postnikov tower of $A \in sk\text{-Alg}$ to

$$F(A) \to \cdots F(A_{\le n}) \to F(A_{\le n-1}) \to \cdots \to F(A_0) = F(\pi_0(A))$$

which one can rewrite as

$$F(A) \to \cdots F_n(A) \to F_{n-1}(A) \to \cdots \to F_0(A) = F(\pi_0(A)).$$

In Ho (Set_Δ), this reads

$$\mathbb{R}F(A) \to \cdots \mathbb{R}F_n(A) \to \mathbb{R}F_{n-1}(A) \to \cdots \to \mathbb{R}F_0(A) = F(\pi_0(A))$$

since $F(A) \cong \mathbb{R}F(A)$ in Ho (Set_Δ), where $\mathbb{R}F(A) = \mathbb{R}_\tau\underline{\mathrm{Hom}}(\mathbb{R}\mathrm{Spec}\, A, F)$, $\mathbb{R}_\tau\underline{\mathrm{Hom}}$ the derived simplicial Hom for $\text{k-D}^{-}\text{Aff}^{\sim,\text{ét.}}$, which, because F is a stack, can also be written using $\mathbb{R}\underline{\mathrm{Hom}}$, the derived simplicial hom of $\text{k-D}^{-}\text{Aff}^{\wedge}$.

Now if F is an n-geometric D^{-}-stack, then $\mathbb{R}F(A) \cong \mathrm{holim}_n\, \mathbb{R}F_n(A)$, which one can view, in the spirit of functor calculus, as a convergence of the above tower for F at the point A.

Another hint that there may be a link between those Taylor towers, whether they be for a geometric stack or for a homotopy functor, comes from the following considerations. As pointed out in [79], the homotopy fiber of the morphism $A_{\le n} \to A_{\le n-1}$ in the Postnikov tower of $A \in sk\text{-Alg}$ is isomorphic to $S^n \otimes_k \pi_n A$ in Ho $(sk\text{-Mod})$, where the notion of homotopy fiber is as defined in [40]; for a functorial factorization $A_{\leq n} \to E(A_{\leq n}) \to A_{n-1}$, a trivial cofibration followed by a fibration, the homotopy fiber of $A_{\leq n} \to A_{\leq n-1}$ is defined by $P_n = * \times_{A_{n-1}} E(A_{\leq n})$, and thus the following diagram in particular is a homotopy Cartesian square:

$$\begin{array}{ccc} P_n & \longrightarrow & A_{\leq n} \\ \downarrow & & \downarrow \\ * & \longrightarrow & A_{\leq n-1}. \end{array}$$

Suppose now $F \in$ k-D$^-$Aff$^{\sim,\text{ét.}}$ preserves homotopy Cartesian squares. It follows that

$$\begin{array}{ccc} F(P_n) & \longrightarrow & F(A_{\leq n}) \\ \downarrow & & \downarrow \\ F(*) & \longrightarrow & F(A_{\leq n-1}) \end{array}$$

is a homotopy Cartesian square as well, that is,

$$F(P_n) \xrightarrow{\simeq} F(*) \times_{F(A_{\leq n-1})} E(FA_{\leq n}).$$

Let $\gamma : \mathrm{Set}_\Delta \to \mathrm{Ho}\,(\mathrm{Set}_\Delta)$ be the canonical projection. It follows that

$$\gamma F(P_n) \xrightarrow{\cong} \gamma F(*) \times_{\gamma F(A_{\leq n-1})} \gamma E(FA_{\leq n}).$$

Suppose F is reduced: $F(*) = *$, so that $\gamma F(*) = \gamma * = *$. This reads

$$\gamma F(P_n) \xrightarrow{\cong} * \times_{\gamma F(A_{\leq n-1})} \gamma E(FA_{\leq n}).$$

Now $\gamma F : sk\text{-Alg} \to \mathrm{Ho}\,(\mathrm{Set}_\Delta)$ factors uniquely through $\mathrm{Ho}\,(sk\text{-Alg})$: $\mathrm{Ho}\,(\gamma F) : \mathrm{Ho}\,(sk\text{-Alg}) \to \mathrm{Ho}\,(\mathrm{Set}_\Delta)$, which we denote again by γF. Using the fact that the homotopy fiber P_n of $A_{\leq n} \to A_{\leq n-1}$ is isomorphic to $S^n \otimes_k \pi_n A$, which we denote by ΩM, it follows that

$$\gamma F(\Omega M) \cong * \times_{\gamma F(A_{\leq n-1})} \gamma E(FA_{\leq n}).$$

We have an equivalence $F(A) \xrightarrow{\simeq} \mathbb{R}F(A)$ in Set_Δ for $A \in sk\text{-Alg}$ since F is a stack (hence an isomorphism $\gamma F(A) \cong \mathbb{R}F(A)$ in $\mathrm{Ho}\,(\mathrm{Set}_\Delta)$), so that we have an equivalence $E(FA) \simeq E(\mathbb{R}FA)$ by the 2–3 property, and consequently an isomorphism $\gamma E(FA) \cong \gamma E(\mathbb{R}FA)$ in $\mathrm{Ho}\,(\mathrm{Set}_\Delta)$. It follows that

$$\mathbb{R}F(\Omega M) \cong * \times_{\mathbb{R}F(A_{\leq n-1})} \gamma E(\mathbb{R}F(A_{\leq n})).$$

Since we work in Ho (Set_Δ), we can drop γ. This now reads

$$\mathbb{R}F(\Omega M) \cong * \times_{\mathbb{R}F(A_{\leq n-1})} E(\mathbb{R}F(A_{\leq n})).$$

Suppose now F is a monoidal functor: $F(A) \otimes_{\mathrm{Set}_\Delta} F(B) \xrightarrow{\cong} F(A \otimes_k B)$. Using the fact that $\Omega M = S^n \otimes_k \pi_n A$, this reads

$$\mathbb{R}F(S^n) \otimes_{\mathrm{Set}_\Delta} \mathbb{R}F(\pi_n A) \cong * \times_{\mathbb{R}F(A_{\leq n-1})} E(\mathbb{R}F(A_{\leq n}))$$

which means $\mathbb{R}F(S^n) \otimes \mathbb{R}F(\pi_n A)$ is the homotopy fiber of $\mathbb{R}F_n(A) \to \mathbb{R}F_{n-1}(A)$ in Ho (Set_Δ).

Now recall from [31] that for a homotopy functor F, the homotopy fiber of $P_n F(X) \to P_{n-1}F(X)$, denoted $D_n F(X)$, is, for spectra-valued functors, of the form $(C_n \wedge (X^{\wedge n}))_{h\Sigma_n}$ for some spectrum C_n with a Σ_n action. This is surprisingly similar to what we have for our stack. The correspondence is given by associating C_n, which Goodwillie calls the nth derivative of F at x and which he denotes by $\partial^{(n)}F(x)$, with $F(S^n)$ and by associating $X^{\wedge n}$ with $F(\pi_n A)$. The corresponding action of the symmetric group Σ_n is given by permuting the entries of $[n]$ in $\Delta^n = \mathrm{Hom}(-, [n])$. Note that here X is regarded as a space over the one point space $*$, so likewise $A \in sk$-Alg is regarded as an algebra over 1. Thus in addition to having convergence of Taylor towers for ρ-analytic functors and for n-geometric D^--stacks, incremental morphisms in such towers have homotopy fibers that are formally the same if in addition we focus on those stacks that preserve homotopy Cartesian diagrams and are reduced and monoidal.

22.4.2 Analyticity and obstruction theory: a study

Now that we have seen we have comparable phenomena at the level of towers, we study this correspondence more carefully. We will now see that what makes F an n-geometric stack in $D^-\mathrm{St}(k)$ is very close indeed to the conditions that Goodwillie put on a homotopy functor to have convergence of its Taylor tower. Coming back to the condition that F preserves homotopy Cartesian squares, observe that this is a scaled down condition on a stack F relative to what Goodwillie asked his functors to satisfy for convergence, namely that they map strongly coCartesian squares to Cartesian squares.

It turns out that this is the only such condition of the sort we will impose on our stacks, and for this reason, we refer to it as the *op-excisive* condition.

In the representability theorem for stacks Theorem C.0.9 we are considering, F has an obstruction theory. We will show this is morally the same for stacks as asking a homotopy functor to be analytic. For a stack F to have an obstruction theory necessitates that it be infinitesimally Cartesian and that it has a global cotangent complex. Recall what infinitesimally Cartesian means: given the homotopy Cartesian square

$$\begin{array}{ccc} A \oplus_d \Omega M & \longrightarrow & A \\ \downarrow & & \downarrow \\ A & \longrightarrow & A \oplus M \end{array}$$

the following square is homotopy Cartesian as well:

$$\begin{array}{ccc} \mathbb{R}F(A \oplus_d \Omega M) & \longrightarrow & \mathbb{R}F(A) \\ \downarrow & & \downarrow \\ \mathbb{R}F(A) & \longrightarrow & \mathbb{R}F(A \oplus M). \end{array}$$

This is automatic if F is op-excisive. Thus asking that an op-excisive stack F has an obstruction theory boils down to showing it has a cotangent complex, which we will show corresponds in a sense to the stably excisive condition, needed for the convergence of Taylor towers of homotopy functors.

To ask that F has a cotangent complex means we have a $(-n)$-connective stable module $\mathbb{L}_{F,x}$ such that $\mathbb{D}\mathrm{er}_F(X,-) \cong \mathbb{R}\underline{h}_s^{\mathbb{L}_{F,x}}$ as we have seen, where $X = \mathbb{R}\underline{\mathrm{Spec}\,A}$ and $x : X \to F$. The derivation from F functor $\mathbb{D}\mathrm{er}_F(X,-)$ has a lift to a functor A-Mod $\to \mathrm{Set}_\Delta$, which with M associates the homotopy fiber at $x \in F(A)$ of the map $F(A \oplus M) \to F(A)$ for any $M \in A$-Mod. This is the image under F of the map $A \oplus M \to A$. Collectively those maps can be represented by a functor $\mathcal{X} : A\text{-Mod} \to sk\text{-Alg}$, which with M associates $\mathcal{X}(M) = A \oplus M$ in such a manner that $A \oplus M \to A$ reads $\mathcal{X}(M) \to \mathcal{X}(\emptyset)$. Now observe that for Goodwillie, being ρ-analytic for a homotopy functor F means that it is n-stably excisive, meaning it satisfies $E(n,c,K)$, which we reproduce here: if $\mathcal{X} : \mathcal{P}(S) \to \mathcal{C}$ is a strongly coCartesian $(n+1)$-cube such that $\mathcal{X}(\emptyset) \to \mathcal{X}(s)$ is k_s-connected for all $s \in S$, with $k_s \geq K$, then $F(\mathcal{X})$ is $(-c + \sum k_s)$-Cartesian. Thus on the one hand one considers maps of the form $\mathcal{X}(\emptyset) \to \mathcal{X}(s)$ in the calculus of functors, whereas for stacks, one considers maps of the form $A \oplus M \to A \oplus \emptyset = A$ instead, which read $\mathcal{X}(M) \to \mathcal{X}(\emptyset)$. The fact that we use "wrong" direction maps is explained by the fact that we ask that F preserves homotopy Cartesian diagrams, as opposed to mapping homotopy coCartesian diagrams to homotopy Cartesian diagrams in the case of functor calculus. However, the notion of cubical diagram here is moot for stacks; A-Mod, the indexing set for $\mathcal{X}$, is not finite. There is no need to work with strongly coCartesian diagrams either, since our stack F is assumed to be op-excisive. In the last section of this work, we will see that picking $\mathcal{X} = A \oplus -$ is not simply an ad hoc identification; the convergence of the Taylor tower of a homotopy functor F at some space X over Y hinges on the choice of $\mathcal{X} = X *_Y -$ as a cubical diagram, which we identify with $A \oplus -$ in Algebraic Geometry.

Note that whether it be a Taylor tower or a cotangent complex, what we have is very much deformation-theoretic, and thus we will put an emphasis on connectivity rather than on connectedness. This choice is further reinforced by the fact that if $\mathcal{X}(\emptyset) \to \mathcal{X}(s)$ is k_s-connected, then $F(\mathcal{X})$ is $(-c + \sum k_s)$-Cartesian, i. e., by Definition 1.3 of [30] the homotopy fibers of $F \circ \mathcal{X}(\emptyset) \simeq \operatorname{holim} F \circ \mathcal{X} \to \operatorname{holim}(F \circ X|_{\mathcal{P}_0(S)})$ are $(-c + \sum k_s - 1)$-connected. Collectively we have preserved the connectedness of each map. We will show that the same is true for connectivity in the case of stacks.

Now that we have defined diagrams for stacks, we study more carefully what the assumptions and claims made on diagrams in the calculus of functors translate to in the language of derived stacks. For Goodwillie, $\mathcal{X}(\emptyset) \to \mathcal{X}(s)$ being k_s connected means that its homotopy fibers are $(k_s - 1)$-connected, and here we are considering homotopy (limit) fibers. In the case of stacks, $\mathcal{X} = A \oplus -$, the homotopy fibers of morphisms $A \oplus M \to A$ depend on M. Thus one defines the connectivity of the map $A \oplus M \to A$ to be the one for M, where we say M is $(-p)$-connective if it is of the form $M = \Omega^p M'$ for some 0-connective $M' \in A$-Mod. Because we have such maps, they functorially map to $F(A \oplus M) \to F(A)$, the collection of which forms $F(\mathcal{X})$. For Goodwillie, $F\mathcal{X}$ being $(-c + \sum k_s)$-Cartesian means that the homotopy fibers of $F\mathcal{X}(\emptyset) \to \operatorname{holim}(F|_{\mathcal{P}_0(S)})$ are $(-c + \sum k_s - 1)$-connected. Instead of looking at the fibers of the map $\operatorname{holim} F(A \oplus M) \to F(A)$ to have a statement mirroring the one made by Goodwillie for $F(\mathcal{X})$, we consider the homotopy fibers of each individual map $F(A \oplus M) \to F(A)$ instead, all of which, as pointed out above, are functorially represented by $\mathbb{D}\mathrm{er}_F(X, -)$ with $X = \underline{\mathbb{R}\mathrm{Spec}\, A}$. Thus for the calculus of functors, the n-stably excisive condition for functors F focuses on $F\mathcal{X}$, whereas for stacks, one looks at $\mathbb{D}\mathrm{er}_F(X, -)$ instead. For stacks, the observation made above regarding the preservation of the connectedness of objects in the calculus of functors (captured by the presence of $\sum k_s$) should translate into the preservation of connectivity as we argued. Indeed, for F an n-geometric D^--stack, $\mathbb{D}\mathrm{er}_F(X, \Omega M) \cong \Omega \mathbb{D}\mathrm{er}_F(X, M)$, and thus if $A \oplus M \to A$ has connectivity p, meaning M is a p-loop object, then so does $\mathbb{D}\mathrm{er}_F(X, M)$, whose lift is the homotopy fiber of $F(A \oplus M) \to F(A)$, so F preserves the connectivity of each map. To address the occurrence of $-c$ in the n-stably excisive condition, observe that $\mathbb{D}\mathrm{er}_F(X, -)$ is corepresented by $\mathbb{L}_{F,x}$, which is connective.

Regarding the connectivity of $\mathbb{L}_{F,x}$, by Lemma 1.4.3.10 of [79] an n-geometric D^--stack F is infinitesimally Cartesian, so by the proof of Proposition 1.4.2.7 of the same reference it satisfies the conditions of Proposition 1.4.1.11, from which it follows that F has a global cotangent complex, which is $(-n)$-connective with the same index n.

Thus we have shown that for an op-excisive (reduced, monoidal) n-geometric D^--stack F, ρ-analyticity translates into F having an obstruction theory.

22.5 Parallel results

22.5.1 Polynomial approximations

At this point, we can draw various comparisons between both theories. On the one hand, for a homotopy functor $F : \mathcal{C}_Y \to \mathcal{D}$ and $X \in \mathcal{C}_Y$, we have

$$P_n F(X) = \operatorname*{hocolim}_k T_n^k F(X)$$

$$\cong \operatorname*{hocolim}_{k}\Big(\operatorname*{holim}_{(U_1,\dots,U_k)\in\mathcal{P}_0(\underline{n+1})^k} F(X *_Y (U_1 * \cdots * U_k))\Big)$$

with $X *_Y U = \operatorname{hocolim}(X \leftarrow X \times U \to Y \times U)$. We compare this with $\mathbb{R}F_n(A) = \mathbb{R}F(A_{\leq n})$, $\pi_i(A_{\leq n}) = 0$ if $i > n$. From [28], $A_{\leq n} = \Delta \downarrow A_{/\sim_n}$ where $\alpha \sim_n \beta$ for $\alpha, \beta : \Delta^q \to A$ if their restrictions to $\mathrm{sk}_n(\Delta^q)$ coincide. One can write

$$A_{\leq n} = \cup_{q\geq 0} \coprod_{\alpha\in\Delta^q\downarrow A} [\alpha]$$

with $[\alpha] = \{\beta \in \Delta^q \downarrow A \,|\, \alpha \sim_n \beta\}$.

If $\alpha, \beta \in \Delta^q \downarrow A$, then define $\alpha \star_A^q \beta$ by $\alpha \cup \beta$ if $\alpha \sim_n \beta$ and $\emptyset$ otherwise. With this, if $F \in \text{k-D}^-\mathrm{Aff}^{\sim,\text{ét.}}$, then

$$\begin{aligned}
\mathbb{R}F_n(A) &= \mathbb{R}F\Big(\cup_{q\geq 0} \coprod_{\alpha} \cup_{k\geq 0} \alpha \star_A^q (\beta_1 \star_A^q \cdots \star_A^q \beta_k)\Big) \\
&= \mathbb{R}F\Big(\cup_{k\geq 0} \cup_{q\geq 0} \coprod_{\alpha\in\Delta^q\downarrow A} \bigcup_{\substack{\beta_l\sim_n\alpha \\ 1\leq l\leq k}} \alpha \star_A^q (\beta_1 \star_A^q \cdots \star_A^q \beta_k)\Big) \\
&= \mathbb{R}F\Big(\cup_{k\geq 0} \cup_{q\geq 0} \bigcup_{\substack{\beta_l\in\Delta^q\downarrow A \\ 1\leq l\leq k}} (\Delta^q \downarrow A) \star_A^q (\beta_1 \star_A^q \cdots \star_A^q \beta_k)\Big) \\
&= \mathbb{R}F\Big(\operatorname*{colim}_{k} \cup_{q\geq 0} \operatorname*{colim}_{(\beta_1,\dots,\beta_k)\in \mathrm{Sk}_{0,q}(n)^k} \Delta^q \downarrow A \star_A^q (\beta_1 \star_A^q \cdots \star_A^q \beta_k)\Big)
\end{aligned}$$

where $\mathrm{Sk}_{0,q}(n) = \{\beta \in (\Delta^q \downarrow A) - \emptyset \,|\, \alpha \sim_n \beta\}$. With $\Delta^q \downarrow A \cong A_q$, this reads

$$\mathbb{R}F_n(A) = \mathbb{R}F\Big(\operatorname*{colim}_{k} \cup_{q\geq 0} \operatorname*{colim}_{(\beta_1,\dots,\beta_k)\in \mathrm{Sk}_{0,q}(n)^k} A_q \star_A^q (\beta_1 \star_A^q \cdots \star_A^q \beta_k)\Big).$$

Denoting $\mathrm{Sk}_0(n)^k = \cup_{q\geq 0}\mathrm{Sk}_{0,q}(n)^k$, write

$$\begin{aligned}
&\operatorname*{colim}_{(\beta_1,\dots,\beta_k)\in \mathrm{Sk}_0(n)^k} A \star_A (\beta_1 \star_A \cdots \star_A \beta_k) \\
&\qquad = \cup_{q\geq 0} \operatorname*{colim}_{(\beta_1,\dots,\beta_k)\in \mathrm{Sk}_{0,q}(n)^k} A_q \star_A^q (\beta_1 \star_A^q \cdots \star_A^q \beta_k)
\end{aligned}$$

where $\star_A$ is a levelwise product, so that

$$\mathbb{R}F_n(A) = \mathbb{R}F\Big(\operatorname*{colim}_{k} \operatorname*{colim}_{(\beta_1,\dots,\beta_k)\in \mathrm{Sk}_0(n)^k} A \star_A (\beta_1 \star_A \cdots \star_A \beta_k)\Big)$$

which one can compare with

$$P_nF(X) \cong \operatorname*{hocolim}_{k} \operatorname*{holim}_{(U_1,\dots,U_k)\in\mathcal{P}_0(\underline{n+1})^k} F(X *_Y (U_1 * \cdots * U_k)).$$

22.5.2 Pointwise approximations

We now consider Proposition 1.6 of [31], which states in particular that if F and G are two stably n-excisive homotopy functors and $u : F \Rightarrow G$, then F and G satisfy the following condition: there are constants c and K such that for all $k \geq K$ and for all $X \in \mathcal{C}_Y$ for which $X \to Y$ is k-connected, the map $F(X) \to G(X)$ is $(-c+(n+1)k)$-connected. The fact that $F(X) \to G(X)$ is $(-c+(n+1)k)$-connected for all X shows that there is an intrinsic connective object to u. Now being n-stably excisive is a property of ρ-analytic functors, which we associate with n-geometric D^--stacks in the theory of stacks. Thus to show for stacks a result similar to Proposition 1.6, we start with two n-geometric D^--stacks F and G and a morphism of stacks $f : F \to G$. We consider maps of the form $f_A : F(A) \to G(A)$ for $A \in sk$-Alg, and we aim to show that all such maps are inherently connective with the same connectivity in a sense to be specified. Since we work deformation-theoretically, we will lift those maps to all diagrams of the form

$$\begin{array}{ccc} F(A \oplus M) & \dashrightarrow & G(A \oplus M) \\ \downarrow & & \downarrow \\ F(A) & \longrightarrow & G(A) \end{array} \tag{22.1}$$

indexed by $M \in A$-Mod, whose individual homotopy fibers are functorially represented by derivation functors. In lieu of considering (22.1), we therefore focus on the following natural transformation:

$$\mathbb{D}\mathrm{er}_F(X,-) \Rightarrow \mathbb{D}\mathrm{er}_G(X,-)$$

with $X = \mathbb{R}\underline{\mathrm{Spec}}\,A$, in the same manner that we had $u_X : F(X) \to G(X)$ in Proposition 1.6. To be precise, for an A-point $x : X = \mathbb{R}\underline{\mathrm{Spec}}\,A \to F$ of F, we have an A-point $f \circ x : \mathbb{R}\underline{\mathrm{Spec}}\,A \to G$ of G, and the natural transformation above appears as an induced map

$$f_* = df : \mathbb{D}\mathrm{er}_F(X,-) \Rightarrow \mathbb{D}\mathrm{er}_G(X,-)$$

where $\mathbb{D}\mathrm{er}_G(X,-)$ is taken at the point $f(x)$. Since in Proposition 1.6 $u_X : F(X) \to G(X)$ is $(-c+(n+1)k)$-connected, its homotopy fibers are $(-c+(n+1)k-1)$-connected. This motivates the introduction of $\mathbb{D}\mathrm{er}_{F/G}(X,-)$, the homotopy fiber of df ([79]), defined by

$$\mathbb{D}\mathrm{er}_{F/G}(X,M) = \mathrm{Map}_{X/k\text{-}D^-\mathrm{Aff}^{\sim,\text{ét}}/G}(X[M],F)$$

and we want to show it has an intrinsic connective object. To that end, observe that f has a cotangent complex, which is connective, as we now explain. By Definition 1.4.1.14 of [79], $f : F \to G$ has a *relative cotangent complex at* x relative to sk-Alg if there is a $(-n)$-connective stable A-module $\mathbb{L}_{F/G,x} \in \mathrm{Ho}(\mathrm{Sp}(A\text{-Mod}))$ and an isomorphism $\mathbb{D}\mathrm{er}_{F/G}(X,-) \cong \mathbb{R}\underline{h}_s^{\mathbb{L}_{F/G,x}}$. If this is true, then the homotopy fiber of $\mathbb{D}\mathrm{er}_F(X,-) \Rightarrow$

$\mathbb{D}\mathrm{er}_G(X,-)$ would be connective. However, note that this is only for fixed A, whereas we have $F(X) \to G(X)$ $(-c + (n+1)k)$-connected for all $X \in \mathcal{C}_Y$ in the calculus of functors. This means that the relative cotangent complexes we just introduced must be found for all points $A \in sk$-Alg and exhibit the same connectivity. Consequently, we need Definition 1.4.1.15 of [79]: we say f has a *cotangent complex relative to* (sk-Mod, sk-Mod, sk-Alg) if for all $A \in sk$-Alg and $x : \underline{\mathbb{R}\mathrm{Spec}\,A} \to F$, f has a cotangent complex $\mathbb{L}_{F/G,x}$ at x, and additionally for all $u : A \to B$ in sk-Alg, any morphism

$$\begin{array}{ccc} \underline{\mathbb{R}\mathrm{Spec}\,B} & \longrightarrow & \underline{\mathbb{R}\mathrm{Spec}\,A} \\ & {\scriptstyle y}\searrow \quad \swarrow{\scriptstyle x} & \\ & F & \end{array}$$

in Ho (k-D^{-}Aff$^{\sim,\text{ét.}}$) induces an isomorphism $\mathbb{L}_{F/G,x} \otimes^{\mathbb{L}}_A B \xrightarrow{\cong} \mathbb{L}_{F/G,y}$ in Ho (Sp(B-Mod)), a sort of coherence condition. But by Lemma 1.4.1.16 of [79], if both F and G have a cotangent complex, so does f. However, by Theorem C.0.9, if both F and G are n-geometric D^{-}-stacks, then they have an obstruction theory; in particular, they have a cotangent complex, and hence so does f. In other terms, lifting each morphism $F(A) \to G(A)$ to the level of derivations, we obtain a natural transformation $\mathbb{D}\mathrm{er}_F(X,-) \Rightarrow \mathbb{D}\mathrm{er}_G(X,-)$ with $X = \underline{\mathbb{R}\mathrm{Spec}\,A}$, whose homotopy fiber $\mathbb{D}\mathrm{er}_{F/G}(X,-)$ at x has a (connective) cotangent complex, and for all $u : A \to B$ in sk-Alg, $x : \underline{\mathbb{R}\mathrm{Spec}\,A} \to F$, and $y : \underline{\mathbb{R}\mathrm{Spec}\,B} \to F$, those cotangent complexes are compatible in the following sense: $\mathbb{L}_{F/G,x} \otimes^{\mathbb{L}}_A B \xrightarrow{\cong} \mathbb{L}_{F/G,y}$, which is our analog for stacks of Proposition 1.6 for homotopy functors.

22.5.3 Reconstruction theorem

We have the following result from [2, Theorem 2.9], a reconstruction theorem for the Taylor tower of a homotopy functor $F : \mathrm{Sp} \to \mathrm{Sp}$, which states that there is a natural homotopy Cartesian square of the form

$$\begin{array}{ccc} P_nF(X) & \longrightarrow & (\partial_n F \wedge X^{\wedge n})^{h\Sigma_n} \\ \downarrow & & \downarrow \\ P_{n-1}F(X) & \longrightarrow & (\partial_n F \wedge X^{\wedge n})^{t\Sigma_n} \end{array}$$

where if X is a spectrum with an action of a finite group G, and X^G is the spectrum of G-invariants of X, then X^{tG} is the cofiber of the norm map $X/G \to X^G$. For our part, by Lemma 2.2.1.1. of [79] there is a unique derivation $d_n \in \pi_0(\mathbb{D}\mathrm{er}_k(A_{\le n-1}, S^{n+1} \otimes_k \pi_n A))$ for which we have an isomorphism in Ho (sk-Alg$_{/A_{\le n-1}}$) between the natural projection $A_{\le n-1} \oplus_{d_n} (S^n \otimes_k \pi_n A) \to A_{\le n-1}$ and $A_{\le n} \to A_{\le n-1}$. For this particular derivation, we have a homotopy Cartesian square

$$\begin{array}{ccc} A_{\leq n-1} \oplus_{d_n} \Omega(S^{n+1} \otimes_k \pi_n A) & \longrightarrow & A_{\leq n-1} \\ \downarrow & & \downarrow{\scriptstyle d_n} \\ A_{\leq n-1} & \xrightarrow[s]{} & A_{\leq n-1} \oplus (S^{n+1} \otimes_k \pi_n A). \end{array}$$

Let $M = S^{n+1} \otimes_k \pi_n A$. The reasoning now is the same as for showing homotopy fibers of connecting maps in towers are similar in form. Because F is op-excisive, we have a homotopy Cartesian square

$$\begin{array}{ccc} F(A_{\leq n-1} \oplus_{d_n} \Omega M) & \longrightarrow & F(A_{\leq n-1}) \\ \downarrow & & \downarrow \\ F(A_{\leq n-1}) & \longrightarrow & F(A_{\leq n-1} \oplus M). \end{array}$$

Thus $F(A_{\leq n-1} \oplus_{d_n} \Omega M) \xrightarrow{\simeq} FA_{\leq n-1} \times_{F(A_{\leq n-1} \oplus M)} E(FA_{\leq n-1})$. By Proposition 1.3.3.8 of [40], for objects A, B, C of some right proper model category $\mathcal{C}$, $A \times_C E(B)$ is weakly equivalent to $E(A) \times_C E(B) = A \times^h_C B$, hence an isomorphism in $\operatorname{Ho}(\mathcal{C})$. It follows that $\gamma F(A_{\leq n-1} \oplus_{d_n} \Omega M) \xrightarrow{\cong} \gamma FA_{\leq n-1} \times^h_{\gamma F(A_{\leq n-1} \oplus M)} \gamma FA_{\leq n-1}$. Now $\gamma F : sk\text{-Alg} \to \operatorname{Ho}(\mathrm{Set}_\Delta)$ has a unique factoring map $\operatorname{Ho}(\gamma F) : \operatorname{Ho}(sk\text{-Alg}) \to \operatorname{Ho}(\mathrm{Set}_\Delta)$, also denoted γF for simplicity. Further, using the fact that $\Omega M \cong S^n \otimes_k \pi_n A$ and the fact that the map $A_{\leq n-1} \oplus_{d_n} S^n \otimes_k \pi_n A \to A_{\leq n-1}$ is isomorphic to $A_{\leq n} \to A_{\leq n-1}$, it follows that we have the following isomorphism in $\operatorname{Ho}(\mathrm{Set}_\Delta)$:

$$\gamma F(A_{\leq n}) \xrightarrow{\cong} \gamma F(A_{\leq n-1}) \times^h_{\gamma F(A_{\leq n-1} \oplus M)} \gamma FA_{\leq n-1}.$$

Finally, for $A \in sk\text{-Alg}$, $F(A) \xrightarrow{\simeq} \mathbb{R}F(A)$, hence an isomorphism $\gamma F(A) \xrightarrow{\cong} \mathbb{R}F(A)$, from which it follows that we have a homotopy pullback square in $\operatorname{Ho}(\mathrm{Set}_\Delta)$:

$$\begin{array}{ccc} \mathbb{R}F(A_{\leq n}) & \longrightarrow & \mathbb{R}F(A_{\leq n-1}) \\ \downarrow & & \downarrow \\ \mathbb{R}F(A_{\leq n-1}) & \longrightarrow & \mathbb{R}F(A_{\leq n-1} \oplus (S^{n+1} \otimes_k \pi_n A)) \end{array}$$

which provides a reconstruction of the tower for $\mathbb{R}F(A)$ as well.

22.6 Taking stock – general theory

22.6.1 Convergence of the Taylor tower

Going over the proof of Theorem 1.13 of [31], let F be ρ-analytic, so that it is n-stably excisive. In the definition of $E(n, c, K)$, we pick $\mathcal{X} = X *_Y -$ as a cubical diagram from $\mathcal{P}(\underline{n+1})$

into $\mathcal{C}$. If $s \in \underline{n+1}$, then $\mathcal{X}(\emptyset) \to \mathcal{X}(s)$ reads $X \to X *_Y s$. Now for the convergence of the Taylor tower, we suppose $X \to Y$ is k-connected. By Remark 1.1 of [31] it follows that $X *_Y s \to Y$ is $(k+1)$-connected, and by Proposition 1.5 of [30] that $X \to X *_Y s$ is k-connected for all $s \in \underline{n+1}$, so that $F\mathcal{X}$ is $(-c + (n+1)k)$-Cartesian since F satisfies $E(n, c, K)$, which exactly means that $F\mathcal{X}(\emptyset) \to \operatorname{holim}(F\mathcal{X}|_{\mathcal{P}_0(\underline{n+1})})$ is $(-c + (n+1)k)$-connected. This map reads $F(X) \to \operatorname{holim}_{U \in \mathcal{P}_0(\underline{n+1})} F(X *_Y U) = T_n F(X)$. By Definition 1.2 this is referred to as saying that $F \to T_n F$ satisfies $\mathcal{O}(n, c, K)$. In the statement of Theorem 1.13, $X \to Y$ is $(\rho+1)$-connected, where $k = \rho + 1$ is that particular value that allows us to have convergence of the Taylor tower. Thus a key step in proving the convergence of the Taylor tower for $F(X)$ with F ρ-analytic and $X \to Y$ $(\rho+1)$-connected consists in first picking $\mathcal{X} = X *_Y -$, from which we have $F \to T_n F$ satisfying $\mathcal{O}(n, c, K)$ for some c and K, which is interpreted as saying that F and $T_n F$ agree to order n. Aside from the fact that $T_n F$ will be useful in constructing $P_n F$, this degree n approximation is essential to prove the convergence of the Taylor tower for $F(X)$.

Note that it is at this point that we took the analogous diagram for stacks to be represented by $A \oplus M \to A$, which amounts to identifying $X *_Y -$ with $A \oplus -$.

22.6.2 Convergence of the tower for stacks

Presently, we focus on the convergence of the tower involving $\mathbb{R}F(A_{\leq n})$. Essentially, we have $\mathbb{R}F(A) \cong \operatorname{holim}_n \mathbb{R}F(A_{\leq n})$ by virtue of the fact that $A_{\leq n}$ can be expressed as a square zero extension. More specifically, following the proof of Theorem C.0.9 of [79], to show that we have such an isomorphism in the homotopy category of simplicial sets, we show that we have an equivalence in Set_Δ or, equivalently, that for some $x \in \pi_0(\operatorname{holim}_n \mathbb{R}F(A_{\leq n}))$, the homotopy fiber of the natural map $\mathbb{R}F(A) \to \operatorname{holim}_n \mathbb{R}F(A_{\leq n})$ is contractible. To achieve this, one uses induction. If $x_n \in \mathbb{R}F(A_{\leq n})$ is the projection of x, one argues there is a representable D^--stack U along with a smooth morphism $U \to F$ such that one has $y_n \in \pi_0(\mathbb{R}U(A_{\leq n}))$ mapping to x_n. Because $A_{\leq n}$ is a square zero extension of $A_{\leq n-1}$, one finds that the morphism $\mathbb{R}U(A_{\leq n+1}) \to \mathbb{R}U(A_{\leq n}) \times^h_{\mathbb{R}F(A_{\leq n})} \mathbb{R}F(A_{\leq n+1})$ is surjective, which allows us to lift y_n to some element $y_{n+1} \in \pi_0(\mathbb{R}U(A_{\leq n+1}))$, and from there one obtains an element $y \in \pi_0(\operatorname{holim} \mathbb{R}U(A_{\leq n}))$ mapping to x. Thus we get a commutative diagram

$$\begin{array}{ccc} \mathbb{R}U(A) & \longrightarrow & \operatorname{holim} \mathbb{R}U(A_{\leq n}) \\ \downarrow & & \downarrow \\ \mathbb{R}F(A) & \longrightarrow & \operatorname{holim} \mathbb{R}F(A_{\leq n}). \end{array}$$

By induction this is a homotopy Cartesian square, and the top map is an equivalence, which allows us to conclude. All the results necessary to conduct this proof are based on the fact that F has an obstruction theory and the fact that the Postnikov tower for $A \in$

sk-Alg is constructed from square zero extensions, and thus the underlying reasoning is purely deformation-theoretic.

To emphasize this last point, by Lemma 2.2.1.1 there is a unique derivation $d_n \in \pi_0(\mathbb{D}\mathrm{er}_k(A_{\leq n-1}, S^{n+1} \otimes_k \pi_n A))$ such that $A_{\leq n} \to A_{\leq n-1}$ is isomorphic to $A_{\leq n-1} \oplus_{d_n} S^n \otimes_k \pi_n A \to A_{\leq n-1}$. Thus the tower for $F_n(A) = F(A_{\leq n})$ reads

$$\cdots \to \mathbb{R}F(A_{\leq n-1} \oplus_{d_n} (S^n \otimes_k \pi_n A)) \to \mathbb{R}F(A_{\leq n-1}) \to \cdots.$$

It follows that starting from an element of $\pi_0 F(A)$, to climb up in the tower for $\mathbb{R}F(A)$, one has to iterate moves of the form $\mathbb{R}F(A) \to \mathbb{R}F(A \oplus_d \Omega M)$ for $M \in A\text{-Mod}_1$.

Using the fact that $\mathbb{R}F(A) \cong \mathbb{R}\underline{\mathrm{Hom}}(\mathbb{R}\underline{\mathrm{Spec}}\, A, F)$, this amounts to having a lift of x to x' in the following diagram:

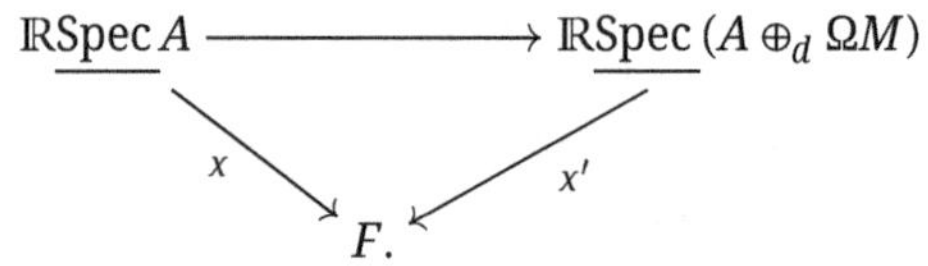

However, by Proposition 1.4.2.5 we have such a lift if we have a vanishing obstruction class in $\mathrm{Map}_{\mathrm{Sp}(A\text{-Mod})}(\mathbb{L}_{F,x}, M)$. Thus we obtain a construction that is inherently deformation-theoretic.

22.6.3 Abstraction

For Goodwillie calculus, we used the cubical diagram $X *_Y -$ to obtain that $F(X) \to \operatorname{holim} F(X *_Y U)$ is $(-c + (n+1)k)$-connected. This is a pivotal statement for proving the convergence of the Taylor tower of $F(X)$. Note that assuming F is ρ-analytic, if X is close enough to Y in the sense that $X \to Y$ is $\rho + 1$-connected, then we have the convergence of the Taylor tower as n goes to infinity in $F(X) \to P_n F(X)$. Essentially, we have convergence in a neighborhood of Y.

For stacks, we took our analogous diagram to be given by $\mathcal{X} = A \oplus -$, and in lieu of dealing with $F\mathcal{X}$, we considered all maps $F(A \oplus M) \to F(A)$ and argued that functorially their homotopy fibers are a lift of the functor $\mathbb{D}\mathrm{er}_F(X, -)$, which because F has an obstruction theory, can be corepresented by $\mathbb{L}_{F,x}$. Having a cotangent complex $\mathbb{L}_{F,x}$ allows us to use obstruction theory arguments to eventually prove the convergence of $\mathbb{R}F(A_{\leq n})$ to $\mathbb{R}F(A)$.

So far both theories are built upon the choice of a deformation theoretic diagram, whether it be $X *_Y -$ or $A \oplus -$, and a pivotal result, $F \to T_n F$ agreeing to order n in Goodwillie's calculus and the existence of $\mathbb{L}_{F,x}$ for stacks. We also had a notion of neighborhood for the calculus of functors. We now investigate such a notion for stacks.

To do so, observe that the convergence of the Postnikov tower for $\mathbb{R}F(A)$ hinges on Lemma C.0.10, in the proof of which by using a local argument one argues that one can find a representable D^{-}-stack U along with a smooth morphism $U \to F$ such that we have the following summarizing diagram:

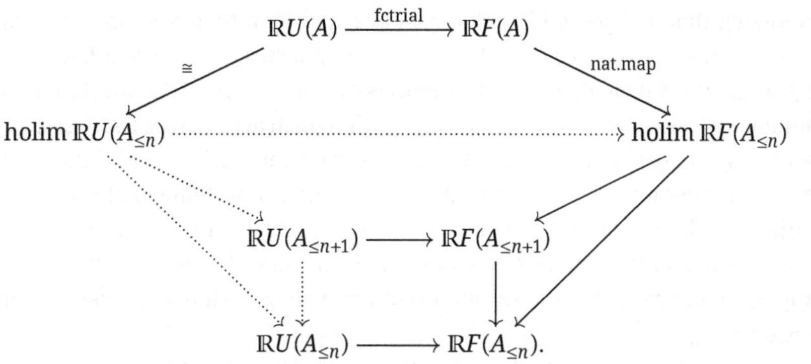

The argument goes as follows: one can find U such that $x_n \in \pi_0(\mathbb{R}F(A_{\leq n}))$ comes from some $y_n \in \pi_0(\mathbb{R}U(A_{\leq n}))$, which gives us the bottom two horizontal maps. Using the fact that the Postnikov tower for A is built from square zero extensions, one can show that for large n, the morphism $\mathbb{R}U(A_{\leq n+1}) \to \mathbb{R}U(A_{\leq n}) \times^{h}_{\mathbb{R}F(A_{\leq n})} \mathbb{R}F(A_{\leq n+1})$ is surjective, which provides us with the bottom square, hence the dotted maps from holim $\mathbb{R}U(A_{\leq n})$, as well as the horizontal dotted map between homotopy limits. The map $\mathbb{R}U(A) \to \text{holim}\, \mathbb{R}U(A_{\leq n})$ is an equivalence since U is representable, the top horizontal map we have by functoriality, and the right diagonal map is the natural map, which we aim to show is an equivalence. By an inductive argument, which is the statement of Lemma C.0.10, the top square is homotopy Cartesian, which allows us to conclude. Thus it seems everything hinges on the local existence of U.

This we have by virtue of Corollary 2.2.2.9, which essentially says that the convergence of the tower for $\mathbb{R}F(A)$ can be studied locally on the small étale site of A as observed in [79], which is an analogous statement to having $X \to Y$ ρ+1-connected in Goodwillie calculus. Both statements provide us with a notion of a neighborhood. Further in the above diagram, one can climb up in the Postnikov tower of $\mathbb{R}U(A)$ for large values of n by Corollary 2.2.5.3 of [79], the argument being that some fibers are simply connected, which immediately gives us the desired convergence, and this also mirrors the fact that the connectedness of $F(X) \to P_nF(X)$ tends to zero as n goes to infinity for the convergence of the Taylor tower of homotopy functors. Both theories use deformation-theoretic arguments, the existence of a cotangent complex for stacks, the fact that the Postnikov tower for $A \in sk$-Alg is built up from iterated square zero extensions, and the connectedness arguments in the case of Goodwillie calculus, along with repeated joins from X, providing us with $T^{i}_{n}F(X)$, the homotopy limit of which is $P_nF(X)$.

Thus heuristically an overarching theory would go like so: let $\mathcal{C}$ be a category with a notion of neighborhood, let $\mathcal{D}$ be a category with a notion of equivalence, let $F : \mathcal{C} \to \mathcal{D}$ be a functor, let $X \in \text{Ob}\, \mathcal{C}$, and let $\cdots \to X_n \to X_{n-1} \to \cdots$ be a deformation-theoretic tower induced by X. If F satisfies some deformation theoretic constraint and X lies in a certain neighborhood, then one has $F(X) \cong \operatorname{holim} F(X_n)$ in $\operatorname{Ho}(\mathcal{D})$. We denote the abstraction of such a formalism by $\mathbb{T}_0$.

Applying this to Goodwillie calculus, governed by a theory $\mathbb{G}$, $\mathcal{C}$ is a category of spaces over some fixed space Y, and $\mathcal{D}$ is a category of spaces or spectra. On $\mathcal{C}$ the notion of neighborhood is provided by connectedness: $X \to Y$ is $(\rho+1)$-connected. $F : \mathcal{C} \to \mathcal{D}$ is a homotopy functor. The tower induced by X is constructed using the join $X *_Y U$ and its iterations, giving rise first to $T_n F(X)$, then to its powers $T_n^i F(X)$, and finally to $P_n F(X)$. The constraint put on F is ρ-analytic and controls the connectedness of the image of cubical diagrams by F and, in particular, those of the form $F(X) \to T_n F(X)$. That we use connectedness is natural since F is valued in a category of spaces, on which weak homotopy equivalence is the natural notion of equivalence. Thus $\mathbb{G}$ presents itself as an expansion of $\mathbb{T}_0$.

For geometric stacks in derived Algebraic Geometry, governed by a theory $\mathbb{S}$, $\mathcal{C} = sk\text{-Alg}$, and $\mathcal{D} = \text{Set}_\Delta$. However, the functor F is a stack, object of $\text{k-D}^-\text{Aff}^{\sim,\text{ét.}}$, and thus the notion of a neighborhood on $\mathcal{C}$ is really given by the fact that we have an étale model topology on $\text{k-D}^-\text{Aff}$. This is illustrated in Corollary 2.2.2.9 of [79], which shows that one can work locally on the small étale site of A. That provides us with a neighborhood of A. The tower for A is given by its Postnikov tower in sk-Alg. It is deformation-theoretic in nature since one can show that it is built from square zero extensions. The constraint we put on F is that it is an op-excisive n-geometric D^--stack, which implies in particular that it has an obstruction theory, which we saw is the analog of being ρ-analytic for homotopy functors. To have an obstruction theory means in particular that F has a cotangent complex. This is necessary to show that, along with the fact that locally on the small étale site of A (which guarantees we have a smooth morphism $U \to F$ from some representable stack U), the homotopy fiber of $\mathbb{R}U(A_{\leq n+1}) \to \mathbb{R}U(A_{\leq n}) \times^h_{\mathbb{R}F(A_{\leq n})} \mathbb{R}F(A_{\leq n+1})$ is simply connected for large values of n, we have the isomorphism $\mathbb{R}F(A) \cong \operatorname{holim} \mathbb{R}F(A_{\leq n})$. Thus $\mathbb{S}$ presents itself as an expansion of $\mathbb{T}_0$ as well.

23 Spectral "schematization" of homotopy types

23.1 Introduction

It is important to note that the present work is in no way an addendum, nor does it supplant Toen's work on the schematization of homotopy types, which is very complete. Our aim is modest; we wanted to develop a formalism in which a schematization of homotopy types would make sense, along with an accompanying notion of homotopy groups from which such homotopy types would be defined, starting from very simple considerations. In doing so, we arrived at a notion of homotopy groups of objects in a comma category, generalizing the classical homotopy groups of spaces, something that we found to be of interest to algebraic topologists. For this reason, after discussing a schematization of homotopy types proper, we study those generalized homotopy groups in the first part of the present section. In the second part, we systematize such an approach, and we find ourselves abandoning the base model category k-CAlg in favor of the model category of symmetric spectra Sp^{Σ}. Thus what we obtain at the onset should really be called a spectralization of homotopy types.

Grothendieck started with the observation in [35] that homotopy types are essentially discrete structures, so one may as well start our investigation with a free $\mathbb{Z}$-module M of finite type as a model of a homotopy type. It gives rise to a vector bundle over $S_0 = \operatorname{Spec}\mathbb{Z}$, which he denoted by $W(M)$, whose $\mathbb{Z}$-module of sections is M itself. The functor $M \mapsto W(M)$ is fully faithful, the point being that this provides an embedding of homotopy types into a category of geometric objects such as bundles. He then generalized that to any $\mathbb{Z}$-module M, i. e., an abelian group. In that case, our vector bundle should be defined as $W(M) : k \mapsto M \otimes_{\mathbb{Z}} k$, a functor on the category of commutative $\mathbb{Z}$-algebras, still very close to being a vector bundle. The resulting functor $M \mapsto W(M)$ is still faithful. Grothendieck pulled back this structure to spaces via torsors. What he thought about constructing then is an auxiliary category $U(k)$ of unipotent bundles over k, not quite schemes over k, given by functors $k\text{-CAlg} = (k\text{-Aff})^{\mathrm{op}} \to \mathrm{Set}$, i. e., presheaves on k-Aff. If we denote by τ the fpqc topology on k-Aff, then we can consider the model category of sheaves on k-Aff that Toen denotes by $k\text{-Aff}^{\sim,\tau}$, and we are looking for a submodel category

$$U(k) \subset k\text{-Aff}^{\sim,\tau} \subset \widehat{k\text{-Aff}}.$$

Grothendieck defined $U(k)$ as the union of an increasing sequence of subcategories $U_n(k)$. He defined $U_{n+1}(k)$ as follows: for any k-module M, the corresponding vector bundle is defined via $W(M)(k') = M \otimes_k k'$, and $X_{n+1} \in U_{n+1}(k)$ if and only if there are some $X_n \in U_n(k)$ and a k-module M such that X_{n+1} is isomorphic to a torsor over X_n with group $W(M)$. In other terms, the initial embedding of a discrete structure M into the category of schemes, or vector bundles as he saw them, is pulled back to $U(k)$, thereby

https://doi.org/10.1515/9783111334073-023

endowing the latter with the flavor of a *schematic homotopy type*. Later in his same work, Grothendieck observes that the modules M have to be projective.

Recall that the sections functor on $W(M)$ gives us back M, which is supposed to represent a homotopy type. Thus we would like to have a sections functor in the present setting as well to extract the homotopy type information from objects X of the category $U(k)$. It turns out that we do have a sections functor $U(k) \to \mathrm{Set} : X \mapsto X(k)$ that induces a map

$$\mathrm{Hom}(\Delta^{\mathrm{op}}, U(k)) \to \mathrm{Hom}(\Delta^{\mathrm{op}}, \mathrm{Set}).$$

Grothendieck initially argued that since the right-hand side should model homotopy types, this makes simplicial objects of $U(k)$ into homotopy theoretic objects, which means that we have an implicit notion of weak equivalence in $sU(k)$ and hence formally a corresponding homotopy category. More specifically, he aimed for a subcategory $\mathcal{M}(k)$ of $U(k)$ whose objects have homotopy groups endowed with a k-module structure, something that he will drop later, but nevertheless the idea remains: one should aim for a subcategory of $sU(k)$ with the desired homotopy theoretic attributes. Then the notion of weak equivalence we have above should lead to a homotopy category $\mathrm{Hot}_n(k)$ of n-connected schematic homotopy types over k, homotopy types because of a yet to be specified notion of weak equivalence, schematic because objects in $U(k)$ have the flavor of schemes insofar as they are torsors with groups $W(M)$ that initially were viewed as scheme-like objects. This is the picture we have to keep in mind. Much of the work Grothendieck did from there revolved around defining a proper notion of homotopy. The first hint that this preliminary picture may not work comes from the very definition of $\mathcal{M}(k)$; how could one construct homotopy groups endowed with a nontrivial k-module structure? Looking elsewhere, Grothendieck pointed out that we do have a classical abelianization functor for homotopy types $\mathrm{Hot}_n \to D_{\bullet}(\mathrm{Ab})$.

One should then have a commutative diagram

$$\begin{array}{ccc}
\mathrm{Hot}_n(k) & \longrightarrow & D_{\bullet}(\mathrm{Ab}_k) \\
\downarrow & & \downarrow \\
\mathrm{Hot}_n & \longrightarrow & D_{\bullet}(\mathrm{Ab})
\end{array}$$

and hence we need an abelianization, or linearization functor $\mathrm{Hot}_n(k) \to D_{\bullet}(\mathrm{Ab}_k)$, a localization of a functor $\mathcal{M}_n(k) \to \mathrm{ch}_{\bullet}(\mathrm{Ab}_k)$. Note, incidentally, that at this point, Eilenberg–MacLane spaces make their appearance, essentially as linearized, schematic homotopy types. Ultimately, Grothendieck was aiming for a k-linearization functor L : $U(k) \to \mathrm{Ab}_k, X \simeq W(M) \mapsto L(X)$. Defining $L(X)$ necessitates taking completions, something he wanted to avoid, so he considered pointed homotopy types instead, something that is quite natural. Hence he considered a category $U(k)^{\bullet}$ of pointed homotopy types instead of $U(k)$, with a corresponding k-linearization functor L_{pt} that is well behaved with regards to ring extensions. For a simplicial object $X_* \in sU(k)^{\cdot}$ in $U(k)^{\cdot}$, one defines its homology by $LH_{\cdot}(X_*) = L_{\mathrm{pt}}(X_*)$, and one says $X_* \to X'_*$ is a weak equivalence if the total homology functor $LH_{\cdot}$ transforms this map into a quasi-isomorphism. Later Grothendieck defines $H_i(X_*) := \pi_i(L_{\mathrm{pt}}(X_*)) = H_i(LH_{\cdot}(X_*))$, something that he regarded as a more precise way of defining a notion of weak equivalence than the one alluded to above. At some point later, he introduced a Lie functor $\mathrm{Lie} : U(k)^{\cdot} \to \mathrm{Ab}_k, X = W(M) \mapsto M$ with the usual homotopy groups $\pi_i(X_*) = \pi_i(X_*(k))$ satisfying $\pi_i(X_*) \simeq \pi_i(\mathrm{Lie}(X_*))$, which he saw as leading to a natural definition of homotopy invariance of X_* and hence to a notion of weak equivalence defined as producing quasi-isomorphisms for the corresponding Lie chain complexes. Even though this is yet another notion of weak equivalence, which he considered as a cornerstone of a theory of schematic homotopy types, he will admit later that a notion of weak equivalence is not agreed upon. Thus the very definition of schematic homotopy type is not settled.

Regarding homotopy groups, them being k-modules will be dropped quite fast by Grothendieck, but one thing he will keep is that higher homotopy groups π_n for $n > 1$ are endowed with an action of π_1, hinting at the fact that homotopy groups themselves may possibly have to be upgraded to something more general. This is also apparent in the fact that since the beginning of his work on the schematization problem, he wanted to use Postnikov devissage to define schematic homotopy types, something he will find not to be too useful, yet not completely abandoned. He later admitted that this was constantly on his mind and at times led him astray. This also points to the fact that homotopy groups carry an additional structure. It needs to be mentioned as well that Grothendieck at some point considered π_1 to be a group object in $U(k)$.

From this short synopsis one gets the flavor of what Grothendieck had in mind regarding schematic homotopy types, even though the very definition of weak equivalence to develop such a formalism remains elusive, and one cannot help but notice how homotopy groups themselves should probably be generalized.

One may summarize the above approach as working with schematic objects, containing some homotopy theoretic information, which is extracted in some sense with a sections functor, and whose intrinsic homotopic information is revealed via a notion of equivalence, the latter being defined as inducing quasi-isomorphisms for a well-chosen homology theory. Instead of using a sections functor, we will work with covers of geometric objects X by spectra of algebraic objects of finite type, a sort of probing procedure, thereby giving us a sense in which X can be regarded as being schematic, as well as a homotopy theoretic object. The homotopy types proper will be defined exactly as Grothendieck defined them, that is, as being dependent upon a notion of π_*-equivalence.

To talk about homotopy type, one needs to have an ambient model category in mind or, at the very least, a category with a subclass of weak equivalences, and if we are working with spaces, then we need a homotopy category of spaces. To speak of schematic homotopy type over a ground ring k while working with spaces as well means, from our perspective, putting any given space X and k on a same footing for them to be comparable, and hence we first have to explain how k could be seen as an object of a given homotopy category of spaces. This is fairly immediate: we consider the spectrum Spec k of a ring k, appropriately topologized. Now given X, looking for a notion of schematic homotopy type over k, we consider morphisms $\operatorname{Spec} R \to X$ in Top for $R \in k$-CAlg as providing such a notion, a schematic cover of X in a sense. Now recall that the modules M Grothendieck started with are supposed to be homotopy types, and for this reason, they are taken to be projective modules of finite type. Thus in the present context, $R \in k$-Mod is projective of finite type. Let $k\text{-CAlg}_{\text{pft}} = \text{Comm}(k\text{-Mod}_{\text{pft}})$, where pft stands for projective of finite type. Our cover can be formally represented as a functor

$$\text{Hom}_{\text{Top}}(\text{Spec}\,(-), X) : k\text{-CAlg}_{\text{pft}} \to \text{Set}$$

thereby providing a perception of X by schematic objects that themselves carry a homotopy type. In a sense, probing X by such objects should produce the schematic homotopy type of X. Practically, for extracting the homotopy-theoretic information in such covers, we introduce a notion of homotopy group of coverings by defining

$$\pi_n(\operatorname{Spec} R \xrightarrow{f} X) = f_* \pi_n(\operatorname{Spec} R).$$

This presupposes that we work with pointed spaces, and this will be implied throughout this work. Functorially, this can be represented by a graded functor $\pi_*(\text{Spec}\,(-) \to X) = \pi^k_{*,X}$, and weak equivalences are defined as π_*-equivalences, namely $X \sim X'$ iff $\pi^k_{*,X} \cong \pi^k_{*,X'}$. In the first part of the present section, we will derive a few properties of such homotopy groups. In the second part, we will generalize those homotopy groups; we have $\pi^k_{n,X}(R) = \pi_n(\operatorname{Spec} R \xrightarrow{f} X) = f_*\pi_n(\operatorname{Spec} R)$ for $R \in k\text{-CAlg}_{\text{pft}}$. Focusing on $\pi_n(\operatorname{Spec} R)$, if we want this group to be representable, then we look for some $R_0 \in k\text{-CAlg}_{\text{pft}}$ such that $\pi_n = \text{Hom}_{k\text{-CAlg}_{\text{pft}}}(-, R_0)_n = \text{Hom}_{k\text{-Aff}_{\text{pft}}}(\operatorname{Spec} R_0, -)_n$ with $k\text{-Aff}_{\text{pft}} = k\text{-CAlg}^{\text{op}}_{\text{pft}}$. Written differently,

$$[S^n, \operatorname{Spec} R] = \pi_n(\operatorname{Spec} R) = \text{Hom}_{k\text{-Aff}_{\text{pft}}}(\operatorname{Spec} R_0, \operatorname{Spec} R)_n$$

which can be implemented by taking R_0 to be a graded object, and the hom object to be in some homotopy category. One can immediately think of spectra with the sphere spectrum $\Sigma^\infty S^0 = \mathbb{S}$ playing the role of $\operatorname{Spec} R_0$, and $\operatorname{Spec} R$ being replaced by an E_∞-ring. Thus the functor $\pi_* = [S^*, -]$ should be replaced by $\mathbb{R}\underline{\text{Hom}}_{\text{Comm}(\text{Sp}^\Sigma)}(\mathbb{S}, -) = \mathbb{R}\underline{\text{Hom}}_{S\text{-Aff}}(-, \operatorname{Spec}\mathbb{S})$, where $S\text{-Aff} = (\text{Comm}(\text{Sp}^\Sigma))^{\text{op}}$.

Note that at this point, there is a clear departure from working with objects $R \in k$-CAlg, since the most basic objects are symmetric spectra. Instead of topological spaces, we will consider simplicial sets K and their symmetric suspension spectrum $\Sigma^\infty K = X$. Homotopy types will be represented by symmetric spectra of finite type. We consider coverings of finite type $\{\mathbf{H} \to X\}$ for $\mathbf{H} \in \text{Sp}^\Sigma_{\text{ft}}$, functorially represented as $\mathbb{R}\underline{\text{Hom}}(-, X) : \text{Ho}\,(\text{Sp}^\Sigma_{\text{ft}}) \to \text{Ho}\,(\text{Sp}^\Sigma)$, a perception of finite type, where Sp^Σ is endowed with the positive model structure of [70], $\underline{\text{Hom}}$ is the internal Hom of Sp^Σ, and $\mathbb{R}\underline{\text{Hom}}$

its derived Hom. For $f : \mathbf{H} \to X$, we have an induced map $\mathrm{Ho}(f_*) : \mathbb{R}\underline{\mathrm{Hom}}(\mathbb{S}, \mathbf{H}) \to \mathbb{R}\underline{\mathrm{Hom}}(\mathbb{S}, X)$, and the collection of subobjects of the form $\mathrm{Ho}(f_*)\mathbb{R}\underline{\mathrm{Hom}}(\mathbb{S}, \mathbf{H})$ defines a functor Π_X, in the same manner that earlier $f_*\pi_n(\mathrm{Spec}\, R) = f_*[S^n, \mathrm{Spec}\, R]$ defined $\pi^k_{n,X}(R)$. Then one defines a weak equivalence in Sp^Σ as follows: $X \sim X'$ if and only if $\Pi_X \cong \Pi_{X'}$. This gives us a notion of homotopy type, defined via symmetric spectra of finite type covering X (resp. X').

Thus we have a spectral interpretation of the schematization of homotopy types.

Notation: In spite of the fact that we typically use the notation Set_Δ to denote the category of simplicial sets, in this chapter only, we will use the more compact notation $\mathcal{S}^\Sigma_*$ to denote the category of symmetric sequences instead of the more cumbersome $\mathrm{Set}^\Sigma_{\Delta,*}$.

23.2 Part I: Algebraic Topology

23.2.1 Generalized homotopy groups

As pointed out in the introduction, for a given topological space X and a commutative ring k, to talk about the homotopy type of X over k, one would want to put X and k on a same footing to be able to define invariants of one object relative to the other. Thus it is natural to regard X as a space over $\mathrm{Spec}\, k$ and to consider morphisms $\mathrm{Spec}\, R \to X$ for $R \in k\text{-CAlg}$. In addition, we ask that those k-algebras be projective of finite type. The image of that morphism tells us what is X relative to R. The collection of all such maps provides an affine covering family $\{\mathrm{Spec}\, R \to X \mid R \in k\text{-CAlg}_{\mathrm{pft}}\}$. This can be further formalized by considering the functor

$$\mathrm{Hom}_{\mathrm{Top}}(\mathrm{Spec}\,(-), X) : k\text{-CAlg}_{\mathrm{pft}} \to \mathrm{Set}$$

which we defined as the affine perception in Spectral Algebraic Geometry.

A morphism $\mathrm{Spec}\, R \to X$ is regarded as an object of $\mathrm{Top}_{/X}$, and we generalize homotopy groups of spaces to homotopy groups of morphisms of spaces as follows. Consider the (local system) functor

$$\begin{aligned} \pi_n : \mathrm{Top} &\to \mathrm{Grp}^{\mathrm{Top}/\cdot}, \\ X &\mapsto \pi_n(X) := \pi_{nX} \end{aligned}$$

where

$$\begin{aligned} \pi_{nX} : \mathrm{Top}_{/X} &\to \mathrm{Grp}, \\ (Y \xrightarrow{f} X) &\mapsto \pi_{nX}(Y \xrightarrow{f} X) = f_*\pi_n(Y). \end{aligned}$$

We drop the subscript X presently for notation's sake. Observe that the classical homotopy group is recovered as $\pi_n(X \xrightarrow{\text{id}} X) = \text{id}_*\pi_n(X) = \pi_n(X)$, where $X \xrightarrow{\text{id}} X$ is the terminal object in $\text{Top}_{/X}$, so the classical picture can be seen as being embedded in this more general formalism. Observe that one may further generalize this by letting Y itself in $\pi_n(Y \xrightarrow{f} X) = f_*\pi_n(Y)$ be an object of a comma category, say $Y = A \xrightarrow{f} B$, with the present case above being recovered by letting $Y = Y$. Indeed,

$$\begin{aligned}\pi_n(A \xrightarrow{f} B \xrightarrow{g} X) &= \pi_n(A \xrightarrow{g\circ f} X)\\ &= (g \circ f)_*\pi_n(A)\\ &= g_* \circ f_*\pi_n(A)\\ &= g_*\pi_n(A \xrightarrow{f} B)\end{aligned}$$

an immediate generalization of $\pi_n(Y \xrightarrow{g} X) = g_*\pi_n(Y)$.

We will be interested in the affine part of this functor, namely $\pi_{n,X} = \pi_n(\text{Spec}\,(-) \to X)$, if we regard it as a functor on k-CAlg. However, we will regard $\pi_{n,X}$ as a functor on k-Aff. Observe that X is not necessarily affine, so one has to be careful in using the notation $(k\text{-Aff}_{\text{pft}})_{/X}$. To be precise, if we consider the projection on the domain $\text{pr}_1(X \to Y) = X$ and the inclusion of $k\text{-Aff}_{\text{pft}}$ into Top, then $\pi_{n,X}$ appears as a functor on $k\text{-Aff}_{\text{pft}} \times_{\text{Top}} \text{Top}_{/X}$, which we denote by $(k\text{-Aff}_{\text{pft}})_{/X}$, and whose objects are pairs $(\text{Spec}\,R, \text{Spec}\,R \xrightarrow{f} X)$. A morphism $(\text{Spec}\,R, f) \to (\text{Spec}\,R', f')$ in this category is a morphism $u : \text{Spec}\,R \to \text{Spec}\,R'$ that makes the following diagram commute:

$$\begin{array}{ccc} \text{Spec}\,R & & \\ \downarrow u & \searrow^{f} & \\ & & X \\ & \nearrow_{f'} & \\ \text{Spec}\,R'. & & \end{array}$$

This induces the following commutative diagram of homotopy groups:

$$\begin{array}{ccc} \pi_n\text{Spec}\,R & & \\ \downarrow u_* & \searrow^{f_*} & \\ & & \pi_n X \\ & \nearrow_{f'_*} & \\ \pi_n\text{Spec}\,R'. & & \end{array}$$

Thus we have a covariant functor

$$\pi_{n,X} : (k\text{-Aff}_{\text{pft}})_{/X} \to \text{Grp},$$
$$(\operatorname{Spec} R, \operatorname{Spec} R \xrightarrow{f} X) \mapsto f_*\pi_n \operatorname{Spec} R.$$

Thus defined, $\pi_{*,X}$ gives us a schematic probe of the homotopic character of X. Define two spaces X and X' to be equivalent if they are π_*-*equivalent*,that is, $\pi_{*,X} \cong \pi_{*,X'}$. This provides us with a notion of weak equivalence, based on schematic objects, and hence with a notion of schematic homotopy type.

23.2.2 Properties of homotopy groups of maps

We use mainly [61] as a resource to develop our formalism. In terms of loop spaces, one can write

$$\pi_n(A \xrightarrow{f} X) = f_*\pi_n(A) = f_*\pi_{n-1}\Omega A = \cdots = f_*\pi_0(\Omega^n A).$$

For having long exact sequences of homotopy groups, we consider, for $A \xrightarrow{f} X$, the following homotopy fiber [61]:

$$Ff = A \times_f PX = \{(a, \xi) \mid f(a) = \xi(1)\} \subset A \times PX$$

with $PX = F(I, X)$, I based at 0, and a fixed point $*$ of X, so Ff consists of paths $\gamma : * \to f(A) \subset X$, i. e., $Ff = P(X; *, f(A))$, the space of paths in X starting at the fixed point $*$ and ending in $f(A)$. We have a natural projection $\pi : Ff \to A$, $(x, \chi) \mapsto x$. We also have an inclusion $\iota : \Omega X \to Ff$ specified by $\iota(\xi) = (*, \xi)$, $* \in f(A)$. This gives us a fiber sequence generated by f as given in [61]:

$$\cdots \to \Omega^2 A \xrightarrow{\Omega^2 f} \Omega^2 X \xrightarrow{-\Omega\iota} \Omega Ff \xrightarrow{-\Omega\pi} \Omega A \xrightarrow{-\Omega f} \Omega X \xrightarrow{\iota} Ff \xrightarrow{\pi} A \xrightarrow{f} X$$

where as defined in [61], $(-\Omega f)(\eta)(t) = (f \circ \eta)(1-t)$ for $\eta \in \Omega A$. From there we get a long exact sequence of pointed sets for any based space Z:

$$\cdots \to [Z, \Omega Ff] \to [Z, \Omega A] \to [Z, \Omega X] \to [Z, Ff] \to [Z, A] \to [Z, X].$$

Then we let $Z = S^n$ and $[S^n, -] = \pi_n$ to arrive at

$$\begin{array}{ccccccccccccc}
\cdots \longrightarrow & \pi_n\Omega Ff & \longrightarrow & \pi_n\Omega A & \longrightarrow & \pi_n\Omega X & \longrightarrow & \pi_n Ff & \longrightarrow & \pi_n A & \longrightarrow & \pi_n X \\
& \| & & \| & & \| & & \| & & & & \\
& \pi_{n+1}Ff & & \pi_{n+1}A & & \pi_{n+1}X & & \pi_n P(X;*,f(A)) & & & & \\
& \| & & & & & & \| & & & & \\
& \pi_{n+1}P(X;*,f(A)) & & & & & & \pi_{n+1}(X,f(A)) & & & & \\
& \| & & & & & & & & & & \\
& \pi_{n+2}(X,f(A)) & & & & & & & & & &
\end{array}$$

or, in simplified form,

$$\cdots \to \pi_n A \xrightarrow{f_*} \pi_n X \to \pi_n(X,fA) \xrightarrow{\partial} \pi_{n-1}A \to \cdots \to \pi_0 A \to \pi_0 X.$$

To show that π_n is functorial on morphisms, we will need the following lemma.

Lemma 23.2.2.1.

$$\pi_n A/\partial\pi_{n+1}(X,fA) \cong \pi_n(A \xrightarrow{f} X). \tag{23.1}$$

Proof. $\pi_n(A \xrightarrow{f} X) = f_*\pi_n A$ is the image of $\pi_n A \xrightarrow{f_*} \pi_n X$, so $\pi_n A \twoheadrightarrow \pi_n(A \xrightarrow{f} X)$, and because we have a long exact sequence above, the result follows. □

We would like now to answer the question whether a map $A \xrightarrow{\tau} B$ induces a map $\pi_n(A \xrightarrow{f} X) \to \pi_n(B \xrightarrow{g} X)$. More generally, for two objects $A \xrightarrow{f} X$ and $B \xrightarrow{g} Y$, with morphism between them given by a commutative diagram (τ, h) of the form

$$\begin{array}{ccc}
A & \xrightarrow{\tau} & B \\
{\scriptstyle f}\downarrow & & \downarrow{\scriptstyle g} \\
X & \xrightarrow[h]{} & Y.
\end{array} \tag{23.2}$$

Proposition 23.2.2.2. *Given $A \xrightarrow{f} X$ and $B \xrightarrow{G} Y$, a map $(\tau, h) : (A, X) \to (B, Y)$ as in (23.2) induces a morphism $\pi_n(A \xrightarrow{f} X) \xrightarrow{(\tau,h)_*} \pi_n(B \xrightarrow{g} Y)$.*

Proof. By the commutativity of (23.2) we have the following commutative diagram:

$$\begin{array}{ccccc}
\pi_{n+1}(X,fA) & \xrightarrow{\partial} & \pi_n A & \xrightarrow{f_*} & \pi_n X \\
& & {\scriptstyle \tau_*}\downarrow & & \downarrow{\scriptstyle h_*} \\
\pi_{n+1}(Y,gB) & \xrightarrow[\partial]{} & \pi_n B & \xrightarrow[g_*]{} & \pi_n Y
\end{array}$$

from which we have

$$\begin{aligned}\partial\pi_{n+1}(X,fA) = \operatorname{Ker} f_* \subset \operatorname{Ker} h_* \circ f_* &= \operatorname{Ker} g_* \circ \tau_* \\ &= \{u \in \pi_n A,\ \tau_* u \in \operatorname{Ker} g_*\} \\ &= \{u \in \pi_n A,\ \tau_* u \in \partial\pi_{n+1}(Y,gB)\} \\ &= \tau_*^{-1}\partial\pi_{n+1}(Y,gB) := H\end{aligned}$$

that is, $\partial\pi_{n+1}(X,fA) \subset H$. Now we have a map

$$\pi_n A \to \pi_n B \twoheadrightarrow \pi_n B/\partial\pi_{n+1}(Y,gB)$$

and $H = \tau_*^{-1}\partial\pi_{n+1}(Y,gB) \lhd \pi_n A$, so there is a map

$$\pi_n A/H \to \pi_n B/\partial\pi_{n+1}(Y,gB)$$

and since we have shown $\partial\pi_{n+1}(X,fA) \subset H$, it follows that we also have a map

$$\pi_n A/\partial\pi_{n+1}(X,fA) \to \pi_n A/H$$

so that by composition we get a map

$$\pi_n A/\partial\pi_{n+1}(X,fA) \to \pi_n B/\partial\pi_{n+1}(Y,gB)$$

that is, a morphism (τ, h) from $A \xrightarrow{f} X$ to $B \xrightarrow{g} Y$ induces a map $\pi_n(A \xrightarrow{f} X) \to \pi_n(B \xrightarrow{g} Y)$ by (23.1). □

Proposition 23.2.2.3. *If X is contractible for any $A \xrightarrow{f} X$, then $\pi_n(A \xrightarrow{f} X) = 0$.*

Proof. $\pi_n(X) = 0$ for all $n \geq 0$, and $\pi_n(A \xrightarrow{f} X) = f_*\pi_n A \subset \pi_n X$, so that $\pi_n(A \xrightarrow{f} X) = 0$ for all $n \geq 0$. □

Proposition 23.2.2.4. *If X is discrete for any $A \xrightarrow{f} X$, then $\pi_n(A \xrightarrow{f} X) = 0$.*

Proof. X being discrete, $\pi_n X = 0$, and the rest of the proof is identical to that of the previous result. □

Proposition 23.2.2.5. $\pi_n((A \xrightarrow{f} X) \times (B \xrightarrow{g} Y)) \cong \pi_n(A \xrightarrow{f} X) \times \pi_n(B \xrightarrow{g} Y)$.

Proof. It suffices to write

$$\begin{aligned}\pi_n((A \xrightarrow{f} X) \times (B \xrightarrow{g} Y)) &= \pi_n(A \times B \xrightarrow{f\times g} X \times Y) \\ &= (f \times g)_*\pi_n(A \times B) \\ &\cong (f \times g)_*\pi_n A \times \pi_n B \\ &= f_* \times g_*\pi_n A \times \pi_n B\end{aligned}$$

$$= f_*\pi_n A \times g_*\pi_n B$$
$$= \pi_n(A \xrightarrow{f} X) \times \pi_n(B \xrightarrow{g} Y). \qquad \square$$

Proposition 23.2.2.6. *$\pi_n(A \xrightarrow{f} X)$ does not depend on the basepoint.*

Proof. For classical homotopy groups, this is proved in [61] and is stated as follows: if $f : (X,A) \to (Y,B)$ is a map of pairs and $\alpha : I \to A$ is a path from a to a' in A, then the following diagram commutes:

$$\begin{array}{ccc} \pi_n(X,A,a) & \xrightarrow{f_*} & \pi_n(Y,B,f(a)) \\ \cong\downarrow & & \downarrow\cong \\ \pi_n(X,A,a') & \xrightarrow[f_*]{} & \pi_n(Y,B,f(a')). \end{array} \tag{23.3}$$

In our setting, this would read as follows: for a map $(A \xrightarrow{f} X) \xrightarrow{(\tau,h)} (B \xrightarrow{g} Y)$ and a path $\alpha : I \to A$ from a to a' in A, we prove that the following diagram is commutative:

$$\begin{array}{ccc} \pi_n(A \xrightarrow{f} X, a) & \xrightarrow{(\tau,h)_*} & \pi_n(B \xrightarrow{g} Y, \tau(a)) \\ \cong\downarrow & & \downarrow\cong \\ \pi_n(A \xrightarrow{f} X, a') & \xrightarrow[(\tau,h)_*]{} & \pi_n(B \xrightarrow{g} Y, \tau(a')) \end{array}$$

which can be rewritten in the following form:

$$\begin{array}{ccc} f_*\pi_n(A,a) & \longrightarrow & g_*\pi_n(B,\tau(a)) \\ \downarrow & & \downarrow \\ f_*\pi_n(A,a') & \longrightarrow & g_*\pi_n(B,\tau(a')). \end{array}$$

This is what we prove. We start from

$$\begin{array}{ccc} A & \xrightarrow{\tau} & B \\ f\downarrow & & \downarrow g \\ X & \xrightarrow[h]{} & Y \end{array}$$

which produces the following CD:

$$\begin{array}{ccc} \pi_n(A,a) & \xrightarrow{\tau_*} & \pi_n(B,\tau a) \\ f_*\downarrow & & \downarrow g_* \\ \pi_n(X,fa) & \xrightarrow[h_*]{} & \pi_n(Y,g\tau a). \end{array} \tag{23.4}$$

It follows that along with (23.3), we have:

$$\begin{array}{ccccccc}
f_*\pi_n(A,a) & & & \overset{h_*}{\dashrightarrow} & & & g_*\pi_n(B,\tau a)\\
 & \nwarrow^{f_*} & & & & \nearrow^{g_*} & \\
 & & \pi_n(A,a) & \xrightarrow{\tau_*} & \pi_n(B,\tau a) & & \\
\cong\downarrow & & \cong\downarrow & & \downarrow\cong & & \downarrow\cong\\
 & & \pi_n(A,a') & \xrightarrow[\tau_*]{} & \pi_n(B,\tau a') & & \\
 & \swarrow_{f_*} & & & & \searrow_{g_*} & \\
f_*\pi_n(A,a') & & & \underset{h_*}{\dashrightarrow} & & & g_*\pi_n(B,\tau a').
\end{array}$$

Now the center square commutes by (23.3), the side squares commute by the functoriality of f_* and g_*, and the top and bottom squares commute by (23.4), making the outside square commute, which means exactly

$$\begin{array}{ccc}
f_*\pi_n(A,a) & \longrightarrow & g_*\pi_n(B,\tau a)\\
\cong\downarrow & & \downarrow\cong\\
f_*\pi_n(A,a') & \longrightarrow & g_*\pi_n(B,\tau a')
\end{array}$$

which completes the proof. □

23.2.3 Colimits

We consider now the following behavior under colimits as covered in [61]: if $X_i \hookrightarrow X_{i+1}$ are inclusions of spaces with colimit X, then the natural map $\operatorname{colim}_i \pi_n(X_i) \to \pi_n X$ is an isomorphism for each n. In our setting, if $X_i \hookrightarrow X_{i+1}$ are inclusions (with colimit X), so are $h_i : X_i \hookrightarrow X$. Consider CDs of the form

$$\begin{array}{ccc}
A_i & \xrightarrow{\tau_i} & A\\
{\scriptstyle f_i}\downarrow & & \downarrow{\scriptstyle f}\\
X_i & \xrightarrow[h_i]{} & X.
\end{array}$$

We ask that $A_i \hookrightarrow A_{i+1}$ be inclusions as well, so that τ_i itself is an inclusion. Hence we are looking at morphisms $(A_i \xrightarrow{f_i} X_i) \hookrightarrow (A_{i+1} \xrightarrow{f_{i+1}} X_{i+1})$ (induced by inclusions on domains and codomains), with $\operatorname{colim}_i(A_i \xrightarrow{f_i} X_i) = A \xrightarrow{f} X$, $A = \operatorname{colim} A_i$. The morphism $f : A \to X$ is defined as f_i on A_i. Note that this can be generalized to other situations where A_i and

X_i are objects of categories other than Top and are both ascending chains of objects by inclusion. We have the following:

Proposition 23.2.3.1. *If* $A \xrightarrow{f} X = colim_n(A_n \xrightarrow{f_n} X_n)$*, then the natural map*

$$\underset{n}{colim}\, \pi_*(A_n \xrightarrow{f_n} X_n) \to \pi_*(A \xrightarrow{f} X)$$

is an isomorphism.

Proof. Because $A_i \hookrightarrow A_{i+1}$ (resp., $X_i \hookrightarrow X_{i+1}$), we also have inclusions of groups $\pi_*(A_i) \hookrightarrow \pi_*(A_{i+1})$ (resp. $\pi_*(X_i) \hookrightarrow \pi_*(X_{i+1})$). We aim to prove that $\mathrm{colim}_n f_{n*}\pi_k(A_n) \cong f_*\pi_k(A)$ for all $k \geq 0$. Consider the ascending inclusions of groups and morphisms induced by the inclusions $A_i \hookrightarrow A_{i+1}$ and $X_i \hookrightarrow X_{i+1}$:

$$\begin{array}{ccccccc}
\cdots & \longrightarrow & \pi_*(A_i) & \hookrightarrow & \pi_*(A_{i+1}) & \hookrightarrow & \cdots \\
 & & \downarrow{\scriptstyle f_{i*}} & & \downarrow{\scriptstyle f_{i+1,*}} & & \\
\cdots & \longrightarrow & \pi_*(X_i) & \hookrightarrow & \pi_*(X_{i+1}) & \hookrightarrow & \cdots.
\end{array}$$

The colimit of such a diagram by the classical result as stated in [61] is provided by $\pi_*(A) \xrightarrow{f_*} \pi_*(X)$. However, we wish to show that the colimit of the images of $\pi_*(A_i)$ in $\pi_*(X_i)$ under f_{i*} is the image of $\pi_*(A)$ in $\pi_*(X)$ under f_*. Observe that those images in $\pi_*(X_i)$ form an ascending chain under inclusion, with their union as colimit, that is,

$$\begin{aligned}
\underset{n}{\mathrm{colim}}\, \pi_k(A_n \xrightarrow{f_n} X_n) &= \underset{n}{\mathrm{colim}}\, f_{n*}\pi_k(A_n) \\
&= \underset{n}{\mathrm{colim}}\, \mathrm{Im} f_n|_{\pi_k(A_n)} \\
&= \bigcup \mathrm{Im} f_n|_{\pi_k(A_n)} \\
&= \mathrm{Im} f|_{\pi_k(A)} \\
&= f_*\pi_k(A) = \pi_k(A \xrightarrow{f} X)
\end{aligned}$$

for all $k \geq 0$. □

23.2.4 Generalized Freudenthal suspension theorem

Recall from [61], for example, that for a based space X, we have the suspension homomorphism $\Sigma : \pi_n X \to \pi_{n+1}(\Sigma X)$ defined by $\Sigma f = f \wedge \mathrm{id} : S^{n+1} \cong S^n \wedge S^1 \to X \wedge S^1 = \Sigma X$, giving rise to the following result: if X is $(n-1)$-connected and nondegenerately based, then $\Sigma : \pi_q(X) \to \pi_{q+1}\Sigma X$ is a bijection for $q < 2n - 1$ and a surjection for $q = 2n - 1$. We prove the following:

Theorem 23.2.4.1. *If X and A are $(n-1)$-connected $(n \geq 1)$ nondegenerately based spaces, then*

$$\Sigma : \pi_q(A \xrightarrow{f} X) \to \pi_{q+1}(\Sigma A \xrightarrow{f \wedge \mathrm{id}} \Sigma X)$$

is a bijection for $q < 2n-1$ and a surjection for $q = 2n-1$.

Proof. By the functoriality of Σ we have the CD

$$\begin{array}{ccc} A & \xrightarrow{\Sigma} & \Sigma A \\ {\scriptstyle f}\downarrow & & \downarrow{\scriptstyle \Sigma f = f \wedge \mathrm{id}} \\ X & \xrightarrow[\Sigma]{} & \Sigma X \end{array}$$

which induces the CD

$$\begin{array}{ccc} \pi_q A & \xrightarrow{\Sigma} & \pi_{q+1}\Sigma A \\ {\scriptstyle f_*}\downarrow & & \downarrow{\scriptstyle (f \wedge \mathrm{id})_*} \\ f_*\pi_q A & \dashrightarrow & (f \wedge \mathrm{id})_*\pi_{q+1}\Sigma A \\ \cap & & \cap \\ \pi_q X & \xrightarrow[\cong]{\Sigma} & \pi_{q+1}\Sigma X. \end{array}$$

To fix ideas, consider $\gamma \in \pi_q A$. An element of $f_*\pi_q A \subset \pi_q X$ is of the form $f_*\gamma$, which is represented by $S^q \xrightarrow{\gamma} A \xrightarrow{f} fA$. Under Σ on $\pi_q X$, it maps to $S^q \wedge S^1 \xrightarrow{\gamma \wedge \mathrm{id}} A \wedge S^1 \xrightarrow{f \wedge S^1} fA \wedge S^1$, which reads $(f \wedge \mathrm{id})_*\Sigma\gamma$, and this last object is in $(f \wedge \mathrm{id})_*\pi_{q+1}\Sigma A$, thereby the restriction of Σ on $\pi_q X$ to $f_*\pi_q A$ does provide the desired dotted map. Because the bottom horizontal map is the suspension map on classical homotopy groups for which we have the Freudenthal suspension theorem, by virtue of our commutative diagram it follows that

$$\Sigma : f_*\pi_q A \to (f \wedge \mathrm{id})_*\pi_{q+1}\Sigma A$$

satisfies the same result as its classical counterpart, that is,

$$\Sigma : \pi_q(A \xrightarrow{f} X) \to \pi_{q+1}(\Sigma A \xrightarrow{f \wedge \mathrm{id}} \Sigma X)$$

is a bijection for $q < 2n-1$ and a surjection for $q = 2n-1$. □

23.2.5 Base change

Finally, we discuss base change. We use the notations

$$\pi^k_{nX} : (k\text{-Aff}_{\text{pft}})_{/X} \to \text{Grp},$$
$$\{\operatorname{Spec} k \leftarrow \operatorname{Spec} R \to X\} \mapsto \pi_n(\operatorname{Spec} R \to X).$$

Proposition 23.2.5.1. *π^k_{nX} is functorial in k, that is, a morphism of rings $k \to k'$ induces a natural transformation $\pi^k_{nX} \to \pi^{k'}_{nX}$.*

Proof. Suppose we have a morphism of base rings $b : k \to k'$. It induces $b^{\text{op}} : \operatorname{Spec} k \leftarrow \operatorname{Spec} k'$. Hence given $\{\operatorname{Spec} k' \leftarrow \operatorname{Spec} R \to X\}$, we get

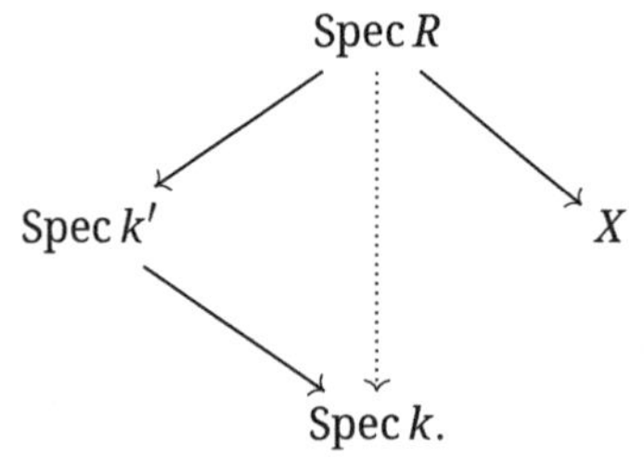

Thus b^{op} induces a map

$$(b^{\text{op}})_* : (k'\text{-Aff}_{\text{pft}})_{/X} \to (k\text{-Aff}_{\text{pft}})_{/X}$$

and finally by composition with π^k_{nX} we get a map

$$\pi^k_{nX} \circ (b^{\text{op}})_* : (k'\text{-Aff}_{\text{pft}})_{/X} \to \text{Grp}$$

that is, we have a map $((b^{\text{op}})_*)^* : \pi^k_{nX} \to \pi^{k'}_{nX}$ induced by a base change morphism $b : k \to k'$. □

23.3 Part II: Algebraic Geometry

Homotopy groups of schematic homotopy types are endowed with some additional structure; initially, they are asked to be k-modules, something that will be dropped later in [35]. π_1 is asked to operate on the π_i for $i > 1$, and π_1 itself is regarded as a group object in $U(k)$. As the next step towards defining such a notion of homotopy groups, we further generalize the functor $\pi_{*,X}$ introduced in the previous section.

We start by just considering $\pi_n : k\text{-CAlg}^{\text{op}} \to \text{Set}$, $R \mapsto \pi_n(\operatorname{Spec} R)$. If we want π_n itself to be a sheaf on k-CAlg, we can make it representable by an object R_0 of k-CAlg, and hence

$$\pi_n(\operatorname{Spec} R) = \operatorname{Hom}_{k\text{-CAlg}}(R, R_0) = \operatorname{Hom}_{k\text{-Aff}}(\operatorname{Spec} R_0, \operatorname{Spec} R)$$

or, in other terms,

$$[S^n, \operatorname{Spec} R] = \operatorname{Hom}_{k\text{-Aff}}(\operatorname{Spec} R_0, \operatorname{Spec} R).$$

This points to the fact that the Hom on the right-hand side is already taken in a homotopy category to be specified and that probably $\operatorname{Spec} R_0$ should be regarded as a component of a semisimplicial object, or even a spectrum. We are looking for an object that would play the role of spheres in a homotopy category. We consider the category of symmetric spectra Sp^Σ as covered in [43] and [70]. We choose to represent spheres by the sphere spectrum $\Sigma^\infty S^0 = \mathbb{S}$. Considering the sphere spectrum is suitable since it is the initial object in the ∞-category of ring spectra. Spaces are generalized to the setting of (topological) symmetric spectra via

$$X \simeq S^0 \wedge X \xrightarrow{\Sigma} S^1 \wedge X \to \cdots$$

i. e., $X \mapsto \Sigma^\infty X$. Homotopically, we can work with simplicial sets instead. For a simplicial set K, we therefore work with $\Sigma^\infty K$, the symmetric suspension spectrum of K. Therefore, what was considered as $\operatorname{Spec} R$ in classical homotopy theory should be replaced by $A \in \mathrm{Sp}^\Sigma$.

Note the change of paradigm; we no longer work with $k\text{-CAlg}_{\mathrm{pft}}$ as our base model category, but rather with Sp^Σ. Thus strictly speaking, we no longer work with schematic objects, but rather with spectralizations thereof.

We will follow our classical study of schematic homotopy types as much as possible. One important note at this point: since we work with spectra, homotopy groups of such objects are typically defined using stable homotopy groups. We purposefully deviate from that norm since we are aiming for a far more rudimentary object that would encapsulate the notion of homotopy group. To define such an object, we first define symmetric sequences, which are sequences of pointed simplicial sets with a basepoint preserving left action of Σ_n on their nth component. We consider $\Sigma = \coprod_{n\geq 0} \Sigma_n$ as a category with objects finite sets $\overline{n} = \{1, 2, \ldots, n\}$ with automorphisms as maps, and we denote by $\mathcal{S}_*^\Sigma$ the category of symmetric sequences. We have a forgetful functor

$$\begin{aligned} \mathrm{Sp}^\Sigma &\to \mathcal{S}_*^\Sigma, \\ A &\mapsto \underline{A} \end{aligned}$$

with $A_n = (\underline{A})_n$. We then regard $\operatorname{Hom}_{\mathcal{S}_*^{\Sigma_n}}(S^n, A_n)$ as the object corresponding to $\pi_n(A_n)$ in this setting. Taking the product of all such sets, we get

$$\prod_p \operatorname{Hom}_{\mathcal{S}_*^{\Sigma_p}}(S^p, A_p) = \operatorname{Hom}_{\mathcal{S}_*^\Sigma}(\underline{\mathbb{S}}, \underline{A})$$

for $A \in \mathrm{Sp}^{\Sigma}$, as shown in [43]. Turning those into simplicial sets by using mapping spaces $\mathrm{Map}(X, Y) = \mathrm{Hom}(X \wedge \Delta[-]_{+}, Y)$ with $\Delta[-] : \Delta \to \mathrm{Set}_{\Delta}$, $\Delta[-]_{+} = \Delta[-] \coprod \Delta[0]$, we still have $\mathrm{Map}_{\mathcal{S}^{\Sigma}_{*}}(\underline{\mathbb{S}}, \underline{A}) = \prod_p \mathrm{Map}_{\Sigma_p}(S^p, A_p)$. To lift this up to the level of spectra, we use the tensor product of symmetric sequences as introduced in [43], and we have

$$\pi_{*}(A) := \mathrm{Map}_{\mathcal{S}^{\Sigma}_{*}}(\underline{\mathbb{S}}, \underline{A}) \cong \mathrm{Map}_{\mathrm{Sp}^{\Sigma}}(\mathbb{S} \otimes \underline{\mathbb{S}}, A)$$

for $A \in \mathrm{Sp}^{\Sigma}$, where we used the fact that $\mathbb{S} \otimes - : \mathcal{S}^{\Sigma}_{*} \to \mathrm{Sp}^{\Sigma}$ is the left adjoint of the forgetful functor $\mathrm{Sp}^{\Sigma} \to \mathcal{S}^{\Sigma}_{*}$.

The role played earlier by schematic coverings of the form $\{\mathrm{Spec}\, R \to X\}$ for $R \in k\text{-}\mathrm{CAlg}_{\mathrm{pft}}$ or, equivalently, by the affine perception functor $\mathrm{Hom}(\mathrm{Spec}\,(-), X)$ is now played by the functor $\mathbb{R}\underline{\mathrm{Hom}}(-, A) : \mathrm{Ho}\,(\mathrm{Sp}^{\Sigma}_{\mathrm{ft}}) \to \mathrm{Ho}\,(\mathrm{Sp}^{\Sigma})$, where $\mathrm{Sp}^{\Sigma}_{\mathrm{ft}}$ corresponds to the category of *symmetric spectra of finite type* (simply defined as spectra with finite simplicial components), which we regard as homotopy type objects, and we put the stable model structure of [70] on Sp^{Σ}. Recall that earlier if we had a morphism $\mathrm{Spec}\, R \xrightarrow{f} X$, then this gave rise to a subgroup $f_{*}\pi_n \mathrm{Spec}\, R$, the collection of whose objects we turned into a functor. In the present context, for $\mathbf{H} \in \mathrm{Sp}^{\Sigma}_{\mathrm{ft}}$ and a given morphism $f : \mathbf{H} \to A$, we have an induced map of simplicial sets in $\mathrm{Ho}\,(\mathrm{Set}_{\Delta})$,

$$\pi_{*}\mathbf{H} = \mathrm{Map}_{\mathrm{Sp}^{\Sigma}}(\mathbb{S} \otimes \mathbb{S}, \mathbf{H}) \xrightarrow{\mathrm{Ho}\,(f_{*})} \mathrm{Map}_{\mathrm{Sp}^{\Sigma}}(\mathbb{S} \otimes \mathbb{S}, A) = \pi_{*}A.$$

Then we define the homotopy group of the cover $\mathbf{H} \to A$ to be given by $\mathrm{Ho}\,(f_{*})\mathrm{Map}(\mathbb{S} \otimes \mathbb{S}, \mathbf{H})$. Functorially, the spectral generalization of the homotopy group functor $\pi_{*,X}$ for a topological space X is given by the functor

$$\begin{aligned} \Pi_A : \mathrm{Sp}^{\Sigma} \times_{\mathrm{Sp}^{\Sigma}_{\mathrm{ft}}} A^{\mathrm{Sp}^{\Sigma}} &\to \mathrm{Ho}\,(\mathrm{Set}_{\Delta}), \\ (\mathbf{H}, \mathbf{H} \xrightarrow{f} A) &\mapsto \mathrm{Ho}\,(f_{*})\mathrm{Map}_{\mathrm{Sp}^{\Sigma}}(\mathbb{S} \otimes \mathbb{S}, \mathbf{H}). \end{aligned}$$

Thus defined, we probe A with homotopy types $\mathbf{H}$. Thus using this notion of homotopy group to define π_{*}-equivalences will yield spectral homotopy types: we define the following equivalence in Sp^{Σ}: $A \sim A'$ if and only if $\Pi_A \cong \Pi_{A'}$. This constitutes a spectralization of the schematic homotopy type we constructed in the first part of this work.

24 Higher Galois for Segal topos

24.1 Introduction

From the perspective of shape theory one can define the shape of a Segal topos T, as Toen and Vezzosi did in [77], as being given by $H_T = |\mathbb{R}\underline{\mathrm{Hom}}^{\mathrm{geom}}(T, -)|$. Here $|\cdot|$ denotes the realization of a Segal category as in [41], left adjoint of the fundamental groupoid functor Π_∞. In [77], it is proven that for a CW complex X, if $T = \mathrm{Loc}(*) = \mathrm{Top}$, then $X \simeq |\mathbb{R}\underline{\mathrm{Hom}}^{\mathrm{geom}}(\mathrm{Top}, \mathrm{Loc}(X))|$ or, equivalently, $\Pi_\infty(X) \simeq \mathbb{R}\underline{\mathrm{Hom}}^{\mathrm{geom}}(\mathrm{Top}, \mathrm{Loc}(X))$. However, in SGA 1, Grothendieck suggests that the category of fiber functors from $\mathcal{X}$ to T in SeT (the category of Segal topos), defined by $\mathbb{R}\underline{\mathrm{Hom}}^*_{\mathrm{SeT}}(\mathcal{X}, T) = \mathbb{R}\underline{\mathrm{Hom}}^{\mathrm{geom}}(T, \mathcal{X})$, could be called a fundamental groupoid, thereby providing a generalization of the above result from [77]. This can be made precise, as we will do in this work, but this will not be the fundamental ∞-groupoid of $\mathcal{X}$ itself, rather it will be regarded as the fundamental ∞-groupoid of a Segal category of functors instead.

To see how this comes about, recall that for two Segal topos T and $\mathcal{X}$ or, more generally, two topos, a geometric morphism f from T to $\mathcal{X}$ consists of a pair of adjoint functors $f_* : T \to \mathcal{X}$ and $f^* : \mathcal{X} \to T, f^* \dashv f_*, f^*$ left exact. Thus if SeT denotes the *category of Segal topos*, then we adopt the notation $\mathbb{R}\underline{\mathrm{Hom}}_{\mathrm{SeT}}(T, \mathcal{X})$ for $\mathbb{R}\underline{\mathrm{Hom}}^{\mathrm{geom}}(T, \mathcal{X})$, which one can define as $\mathbb{R}\underline{\mathrm{Hom}}^*_{\mathrm{SeT}}(\mathcal{X}, T) \simeq \mathbb{R}\underline{\mathrm{Hom}}_{*,\mathrm{SeT}}(T, \mathcal{X})^{\mathrm{op}}$ depending on whether we fix our attention on left or right adjoints. Now it turns out that fiber functors are left exact, and working with Segal topos, it is therefore natural, in the spirit of Grothendieck, to expect the groupoid of fiber functors in the present setting to be given by $\mathbb{R}\underline{\mathrm{Hom}}^*_{\mathrm{SeT}}(\mathcal{X}, T)$, sub-Segal category of $\mathbb{R}\underline{\mathrm{Hom}}(\mathcal{X}, T) = \widehat{\mathcal{X}^{\mathrm{op}}}$. We prove, using Toen's work [77], that $\mathbb{R}\underline{\mathrm{Hom}}^*_{\mathrm{SeT}}(\mathcal{X}, T)$ is a Segal groupoid for any T. Hence we adopt the suggestive notation $\mathbb{R}\underline{\mathrm{Hom}}^*_{\mathrm{SeT}}(\mathcal{X}, T) = \Pi_{\infty,T}\widehat{\mathcal{X}^{\mathrm{op}}}$.

Considering local systems $i_T = \mathbb{R}\underline{\mathrm{Hom}}(-, T)$ on it, we show that we have a unique, universal map $\mathcal{X} \to i_T\Pi_{\infty,T}\widehat{\mathcal{X}^{\mathrm{op}}}$. In particular, if $T = \mathcal{X}$, then for any Segal groupoid $\mathcal{Y}$, we have a morphism of Segal categories from $i_\mathcal{X}\Pi_{\infty,\mathcal{X}}\widehat{\mathcal{X}^{\mathrm{op}}}$ to the internal perception $\mathbb{R}\underline{\mathrm{Hom}}(\mathcal{Y}, \mathcal{X})$ of $\mathcal{X}$ relative to $\mathcal{Y}$. In other terms the perception $\mathbb{R}\underline{\mathrm{Hom}}(-, \mathcal{X})$ of $\mathcal{X}$ originates in part from $i_\mathcal{X}\Pi_{\infty,\mathcal{X}}\widehat{\mathcal{X}^{\mathrm{op}}}$.

If we apply this to a description of all natural phenomena, then one can envision that the laws of nature are modeled by simplicial algebras, objects in sk-CAlg for a commutative ring k, realized via derived stacks, the collection of which is a Segal topos dSt(k), to which we can apply all this formalism. One advantage of using stacks to model algebraic realizations of natural laws is that one can completely bypass the use of “target spaces” and “fields” living on such spaces, whose dynamics is given by equations of motion derived from an independent “Lagrangian”, notwithstanding the fact that such target spaces for the most part would be obtained from compactification(s). Here dynamics is dictated by the functoriality of stacks themselves, which we interpret as providing coherent manifestations of natural laws. All the derived stacks $F : sk\text{-CAlg} \to \mathrm{Set}_\Delta$ together

https://doi.org/10.1515/9783111334073-024

form a Segal topos $\mathcal{X} = \mathrm{dSt}(k)$, which therefore corresponds to the collection of all natural phenomena. A holistic picture corresponds to working with $\mathcal{X}$, which is static, along with its dynamics, which we regard as given by the Segal category $\mathbb{R}\underline{\mathrm{Hom}}^*_{\mathrm{SeT}}(\mathcal{X}, \mathcal{X})$. Observe that the latter contains $\mathcal{X}$ itself since $\mathrm{id}_{\mathcal{X}} \in \mathbb{R}\underline{\mathrm{Hom}}^*_{\mathrm{SeT}}(\mathcal{X}, \mathcal{X})$. Thus a complete picture is provided by just considering $\mathbb{R}\underline{\mathrm{Hom}}^*_{\mathrm{SeT}}(\mathcal{X}, \mathcal{X})$. This approach has the advantage of repackaging all of Physics in a purely algebro-geometric object such as a Segal topos, in the spirit of Kontsevich's take on the mirror symmetry problem by introducing the homological mirror symmetry formalism [50].

Our references for Segal categories are standard: [77, 74, 75, 66, 41, 71]. For derived stacks, we use the foundational papers [76, 77, 79, 81, 74, 75].

24.2 Grothendieck's take on Galois theory

We briefly remind the reader of the following fundamental result of Galois theory: for a finite Galois extension K of a field k, $G = \mathrm{Gal}(K/k)$, we have a one-to-one correspondence between subfield extensions $k \subset E \subset K$ and subgroups $H \subset G$ given by $E \mapsto \mathrm{Gal}(K/E)$ and $H \mapsto K^H$. It was Grothendieck's idea in SGA 1 to categorize this result, as clearly recounted in [14], by first using the well-known fact that there is a one-to-one correspondence between conjugacy classes of subgroups of G and isomorphism classes of transitive G-sets, whose category we denote tSets^G, and by regarding field extensions as a category $\mathcal{C}^{\mathrm{op}}$ whose objects are fields, k then becoming the terminal object of $\mathcal{C}$. One then moves into the categorical realm. An object A of $\mathcal{C}$ being fixed, under the assumption that for all objects X of $\mathcal{C}$, there is a morphism $A \to X$, which is further a strict epi, meaning the joint coequalizer of all the parallel pairs that it coequalizes, if we assume we have a notion of categorical quotient $A \to A/H$, preserved by $\mathrm{Hom}_{\mathcal{C}}(A, -) = [A, -]$, and finally if we assume that $\mathrm{End}(A) = \mathrm{Aut}(A)$, then under those assumptions, we have an adjoint equivalence

$$[A, -] : \mathcal{C} \rightleftarrows \mathrm{tSets}^G : A \times_G - \tag{24.1}$$

thereby providing an abstraction of the classical Galois statement. Grothendieck's idea then was to observe that $[A, -]$ being a map from $\mathcal{C}$ to Set, one may as well start from a fiber functor $F : \mathcal{C} \to \mathrm{fSets}$, left exact among other things, and under mild conditions on $\mathcal{C}$ and on F itself, one obtains a generalization of the above result [33, 14]. One would first show that F is procorepresentable, $F = [P, -]$, from which one would make a transition from the result given in (24.1) to the following adjoint equivalence:

$$F = \mathrm{Hom}_{\mathcal{C}}(P, -) = [P, -] : \mathcal{C} \rightleftarrows \mathrm{fSets}^{\pi} : P \times_{\pi} -$$

with a profinite group $\pi = \mathrm{Aut}(P)^{\mathrm{op}}$. Grothendieck then observed that this depended of course on F and that a way to make this independent of the choice of a fiber functor was to consider $\Gamma = \{F : \mathcal{C} \to \mathrm{fSets}\}$, which he argued is a groupoid. Considering local systems on Γ, meaning introducing $i = \mathrm{Hom}_{\mathcal{C}}(-, \mathrm{fSets})$, he then proved in a few lines that $i\Gamma \simeq \mathcal{C}$.

24.3 Grothendieck's Galois theory for Segal topos

We now promote $\mathcal{C}$ and fSets to the status of Segal topoi as introduced in [77]. Thus our base category $\mathcal{C}$ becomes a fixed Segal topos $\mathcal{X}$, and we consider functors valued in any Segal topos T. Fiber functors in the Segal topos setting would be represented by left exact left adjoints. This justifies the definition

$$\mathbb{R}\underline{\mathrm{Hom}}^{*}_{\mathrm{SeT}}(\mathcal{X}, T) = \mathbb{R}\underline{\mathrm{Hom}}_{\mathrm{SeT}}(T, \mathcal{X})$$

as given in [77], where $\mathbb{R}\underline{\mathrm{Hom}}_{\mathrm{SeT}}$ is denoted $\mathbb{R}\underline{\mathrm{Hom}}^{\mathrm{geom}}$ instead, and $\mathbb{R}\underline{\mathrm{Hom}}^{*}_{\mathrm{SeT}}(\mathcal{X}, T)$ refers to the sub-Segal category of $\mathbb{R}\underline{\mathrm{Hom}}(\mathcal{X}, T)$ spanned by left exact left adjoints. Recall from the same paper that SePC, the category of Segal pre-categories, is a symmetric monoidal model category [42] with the direct product as monoidal product, and hence Ho (SePC) has an internal Hom object [42] denoted $\mathbb{R}\underline{\mathrm{Hom}}(A, B) \in \mathrm{Ho}\,(\mathrm{SePC})$ for $A, B \in \mathrm{Ho}\,(\mathrm{SePC})$, where $\mathbb{R}\underline{\mathrm{Hom}}(A, B) \cong \underline{\mathrm{Hom}}(A, RB)$, $\underline{\mathrm{Hom}}$ the internal Hom object in SePC. However, we also have an anti-equivalence $\mathbb{R}\underline{\mathrm{Hom}}^{*}_{\mathrm{SeT}}(\mathcal{X}, T) \simeq \mathbb{R}\underline{\mathrm{Hom}}_{*,\mathrm{SeT}}(T, \mathcal{X})^{\mathrm{op}}$. That this is an anti-equivalence follows from the commutative diagram below; for two Segal topos $\mathcal{X}$ and $\mathcal{Y}$, two geometric morphisms $f, g : \mathcal{X} \to \mathcal{Y}$, a morphism $\alpha : f^* \Rightarrow g^*$ in $\mathbb{R}\underline{\mathrm{Hom}}^{*}_{\mathrm{SeT}}(\mathcal{Y}, \mathcal{X})$, $x \in \mathcal{X}_0$, and $y \in \mathcal{Y}_0$, there corresponds a morphism $\beta : g_* \Rightarrow f_*$ in $\mathbb{R}\underline{\mathrm{Hom}}_{*,\mathrm{SeT}}(\mathcal{X}, \mathcal{Y})$ as in

$$\begin{array}{ccc}
\mathcal{X}_{(g^* y, x)} & \xrightarrow{\cong} & \mathcal{Y}_{(y, g_* x)} \\
{\scriptstyle \alpha_y^*}\big\downarrow & & \big\downarrow{\scriptstyle \beta_{x,*}} \\
\mathcal{X}_{(f^* y, x)} & \xrightarrow[\cong]{} & \mathcal{Y}_{(y, f_* x)}.
\end{array}$$

Hence we can also define $\mathbb{R}\underline{\mathrm{Hom}}_{\mathrm{SeT}}(T, \mathcal{X})$ as

$$\mathbb{R}\underline{\mathrm{Hom}}_{\mathrm{SeT}}(T, \mathcal{X}) = \mathbb{R}\underline{\mathrm{Hom}}_{*,\mathrm{SeT}}(T, \mathcal{X})^{\mathrm{op}}.$$

Theorem 24.3.1. *For two Segal topos $\mathcal{X}$ and T, $\mathbb{R}\underline{\mathrm{Hom}}_{\mathrm{SeT}}(T, \mathcal{X})$ is a Segal groupoid.*

Proof. $\mathcal{X}$ being a Segal topos, by definition there is a fully faithful map $i : \mathcal{X} \to \hat{B} = \mathbb{R}\underline{\mathrm{Hom}}(B^{\mathrm{op}}, \mathrm{Top})$, where B is a Segal category, and $\mathrm{Top} = \mathrm{Loc}(*) = L(\mathrm{Set}_{\Delta})$. This map induces a fully faithful morphism of Segal categories $i_* : \mathbb{R}\underline{\mathrm{Hom}}(T, \mathcal{X}) \to \mathbb{R}\underline{\mathrm{Hom}}(T, \hat{B})$. As a matter of fact, since right adjoints with a left exact left adjoint map to right adjoints with a left exact left adjoint, $i_* : \mathbb{R}\underline{\mathrm{Hom}}_{*,\mathrm{SeT}}(T, \mathcal{X}) \to \mathbb{R}\underline{\mathrm{Hom}}_{*,\mathrm{SeT}}(T, \hat{B})$ is also fully faithful (this is the object of the lemmas that follow). Now using the adjunction formula

$\mathbb{R}\underline{\mathrm{Hom}}(A \times B, C) \simeq \mathbb{R}\underline{\mathrm{Hom}}(A, \mathbb{R}\underline{\mathrm{Hom}}(B, C))$ of [77] restricted to right adjoints with a left exact left adjoint, we have

$$\begin{aligned}\mathbb{R}\underline{\mathrm{Hom}}_{*,\mathrm{SeT}}(T, \hat{B}) &= \mathbb{R}\underline{\mathrm{Hom}}_{*,\mathrm{SeT}}(T, \mathbb{R}\underline{\mathrm{Hom}}(B^{\mathrm{op}}, \mathrm{Top}))\\ &\cong \mathbb{R}\underline{\mathrm{Hom}}_{*,\mathrm{SeT}}(T \times^{\mathbb{L}} B^{\mathrm{op}}, \mathrm{Top})\\ &= \mathbb{R}\underline{\mathrm{Hom}}(B^{\mathrm{op}}, \mathbb{R}\underline{\mathrm{Hom}}_{*,\mathrm{SeT}}(T, \mathrm{Top}))\\ &= \mathbb{R}\underline{\mathrm{Hom}}(B^{\mathrm{op}}, \mathbb{R}\underline{\mathrm{Hom}}_{\mathrm{SeT}}(T, \mathrm{Top})^{\mathrm{op}})\end{aligned}$$

where on the second line, we just have a left exact left adjoint between T and Top. We then use the fact that $\mathbb{R}\underline{\mathrm{Hom}}_{\mathrm{SeT}}(T, \mathrm{Top})$ is a Segal groupoid as proved in [77], so that $\mathbb{R}\underline{\mathrm{Hom}}(B^{\mathrm{op}}, \mathbb{R}\underline{\mathrm{Hom}}_{\mathrm{SeT}}(T, \mathrm{Top})^{\mathrm{op}})$ itself is a Segal groupoid, and hence so is $\mathbb{R}\underline{\mathrm{Hom}}_{*,\mathrm{SeT}}(T, \hat{B})$, and $\mathbb{R}\underline{\mathrm{Hom}}_{*,\mathrm{SeT}}(T, \mathcal{X})$ faithfully maps into it, so is a Segal groupoid, or, equivalently, $\mathbb{R}\underline{\mathrm{Hom}}_{\mathrm{SeT}}(T, \mathcal{X})$ is a Segal groupoid. □

In the proof, we used the fact that for $A, B \in \mathrm{Ho}\,(\mathrm{SeCat})$, $\mathbb{R}\underline{\mathrm{Hom}}(A, B) \in \mathrm{Ho}\,(\mathrm{SeCat})$. Indeed, by [74] $B \in \mathrm{SeCat}$ implies we have a fully faithful embedding $B \to LM$ for some model category M. It follows that we have a fully faithful embedding $\mathbb{R}\underline{\mathrm{Hom}}(A, B) \to \mathbb{R}\underline{\mathrm{Hom}}(A, LM) \cong L(M^A)$, where we used the *strictification theorem*. Note that this holds for $A \in \mathrm{Cat}_\Delta$, the category of simplicial categories. However, $\mathrm{Ho}\,(\mathrm{SeCat}) \simeq \mathrm{Ho}\,(\mathrm{Cat}_\Delta)$, so we can regard A as a simplicial category. Further, $L(M^A) \in \mathrm{Ho}\,(\mathrm{Cat}_\Delta) \simeq \mathrm{Ho}\,(\mathrm{SeCat})$, and hence we regard $\mathbb{R}\underline{\mathrm{Hom}}(A, B)$ as a Segal category.

We can also mention in passing that for $A, B \in \mathrm{Ho}\,(\mathrm{SeCat})$, a morphism $f : A \to B$ is regarded as an object of $\mathbb{R}\underline{\mathrm{Hom}}(A, B)$. To see this, observe as in [71] that if $[0]$ is the category with one object 0, regarded as a Segal category, then by adjunction $\mathrm{Hom}_{\mathrm{Ho}\,(\mathrm{SeCat})}(A \times [0], B) \cong \mathrm{Hom}_{\mathrm{Ho}\,(\mathrm{SeCat})}([0], \mathbb{R}\underline{\mathrm{Hom}}(A, B))$. An object in the left-hand side Hom corresponds to a map of bisimplicial sets, whereas an object in the right-hand side Hom corresponds to picking an object of $\mathbb{R}\underline{\mathrm{Hom}}(A, B)$. In the same manner, if $[1]$ denotes the category with two objects 0 and 1 and a unique morphism $0 \to 1$, then for $f, g : A \to B$, a natural transformation $\alpha : f \Rightarrow g$ can be seen as an object of $\mathrm{Hom}_{\mathrm{Ho}\,(\mathrm{SeCat})}(A \times [1], B)$ or by adjunction as an object of $\mathrm{Hom}_{\mathrm{Ho}\,(\mathrm{SeCat})}([1], \mathbb{R}\underline{\mathrm{Hom}}(A, B))$, that is, an object of $\mathbb{R}\underline{\mathrm{Hom}}(A, B)_{(f,g)}$.

We now go over this claim used in the proof of the above theorem that $\mathbb{R}\underline{\mathrm{Hom}}_{*,\mathrm{SeT}}(T, \mathcal{X}) \to \mathbb{R}\underline{\mathrm{Hom}}_{*,\mathrm{SeT}}(T, \hat{B})$ is fully faithful. First, we have to understand what is a morphism of adjunctions. We follow [59]. Given two adjunctions $F : X \rightleftarrows A : G$ and $F' : X \rightleftarrows A : G'$, $\sigma : F \Rightarrow F'$ and $\tau : G' \Rightarrow G$ are said to be conjugate when the following diagram commutes for all $x \in X$ and $a \in A$:

$$\begin{array}{ccc} A(F'x, a) & \xrightarrow{\cong} & X(x, G'a) \\ {\scriptstyle (\sigma_x)^*}\downarrow & & \downarrow{\scriptstyle (\tau_a)_*} \\ A(Fx, a) & \xrightarrow[\cong]{} & X(x, Ga). \end{array}$$

We regard a conjugate pair (σ, τ) of natural transformations as a map of adjunctions from (F, G) to (F', G'). Now recall that we want to show that $\mathbb{R}\underline{\mathrm{Hom}}_{*,\mathrm{SeT}}(T, \mathcal{X}) \xrightarrow{i_*} \mathbb{R}\underline{\mathrm{Hom}}_{*,\mathrm{SeT}}(T, \hat{B})$ is fully faithful, so we consider F, G in $\mathbb{R}\underline{\mathrm{Hom}}_{*,\mathrm{SeT}}(T, \mathcal{X})$ and $\varphi : F \Rightarrow G$ in $\mathbb{R}\underline{\mathrm{Hom}}_{*,\mathrm{SeT}}(T, \mathcal{X})$. By the previous discussion, φ should be part of a conjugate pair of natural transformations. To be precise, we consider a map of adjunctions $(G', G) \to (F', F)$ with $F' : \mathcal{X} \rightleftarrows T : F$ and $G' : \mathcal{X} \rightleftarrows T : G$, with F' and G' left exact, $\varphi' : G' \Rightarrow F'$, and $\varphi : F \Rightarrow G$. Diagrammatically, this means that the following diagram is commutative:

$$\begin{array}{ccc} T_{(F'x,t)} & \xrightarrow{\cong} & \mathcal{X}_{(x,Ft)} \\ {\scriptstyle (\varphi'_x)^*}\downarrow & & \downarrow{\scriptstyle (\varphi_t)_*} \\ T_{(G'x,t)} & \xrightarrow[\cong]{} & \mathcal{X}_{(x,Gt)} \end{array}$$

which using compositions in Segal categories should read

$$\begin{array}{ccc} T_{(G'x,F'x)} \times T_{(F'x,t)} & \xrightarrow{\cong} & \mathcal{X}_{(x,Ft)} \times \mathcal{X}_{(Ft,Gt)} \\ \downarrow & & \downarrow \\ T_{(G'x,t)} & \xrightarrow[\cong]{} & \mathcal{X}_{(x,Gt)}. \end{array}$$

Knowing that both F' and G' are left exact, and $\varphi' : G' \Rightarrow F'$ (that is, for all x, $\varphi'_x : G'x \to F'x$), for a diagram x^α in $\mathcal{X}$ used to prove left exactness, we have

$$\begin{aligned} \varphi'_{\lim x^\alpha} &= G'(\lim x^\alpha) \to F'(\lim x^\alpha) \\ &= \lim G'x^\alpha \to \lim F'x^\alpha \\ &= \lim(G'x^\alpha \to F'x^\alpha) \\ &= \lim \varphi'_{x^\alpha} \end{aligned}$$

and thus φ' is left exact as well. To conclude, if $\varphi : F \Rightarrow G$ and $\varphi' : G' \Rightarrow F'$ form a conjugate pair in $\mathbb{R}\underline{\mathrm{Hom}}(T, \mathcal{X})$, then φ' is left exact.

Lemma 24.3.2. *In the setting of the above theorem,*

$$i_* : \mathbb{R}\underline{\mathrm{Hom}}_{*,\mathrm{SeT}}(T, \mathcal{X}) \to \mathbb{R}\underline{\mathrm{Hom}}_{*,\mathrm{SeT}}(T, \hat{B})$$

is well-defined.

Proof. Let $F \in \mathbb{R}\underline{\mathrm{Hom}}_{*,\mathrm{SeT}}(T, \mathcal{X})$. We first show that $i_* F$ is a right adjoint with a left exact left adjoint. Since $\mathcal{X} \in \mathrm{SeT}$, there is a fully faithful map $i : \mathcal{X} \to \hat{B}$ as we saw in the proof of the above theorem, but we also have a left exact left adjoint $j : \hat{B} \to \mathcal{X}$. F itself has a left exact left adjoint, which we denote by F'. We consider the following compositions:

$$T \underset{F'}{\overset{F}{\rightleftarrows}} \mathcal{X} \underset{j}{\overset{i}{\rightleftarrows}} \hat{B}.$$

We argue that $j^*F' \dashv i_*F$ in $\mathbb{R}\underline{\mathrm{Hom}}(T, \hat{B})$, and j^*F' is left exact. We first show the adjointness. It suffices to write

$$\begin{aligned} T_{(F' \circ jb, t)} &\cong \mathcal{X}_{(jb, Ft)} \\ &\cong \hat{B}_{(b, iFt)}. \end{aligned}$$

We now show that j^*F' is left exact. Consider a diagram in $\hat{B}$. We have $j^*F'(\lim b^\alpha) = F' \circ j(\lim b^\alpha) = F'(\lim jb^\alpha)$ since j is left exact, and this is further equal to $\lim(F' \circ jb^\alpha)$ since F' is left exact as well. But this is $\lim j^*F'(b^\alpha)$. Thus i_* maps objects of $\mathbb{R}\underline{\mathrm{Hom}}_{*,\mathrm{SeT}}(T, \mathcal{X})$ to objects of $\mathbb{R}\underline{\mathrm{Hom}}_{*,\mathrm{SeT}}(T, \hat{B})$.

Finally, we show that i_* maps morphisms in $\mathbb{R}\underline{\mathrm{Hom}}_{*,\mathrm{SeT}}(T, \mathcal{X})$ to morphisms of $\mathbb{R}\underline{\mathrm{Hom}}_{*,\mathrm{SeT}}(T, \hat{B})$. Let F, G in $\mathbb{R}\underline{\mathrm{Hom}}_{*,\mathrm{SeT}}(T, \mathcal{X})$ with respective left exact left adjoints F', G' in $\mathbb{R}\underline{\mathrm{Hom}}^*_{\mathrm{SeT}}(\mathcal{X}, T)$, and let $\varphi : F \Rightarrow G$ and $\varphi' : G' \Rightarrow F'$ (left exact) be a conjugate pair for those adjoints. We need to show that $\phi = i_*\varphi : i_*F \Rightarrow i_*G$ has a left exact conjugate in $\mathbb{R}\underline{\mathrm{Hom}}^*_{\mathrm{SeT}}(\hat{B}, T)$.

We have the following compositions:

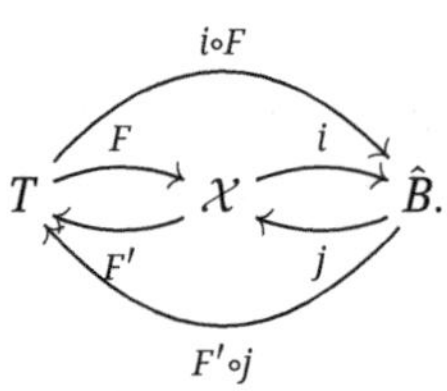

We have $i_*\varphi : i_*F \Rightarrow i_*G$ and $j^*\varphi' : j^*G' \Rightarrow j^*F'$. We claim that $(j^*\varphi', i_*\varphi)$ is a conjugate pair with $j^*\varphi'$ left exact. We want the following diagram to commute:

$$\begin{array}{ccc} T_{(j^*F'b,t)} & \xrightarrow{\cong} & \hat{B}_{(b,i_*Ft)} \\ \downarrow{\scriptstyle (j^*\varphi'_b)^*} & & \downarrow{\scriptstyle (i_*\varphi_t)_*} \\ T_{(j^*G'b,t)} & \xrightarrow[\cong]{} & \hat{B}_{(b,i_*Gt)}. \end{array}$$

We have

$$\begin{array}{ccccccc} T_{(j^*F'b,t)} & = & T_{(F'jb,t)} & \xrightarrow{\cong} & \mathcal{X}_{(jb,Ft)} & \xrightarrow{\cong} & \hat{B}_{(b,iFt)} \\ \downarrow{\scriptstyle (j^*\varphi'_b)^*} & & \downarrow{\scriptstyle (\varphi'_{jb})^*} & & \downarrow{\scriptstyle (\varphi_t)_*} & & \downarrow{\scriptstyle (i_*\varphi_t)_*} \\ T_{(j^*G'b,t)} & = & T_{(G'jb,t)} & \xrightarrow[\cong]{} & \mathcal{X}_{(jb,Gt)} & \xrightarrow[\cong]{} & \hat{B}_{(b,iGt)}. \end{array}$$

The two squares on the right commute by adjunction, which gives us our desired commutativity, i. e., $j^*\varphi'$ and $i_*\varphi$ form a conjugate pair. We also have that $j^*\varphi'$ is left exact as a composite of left exact functors. To summarize, for a morphism φ in $\mathbb{R}\underline{\mathrm{Hom}}_{*,\mathrm{SeT}}(T,\mathcal{X})$, part of a conjugate pair (φ',φ) with φ' left exact, $i_*\varphi$ is part of a conjugate pair $(j^*\varphi', i_*\varphi)$ in $\mathbb{R}\underline{\mathrm{Hom}}(T,\hat{B})$, that is, morphisms in $\mathbb{R}\underline{\mathrm{Hom}}_{*,\mathrm{SeT}}(T,\mathcal{X})$ map to morphisms in $\mathbb{R}\underline{\mathrm{Hom}}_{*,\mathrm{SeT}}(T,\hat{B})$. □

Lemma 24.3.3. *In the setting of the above theorem,*

$$i_* : \mathbb{R}\underline{\mathrm{Hom}}_{*,\mathrm{SeT}}(T,\mathcal{X}) \to \mathbb{R}\underline{\mathrm{Hom}}_{*,\mathrm{SeT}}(T,\hat{B})$$

is fully faithful.

Proof. Let $F, G \in \mathbb{R}\underline{\mathrm{Hom}}_{*,\mathrm{SeT}}(T,\mathcal{X})$, and let $\phi : i_*F \Rightarrow i_*G$ in $\mathbb{R}\underline{\mathrm{Hom}}_{*,\mathrm{SeT}}(T,\hat{B})$. We show we have a morphism $\varphi : F \Rightarrow G$ in $\mathbb{R}\underline{\mathrm{Hom}}_{*,\mathrm{SeT}}(T,\mathcal{X})$. Note that ϕ is a morphism in $\mathbb{R}\underline{\mathrm{Hom}}(T,\hat{B})$, so $i_* : \mathbb{R}\underline{\mathrm{Hom}}(T,\mathcal{X}) \to \mathbb{R}\underline{\mathrm{Hom}}(T,\hat{B})$ being fully faithful, there is some morphism $\varphi : F \Rightarrow G$ in $\mathbb{R}\underline{\mathrm{Hom}}(T,\mathcal{X})$. We show it has a left exact conjugate φ'. We use the fact that F and G have left exact left adjoints, which we denote by F' and G', respectively. Since $j \dashv i$, we have from earlier work that $j^*F' \dashv i_*F$ and $j^*G' \dashv i_*G$ in $\mathbb{R}\underline{\mathrm{Hom}}(T,\hat{B})$. Since ϕ is a morphism in $\mathbb{R}\underline{\mathrm{Hom}}_{*,\mathrm{SeT}}(T,\hat{B})$ from i_*F to i_*G, it has a left exact conjugate $\phi' : j^*G' \Rightarrow j^*F'$. At the level of adjoints, we have $T_{(j^*F'b,t)} \cong \hat{B}_{(b,i_*Ft)}$, which we can write as $T_{(F'jb,t)} \cong \hat{B}_{(b,iFt)}$. Letting $b = ix$, this reads

$$T_{(F'jix,t)} \cong \hat{B}_{(ix,iFt)} \cong \mathcal{X}_{(x,Ft)}$$

where we used the fact that i is fully faithful. It follows that $T_{(F'jix,t)} \cong \mathcal{X}_{(x,Ft)}$, and hence $F'ji \simeq F'$ up to equivalence. Since we work in Ho (SeCat), we can therefore identify $F'ji$ with F', and $G'ji$ with G'. We have the following commutative diagram:

$$\begin{array}{ccc} T_{(j^*F'ix,t)} = T_{(F'jix,t)} = T_{(F'x,t)} & \xrightarrow{\cong} & \mathcal{X}_{(x,Ft)} \\ \Big\downarrow {\scriptstyle (\phi'_{ix})^*} & & \Big\downarrow {\scriptstyle (\varphi_t)_*} \\ T_{(j^*G'ix,t)} = T_{(G'jix,t)} = T_{(G'x,t)} & \xrightarrow[\cong]{} & \mathcal{X}_{(x,Gt)} \end{array}$$

that is, $\phi'_{ix} : G'x \to F'x$, or $i_*\phi' = \varphi' : G' \Rightarrow F'$ is conjugate to φ and also left exact by composition, since i right adjoint preserves limits, and ϕ' is left exact. Thus i_* is full. For faithfulness, let $\varphi_1, \varphi_2 : F \Rightarrow G$ in $\mathbb{R}\underline{\mathrm{Hom}}_{*,\mathrm{SeT}}(T,\mathcal{X})$. Suppose $i_*\varphi_1 = i_*\varphi_2 = \phi : i_*F \Rightarrow i_*G$. Pointwise this means $\phi_t : iFt \to iGt$, and i being faithful, we have a unique map $\varphi_t : Ft \to Gt$ for any t, and hence $\varphi_1 = \varphi_2$. □

Regarding notations, since $\widehat{\mathcal{X}^{\mathrm{op}}} = \mathbb{R}\underline{\mathrm{Hom}}(\mathcal{X},T)$ relative to T, and given that $\mathbb{R}\underline{\mathrm{Hom}}^*_{\mathrm{SeT}}(\mathcal{X},T)$ is a full sub-Segal category thereof [77], we will write $\Pi_{\infty,T}\widehat{\mathcal{X}^{\mathrm{op}}}$ for $\mathbb{R}\underline{\mathrm{Hom}}^*_{\mathrm{SeT}}(\mathcal{X},T) = \mathbb{R}\underline{\mathrm{Hom}}_{\mathrm{SeT}}(T,\mathcal{X})$, since we have just proved that it is a Segal groupoid. Then for any Segal topos T, this defines a functor

$$\Pi^{\wedge}_{\infty,T} : \text{Ho}\,(\text{SeT})^{\text{op}} \to \text{Ho}\,(\text{SeGpd}),$$
$$\mathcal{X} \mapsto \Pi_{\infty,T}\widehat{\mathcal{X}^{\text{op}}}$$

where SeGpd denotes the category of Segal groupoids. We now define

$$i_T : \text{Ho}\,(\text{SeGpd})^{\text{op}} \to \text{Ho}\,(\text{SeCat}),$$
$$A \mapsto i_T(A) = \mathbb{R}\underline{\text{Hom}}(A, T).$$

We claim that i_T is actually valued in Ho (SeT) if T is a Segal topos:

Lemma 24.3.4. *If $T \in \text{SeT}$ and $A \in \text{SeCat}$, then $i_T A = \mathbb{R}\underline{\text{Hom}}(A, T) \in \text{Ho}\,(\text{SeT})$.*

Proof. Since $T \in \text{SeT}$, we know there is a fully faithful embedding $i : T \to \hat{B} = \mathbb{R}\underline{\text{Hom}}(B^{\text{op}}, \text{Top})$ with a left exact left adjoint j. The map i induces a fully faithful map

$$\begin{aligned} i_* : \mathbb{R}\underline{\text{Hom}}(A, T) \to \mathbb{R}\underline{\text{Hom}}(A, \hat{B}) \\ &= \mathbb{R}\underline{\text{Hom}}(A, \mathbb{R}\underline{\text{Hom}}(B^{\text{op}}, \text{Top})) \\ &\cong \mathbb{R}\underline{\text{Hom}}(A \times^{\mathbb{L}} B^{\text{op}}, \text{Top}). \end{aligned}$$

We now determine whether i_* has a left exact left adjoint. We show it is $j_* : i_{\hat{B}}A \to i_T A$. We want $\text{Hom}_{i_T A}(j_* F, G) \cong \text{Hom}_{i_{\hat{B}}A}(F, i_* G)$. Consider

$$\begin{aligned} \text{Hom}_{i_T A}(j_* F, G) &= \text{Hom}_{i_T A}(j \circ F, G) \\ &= \coprod_{a \in A} \text{Hom}_T((j \circ F)(a), Ga) \\ &= \coprod_{a} \text{Hom}_T(j(Fa), Ga) \\ &\cong \coprod_{a} \text{Hom}_{\hat{B}}(Fa, i(Ga)) \\ &= \text{Hom}_{i_{\hat{B}}A}(F, i_* G) \end{aligned}$$

where we have used the fact that natural transformations between functors are defined pointwise. Thus $j_* \dashv i_*$. We now argue that j_* is left exact.

A diagram of objects in $i_{\hat{B}}A$ is defined pointwise over A, and hence

$$\begin{aligned} j_* \lim F^a &= j \circ \lim F^a \\ &= \coprod_{a \in A} j \circ \lim F^a(a) \\ &= \coprod_{a} \lim j \circ F^a(a) \\ &= \lim j_* F^a \end{aligned}$$

where we used the fact that j is left exact. Thus j_* is left exact, which completes the proof. □

Theorem 24.3.5. *For $T \in \mathrm{SeT}$, we have an adjunction*

$$i_T : \mathrm{Ho}\,(\mathrm{SeGpd})^{\mathrm{op}} \rightleftarrows \mathrm{Ho}\,(\mathrm{SeT}) : (\Pi^{\wedge}_{\infty,T})^{\mathrm{op}},$$
$$(\Pi^{\wedge}_{\infty,T})^{\mathrm{op}} \dashv i_T.$$

It follows that for all Segal topos $\mathcal{X}$, we have a unique unit map up to homotopy:

$$\mathcal{X} \to i_T \Pi_{\infty,T} \widehat{\mathcal{X}^{op}}.$$

Proof. It suffices to write, for $A, T \in \mathrm{Ho}\,(\mathrm{SeT})$, $B \in \mathrm{Ho}\,(\mathrm{SeGpd})$,

$$\begin{aligned}
\mathrm{Hom}^*_{\mathrm{Ho}\,(\mathrm{SeT}),\mathrm{lex}}(A, i_T B) &= \mathrm{Hom}^*_{\mathrm{Ho}\,(\mathrm{SeT}),\mathrm{lex}}(A, \mathbb{R}\underline{\mathrm{Hom}}(B, T)) \\
&\cong \mathrm{Hom}^*_{\mathrm{Ho}\,(\mathrm{SeT}),\mathrm{lex}}(A \times^{\mathbb{L}} B, T) \\
&\cong \mathrm{Hom}_{\mathrm{Ho}\,(\mathrm{SeGpd})}(B, \mathbb{R}\underline{\mathrm{Hom}}^*_{\mathrm{SeT}}(A, T)) \\
&= \mathrm{Hom}_{\mathrm{Ho}\,(\mathrm{SeGpd})}(B, \Pi_{\infty,T} \widehat{A^{\mathrm{op}}}) \\
&= \mathrm{Hom}_{\mathrm{Ho}\,(\mathrm{SeGpd})^{\mathrm{op}}}(\Pi_{\infty,T} \widehat{A^{\mathrm{op}}}, B)
\end{aligned}$$

where on the second line the left exactness is between A and T. □

It follows that for $T \in \mathrm{SeT}$, $\mathcal{X} \to i_T$ is a universal arrow. In particular, for $T = \mathcal{X}$, we have a universal arrow from $\mathcal{X}$ to its homotopical, internal perception $i_{\mathcal{X}} = \mathbb{R}\underline{\mathrm{Hom}}(-, \mathcal{X})$. Recall that we regard a perception $\mathbb{R}\underline{\mathrm{Hom}}(-, X)$ as providing a representation of the object X. That the morphism $\mathcal{X} \to \mathbb{R}\underline{\mathrm{Hom}}(-, \mathcal{X})$ is a universal arrow supports this concept. Now by universality, for any $\mathcal{Y} \in \mathrm{Ho}\,(\mathrm{SeGpd})$, we have a unique map $\mathcal{Y} \to \Pi_{\infty,\mathcal{X}} \widehat{\mathcal{X}^{\mathrm{op}}}$ in $\mathrm{Ho}\,(\mathrm{SeGpd})^{\mathrm{op}}$ making the following diagram commutative:

$$\begin{array}{ccc}
\mathcal{X} & \longrightarrow & \mathbb{R}\underline{\mathrm{Hom}}(\Pi_{\infty,\mathcal{X}} \widehat{\mathcal{X}^{\mathrm{op}}}, \mathcal{X}) \\
& \searrow & \vdots \\
& & \mathbb{R}\underline{\mathrm{Hom}}(\mathcal{Y}, \mathcal{X}).
\end{array}$$

In such a diagram, $\mathbb{R}\underline{\mathrm{Hom}}(\mathcal{Y}, \mathcal{X})$ is the internal perception of $\mathcal{X}$ relative to $\mathcal{Y} \in \mathrm{Ho}\,(\mathrm{SeGpd})$. The morphism $\mathcal{X} \to \mathbb{R}\underline{\mathrm{Hom}}(\mathcal{Y}, \mathcal{X})$ is a transition morphism from $\mathcal{X}$ to its perception relative to $\mathcal{Y}$. Since the above diagram is commutative, $\mathbb{R}\underline{\mathrm{Hom}}(\mathcal{Y}, \mathcal{X})$ ultimately comes from $i_{\mathcal{X}} \Pi_{\infty,\mathcal{X}} \widehat{\mathcal{X}^{\mathrm{op}}}$.

24.4 Natural phenomena

We regard natural phenomena as following some basic rules, or fundamental laws, encoded in algebras. Coherent manifestations of such laws will be modeled by stacks valued in simplicial sets, and according to the philosophy of derived algebraic geometry, it is natural then to consider simplicial algebras. Hence we start with a commutative ring k, and we let sk-Mod be the category of simplicial k-modules, commutative monoids of which form sk-CAlg the category of simplicial k-algebras. Its opposite category $\mathrm{d}k\text{-Aff} = L(sk\text{-CAlg})^{\mathrm{op}}$ is referred to as the Segal category of derived affine stacks [78, 76, 79, 74]. We put the ffqc topology on this Segal category. The category $\mathcal{X} = \mathrm{dSt}(k) = \mathrm{d}k\text{-Aff}^{\sim,\mathrm{ffqc}}$ of affine stacks is a localization of the Segal category of prestacks $\widehat{\mathrm{d}k\text{-Aff}} = \mathbb{R}\underline{\mathrm{Hom}}(\mathrm{d}k\text{-Aff}^{\mathrm{op}}, \mathrm{Top})$. The former is a Segal topos [74], and hence all the above formalism applies to $\mathrm{dSt}(k)$. In particular, by universality of the unit map $\mathcal{X} \to i_{\mathcal{X}}\Pi_{\infty,\mathcal{X}}\widehat{\mathcal{X}^{\mathrm{op}}}$ we have a morphism $i_{\mathcal{X}}\Pi_{\infty,\mathcal{X}}\widehat{\mathcal{X}^{\mathrm{op}}} \to \mathbb{R}\underline{\mathrm{Hom}}(\mathcal{Y},\mathcal{X})$ in Ho (SeT) to the homotopy internal perception of $\mathcal{X}$ relative to $\mathcal{Y}$ for any $\mathcal{Y} \in$ Ho (SeGpd).

For a Segal category $\mathcal{X}$, we know there is a Segal category $\mathcal{U}$ such that $\mathcal{U} = \mathbb{R}\underline{\mathrm{Hom}}^*_{\mathrm{SeT}}(\mathcal{X},\mathcal{X}) = \Pi_{\infty,\mathcal{X}}\widehat{\mathcal{X}^{\mathrm{op}}}$. In contrast to just having the Segal category $\mathcal{X}$, with its objects and morphisms, which is very much static, $\mathcal{U}$ provides a functorial, dynamic presentation of $\mathcal{X}$ in the following sense. Consider a geometric morphism $\psi : \mathcal{X} \to \mathcal{X}$ in $\mathcal{U}$, which we regard as a perturbation of $\mathcal{X}$, and let $F, G \in \mathcal{X}$. Then we have an induced morphism of simplicial sets $\psi : \mathcal{X}_{(F,G)} \to \mathcal{X}_{(\psi F,\psi G)}$, a local perturbation on morphisms of $\mathcal{X}$. Further, $\psi = \mathrm{id}_{\mathcal{X}}$ is clearly an element of $\mathcal{U}$. However, $\mathrm{id}_{\mathcal{X}}$ just reproduces $\mathcal{X}$. Thus we can morally see $\mathcal{X}$ as an object of $\mathcal{U}$ or $\mathcal{U}$ as an enlargement of $\mathcal{X}$. It follows that a comprehensive view of $\mathcal{X}$ occurs in a higher Segal category $\mathcal{U}$. Observe that this can be repeated: $\mathbb{R}\underline{\mathrm{Hom}}^*_{\mathrm{SeT}}(\mathcal{U},\mathcal{U})$ is yet another Segal category, in which $\mathcal{U}$ would take its full meaning.

Recall that one representation of $\mathcal{X}$ is provided by its homotopical internal perception $\mathbb{R}\underline{\mathrm{Hom}}(-,\mathcal{X})$. By the universality of $\mathcal{X} \to i_{\mathcal{X}}\Pi_{\infty,\mathcal{X}}\widehat{\mathcal{X}^{\mathrm{op}}}$, for any morphism $f : \mathcal{X} \to \mathbb{R}\underline{\mathrm{Hom}}(\mathcal{Y},\mathcal{X})$ for $\mathcal{Y} \in$ Ho (SeGpd), which corresponds to focusing on one facet of $\mathcal{X}$ relative to $\mathcal{Y}$, we have a unique morphism $\phi : \mathcal{Y} \to \Pi_{\infty,\mathcal{X}}\widehat{\mathcal{X}^{\mathrm{op}}} = \mathcal{U}$ such that the following diagram is commutative:

$$\begin{array}{ccc} \mathcal{X} & \longrightarrow & \mathbb{R}\underline{\mathrm{Hom}}(\mathcal{U},\mathcal{X}) \\ & \searrow & \vdots \\ & & \mathbb{R}\underline{\mathrm{Hom}}(\mathcal{Y},\mathcal{X}). \end{array}$$

The morphism ϕ provides us with the information of $\mathcal{U}$ originating from $\mathcal{Y}$. The commutative diagram tells us that the perception of $\mathcal{X}$ relative to $\mathcal{Y}$ factors through the perception of $\mathcal{X}$ relative to $\mathcal{U}$. Since the latter corresponds to the dynamics of $\mathcal{X}$, this perception reveals the static part of $\mathcal{X}$ relative to itself. On the other hand, Segal groupoids present a dynamic improvement upon simple Segal categories such as $\mathcal{X}$. It follows that an object such as $\mathbb{R}\underline{\mathrm{Hom}}(\mathcal{Y},\mathcal{X})$ provides us with a perception of $\mathcal{X}$ by a dynamic object. The fact that any perception of $\mathcal{X}$ by Segal groupoids factors through $i_{\mathcal{X}}\Pi_{\infty,\mathcal{X}}\widehat{\mathcal{X}^{\mathrm{op}}}$ therefore means that $i_{\mathcal{X}}\mathcal{U}$ is initial in all such perceptions. Note however that it is the morphism ϕ above that is important, not so much the CD. In that regard, it would be interesting to have a cover of $\mathcal{U}$ by such morphisms ϕ. Then the collection of accompanying CDs would provide a complete decomposition of a perception of $\mathcal{X}$ relative to its own dynamics.

One can contrast this formalism with the concept of having a field in Physics being defined on a target space, itself resulting from a compactification. Here a field is replaced by a functor on simplicial algebras, a derived stack to be more precise, with Set_{Δ} as target space. Those functors provide a manifestation of natural laws. The collection of all such functors is the Segal Topos $\mathrm{dSt}(k)$, an enlargement of which is provided by $\mathbb{R}\underline{\mathrm{Hom}}^{*}_{\mathrm{SeT}}(\mathrm{dSt}(k), \mathrm{dSt}(k))$, which because it is dynamic would consequently provide a universal theory of physical phenomena. CDs such as the one above would provide a decomposition of the perception of $\mathrm{dSt}(k)$ as a static object relative to its own dynamics.

25 Infinity-topoi and natural phenomena: generation

25.1 Introduction

In Chapter 24, we argued that one could model natural laws by simplicial commutative algebras and that representations of such laws could be implemented by considering higher stacks, or derived stacks [79, 74] to be more precise. For a commutative ring k and the simplicial category sk-CAlg of simplicial commutative k-algebras, which after Segal localization [74, 77] gives $\mathrm{d}k\text{-Aff} = (L(sk\text{-CAlg}))^{\mathrm{op}}$, the Segal category of derived affine schemes, one can first consider the Segal category of derived prestacks on $\mathrm{d}k$-Aff, $\widehat{\mathrm{d}k\text{-Aff}} = \mathbb{R}\underline{\mathrm{Hom}}(\mathrm{d}k\text{-Aff}^{\mathrm{op}}, \mathrm{Top})$. On $\mathrm{d}k$-Aff, we put the ffqc topology, and we define the Segal category of derived stacks $\mathrm{dSt}(k)$ on $\mathrm{d}k$-Aff to be the left Bousfield localization of $\widehat{\mathrm{d}k\text{-Aff}}$ with regards to ffqc-hypercovers. This is a Segal topos [74]. The purpose of the present work is to discuss the existence of subobject classifiers in such a Segal topos. In the setting of ∞-topoi, such a concept is easier to discuss, so we will consider the ∞-topos associated with $\mathrm{dSt}(k)$. Such an association is a fairly well-known fact; as pointed out in [56, 52, 76, 79], the notions of Segal topos and ∞-topos are essentially equivalent. To be precise, we have an adjunction from the category of simplicial categories Cat_Δ to the category of simplicial sets Set_Δ, with right adjoint the simplicial nerve functor $N : \mathrm{Cat}_\Delta \to \mathrm{Set}_\Delta$ [52], originally introduced by Cordier as the coherent nerve functor [11]. We have $N(\mathrm{dSt}(k)) \in \mathrm{Cat}_\infty$. To see this, it suffices to take a fibrant simplicial category equivalent to $\mathrm{dSt}(k)$. Those have Kan complexes for mapping spaces,[1] and hence their simplicial nerve is an ∞-category.[2] We further show that $\mathrm{dSt}(k)_\infty = N(\mathrm{dSt}(k))$ is an ∞-topos. Focusing then on subobject classifiers in ∞-topos and arguing that an object of an ∞-topos is characterized by its subobjects, it follows that one can regard the subobject classifier in the ∞-topos $\mathrm{dSt}(k)_\infty$ as a generator of natural phenomena.

In Chapter 24 the emphasis was on higher Galois, and for this purpose, we considered $\mathbb{R}\underline{\mathrm{Hom}}^*_{\mathrm{SeT}}(\mathcal{X},\mathcal{U})$ for two Segal topos $\mathcal{X}$ and $\mathcal{U}$, and we showed this is a Segal groupoid. From the perspective of the present work it is preferable to have Segal topos Hom objects however, and $\mathbb{R}\underline{\mathrm{Hom}}(\mathcal{X},\mathcal{U})$ would be a more suitable object to consider. The corresponding object in the context of ∞-topoi is $\mathrm{Fun}(\mathcal{X},\mathcal{U})$, an ∞-topos for two ∞-topoi $\mathcal{X}$ and $\mathcal{U}$. In particular, so is $\mathrm{Fun}(\mathcal{X},\mathcal{X})$, and so it has a subobject classifier. From there one investigates whether there is a universal ∞-topos, universal in the sense of having a universal subobject classifier, and we show that indeed there is one, namely the ∞-category of spaces $\mathcal{S}$.

In the present work, whatever technical definition is invoked but not fully used, it will be referenced, and only those concepts that need to be explained will be covered

1 Rmrk 1.1.4.3 of [52].

2 Prop. 1.1.5.10 of [52].

https://doi.org/10.1515/9783111334073-025

for ease of reading. We first go over the necessary background about Segal categories and ∞-categories. This we do in Section 25.2. As an intermission, we discuss localizations (Dwyer–Kan, Segal, Hammock, Bousfield) in Section 25.3, at which point we can come back to Segal categories and ∞-categories to define those localizations, which leads to the notions of Segal topoi and ∞-topoi. We then define the Segal categories of prestacks and stacks with this formalism. Moving on to the ∞-category side of things in Section 25.4, we then use a result of [52], which essentially states that the simplicial nerve of the subcategory of fibrant–cofibrant objects in the model category of prestacks with the local model structure, meaning that the Segal category of stacks is an ∞-topos, which is what we needed to show that $\mathrm{dSt}(k)$ maps to an ∞-topos. From there we discuss subobject classifiers and how relevant ∞-topoi relate to one another.

For notations, Set_Δ is the category of simplicial sets, Δ is the usual category of combinatorial simplices whose objects are linearly ordered sets $[n] = \{0, 1, \dots, n\}$, with morphisms order-preserving maps. For $A : \Delta \to \mathrm{Set}$, we write A_n for $A([n])$. Our main reference for model categories is [42].

25.2 Higher categories

25.2.1 Segal categories

Our main references for Segal categories will be [41, 66, 77]. One can regard a Segal category as a weak form of simplicial category, where compositions are defined up to equivalence only. To be more precise, a Segal category C is first a bisimplicial set, a functor $C : \Delta^{\mathrm{op}} \to \mathrm{Set}_\Delta$, such that C_0 is a discrete set, its set of objects. This makes a bisimplicial set into a Segal pre-category. The simplicial set of morphism $C_1(x,y)$ between objects $x, y \in C_0$ is the preimage of $(x,y) \in C_0 \times C_0$ under the morphism $C_1 \to C_0 \times C_0$. A Segal category among Segal pre-categories is characterized by the fact that the morphisms of simplicial sets $C_n \to C_1 \times_{C_0} \cdots \times_{C_0} C_1$ are equivalences for all $n \in \Delta$. It is in this regard that Segal categories are generalizations of simplicial categories. Indeed,

$$C_n \cong \coprod_{(a_0,\dots,a_n) \in C_0^{n+1}} C_{(a_0,\dots,a_n)}$$

so for all points $(a_0, \dots, a_n) \in C_0^{n+1}$, a Segal category would just provides us with equivalences of simplicial sets:

$$C_{(a_0,\dots,a_n)} \to C_{(a_0,a_1)} \times \cdots \times C_{(a_{n-1},a_n)}$$

whereas if, say, S is a simplicial category [28], then defining

$$S_n = \coprod_{(x_0,\dots,x_n) \in S_0^{n+1}} \underline{\mathrm{Hom}}(x_0, x_1) \times \cdots \times \underline{\mathrm{Hom}}(x_{n-1}, x_n)$$

we have natural isomorphisms of simplicial sets,

$$S_{(x_0,\ldots,x_n)} \to S_{(x_0,x_1)} \times \cdots \times S_{(x_{n-1},x_n)}.$$

To come back to Segal pre-categories, they form a category SePC on which we can put two model categories, and for this, we refer the reader to [77]. One of those is a left Bousfield localization of the other, and the natural inclusion on the corresponding homotopy categories has a left adjoint SeCat that turns Segal pre-categories into Segal categories. The main point here is that SeCat is a left derived functor, a fact we will use later. We need a notion of homotopy category Ho (C) for a Segal category C: it suffices to take a category with C_0 as a set of objects and $\pi_0(C_{(a,b)})$ as a set of morphisms between two objects $a, b \in C_0$. A morphism $f : C \to D$ of Segal pre-categories will be said to be fully faithful if for all $a, b \in C_0$, the morphism $\mathrm{SeCat}(C)_{(a,b)} \to \mathrm{SeCat}(D)_{(f(a),f(b))}$ is an isomorphism in Ho (Set_Δ), and f is said to be essentially surjective if Ho $(\mathrm{SeCat}(C)) \to$ Ho $(\mathrm{SeCat}(D))$ is essentially surjective. As usual, we will define a morphism of Segal pre-categories to be an equivalence if it is both fully faithful and essentially surjective. If we take this as a notion of weak equivalence on SePC and we take cofibrations to be monomorphisms, this defines a model structure on SePC as shown by Pellissier [66, 77]. This model structure is furthermore internal, with an internal hom $\underline{\mathrm{Hom}}$, and Ho (SePC) is Cartesian closed, with internal hom $\mathbb{R}\underline{\mathrm{Hom}}$. When we say $f : C \to D$ is a morphism of Segal categories, we mean it is an element of $\mathbb{R}\underline{\mathrm{Hom}}(C, D)$. The derived adjunction formula

$$\mathbb{R}\underline{\mathrm{Hom}}(A, \mathbb{R}\underline{\mathrm{Hom}}(B, C)) \cong \mathbb{R}\underline{\mathrm{Hom}}(A \times B, C)$$

is also very useful.

25.2.2 ∞-categories

A standard reference for our purposes will be [52]. By an ∞-category we mean an $(\infty, 1)$-category, where an (∞, n)-category is an ∞-category for which morphisms are invertible for $k > n$. We say a simplicial set K is an ∞-category if one can find a dotted arrow as shown below that can make the following diagram commutative for $0 < i < n$, where $\Delta^n = \mathrm{Hom}_{\mathrm{Set}_\Delta}(-, [n])$, and Λ^n_i is the ith horn of Δ^n:

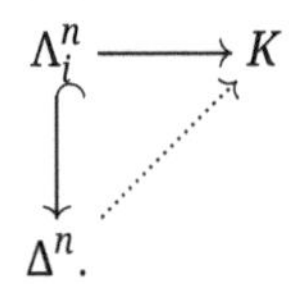

Those simplicial sets that satisfy such an extension property for outer horns as well are referred to as Kan complexes. If $\mathcal{C}$ and $\mathcal{D}$ are ∞-categories, then a functor from $\mathcal{C}$ to $\mathcal{D}$ is simply a map $\mathcal{C} \to \mathcal{D}$ of simplicial sets. We denote $\mathrm{Map}_{\mathrm{Set}_\Delta}(\mathcal{C}, \mathcal{D})$ by $\mathrm{Fun}(\mathcal{C}, \mathcal{D})$, which is the ∞-category of functors from $\mathcal{C}$ to $\mathcal{D}$. We denote by Cat_∞ the ∞-category of small ∞-categories.

25.2.3 Simplicial categories

Set_Δ-enriched categories are referred to as simplicial categories, the category of which will be denoted by Cat_Δ (see [28] as well). Crucial for us will be the simplicial nerve functor N:

$$N : \mathrm{Cat}_\Delta \to \mathrm{Set}_\Delta$$

as originally introduced by Cordier under the name of coherent nerve functor [11], and whose definition is given again in [52] in a format suitable for our purposes. It is defined via the simplicial category functor $\mathfrak{C}$ as presented in [52]. We will not need its full definition, and the reader is referred to that reference for details. One can see $\mathfrak{C} : \mathrm{Set}_\Delta \to \mathrm{Cat}_\Delta$ as a thickening of simplicial sets, through a combinatorial definition. If $\mathcal{C}$ is a simplicial category, then the simplicial nerve $N(\mathcal{C})$ is defined by the adjunction formula

$$\mathrm{Hom}_{\mathrm{Set}_\Delta}(\Delta^n, N(\mathcal{C})) = \mathrm{Hom}_{\mathrm{Cat}_\Delta}(\mathfrak{C}[\Delta^n], \mathcal{C})$$

which one may want to contrast with the usual nerve functor, which for an ordinary category $\mathcal{C}$ satisfies

$$\mathrm{Hom}_{\mathrm{Set}_\Delta}(\Delta^n, N(\mathcal{C})) = \mathrm{Hom}_{\mathrm{Cat}}([n], \mathcal{C}).$$

Note that we have an equivalence of simplicial categories $\mathfrak{C}[\Delta^n] \to [n]$. Kan, the full subcategory of Set_Δ spanned by Kan complexes, is a simplicial category; $\mathcal{S} = N(\mathrm{Kan})$ is referred to as the ∞-category of spaces.

25.3 Localizations

We will use a few different localizations, so to fix ideas, we will go over those we need and provide references for more ample details.

25.3.1 Bousfield localization

Our main reference for *Bousfield localizations* will be [40]. Recall that if M is a model category and S is a class of maps in M, then an object X of M is said to be *S-local* if it is fibrant and if for any $f : A \to B$ in S, the induced map of homotopy function complexes $f^* : \mathrm{Map}(B, X) \to \mathrm{Map}(A, X)$ is a weak equivalence in Set_Δ. Now a map

$g : X \to Y$ in M is an S-*local equivalence* if for all S-local objects Z, the induced map $g^* : \mathrm{Map}(Y, Z) \to \mathrm{Map}(X, Z)$ is a weak equivalence in Set_Δ. We then define the left Bousfield localization $L_S M$ of M to be a model structure on the underlying category of M with S-local equivalences as weak equivalences, the same cofibrations as those of M, and fibrations will be maps with the right lifting property with respect to cofibrations that are also S-local equivalences. We will denote the Bousfield localizations by L_{Bous}, the set S being implied.

25.3.2 Simplicial localizations

Dwyer and Kan [15] introduced a slightly more general localization than the usual Gabriel–Zisman localization [21]; for an ordinary category $\mathcal{C}$ and a subcategory $\mathcal{B}$, the *Gabriel–Zisman localization* of $\mathcal{C}$ with respect to $\mathcal{B}$ is denoted $\mathcal{C}[\mathcal{B}^{-1}]$ (or sometimes $\mathcal{B}^{-1}\mathcal{C}$), obtained by formally inverting morphisms in $\mathcal{B}$. Dwyer and Kan generalized this by defining the *simplicial localization* of $\mathcal{C}$ with respect to $\mathcal{B}$ to be given by $L_{\mathrm{DK}}(\mathcal{C}, \mathcal{B}) = F_*\mathcal{C}[F_*\mathcal{B}^{-1}]$, where $F_*\mathcal{C}$ denotes the simplicial free resolution of the category $\mathcal{C}$ (see [15]). $L_{\mathrm{DK}}(\mathcal{C}, \mathcal{B})$ is a simplicial category, and it is a generalization in the sense that $\mathrm{Ho}\,(L_{\mathrm{DK}}[\mathcal{C}, \mathcal{B}]) = \mathcal{C}[\mathcal{B}^{-1}]$.

One can also generalize this definition to the case where $\mathcal{C}$ is a simplicial category by taking the diagonal of the bisimplicial set constructed using the above procedure levelwise: $L_{\mathrm{DK}}(\mathcal{C}, \mathcal{B}) = \mathrm{diag} F_*\mathcal{C}[F_*\mathcal{B}^{-1}]$.

In [77, 74] a still more general localization is defined using the fact that we have a fully faithful functor $\mathrm{Cat}_\Delta \to \mathrm{SeCat}$. Toen and Vezzosi define a Segal analog of L_{DK} as follows: if C is a Segal category and S is a set of morphisms in $\mathrm{Ho}\,(C)$, then there exists a Segal category $L(C, S)$ with a localization $l : C \to L(C, S)$ characterized by the following universal property: for any Segal category D, the induced morphism

$$l^* : \mathbb{R}\underline{\mathrm{Hom}}(L(C, S), D) \to \mathbb{R}\underline{\mathrm{Hom}}(C, D)$$

is fully faithful and has for essential image those morphisms from C to D that send morphisms in S into equivalences in D. Observe that if C is an ordinary category viewed as a Segal category, then $L(C, S) = L_{\mathrm{DK}}(C, S)$ [41]. If M is a model category and W is its set of equivalences, then we will just denote $L(M, W)$ in the sense above by LM. We will denote both *Segal localizations* and *Dwyer–Kan localizations* by the same letter L, and it should be clear from the context which one is being used.

In [16] yet another simplicial localization is introduced, the *hammock localization* L^H, which is better behaved than the original Dwyer–Kan localization $L_{\mathrm{DK}} = L$. We will not need its explicit definition, so the reader is referred to [16] for details, but for us, it is important that if C denotes an ordinary category, then we have a weak equivalence

$$L^H C \to LC.$$

For a simplicial category C, we can also define its hammock localization as $L^H C = \mathrm{diag} L^H C$, and in this case, we also have a weak equivalence

$$\mathrm{diag} L^H C \to LC.$$

From [17], if M is a simplicial model category and M° is its subcategory of fibrant-cofibrant objects, then we have a weak equivalence

$$M^\circ \to \mathrm{diag} L^H M.$$

Collecting things, if M is a simplicial model category, $LM \simeq \mathrm{diag} L^H M \simeq M^\circ$, so $LM \simeq M^\circ$, as observed in [41]. This is a crucial point that will be important later.

25.3.3 Higher topoi

25.3.3.1 Segal topoi

As defined in [77], a map of Segal categories is said to be *left exact* if it preserves finite limits. A localization of Segal categories is a morphism of Segal categories with a fully faithful right adjoint. A Segal category $\mathcal{X}$ is a *Segal topos* if it is a left exact localization of $\hat{C} = \mathbb{R}\underline{\mathrm{Hom}}(C^{\mathrm{op}}, \mathrm{Top})$ with $\mathrm{Top} = L\mathrm{Set}_\Delta$, the *Segal category of simplicial sets*, and C a small Segal category. This means there exists some $i : \mathcal{X} \to \hat{C}$ fully faithful with a left exact left adjoint. As in classical topos theory [60], Toen and Vezzosi define morphisms between Segal topoi to be geometric morphisms; if $\mathcal{X}$ and $\mathcal{Y}$ are Segal topoi, then one denotes by $\mathbb{R}\underline{\mathrm{Hom}}_{\mathrm{SeT}}(\mathcal{X}, \mathcal{Y}) = \mathbb{R}\underline{\mathrm{Hom}}^*_{\mathrm{SeT}}(\mathcal{Y}, \mathcal{X})$ the Segal category of geometric morphisms from $\mathcal{X}$ to $\mathcal{Y}$, defined as the sub-Segal category of $\mathbb{R}\underline{\mathrm{Hom}}(\mathcal{Y}, \mathcal{X})$ spanned by those morphisms that are left exact with a right adjoint. We denote the 2-Segal category of Segal topoi by SeT.

25.3.3.2 ∞-topoi

As defined in [52], a *localization* is a functor between ∞-categories that has a fully faithful right adjoint. The notion of adjoints for ∞-categories makes use of a correspondence and is an abstraction of the notion of adjunction from ordinary category theory [59]. Nevertheless, this definition, which we will not use, can be recast in a classical format by virtue of the fact that if $f : \mathcal{C} \to \mathcal{D}$ and $g : \mathcal{D} \to \mathcal{C}$ are functors between ∞-categories $\mathcal{C}$ and $\mathcal{D}$, then $f \dashv g$ if and only if there exists a morphism $\eta : \mathrm{id}_{\mathcal{C}} \to g \circ f$ in $\mathrm{Fun}(\mathcal{C}, \mathcal{C})$ such that for all $X \in \mathcal{C}$ and $Y \in \mathcal{D}$, we have an induced equivalence in the homotopy category of spaces $\mathcal{H}$:

$$\mathrm{Map}_{\mathcal{D}}(f(X), Y) \simeq \mathrm{Map}_{\mathcal{C}}(X, g(Y)).$$

If $f : \mathcal{C} \to \mathcal{D}$ is a localization, then $\mathcal{D}$ is said to be a localization of $\mathcal{C}$. If g denotes the right adjoint that makes f a localization, then $L = g \circ f : \mathcal{C} \to \mathcal{C}$ will be referred to as a *localization functor*. For the definition of right (left) exactness, the reader is referred to this reference, but the definition is not very illuminating in our context. The following fact is important: a functor is left exact if and only if it preserves finite limits. $\mathcal{X} \in \mathrm{Cat}_\infty$ is an ∞-topos if there exist a small ∞-category $\mathcal{C}$ and an accessible left exact localization functor $\mathrm{Pr}(\mathcal{C}) \to \mathcal{X}$, where $\mathrm{Pr}(\mathcal{C}) = \mathrm{Fun}(\mathcal{C}^{\mathrm{op}}, \mathcal{S})$. We will not need to check that a localization is accessible for the simple reason that most of the categories we will deal with are accessible. Since a morphism between accessible ∞-categories with an adjoint is accessible [52, 56], we will dispense with this accessibility condition. Nevertheless, we can briefly remind the reader what it means to be accessible (see [52] for more details). We first have to go back to simplicial sets. One defines the join $K \star K'$ of two simplicial sets K and K' as follows: for every nonempty finite linearly ordered set I,

$$(K \star K')(I) = \coprod_{I = J \cup J'} K(J) \times K'(J').$$

With this in hand, one denotes $K \star \Delta^0$ by $K^{\triangleright}$, referred to as the right cone of K. Now if κ is a regular cardinal and $\mathcal{C} \in \mathrm{Cat}_\infty$, then $\mathcal{C}$ is said to be κ-filtered if for any κ-small $K \in \mathrm{Set}_\Delta$ and any map $f : K \to \mathcal{C}$, there is an extension $\overline{f} : K^{\triangleright} \to \mathcal{C}$ of f. With this we have an obvious notion of κ-filtered colimits. For $\mathcal{C} \in \mathrm{Cat}_\infty$ with small κ-filtered colimits we say a functor $f : \mathcal{C} \to \mathcal{D}$ in Cat_∞ is κ-continuous if it preserves κ-filtered colimits. For a small ∞-category $\mathcal{C}$, let $\mathrm{Ind}_\kappa(\mathcal{C})$ be the full subcategory of $\mathrm{Pr}(\mathcal{C})$ spanned by functors $f : \mathcal{C}^{\mathrm{op}} \to \mathcal{S}$ that classify right fibrations $\tilde{\mathcal{C}} \to \mathcal{C}$, where $\tilde{\mathcal{C}}$ is a κ-filtered ∞-category. Finally, $\mathcal{C} \in \mathrm{Cat}_\infty$ is said to be κ-accessible, or simply accessible, if there exist a small ∞-category $\mathcal{C}_0$ and an equivalence $\mathrm{Ind}_\kappa(\mathcal{C}_0) \to \mathcal{C}$. If $\mathcal{C}$ is such a category, then a functor $f : \mathcal{C} \to \mathcal{D}$ is accessible if it is κ-continuous for some regular cardinal κ. Observe that if $K \in \mathrm{Set}_\Delta$ and $\mathcal{C}$ is an accessible ∞-category, then $\mathrm{Fun}(K, \mathcal{C})$ is accessible as well. That gives us a lot of accessible categories.

25.3.4 Local model structures

One model structure we will put on our simplicial presheaves is the *local model structure*, something originally introduced by Jardine [44] and used in [76] and [79] to construct Segal categories of derived stacks. First, it is important to go briefly over those constructions for putting them in perspective, referring the reader to the main reference [76] for more details. In that reference, simplicial categories are referred to as S-categories. Let T be such a category. An S-topology τ on T is nothing but a Grothendieck topology τ on its homotopy category $\mathrm{Ho}\,(T)$. This makes (T, τ) a so-called S-site. On the category of simplicial presheaves on T, $\mathrm{sPr}(T)$, there exists a model structure, the local model structure, where equivalences are local equivalences, also known as π_*-equivalences (see [76] for more details or the section on derived stacks). We denote by $\mathrm{sPr}_\tau(T)$ this new model structure, which turns out to be the left Bousfield localization of sPr(T) along local equivalences. We call $\mathrm{sPr}_\tau(T)$ the model category of stacks on (T, τ). Actually, since $\mathrm{id} : \mathrm{sPr}(T) \to \mathrm{sPr}_\tau(T)$ preserves homotopy fiber products, we call such a localization a *left exact Bousfield localization* of sPr(T). We define a *model topos* to be a model category that is Quillen equivalent to a left exact Bousfield localization of sPr(T), T a simplicial category. In particular, that would make $\mathrm{sPr}_\tau(T)$ a model topos.

We have the following result [77]: if M is a model topos, then LM is a Segal topos. Thus $L(\mathrm{sPr}_\tau(T))$ is a Segal topos.

Independently, we can develop the same formalism of stacks starting not from a simplicial category, but from a Segal category. One defines a Segal topology on a Segal category $\mathcal{X}$ as a Grothendieck topology on its homotopy category Ho ($\mathcal{X}$). One defines a notion of local equivalence on $\hat{\mathcal{X}} = \mathbb{R}\underline{\mathrm{Hom}}(\mathcal{X}^{\mathrm{op}}, \mathrm{Top})$ relative to the topology τ on the Segal site $(\mathcal{X}, \tau)$. A stack F on $\mathcal{X}$ is an element of $\hat{\mathcal{X}}$ such that for any local equivalence $\alpha : G \to H$ in $\hat{\mathcal{X}}$, the induced morphism $\alpha^* : \mathrm{Hom}_{\mathrm{Ho}\,(\hat{\mathcal{X}})}(H, F) \to \mathrm{Hom}_{\mathrm{Ho}\,(\hat{\mathcal{X}})}(G, F)$ is bijective. We denote by $\mathcal{X}^{\sim,\tau}$ the Segal category of stacks on $\mathcal{X}$. It turns out that the inclusion $\mathcal{X}^{\sim,\tau} \to \hat{\mathcal{X}}$ has a left exact left adjoint, making $\mathcal{X}^{\sim,\tau}$ a Segal topos.

Now there is a very important result in the theory of Segal categories called *strictification* [41, 77, 76, 79] that states that for a small simplicial category T and a cofibrantly generated simplicial model category M, if we endow M^T with its projective model structure, then we have an isomorphism in Ho (SePC)

$$L(M^T) \cong \mathbb{R}\underline{\mathrm{Hom}}(T, LM).$$

This is sometimes written in the following format [76]: for a category C with a subset of morphisms S,

$$L(M^{C,S}) \cong \mathbb{R}\underline{\mathrm{Hom}}(L(C, S), LM)$$

where $M^{C,S}$ is the model category of restricted diagrams, a left Bousfield localization of M^C with respect to morphisms based in S. In this isomorphism, we used the Quillen equivalence $M^{C,S} \simeq M^{L(C,S)}$.

Using this, one can show that

$$L(\mathrm{sPr}(T)) \cong \mathbb{R}\underline{\mathrm{Hom}}(T^{\mathrm{op}}, \mathrm{Top})$$

as well as

$$L(\mathrm{sPr}_\tau(T)) \cong T^{\sim,\tau}.$$

Then modulo isomorphisms we have the following commutative diagram that sums up this picture:

$$\begin{array}{ccc} \mathrm{sPr}(T) & \xrightarrow{L_{\mathrm{Bous}}} & \mathrm{sPr}_\tau(T) \\ {\scriptstyle L}\downarrow & & \downarrow{\scriptstyle L} \\ \hat{T} = \mathbb{R}\underline{\mathrm{Hom}}(T^{\mathrm{op}}, \mathrm{Top}) & \xrightarrow[L_{\mathrm{Bous}}]{} & T^{\sim,\tau}. \end{array}$$

For instance,

$$\begin{aligned} L(\mathrm{sPr}((sk\text{-CAlg})^{\mathrm{op}})) &= L(\mathrm{Set}_\Delta^{sk\text{-CAlg}}) \\ &\cong \mathbb{R}\underline{\mathrm{Hom}}(L(sk\text{-CAlg}), L(\mathrm{Set}_\Delta)) \\ &= \mathbb{R}\underline{\mathrm{Hom}}(\mathrm{d}k\text{-Aff}^{\mathrm{op}}, \mathrm{Top}) \\ &= \widehat{\mathrm{d}k\text{-Aff}}. \end{aligned}$$

25.3.5 Segal category of derived stacks

As covered in [78, 76, 74], for a commutative ring k, the category sk-CAlg of simplicial commutative k-algebras, and the Segal category $\mathrm{d}k\text{-Aff} = L(sk\text{-CAlg})^{\mathrm{op}}$ of derived affine schemes, on which we put the ffqc topology, $\mathrm{d}k\text{-Aff}^{\sim,\mathrm{ffqc}} = \mathrm{dSt}(k)$ is a left exact localization of $\widehat{\mathrm{d}k\text{-Aff}} = \mathbb{R}\underline{\mathrm{Hom}}(\mathrm{d}k\text{-Aff}^{\mathrm{op}}, \mathrm{Top})$. This will be the primary object of study for us, the reason being, as briefly discussed in Chapter 24, that one can model natural laws by simplicial algebras, the realizations of which can, for the sake of coherence, be represented as stacks. The collection of all such representations is $\mathcal{X} = \mathrm{dSt}(k)$ and would correspond to the collection of all natural phenomena. Recall also from Chapter 24, or from the section on higher Galois, that $\mathbb{R}\underline{\mathrm{Hom}}^*_{\mathrm{SeT}}(\mathrm{dSt}(k), \mathrm{dSt}(k))$ gives a (dynamic) deformation-theoretic presentation of $\mathrm{dSt}(k)$, thereby providing a universal theory of all physical phenomena.

25.4 From Segal topoi to ∞-topoi

It is clear that the transition from Segal topoi to ∞-topoi will be made using the simplicial nerve functor N. In particular, we will be using the following result from [52] and [56]: if C is a small category with a Grothendieck topology, $A = \mathrm{sPr}(C)$ is the category of simplicial presheaves over C with its local model structure [44], and A° is the subcategory of fibrant–cofibrant objects of A for that local model structure, then $N(A^\circ) \simeq \mathrm{Sh}(N(C))^\wedge$, where $\mathrm{Sh}(\mathcal{D}) \subseteq \mathrm{Pr}(\mathcal{D})$ for a small ∞-category $\mathcal{D}$ with a Grothendieck topology [52], corresponds to a notion of higher category of sheaves [52]. It turns out that $\mathrm{Sh}(\mathcal{D})$ is a *topological localization* of $\mathrm{Pr}(\mathcal{D})$ [52], and those are accessible and left exact, and hence $\mathrm{Sh}(\mathcal{D})$ is an ∞-topos itself. If we further take the Bousfield localization of such an ∞-topos with respect to ∞-connective morphisms [52], then one gets its *hypercompletion* $\mathrm{Sh}(\mathcal{D})^\wedge$, which again is an ∞-topos. This result above starting from an ordinary category holds also in the case where $\mathcal{C}$ is a simplicial model category, and that will be important for us since we are taking $\mathcal{C} = sk\text{-CAlg}$.

The situation can be summarized as follows. If one takes $\mathcal{C} = (sk\text{-CAlg})^{\mathrm{op}}$, then $A = \mathrm{sPr}(\mathcal{C}^{\mathrm{op}})$ is a simplicial model category, on which we can put a local model structure. Actually, on A itself, we first put the injective model structure, which is Quillen equivalent to the projective model structure. Recall that the *injective model structure* is obtained by taking objectwise cofibrations as cofibrations, whereas the projective model structure is obtained by starting instead with objectwise fibrations. Then we have

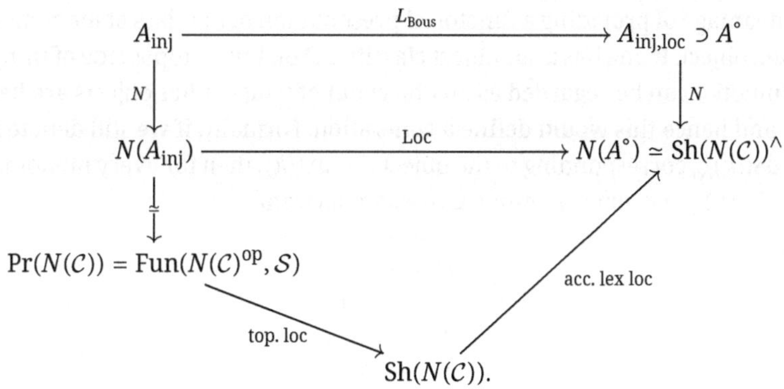

We apply this to the case $\mathcal{C} = (sk\text{-CAlg})^{\mathrm{op}}$, $A = \mathrm{sPr}(\mathcal{C}) = \mathrm{Set}_{\Delta}^{sk\text{-CAlg}}$ with the injective model structure, so that $A_{\mathrm{inj,loc}} = \mathrm{sPr}_{\mathrm{ffqc}}(\mathcal{C})$. Now

$$A^{\circ} \simeq LA_{\mathrm{inj,loc}} = L(\mathrm{sPr}_{\mathrm{ffqc}}((sk\text{-CAlg}))^{\mathrm{op}}) \simeq \mathrm{dSt}(k)$$

so that

$$N(A^{\circ}) \simeq N(\mathrm{dSt}(k)) = \mathrm{dSt}(k)_{\infty}$$

is an ∞-topos since it is equivalent to an ∞-topos $\mathrm{Sh}(N(sk\text{-CAlg}^{\mathrm{op}}))^{\wedge}$. From then on we will work with $\mathrm{dSt}(k)_{\infty}$, the ∞-topos associated with the Segal topos $\mathrm{dSt}(k)$.

25.5 Subobject classifiers

First, a bit of definitions from [52]. If $\mathcal{X}$ is an ∞-category with pullbacks, S is a class of morphisms in $\mathcal{X}$, and $\mathcal{O}_{\mathcal{X}} = \mathrm{Fun}(\Delta^1, \mathcal{X})$, then we denote by $\mathcal{O}_{\mathcal{X}}^{(S)}$ the subcategory of $\mathcal{O}_{\mathcal{X}}$ whose objects are elements of S and morphisms are pullback diagrams. For $\mathcal{X} \in \mathrm{Cat}_{\infty}$ with pullbacks and a collection S of morphisms in $\mathcal{X}$ stable under pullback, we say a morphism $f : X \to Y$ classifies S if it is a final object of $\mathcal{O}_{\mathcal{X}}^{(S)}$, which can equivalently be stated by saying that the object Y classifies S. If for S, we take the class of monomorphisms in $\mathcal{X}$, i. e., those morphisms $W \to Z$ that are (-1)-truncated objects of $\mathcal{X}_{/Z}$, then we define a *subobject classifier* of an ∞-category $\mathcal{X}$ with pullbacks to be an object Ω that classifies the collection of all monomorphisms in $\mathcal{X}$ or, equivalently, a final object of $\mathcal{O}_{\mathcal{X}}^{(\mathrm{mono})}$. Now every ∞-topos has a subobject classifier, and hence so does $\mathrm{dSt}(k)_\infty$.

The importance of such a concept can be traced back to the original motivation for choosing $\mathrm{dSt}(k)$ in the first place; stacks model realizations of natural laws, i. e., natural phenomena. Monomorphisms into a given stack can be seen as constituents of such a stack, and hence $F \in \mathrm{dSt}(k)$ can be seen as being characterized by monomorphisms in $\mathrm{dSt}(k)/F$. In terms of perception, this corresponds to considering the faithful perception $\mathrm{Hom}_{\mathrm{mono}}(-, F)$ of F, spanned by those morphisms into F that are monomorphisms. This has the advantage of providing a functorial presentation of the class of monomorphisms into a given object. If one has a subobject classifier Ω on the ∞-topos side of things, by its very definition it can be regarded as an object out of which other objects are being constructed, and hence this would define a generation. Formally, if we still denote by F the object of $\mathrm{dSt}(k)_\infty$ corresponding to the object $F \in \mathrm{dSt}(k)$, then for every monomorphism $G \to F$ in $\mathrm{dSt}(k)_\infty$, we have a unique Cartesian diagram

$$\begin{array}{ccc} G & \longrightarrow & 1 \\ \downarrow & & \downarrow \\ F & \longrightarrow & \Omega \end{array}$$

where 1 is a universal subobject of Ω.

For later purposes, consider $\mathrm{Fun}(\mathcal{X},\mathcal{U})$ for $\mathcal{U}$ ∞-topoi $\mathcal{X}$. In particular, $\mathcal{U}$ is a left exact localization of $\mathrm{Pr}(\mathcal{C})$ for some small ∞-category $\mathcal{C}$. From [56] we know $\mathrm{Pr}(\mathcal{C})$ is an ∞-topos. Denote by $\pi : \mathrm{Pr}(\mathcal{C}) \to \mathcal{U}$ the left exact left adjoint of this localization. It induces a localization $\pi_* : \mathrm{Fun}(\mathcal{X}, \mathrm{Pr}(\mathcal{C})) \rightleftarrows \mathrm{Fun}(\mathcal{X},\mathcal{U}) : i_*$, again left exact. Now

$$\begin{aligned} \mathrm{Fun}(\mathcal{X}, \mathrm{Pr}(\mathcal{C})) &= \mathrm{Fun}(\mathcal{X}, \mathrm{Fun}(\mathcal{C}^{\mathrm{op}}, \mathcal{S})) \\ &\simeq \mathrm{Fun}(\mathcal{X} \times \mathcal{C}^{\mathrm{op}}, \mathcal{S}) \end{aligned}$$

and the latter is an ∞-topos, hence so is $\mathrm{Fun}(\mathcal{X},\mathcal{U})$ by localization, and in particular so is $\mathrm{Fun}(\mathcal{X}, \mathcal{X})$, where $\mathcal{X} = \mathrm{dSt}(k)_\infty$ for us. This mirrors the object $\mathbb{R}\underline{\mathrm{Hom}}^*_{\mathrm{SeT}}(\mathrm{dSt}(k), \mathrm{dSt}(k)) \in \mathrm{SeT}$ that provided a comprehensive, functorial presentation of $\mathrm{dSt}(k)$.

Coming back to $\mathrm{dSt}(k)_\infty$, which has a subobject classifier Ω, one may ask whether this object itself is universal among all such subobject classifiers in the category of ∞-topoi. Since we are dealing with ∞-topoi, we consider geometric morphisms between them. Let $\mathcal{E}$ be an ∞-topos, and let $\pi : \mathcal{E} \to \mathrm{dSt}(k)_\infty$ be the left exact left adjoint of a geometric morphism from $\mathrm{dSt}(k)_\infty$ to $\mathcal{E}$ with right adjoint i. The ∞-topos $\mathcal{E}$ itself has a subobject classifier G of its own, with $\pi G \in \mathrm{dSt}(k)_\infty$, such that

$$\begin{array}{ccc} S & \longrightarrow & 1 \\ \downarrow & & \downarrow \\ \pi G & \dashrightarrow & \Omega \end{array}$$

for a monomorphism $S \to \pi G$. Now $\pi : \mathcal{E} \rightleftarrows \mathrm{dSt}(k)_\infty : i$ induces

$$\pi_* : \mathrm{Fun}(\Delta^1, \mathcal{E}) \rightleftarrows \mathrm{Fun}(\Delta^1, \mathrm{dSt}(k)_\infty) : i_*.$$

From [52], for an ∞-topos $\mathcal{X}$ and any ∞-category $\mathcal{C}$, $\mathrm{Fun}(\mathcal{C}, \mathcal{X})$ is an ∞-topos as we argued in the previous paragraph, so $\mathcal{O}_\mathcal{X}$ is an ∞-topos. Since π_* is left exact, we have again a geometric morphism

$$\pi_* : \mathcal{O}_\mathcal{E} \rightleftarrows \mathcal{O}_{\mathrm{dSt}(k)_\infty} : i_*.$$

It follows that π_* carries monomorphisms in $\mathcal{E}$ into monomorphisms in $\mathrm{dSt}(k)_\infty$ [52]. Further, π preserves pullbacks since it is left exact, so π_* preserves morphisms between monomorphisms, and hence we have a well-defined morphism

$$\pi_* : \mathcal{O}_\mathcal{E}^{(\mathrm{mono})} \to \mathcal{O}_{\mathrm{dSt}(k)_\infty}^{(\mathrm{mono})}$$

which is left exact and hence preserves final objects and, in particular, the subobject classifiers, so $\pi G \xrightarrow{\simeq} \Omega$. Indeed, for $K \in \mathrm{Set}_\Delta$, an object of K is final if it is a final object of hK, where one defines the homotopy category hK of K [52] as the homotopy category $h\mathfrak{C}[K]$ of the simplicial category $\mathfrak{C}[K]$. In this sense, πG and Ω are isomorphic in $h\mathfrak{C}[\mathcal{O}_{\mathrm{dSt}(k)_\infty}^{(\mathrm{mono})}] = h\mathcal{O}_{\mathrm{dSt}(k)_\infty}^{(\mathrm{mono})}$ and hence equivalent in $\mathrm{dSt}(k)_\infty$. To summarize, if $\pi : \mathcal{E} \rightleftarrows \mathrm{dSt}(k)_\infty : i$ is a geometric morphism of ∞-topoi, then the subobject classifier of $\mathcal{E}$ maps to Ω under π.

Now the question is what would be a universal such ∞-topos $\mathcal{E}$. Recall that if one denotes by $\mathcal{RT}\mathrm{op}$ the category of ∞-topoi and right adjoints of geometric morphisms as morphisms between ∞-topoi, then $\mathrm{Fun}^*(\mathcal{S}, \mathcal{X})$ is contractible for any ∞-topos $\mathcal{X}$, which means exactly that $\mathcal{S}$ is a final object of $\mathcal{RT}\mathrm{op}$. Consequently, there is a geometric morphism $\mathrm{dSt}(k)_\infty \to \mathcal{S}$ with a left exact left adjoint, and by the same argument as above, if $\pi : \mathcal{S} \to \mathrm{dSt}(k)_\infty$ is this left exact left adjoint, then it induces a left exact morphism $\mathcal{O}_\mathcal{S}^{(\mathrm{mono})} \to \mathcal{O}_{\mathrm{dSt}(k)_\infty}^{(\mathrm{mono})}$ mapping a final object to a final object in a weak sense, which would make $\mathcal{S}$ a universal ∞-topos in this regard, since $\mathcal{S}$ is a final object of $\mathcal{RT}\mathrm{op}$, consequently making its subobject classifier the sought after universal subobject classifier.

26 Segal topoi and the universality of physical laws

26.1 Introduction

Lurie [52] showed that starting from the usual Grothendieck construction whereby one obtains a category $\mathcal{C}_f \xrightarrow{p} \mathcal{C}$ cofibered in groupoids over $\mathcal{C}$ from a functor $f : \mathcal{C} \to \mathrm{Gpd}$ valued in groupoids, one gets, via the simplicial nerve functor, a left fibration in Set_Δ: $N(\mathcal{C}_f) \xrightarrow{Np} N(\mathcal{C})$. Generalizing this to the ∞-categorical setting, in particular, by replacing Gpd by Set_Δ, Lurie was led to proving that one has a Quillen adjunction

$$(\mathrm{Set}_\Delta)_{/K} \rightleftarrows (\mathrm{Set}_\Delta)^{\mathcal{C}}$$

for $K \in \mathrm{Set}_\Delta$ (generalizing $N(\mathcal{C})$), $\mathcal{C} \in \mathrm{Cat}_\Delta$, and a morphism $\mathfrak{C}[K] \to \mathcal{C}^{\mathrm{op}}$ in Cat_Δ. Here $(\mathrm{Set}_\Delta)_{/K}$ is endowed with the contravariant model structure (see Section 26.2), and the projective model structure is put on $(\mathrm{Set}_\Delta)^{\mathcal{C}}$. Further, if $\mathfrak{C}[K] \to \mathcal{C}^{\mathrm{op}}$ is an equivalence of simplicial categories, then the above equivalence is a Quillen equivalence.

Since our aim is to apply this formalism to fibered categories, it is preferable as pointed out in [52] to work with marked simplices instead. The bridge between both formalisms is warranted by the fact proved in [52] that we have a Quillen equivalence

$$(\mathrm{Set}_\Delta)_{/K} \rightleftarrows (\mathrm{Set}_\Delta^+)_{/K}$$

for all $K \in \mathrm{Set}_\Delta$, where we have put the contravariant model structure on $(\mathrm{Set}_\Delta)_{/K}$ and the Cartesian model structure on $(\mathrm{Set}_\Delta^+)_{/K}$ (see Section 26.4). In the marked setting, we have a Quillen adjunction

$$(\mathrm{Set}_\Delta^+)_{/K} \rightleftarrows (\mathrm{Set}_\Delta^+)^{\mathcal{C}}$$

for every morphism $f : \mathfrak{C}[K] \to \mathcal{C}^{\mathrm{op}}$ in Cat_Δ. Further, if f is an equivalence, then this adjunction becomes a Quillen equivalence. We prove a partial converse to this result. Towards that end, we prove two results (where L denotes a simplicial localization [15, 74].

Proposition 26.1.1. *For $K \in \mathrm{Set}_\Delta$,*

$$L(\mathrm{Set}_\Delta^+)_{/K} \cong L(\mathrm{Set}_\Delta^+)^{\mathfrak{C}[K]^{\mathrm{op}}}.$$

Proposition 26.1.2. *Let $K \in \mathrm{Set}_\Delta$ and $\mathcal{C} \in \mathrm{Cat}_\Delta$. If a morphism $\mathfrak{C}[K]^{\mathrm{op}} \to \mathcal{C}$ induces an isomorphism $L(\mathrm{Set}_\Delta^+)^{\mathfrak{C}[K]^{\mathrm{op}}} \cong L(\mathrm{Set}_\Delta^+)^{\mathcal{C}}$ in* Ho (SePC), *then its induced localization $L\mathfrak{C}[K]^{\mathrm{op}} \to L\mathcal{C}$ is a local morphism in* SePC *relative to Segal categories of prestacks.*

When combined, these results provide us with the desired result:

Theorem 26.1.3. *If $L(\mathrm{Set}_\Delta^+)_{/K} \cong L(\mathrm{Set}_\Delta^+)^{\mathcal{C}}$ in* Ho (SePC) *for some morphism $\mathfrak{C}[K]^{\mathrm{op}} \to \mathcal{C}$, then $L\mathfrak{C}[K]^{\mathrm{op}} \sim_{\mathrm{SePrStck}} L\mathcal{C}$ in* SePC.

https://doi.org/10.1515/9783111334073-026

In this statement, $\mathcal{C} \sim_{\mathrm{SePrStck}} \mathcal{D}$ means the morphism $\mathcal{C} \to \mathcal{D}$ is local with respect to Segal categories of prestacks. From this perspective, $\mathcal{C}$ and $\mathcal{D}$ are considered to be equivalent relative to such Segal categories, hence the suggestive notation.

To go a little further, it is not a Segal topos of prestacks one should consider if one is interested in modeling natural phenomena. Rather, as argued in Chapters 24 and 25, it is really stacks we should work with. If we endow $L\mathfrak{C}[K]^{\mathrm{op}}$ with a Segal topology and consider morphisms of Segal sites $L\mathfrak{C}[K]^{\mathrm{op}} \to L\mathcal{C}$, taking Bousfield localizations of $L(\mathrm{Set}_\Delta^+)^{\mathcal{C}}$ and $L(\mathrm{Set}_\Delta^+)^{\mathfrak{C}[K]^{\mathrm{op}}}$ with respect to their respective hypercovers, we get isomorphic Segal topoi of stacks. In other terms, $L(\mathrm{Set}_\Delta^+)^{\mathfrak{C}[K]^{\mathrm{op}}} \cong L(\mathrm{Set}_\Delta^+)^{\mathcal{C}}$ implies getting isomorphic representations of natural phenomena after Bousfield localization, and the base Segal categories $L\mathfrak{C}[K]^{\mathrm{op}}$ and $L\mathcal{C}$ giving rise to such phenomena are equivalent relative to prestacks, i. e., correspond to the same intrinsic natural laws. This would point to a universality of physical laws within each class. Further observe that this allows us to consider phenomena in reference to K, via the isomorphism $L(\mathrm{Set}_\Delta^+)_{/K} \cong L(\mathrm{Set}_\Delta^+)^{\mathfrak{C}[K]^{\mathrm{op}}}$. Had we picked another simplicial set, we would have obtained another isomorphism class of phenomena with $L(\mathrm{Set}_\Delta^+)^{\mathfrak{C}[K]^{\mathrm{op}}}$ as representative and hence another homotopy class of base Segal categories with $L\mathfrak{C}[K]$ as representative. In other terms, we have a sort of relativity principle whereby the point of view (being K) dictates what are the physical laws that one should consider. The present work inscribes itself within the vision Grothendieck had as far back as [34], namely that Physics should occur in higher categories.

The foundational results, and statements pertaining to the adjunctions whose converse we prove, are all covered in full in [52] and are just included here for ease of reading. No originality is claimed. We just reorganize the presentation a bit for our purposes, and notations may vary only slightly. Our references for model category theory are standard [42, 40]. For Segal categories, we use [77, 78, 74, 66, 41]. For stacks, we use [76] and [79].

In Section 26.2, we discuss the contravariant model structure on $(\mathrm{Set}_\Delta)_{/K}$. In Section 26.3, we give the unmarked Quillen equivalence $(\mathrm{Set}_\Delta)_{/K} \rightleftarrows (\mathrm{Set}_\Delta)^{\mathcal{C}}$ of [52]. Before making the transition to the marked case, we introduce marked simplicial sets in Section 26.4 and present the Cartesian model structure on $(\mathrm{Set}_\Delta^+)_{/K}$ for $K \in \mathrm{Set}_\Delta$ as done in [52]. In Section 26.5, we give the marked Quillen equivalence of [52]: $(\mathrm{Set}_\Delta^+)_{/K} \rightleftarrows (\mathrm{Set}_\Delta^+)^{\mathcal{C}}$. In Section 26.6, we have to define $\mathrm{Top}^+ = L\mathrm{Set}_\Delta^+$ since we will use the strictification theorem $L(\mathrm{Set}_\Delta^+)^{\mathcal{C}} \cong \mathbb{R}\underline{\mathrm{Hom}}(L\mathcal{C}, L\mathrm{Set}_\Delta^+)$. Finally, in Section 26.7, we give the main results of this chapter, everything before, with the exception perhaps of Section 26.6, being initially covered elsewhere.

26.2 Contravariant model structure on $(\mathrm{Set}_\Delta)_{/K}$

For fixed $K \in \mathrm{Set}_\Delta$, one puts a model structure on $(\mathrm{Set}_\Delta)_{/K}$ that Lurie refers to as the contravariant model structure. To do so, we need a few definitions. Following [52], one defines the homotopy category of spaces by $\mathcal{H} = W^{-1}\mathrm{Set}_\Delta$, where W is the set of weak equivalences in Set_Δ. If $\mathcal{CG}$ denotes the category of compactly generated Hausdorff topological spaces, then one has a functor $[\ ] : \mathcal{CG} \to \mathcal{H}, X \mapsto [X]$, $[X]$ the homotopy class of X. One also has the geometric realization functor $|\ | : \mathrm{Set}_\Delta \to \mathcal{CG}$. Denote the composition by $\gamma = [\ ] \circ |\ | : \mathrm{Set}_\Delta \to \mathcal{H}$. Applying γ to each of the morphism spaces of a simplicial category $\mathcal{C}$ gives an $\mathcal{H}$-enriched category, which one denotes by $h\mathcal{C}$. For $K \in \mathrm{Set}_\Delta$, one defines the homotopy category hK of K by $hK = h\mathfrak{C}[K]$, where $\mathfrak{C}[K] \in \mathrm{Cat}_\Delta$ is defined in [52] (we will not need its construction). To put things in perspective, recall that if one endows Set_Δ with the Joyal model structure, with monomorphisms as cofibrations, and weak equivalences those maps $K \to K'$ in Set_Δ inducing equivalences $\mathfrak{C}[K] \to \mathfrak{C}[K']$ in Cat_Δ, then one has a Quillen equivalence $\mathfrak{C} : \mathrm{Set}_\Delta \rightleftarrows \mathrm{Cat}_\Delta : N$, where N is the simplicial nerve functor [11]. One says $f : K \to L$ in Set_Δ is a *categorical equivalence* if $hf : hK \to hL$ is an equivalence of $\mathcal{H}$-enriched categories. Now to define covariant equivalences, recall the join of two simplicial sets K and L: if I is a nonempty finite linearly ordered set, then

$$K \star L = \coprod_{I=J\cup J'} K(J) \times L(J').$$

In particular, one defines the left cone $K^{\triangleleft}$ as $\Delta^0 \star K$ (and the right cone as $K^{\triangleright} = K \star \Delta^0$). A map $X \to Y$ in $(\mathrm{Set}_\Delta)_{/K}$ is said to be a *covariant equivalence* if it induces a categorical equivalence

$$X^{\triangleleft} \coprod_X K \to Y^{\triangleleft} \coprod_Y K.$$

One says a map in $(\mathrm{Set}_\Delta)_{/K}$ is a *covariant cofibration* if it is a monomorphism of simplicial sets, and it is a *covariant fibration* if it has the right lifting property with respect to weak covariant cofibrations. Covariant fibrations, cofibrations, and equivalences determine a left proper combinatorial model structure on $(\mathrm{Set}_\Delta)_{/K}$ called the *covariant model structure* [52]. The *contravariant model structure* on $(\mathrm{Set}_\Delta)_{/K}$ is obtained by considering the same cofibrations; contravariant equivalences f are those maps such that f^{op} is a covariant equivalence in $(\mathrm{Set}_\Delta)_{/K^{\mathrm{op}}}$, and contravariant fibrations are those maps f such that f^{op} is a covariant fibration in $(\mathrm{Set}_\Delta)_{/K^{\mathrm{op}}}$.

26.3 Unmarked Quillen equivalence

The construction that follows is done in full in [52] and is reproduced here for completeness, albeit with different notations. Let $K \in \mathrm{Set}_\Delta$, $\mathcal{C} \in \mathrm{Cat}_\Delta$, and $f : \mathfrak{C}[K] \to \mathcal{C}^{\mathrm{op}}$. Let $*$ be the cone point of $X^{\triangleright}$, $X \in (\mathrm{Set}_\Delta)_{/K}$. By the definition of X one has an induced map $\mathfrak{C}[X] \to \mathfrak{C}[K]$. Consider the pushout diagram

$$\begin{array}{ccccc} \mathfrak{C}[X] & \longrightarrow & \mathfrak{C}[K] & \xrightarrow{f} & \mathcal{C}^{\mathrm{op}} \\ \downarrow & & & & \downarrow \\ \mathfrak{C}[X^{\triangleright}] & & \longrightarrow & & \Gamma \end{array}$$

where

$$\Gamma = \Gamma[K, \mathcal{C}, f, X] = \mathfrak{C}[X^{\triangleright}] \coprod_{\mathfrak{C}[X]} \mathcal{C}^{\mathrm{op}}$$

which can be seen as a correspondence

$$\Gamma[K, \mathcal{C}, f, X] = \mathrm{Corr}[K, f, X](\mathcal{C}^{\mathrm{op}} \to \{*\})$$

which itself can be viewed as a functor

$$\begin{aligned} \mathrm{St}_f X : \mathcal{C} &\to \mathrm{Set}_\Delta, \\ C &\mapsto \mathrm{Map}_\Gamma(C, *) \end{aligned}$$

an object of $(\mathrm{Set}_\Delta)^{\mathcal{C}}$ for each $X \in (\mathrm{Set}_\Delta)_{/K}$, giving rise to a *straightening functor*

$$\begin{aligned} \mathrm{St}_f : (\mathrm{Set}_\Delta)_{/K} &\to (\mathrm{Set}_\Delta)^{\mathcal{C}}, \\ X &\mapsto \mathrm{St}_f X. \end{aligned}$$

There exists an adjunction

$$\mathrm{St}_f : (\mathrm{Set}_\Delta)_{/K} \rightleftarrows (\mathrm{Set}_\Delta)^{\mathcal{C}} : \mathrm{Un}_f$$

which turns out to be a Quillen adjunction if one endows $(\mathrm{Set}_\Delta)_{/K}$ with the contravariant model structure and $(\mathrm{Set}_\Delta)^{\mathcal{C}}$ with the projective model structure. Further, if f is an equivalence of simplicial categories, then $\mathrm{St}_f \dashv \mathrm{Un}_f$ is a Quillen equivalence. All this is proved in [52].

26.4 Marked simplicial sets

As observed in [52], if one wants to regard objects $X \in (\mathrm{Set}_\Delta)_{/K}$ functorially, then one is led to regarding $X \to K$ as a Cartesian fibration, fibrant object in $(\mathrm{Set}_\Delta)_{/K}$ if this category is endowed with a model structure other than the contravariant model structure, one which necessitates considering marked simplicial sets, which are essentially simplicial sets with a collection of marked edges. Given the interpretation we want to draw from $(\mathrm{Set}_\Delta)_{/K}$, we are led to considering this new model structure, called the *Cartesian model structure*, which builds up on the one we introduced prior.

As proved in [52], we have a Quillen equivalence

$$(\mathrm{Set}_\Delta)_{/K} \rightleftarrows (\mathrm{Set}_\Delta^+)_{/K}$$

for all $K \in \mathrm{Set}_\Delta$, as discussed in the introduction. We can see this result as a first step towards generalizing the work in the unmarked case to the marked one.

For $K, S \in \mathrm{Set}_\Delta$ and a map $p : S \to K$ in Set_Δ, there exists some $K_{/p} \in \mathrm{Set}_\Delta$ defined by

$$\mathrm{Hom}_{\mathrm{Set}_\Delta}(L, K_{/p}) = \mathrm{Hom}(L \star S, K)$$

for all $L \in \mathrm{Set}_\Delta$, where on the right-hand side, we consider only those maps $L \star S \to K$ in Set_Δ whose restriction to S is just p. For $\mathcal{C} \in \mathrm{Cat}_\infty$, if $p : \Delta^n \to \mathcal{C}$ classifies an n-simplex σ of $\mathcal{C}$, then write $\mathcal{C}_{/\sigma}$ for $\mathcal{C}_{/p}$. In particular, if $p = [0] = \Delta^0 \to \mathcal{C}$ is a point x of $\mathcal{C}$, then write $\mathcal{C}_{/p} = \mathcal{C}_{/x}$, and if $p : \Delta^1 \to \mathcal{C}$ is a morphism $f : x \to y$ in $\mathcal{C}$, then write $\mathcal{C}_{/p} = \mathcal{C}_{/f}$.

With those notations, if $p : X \to K$ is an inner fibration of simplicial sets, then an edge $u : x \to y$ of X is said to be *p-Cartesian* if the induced map

$$X_{/u} \to X_{/y} \times_{K_{/p(y)}} K_{/p(u)}$$

is a trivial Kan fibration.

Now one says a map $p : X \to K$ in Set_Δ is a *Cartesian fibration* if it is an inner fibration and if, in addition, for all $u : x \to y$ in K and any vertex y_0 of X such that $p(y_0) = y$, there is a p-Cartesian edge $u_0 : x_0 \to y_0$ of X such that $p(u_0) = u$.

One defines a *marked simplicial set* to be a simplicial set X along with a set $\mathcal{E}$ of edges of X (called marked edges) that contains all degenerate edges. A map $f : (X, \mathcal{E}) \to (X', \mathcal{E}')$ of marked simplicial sets is just a map $f : X \to X'$ of simplicial sets such that $f(\mathcal{E}) \subseteq \mathcal{E}'$. One denotes the category of marked simplicial sets by Set_Δ^+. If $K \in \mathrm{Set}_\Delta$, then we denote by $K^\forall$ the marked simplicial set K for which all its edges are marked. In particular, for $K \in \mathrm{Set}_\Delta$, $(\mathrm{Set}_\Delta^+)_{/K^\forall}$ is denoted $(\mathrm{Set}_\Delta^+)_{/K}$. If $p : X \to K$ is a Cartesian fibration in Set_Δ, then we denote by X^{cart} the marked simplicial set X where we have kept as marked edges its p-Cartesian edges only.

Now Set_Δ^+ is Cartesian closed, so for all $X, Y \in \mathrm{Set}_\Delta^+$, there is an internal object Y^X. We denote by $\underline{Y^X}$ the underlying simplicial set of Y^X and by $\mathrm{Map}^\forall(X, Y)$ the marked simplicial set $(\underline{Y^X})^\forall$. As proved in [52], we have that for $p : X \to Y \in (\mathrm{Set}_\Delta^+)_{/K}$, the following two statements are equivalent: for any Cartesian fibration $L \to K$, the induced map

$$\underline{(L^{\mathrm{cart}})^Y} \to \underline{(L^{\mathrm{cart}})^X}$$

is an equivalence in Cat_∞, and the induced map

$$\mathrm{Map}^\forall(Y, L^{\mathrm{cart}}) \to \mathrm{Map}^\forall(X, L^{\mathrm{cart}})$$

is a homotopy equivalence in Kan, the category of Kan complexes. Finally, a map $p : X \to Y$ in $(\mathrm{Set}_\Delta^+)_{/K}$ satisfying those equivalent conditions will be referred to as a *Cartesian equivalence.*

There is a left proper combinatorial model structure on $(\mathrm{Set}_\Delta^+)_{/K}$, called the *Cartesian model structure* [52], whose cofibrations are those maps in $(\mathrm{Set}_\Delta^+)_{/K}$ whose underlying maps on the underlying simplicial sets are cofibrations, weak equivalences are Cartesian equivalences, and fibrations are as usual those maps that have the right lifting property with respect to weak cofibrations. Observe, as this was our aim, that with this model structure, $X \in (\mathrm{Set}_\Delta^+)_{/K}$ is fibrant if and only if it is equivalent to some L^{cart} for a Cartesian fibration $L \to K$. Note that $(\mathrm{Set}_\Delta^+)_{/K}$ is a simplicial model category.

26.5 Marked Quillen equivalence

We first have to introduce a bit of notation. For $K \in \mathrm{Set}_\Delta$, $X \in (\mathrm{Set}_\Delta)_{/K}$, $f : \mathfrak{C}[K] \to \mathcal{C}^{\mathrm{op}}$, and $\sigma : \Delta^n \to \mathrm{Map}_{\mathcal{C}^{\mathrm{op}}}(C, D)$, denote by σ^* the following induced map:

$$\mathrm{St}_f X(D)_n = \mathrm{Map}_\Gamma(D, *)_n \to \mathrm{Map}_\Gamma(C, *)_n = \mathrm{St}_f X(C)_n$$

where

$$\Gamma = \mathfrak{C}[X^{\triangleright}] \coprod_{\mathfrak{C}[X]} \mathcal{C}^{\mathrm{op}}$$

as introduced in Section 26.3. Indeed, if we have $C \to D \in \mathcal{C}^{\mathrm{op}}$, then this induces $\mathrm{Map}_\Gamma(C, *) \leftarrow \mathrm{Map}_\Gamma(D, *)$. In particular, for $\sigma \in \mathrm{Map}_{\mathcal{C}^{\mathrm{op}}}(C, D)_n$, one has the induced map σ^* as defined above.

Now let $c : \Delta^0 \to X$ for $X \in (\mathrm{Set}_\Delta)_{/K}$. Then consider

$$\begin{array}{ccccc}
\Delta^0 & \xrightarrow{c} & X & \longrightarrow & K \\
 & & \downarrow & & \downarrow \\
 & & \mathfrak{C}[X] & \longrightarrow & \mathfrak{C}[K] & \xrightarrow[f]{} & \mathcal{C}^{\mathrm{op}}.
\end{array}$$

Let C be the object of $\mathcal{C}$ thus obtained. One can extend this map as follows:

$$\begin{array}{ccc}
c \star \mathrm{id}_{\Delta^0} : \Delta^1 & \longrightarrow & X^{\triangleright} \\
 & & \downarrow \\
 & & \mathfrak{C}[X^{\triangleright}] \longrightarrow \Gamma = \{\mathcal{C}^{\mathrm{op}} \to *\}
\end{array}$$

where we have morally identified Γ with the set of maps $\mathcal{C}^{\mathrm{op}} \to *$ to emphasize that $c \star \mathrm{id}$ corresponds to a map $C \to *$, an element of $\mathrm{St}_f X(C)$, which we denote by $\tilde{c}$.

Now consider an edge $u : c \to d$ of X as in

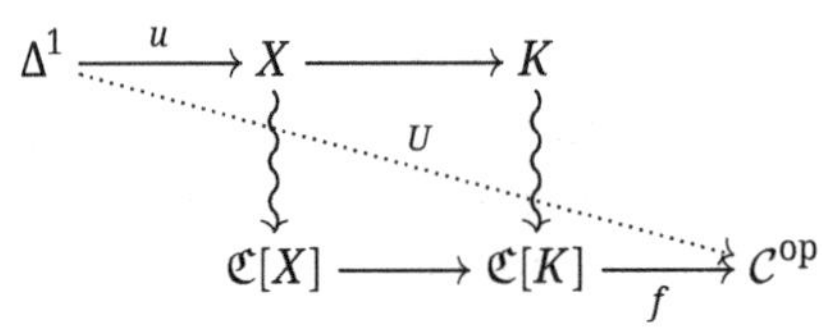

giving rise to $U : C \to D$ in $\mathcal{C}^{\mathrm{op}}$. One can extend this to

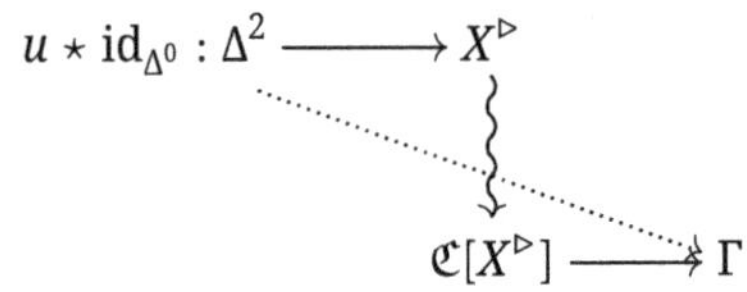

where the dotted map corresponds to

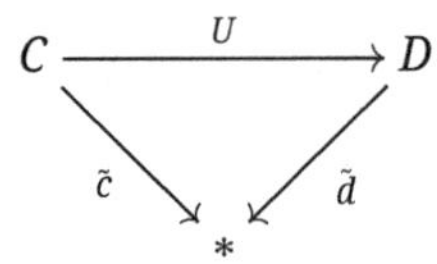

along with a map $\tilde{u} : \tilde{c} \to \tilde{d} \circ U = U^*\tilde{d}$.

Now for an edge $v : d \to e$ of X, one has a corresponding diagram as above:

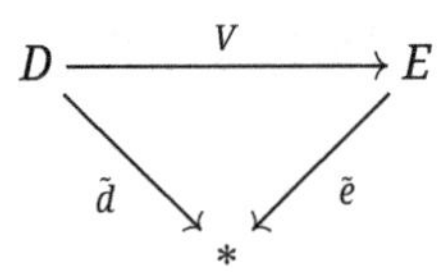

in $\mathrm{St}_f X(D)$, along with a map $\tilde{v} : \tilde{d} \to \tilde{e} \circ V = V^*\tilde{e}$, i. e., $\tilde{v} \in \mathrm{St}_f X(D)_1$. For $\sigma : \Delta^1 \to \mathrm{Map}_{\mathcal{C}^{\mathrm{op}}}(C, D)$, one has an induced map $\sigma^* : \mathrm{St}_f X(D)_1 \to \mathrm{St}_f X(C)_1$. Define $\mathcal{E}_f(C)$ to be the collection of all $\sigma^* \mathrm{St}_f X(D)_1$, that is, the collection of all pullbacks $\sigma^*\tilde{v}$ of all those $\tilde{v}$ in $\mathrm{St}_f X(D)_1$ as defined above for all edges $v : d \to e$ in X and all $\sigma : \Delta^1 \to \mathrm{Map}_{\mathcal{C}^{\mathrm{op}}}(C, D)$. If we limit ourselves to marked edges of $(X, \mathcal{E}) \in \mathrm{Set}_\Delta^+$, then this leads us to defining

$$\begin{aligned} \mathrm{St}_f^+(X, \mathcal{E}) : \mathcal{C} &\to \mathrm{Set}_\Delta^+, \\ C &\mapsto (\mathrm{St}_f X(C), \mathcal{E}_f(C)). \end{aligned}$$

There exists a right adjoint Un_f^+ to St_f^+:

$$\mathrm{St}_f^+ : (\mathrm{Set}_\Delta^+)_{/K} \rightleftarrows (\mathrm{Set}_\Delta^+)^{\mathcal{C}} : \mathrm{Un}_f^+ \tag{26.1}$$

a Quillen adjunction, where $(\mathrm{Set}_\Delta^+)_{/K}$ is endowed with the Cartesian model structure and $(\mathrm{Set}_\Delta^+)^{\mathcal{C}}$ with the projective model structure. This adjunction is furthermore a Quillen equivalence if $f : \mathfrak{C}[K] \to \mathcal{C}^{\mathrm{op}}$ is an equivalence of simplicial categories [52].

26.6 Simplicial localization of Set_Δ^+

With a view towards taking a localization of the above Quillen equivalence (26.1), we need a notion of localization of Set_Δ^+. In [77, 74], Toen and Vezzosi denote the simplicial localization [15] $L\mathrm{Set}_\Delta$ of Set_Δ by Top. We define its marked counterpart. To do so, we start from [15] and adapt their construction to Set_Δ^+.

The free category on Set_Δ^+ is the category $F\mathrm{Set}_\Delta^+ \in \mathrm{Cat}_\Delta$ with a generator $F\phi$ for every nonidentity map $\phi \in \mathrm{Set}_\Delta^+$. This construction comes with two functors $D : F\mathrm{Set}_\Delta^+ \to \mathrm{Set}_\Delta^+$ and $U : F\mathrm{Set}_\Delta^+ \to F^2\mathrm{Set}_\Delta^+$ satisfying the usual comonad conditions. With this in hand, we can define the standard free resolution of Set_Δ^+ as the simplicial object $F_*\mathrm{Set}_\Delta^+$ with $F_k\mathrm{Set}_\Delta^+ = F^{k+1}\mathrm{Set}_\Delta^+ \in \mathrm{Cat}_\Delta$ with face and degeneracy operators given by

$$d_i : F^{k+1}\mathrm{Set}_\Delta^+ \xrightarrow{F^iDF^{k-i}} F^k\mathrm{Set}_\Delta^+$$

and

$$s_i : F^{k+1}\mathrm{Set}_\Delta^+ \xrightarrow{F^iUF^{k-i}} F^{k+2}\mathrm{Set}_\Delta^+.$$

Observe that we have a weak equivalence of bisimplicial categories

$$F_*\mathrm{Set}_\Delta^+ \to \mathrm{Set}_\Delta^+$$

given by

$$F^{k+1}\mathrm{Set}_\Delta^+ \xrightarrow{D^{k+1}} \mathrm{Set}_\Delta^+.$$

To localize $F_*\mathrm{Set}_\Delta^+$, we need a notion of weak equivalence on Set_Δ^+. Recall from [28] the definition of the simplicial category $\Delta \downarrow K$ of a simplicial set K: its objects are maps $\sigma : \Delta^n \to K$ with arrows being commutative diagrams

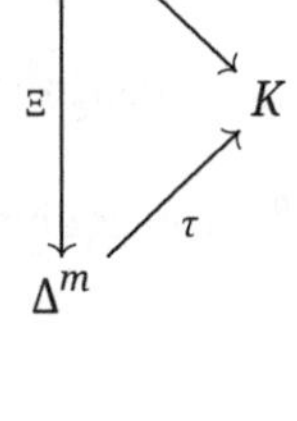

which gives an isomorphism $K \cong \operatorname{colim} \Delta^n$, where the colimit is taken over all morphisms $\Delta^n \to K$ in $\Delta \downarrow K$. This enables one to define the geometric realization of K as $|K| = \operatorname{colim} |\Delta^n| \in \mathrm{Top}$, where the colimit is over the same maps $\Delta^n \to K$. Now for $K^+ = (K, \mathcal{E}) \in \mathrm{Set}_\Delta^+$, define the simplicial category $\Delta \downarrow K^+$ as the category whose objects are maps $\sigma \times \sigma' : \Delta^n \times \Delta^i \to K \times \mathcal{E}$, and morphisms are commutative diagrams

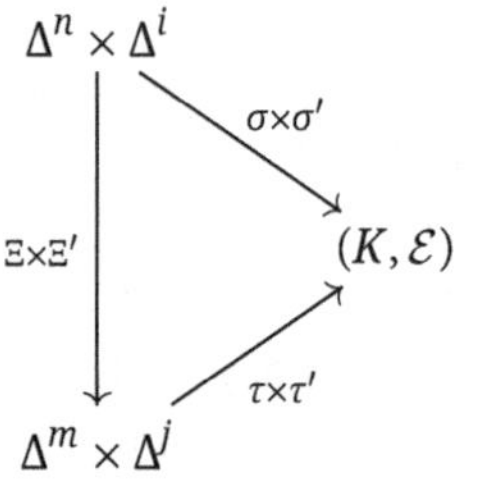

for $i, j = 0, 1$, which allows us to define the geometric realization of K^+ as

$$|(K, \mathcal{E})| = \operatorname{colim} |\Delta^n| \times \operatorname{colim} |\Delta^i| = |K| \times |\mathcal{E}|.$$

Weak equivalences in Set_Δ^+ between marked simplicial sets are then those maps that induce weak homotopy equivalences between their respective realizations. Let W^+ be the set of weak equivalences in Set_Δ^+.

With this notion of equivalence, we have a functor

$$F_* \mathrm{Set}_\Delta^+ \to F_* \mathrm{Set}_\Delta^+[(F_* W^+)^{-1}]$$

defined levelwise. We then define

$$\mathrm{Top}^+ = L(\mathrm{Set}_\Delta^+) = \operatorname{diag} F_* \mathrm{Set}_\Delta^+[(F_* W^+)^{-1}] \in \mathrm{Cat}_\Delta$$

following the localization of simplicial categories as done in [15].

26.7 Equivalence of models for natural phenomena

26.7.1 Isomorphism of Segal categories of prestacks

Applying the Quillen equivalence (26.1) to the case where $\mathcal{C} = \mathfrak{C}[K]^{\mathrm{op}}$, one obtains a Quillen equivalence

$$\mathrm{St}^+ : (\mathrm{Set}_\Delta^+)_{/K} \rightleftarrows (\mathrm{Set}_\Delta^+)^{\mathfrak{C}[K]^{\mathrm{op}}} : \mathrm{Un}^+.$$

In [52], it was proved that if $\mathfrak{C}[K] \to \mathcal{C}^{\mathrm{op}}$ is an equivalence in Cat_Δ, then one has a Quillen equivalence

$$(\mathrm{Set}_\Delta^+)^{\mathfrak{C}[K]^{\mathrm{op}}} \rightleftarrows (\mathrm{Set}_\Delta^+)^{\mathcal{C}}.$$

We wish to prove the converse, namely that if we have such an equivalence, then the two simplicial categories we started from are equivalent in a sense to be specified. We do this after simplicial localization.

According to the strictification theorem [41, 74, 76], we have an equivalence of Segal categories in SePC

$$L(\mathrm{Set}_\Delta^+)^{\mathcal{C}} \simeq \mathbb{R}\underline{\mathrm{Hom}}(L\mathcal{C}, \mathrm{Top}^+) \tag{26.2}$$

and hence an isomorphism in Ho (SePC). We use the isomorphism since Segal categories are objects of Ho (SePC). Here $\mathbb{R}\underline{\mathrm{Hom}}$ is the derived internal Hom in Ho (SePC) with $\underline{\mathrm{Hom}}$ the internal Hom in SePC, the category of Segal pre-categories (see [77] for notations). Suppose now $u : L\mathcal{C}' \to L\mathcal{C}$ induces an equivalence $L(\mathrm{Set}_\Delta^+)^{\mathcal{C}} \simeq L(\mathrm{Set}_\Delta^+)^{\mathcal{C}'}$. We show that this implies $L\mathcal{C} \simeq L\mathcal{C}'$ in SePC in a certain sense. To do so, we use the following result [40]: if $M \in \mathrm{Cat}_\Delta$ and $f : X \to Y$ is a morphism in M, where both X and Y are cofibrant, then f is a weak equivalence if and only if for any fibrant $Z \in M$, the induced map $f^* : \mathrm{Map}(Y, Z) \to \mathrm{Map}(X, Z)$ is a weak equivalence in Set_Δ. Here SePC is a simplicial model category to which we apply this result. Note that morphisms of Segal categories are defined via the derived internal hom $\mathbb{R}\underline{\mathrm{Hom}}$ in Ho (SePC), whereas we denote by Map the mapping space object of SePC. On the one hand, the supposed equivalence $L(\mathrm{Set}_\Delta^+)^{\mathcal{C}} \simeq L(\mathrm{Set}_\Delta^+)^{\mathcal{C}'}$ in SePC induces an isomorphism in Ho (SePC). On the other hand, (26.2) above gives us another isomorphism in Ho (SePC). Combining both, we have

$$\mathbb{R}\underline{\mathrm{Hom}}(L\mathcal{C}, \mathrm{Top}^+) \cong \mathbb{R}\underline{\mathrm{Hom}}(L\mathcal{C}', \mathrm{Top}^+)$$

which means that the Segal categories $L\mathcal{C}$ and $L\mathcal{C}'$ produce isomorphic Segal categories of prestacks in Ho (SePC). We can say more. We have

$$\begin{array}{ccc} \mathbb{R}\underline{\mathrm{Hom}}(L\mathcal{C}, \mathrm{Top}^+) & \xrightarrow[\mathrm{Ho\,(SePC)}]{\cong} & \mathbb{R}\underline{\mathrm{Hom}}(L\mathcal{C}', \mathrm{Top}^+) \\ \| & & \| \\ \underline{\mathrm{Hom}}(QL\mathcal{C}, R\mathrm{Top}^+) & \dashrightarrow & \underline{\mathrm{Hom}}(QL\mathcal{C}', R\mathrm{Top}^+) \end{array}$$

where we have used the definition of the derived Hom $\mathbb{R}\underline{\mathrm{Hom}}$ and the fact that Segal categories are cofibrant in SePC. Hence we have an equivalence in SePC

$$\underline{\mathrm{Hom}}(QL\mathcal{C}, R\mathrm{Top}^+) \simeq \underline{\mathrm{Hom}}(QL\mathcal{C}', R\mathrm{Top}^+)$$

which allows us to write, for $D \in \mathrm{SePC}$,

$$\begin{array}{ccc}
\mathrm{Map}(QD, \underline{\mathrm{Hom}}(QLC, R\mathrm{Top}^+)) & \xrightarrow{\simeq} & \mathrm{Map}(QD, \underline{\mathrm{Hom}}(QLC', R\mathrm{Top}^+)) \\
\downarrow \simeq & & \downarrow \simeq \\
\mathrm{Map}(QD \times QLC, R\mathrm{Top}^+) & & \mathrm{Map}(QD \times QLC', R\mathrm{Top}^+) \\
\downarrow \simeq & & \downarrow \simeq \\
\mathrm{Map}(QLC, \underline{\mathrm{Hom}}(QD, R\mathrm{Top}^+)) & \dashrightarrow & \mathrm{Map}(QLC', \underline{\mathrm{Hom}}(QD, R\mathrm{Top}^+)).
\end{array}$$

Since all objects are cofibrant in SePC, and $u : LC' \to LC$ induces the dotted map above, by the 2–3 property this reads

$$\mathrm{Map}(LC, \hat{E}) \xrightarrow{\simeq} \mathrm{Map}(LC', \hat{E})$$

for any Segal category of prestacks $\hat{E}$. This means $u : LC' \to LC$ is local with respect to such objects. Note that this is different from the original definition of local object in [40] and is just meant here to indicate that relative to Segal categories of prestacks, the Segal categories LC and LC' are considered to be equivalent. We denote this by

$$LC' \underset{\mathrm{SePrStck}}{\sim} LC.$$

Applying this to the case $C' = \mathfrak{C}[K]^{\mathrm{op}}$ in particular, we have the following:

Proposition 26.7.1. *If the morphism* $\mathfrak{C}[K]^{\mathrm{op}} \to C$ *induces* $L(\mathrm{Set}_\Delta^+)^{\mathfrak{C}[K]^{\mathrm{op}}} \simeq L(\mathrm{Set}_\Delta^+)^C$ *in* SePC, *then we have the relation*

$$L\mathfrak{C}[K]^{\mathrm{op}} \underset{\mathrm{SePrStck}}{\sim} LC.$$

26.7.2 Statement of the theorem

Lurie [52] proved that an equivalence $f : \mathfrak{C}[K] \to C^{\mathrm{op}}$ in Cat_Δ induces a Quillen equivalence $(\mathrm{Set}_\Delta^+)_{/K} \rightleftarrows (\mathrm{Set}_\Delta^+)^C$. In this subsection, we prove a partial converse to this statement. We first have the following:

Proposition 26.7.2.1.

$$L(\mathrm{Set}_\Delta^+)_{/K} \xrightarrow{\cong} L(\mathrm{Set}_\Delta^+)^{\mathfrak{C}[K]^{\mathrm{op}}}.$$

Proof. Applying the Quillen equivalence (26.1) to the case where $C = \mathfrak{C}[K]^{\mathrm{op}}$, one obtains a Quillen equivalence

$$\mathrm{St}^+ : (\mathrm{Set}_\Delta^+)_{/K} \rightleftarrows (\mathrm{Set}_\Delta^+)^{\mathfrak{C}[K]^{\mathrm{op}}} : \mathrm{Un}^+.$$

It follows that the hammock localizations of the subcategories of cofibrant objects of such categories are weakly equivalent [17]:

$$\mathrm{diag}L^H[(\mathrm{Set}_\Delta^+)_{/K}]^c \xrightarrow{\simeq} \mathrm{diag}L^H[(\mathrm{Set}_\Delta^+)^{\mathfrak{C}[K]^{\mathrm{op}}}]^c.$$

Given the equivalences

$$\begin{array}{ccc} \mathrm{diag}L^H[(\mathrm{Set}_\Delta^+)_{/K}]^c & \xrightarrow{\simeq} & \mathrm{diag}L^H[(\mathrm{Set}_\Delta^+)^{\mathfrak{C}[K]^{\mathrm{op}}}]^c \\ \downarrow{\scriptstyle\simeq} & & \downarrow{\scriptstyle\simeq} \\ \mathrm{diag}L^H(\mathrm{Set}_\Delta^+)_{/K} & & \mathrm{diag}L^H(\mathrm{Set}_\Delta^+)^{\mathfrak{C}[K]^{\mathrm{op}}} \end{array}$$

we have $\mathrm{diag}L^H(\mathrm{Set}_\Delta^+)_{/K} \xrightarrow{\cong} \mathrm{diag}L^H(\mathrm{Set}_\Delta^+)^{\mathfrak{C}[K]^{\mathrm{op}}}$. Further, if $C \in \mathrm{Cat}_\Delta$, then from [15] we have a roof diagram of equivalences

$$\begin{array}{ccccc} & & \mathrm{diag}L^H F_* C & & \\ & \swarrow{\scriptstyle\simeq} & & \searrow{\scriptstyle\simeq} & \\ \mathrm{diag}L^H C & & & & LC \end{array}$$

leading to an isomorphism in Ho (SePC)

$$L(\mathrm{Set}_\Delta^+)_{/K} \xrightarrow{\cong} L(\mathrm{Set}_\Delta^+)^{\mathfrak{C}[K]^{\mathrm{op}}}. \qquad \square$$

Using this proposition in conjunction with the fact that $L(\mathrm{Set}_\Delta^+)^{\mathfrak{C}[K]^{\mathrm{op}}} \cong L(\mathrm{Set}_\Delta^+)^C$ implies $L\mathfrak{C}[K]^{\mathrm{op}} \sim_{\mathrm{SePrStck}} LC$ in SePC, we have the following:

Theorem 26.7.2.2. *If a morphism* $\mathfrak{C}[K]^{\mathrm{op}} \to C$ *induces an isomorphism* $L(\mathrm{Set}_\Delta^+)_{/K} \cong L(\mathrm{Set}_\Delta^+)^C$ *in* Ho (SePC), *then* $L\mathfrak{C}[K]^{\mathrm{op}} \sim_{\mathrm{SePrStck}} LC$ *in* SePC.

26.7.3 Determining K in $L(\mathrm{Set}_\Delta^+)_{/K}$

In this section, we are interested in the peripheral question of finding what reference simplicial set K would $C \in \mathrm{Cat}_\Delta$ correspond to when we consider the isomorphism $L(\mathrm{Set}_\Delta^+)_{/K} \cong L(\mathrm{Set}_\Delta^+)^C$. Recall from [52] that one has a Quillen adjunction $\mathfrak{C} : \mathrm{Set}_\Delta \rightleftarrows \mathrm{Cat}_\Delta : N$ and hence for any $C \in \mathrm{Cat}_\Delta$, a weak equivalence $\mathbb{L}\mathfrak{C}(N(RC^{\mathrm{op}})) \to RC^{\mathrm{op}}$. Since every object of Set_Δ is cofibrant, we have an equivalence $\mathfrak{C}[K]^{\mathrm{op}} \to RC$ in Cat_Δ with $K = N(RC^{\mathrm{op}})$ [42]. For simplicity, we will consider C fibrant in Cat_Δ. Because we have such an equivalence, by [52] we have a Quillen equivalence $(\mathrm{Set}_\Delta^+)_{/K} \rightleftarrows (\mathrm{Set}_\Delta^+)^C$. Following exactly the same reasoning as in the previous section, it follows that we have an isomorphism of Segal categories $L(\mathrm{Set}_\Delta^+)_{/K} \xrightarrow{\cong} L(\mathrm{Set}_\Delta^+)^C$, showing that $K = N(C^{\mathrm{op}})$ is the desired simplicial set.

26.7.4 Natural phenomena

To push the result of Theorem 26.7.2.2 further, physical phenomena occur in Segal topoi of stacks, not prestacks, so one has to localize those equivalences. One needs to first put a topology on the Segal categories $L\mathfrak{C}[K]^{\mathrm{op}}$ and LC. Recall from [77] that a topology on LC is a Grothendieck topology on its homotopy category $\mathrm{Ho}\,(LC)$. Let τ be one such topology. Denote $(LC)^{\mathrm{op}}$ by $\mathcal{A}\mathrm{ff}_C$, and $\mathcal{A}\mathrm{ff}_C^{\sim,\tau} = L_{\mathrm{Bous}}\mathbb{R}\underline{\mathrm{Hom}}(LC, \mathrm{Top}^+)$, the Segal category of stacks on $\mathcal{A}\mathrm{ff}_C$, where the Bousfield localization is with respect to hypercovers for the topology τ. Now the local equivalence $L\mathfrak{C}[K]^{\mathrm{op}} \sim_{\mathrm{SePrStck}} LC$ in SePC induces a morphism of homotopy categories $\rho : \mathrm{Ho}\,(L\mathfrak{C}[K]^{\mathrm{op}}) \to \mathrm{Ho}\,(LC)$. We are interested in having such a map preserve the Segal categories of stacks on those sites. This happens if ρ is a morphism of sites, or a continuous morphism, as it is sometimes called. If that is the case, then one can take simultaneous Bousfield localizations as in

$$\begin{array}{ccc}
\mathbb{R}\underline{\mathrm{Hom}}(L\mathfrak{C}[K]^{\mathrm{op}}, \mathrm{Top}^+) & \xleftarrow{\cong} & \mathbb{R}\underline{\mathrm{Hom}}(LC, \mathrm{Top}^+) \\
\Big\downarrow{\scriptstyle L_{\mathrm{Bous}}} & & \Big\downarrow{\scriptstyle L_{\mathrm{Bous}}} \\
L\mathfrak{C}[K]^{\sim,\tau'} & \overset{\cong}{\dashleftarrow} & \mathcal{A}\mathrm{ff}_C^{\sim,\tau}.
\end{array}$$

Then from the top isomorphism we have isomorphic phenomena, modeled by Segal topoi of stacks on the bottom row, which correspond to equivalent Segal categories $L\mathfrak{C}[K]^{\mathrm{op}}$ and LC relative to Segal categories of prestacks according to Theorem 26.7.2.2.

Note as an aside that this implies we do not have a unique set of natural laws, but a homotopical class of such laws.

To summarize, we have an isomorphism of Segal topoi of stacks, concurrently with a local equivalence of simplicial sites, both of which follow from an isomorphism of Segal categories of prestacks in Ho (SePC). We regard this formalism as a roof diagram

$$\begin{array}{ccccc}
 & L(\mathrm{Set}_\Delta^+)^{\mathfrak{C}[K]^{\mathrm{op}}} & \xleftarrow{\cong} & L(\mathrm{Set}_\Delta^+)^{C} & \\
 \swarrow{\scriptstyle L_{\mathrm{Bous}}} & & \swarrow \quad \searrow{\scriptstyle L_{\mathrm{Bous}}} & & \searrow \\
(L\mathfrak{C}[K])^{\sim,\tau'} \overset{\cong}{\dashleftarrow} \mathcal{A}\mathrm{ff}_C^{\sim,\tau} & & & & L\mathfrak{C}[K]^{\mathrm{op}} \xrightarrow[\mathrm{SePrStck}]{\sim} LC
\end{array}$$

which we interpret as a weak universality of natural laws, the choice of word universality being meant to indicate that Segal sites in the same class yield equivalent physical phenomena.

27 Internal Bousfield localizations

27.1 Introduction

From [76], recall that one can define a stack in $\mathrm{sPr}(T)$, the model category of simplicial presheaves over a simplicial model category T as an object $F \in \mathrm{sPr}(T)$ satisfying hyperdescent, meaning being local with respect to a certain class of hypercovers or being local with respect to local equivalences, both concepts necessitating to define a topology on T. This is really one localization due to the fact that local equivalences (sometimes called π_*-equivalences; see [76]), maps $F \to G$ in $\mathrm{sPr}(T)$ such that for all $n > 0$, we have an isomorphism of sheaves $\pi_n(F) \to \pi_n(G)$, among other things, are also hypercovers $F \to G$ in $\mathrm{sSPr}(T)$ (viewed as constant simplicial objects), maps such that for all $n \geq 0$,

$$F^{\mathbb{R}\Delta^n} \to F^{\mathbb{R}\partial\Delta^n} \times^h_{G^{\mathbb{R}\partial\Delta^n}} G^{\mathbb{R}\Delta^n}$$

is a τ-covering, τ being a topology on T. Such a map $F \to G$ would be an $\mathbb{S}$-colocal object in the language of [40]. Suppose we consider objects $\tilde{k_0}$ other than the sphere spectrum $\mathbb{S}$ in this definition of a local equivalence, cosimplicial resolutions of some k_0, object of some class K_0. The idea is that we want to consider reference objects that are very general, in keeping with the philosophy of relative Algebraic Geometry. Suppose further we consider functors not valued in Set_Δ, the category of simplicial sets, but in some category $\mathcal{D}$, which for the moment we can suppose is an internal proper cellular model category. Then it would be natural for localization to use internal Hom objects for the definition of homotopy function complexes, as opposed to using Hom sets. As a matter of fact, we will closely follow Hirschhorn's work [40] regarding Bousfield localizations and make the appropriate changes.

We start with a $\mathbb{U}$-small pseudo-model category $(\mathcal{C}, W)$ [76] and a proper cellular symmetric monoidal model category $\mathcal{D}_0$, and we consider the functor category $\mathcal{D}_0^{\mathcal{C}^{\mathrm{op}}}$. Prestacks in this setting are functors $F : \mathcal{C}^{\mathrm{op}} \to \mathcal{D}_0$ mapping equivalences to equivalences; if $x \to y$ is in W, then $Fx \to Fy$ is an equivalence in $\mathcal{D}_0$. If $\mathcal{D}_0 = \mathrm{Set}_\Delta$, then to see equivalence-preserving functors as local objects, one can use the classical Yoneda lemma: $Fx \simeq \mathrm{Hom}(h_x, F)$, $h_x = \mathrm{Hom}_{\mathcal{C}}(-, x)$, $\mathcal{C}$ a simplicial category. For $\mathcal{D}_0$-enriched functors, we must consider enriched Yoneda: $Fx \simeq \int_{y \in \mathcal{C}^{\mathrm{op}}} \underline{\mathrm{Hom}}(h_x(y), Fy)$. Then $Fx \xrightarrow{\simeq} Fy$ will follow from defining a notion of a local object using internal Hom objects. This leads us to defining the internal left Bousfield localization $L_\Gamma \mathcal{D}_0^{\mathcal{C}^{\mathrm{op}}} = \mathcal{D}_0^{\mathcal{C}^{\mathrm{op}}\wedge}$ of $\mathcal{D}_0^{\mathcal{C}^{\mathrm{op}}}$ with respect to $\Gamma = \{h_y \to h_x \mid x \to y \in W^{\mathrm{op}}\}$. However, as will become clear later, for generalizing classical results from [40] to the setting where Hom sets are replaced by internal Hom objects, we denote by Γ^0 the collection we just defined, and we will take a localization of $\mathcal{D}_0^{\mathcal{C}^{\mathrm{op}}}$ with respect to

$$\Gamma := \Gamma^0 \otimes \mathcal{D}_0^{\mathcal{C}^{\mathrm{op}}} = \{h_y \otimes F \to h_x \otimes F \mid x \to y \in W^{\mathrm{op}}, F \in \mathcal{D}_0^{\mathcal{C}^{\mathrm{op}}}\}.$$

https://doi.org/10.1515/9783111334073-027

Our philosophy at this point departs markedly from the standard philosophy of Homotopical Algebraic Geometry [76] in that we have no topology on $\mathcal{C}$, and we limit ourselves to considering only one condition defining π_*-local equivalences, namely $\pi_n(F) \xrightarrow{\simeq} \pi_n(G)$ for all $n > 0$, which we internalize, using not spheres, but arbitrary objects k_0 of some class of objects K_0 of $\mathcal{D}_0^{\mathcal{C}^{\mathrm{op}}}$:

$$\underline{\mathrm{Hom}}(\tilde{k_0}, \hat{F}) \xrightarrow{\simeq} \underline{\mathrm{Hom}}(\tilde{k_0}, \hat{G}) \tag{27.1}$$

where using the sphere spectrum $\mathbb{S}$ for π_*-local equivalences is replaced by using a cosimplicial resolution $\tilde{k_0}$ of a single object k_0. Here $\hat{F}$ is a fibrant approximation to F. (27.1) would define what it means for $F \to G$ to be an internal K_0-colocal equivalence. From there we are naturally led to considering an internal right Bousfield localization $R_{\mathcal{K}_0} L_\Gamma \mathcal{D}_0^{\mathcal{C}^{\mathrm{op}}}$ of $\mathcal{D}_0^{\mathcal{C}^{\mathrm{op}}\wedge}$ with respect to $\mathcal{K}_0$, the class of internal K_0-colocal equivalences.

In Section 27.2, we present the main construction, giving a category of prestacks $\mathcal{D}_0^{\mathcal{C}^{\mathrm{op}}\wedge} = L_\Gamma \mathcal{D}_0^{\mathcal{C}^{\mathrm{op}}}$ from localizing $\mathcal{D}_0^{\mathcal{C}^{\mathrm{op}}}$ along a subset Γ, followed by further taking a right Bousfield localization along a class of colocal equivalences, all concepts being redefined using internal Homs. In Section 27.3, we present foundational results, such as the enriched Yoneda lemma and localization using internal Homs for homotopy function complexes. In Section 27.4, we discuss cardinality arguments needed in the proof of our main result, the existence of a left Bousfield localization using internal Homs instead of Hom sets. In Section 27.5, we present technical results needed to prove the existence of such a localization, which is itself given in Section 27.6. In Section 27.7, we give the results needed to prove the existence of a right Bousfield localization using internal Homs, which is discussed in Section 27.8.

Relation to past work: it was pointed out to the author by J. Gutierrez and D. White that the present work is quite close to past work on the subject. In particular, they both referenced two papers the author was wholly unaware of, namely [6] and [36]. As a matter of fact, Barwick's work is so close to the present one that the original thought of using the Hom from a Quillen adjunction of two variables to define Bousfield localizations must be credited to him. Presently, we discuss localizations of symmetric monoidal model categories, and some work has been done on the subject, albeit in a classical sense, not using internal Homs. This work can be found in [84] and [85], where such localizations are referred to as monoidal Bousfield localizations.

To come back to [84] and [36], there are slight differences to be noted. Barwick works with tractable categories, which are combinatorial and hence cofibrantly generated. We work with cellular model categories, which are also cofibrantly generated. Barwick observes that he does not know of any left proper combinatorial model category that is not tractable. On our part, we do not see how to relate cellular model categories with tractable model categories; our objects of study seem different. Additionally, Barwick works with $\mathcal{V}$-enriched categories $\mathcal{C}$ for a symmetric monoidal model category $\mathcal{V}$, and the "internal" Homs he uses, derived from a Quillen adjunction of two variables due to this enrichment, are objects in $\mathcal{V}$. In particular, he shows that for a small site $\mathcal{C}$ and for a tractable symmetric monoidal model category $\mathcal{V}$ with cofibrant unit, $\mathcal{V}^{\mathcal{C}}$ is a $\mathcal{V}$-model category and so is enriched in $\mathcal{V}$, on which we define an injective local model structure as an enriched left Bousfield localization of $\mathcal{V}^{\mathcal{C}}$ with its injective model structure. Instead, we show that $\mathcal{V}^{\mathcal{C}}$ is a symmetric monoidal model category with an internal Hom, and we use those internal Homs in the definition of our version of Bousfield localizations. For this reason, we call them internal Bousfield localizations, as opposed to enriched Bousfield localizations, which use Hom objects valued in $\mathcal{V}$.

Regarding the Bousfield localization itself, Barwick's construction is an astute one. He does not define a new Bousfield localization. He uses what we refer to as the classical Bousfield localization, that of Hirschhorn in [40]. If H is a set of maps we are localizing with respect to, S is a set representing homotopy classes of H, and I is a generating set of cofibrations with cofibrant domains, then he shows that his enriched Bousfield localization of a tractable left proper model category $\mathcal{C}$ is none other than $L_{I\square S}\mathcal{C}$ in the classical sense and then uses Theorem 17.4.16 of [40] to have his enriched Homs appear. His proof is short, precisely because of his ingenious use of that result, as opposed to ours, which is a tedious reworking of Hirschhorn's work in [40] about left Bousfield localization in the event that Hom sets are replaced by internal Hom objects.

Barwick's work is elegantly generalized in [36] in the larger setting of combinatorial model categories. Gutierrez and Roitzhem consider a Quillen adjunction of two variables between combinatorial model categories, $\mathcal{C} \times \mathcal{D} \to \mathcal{E}$, and they define the left Bousfield localization $L_S\mathcal{E}$ of $\mathcal{E}$ with respect to a set S of morphisms in $\mathcal{C}$, which they call the S-local model structure on $\mathcal{E}$. In the notations of [6] as used above, if we consider a Quillen adjunction of two variables $\mathcal{C} \otimes \mathcal{V} \to \mathcal{C}$ associated with a $\mathcal{V}$-enrichment, then $L_S\mathcal{C} = L_{I\square S}\mathcal{C}$, which corresponds to a $\mathcal{V}$-enriched left Bousfield localization of $\mathcal{C}$ with respect to S as defined by Barwick.

27.2 Construction

We fix a universe $\mathbb{U}$. Let $(\mathcal{C}, W)$ be a $\mathbb{U}$-small pseudo-model category, let $\mathcal{D}_0$ be a proper cellular model category, and let $\mathcal{D}_0^{\mathcal{C}^{op}}$ be the model category of functors from $\mathcal{C}^{op}$ to $\mathcal{D}_0$. It is also a proper cellular model category (Theorem 13.1.14 and Proposition 12.1.5 of [40]). We aim to take left and right Bousfield localizations of $\mathcal{D}_0^{\mathcal{C}^{op}}$ with respect to certain classes of maps in a generalized sense. By this we mean we will use internal Hom objects in the definition of homotopy function complexes instead of Hom sets. We will prepare the ground for operating such localizations. This is what "construction" is in reference to. We will first introduce $\Gamma^0 = \{h_y \to h_x \mid x \to y \in W^{op}\}$, where $h_a = \mathrm{Hom}_{\mathcal{C}}(-, a)$, and W denotes the class of weak equivalences in $\mathcal{C}$. This collection is enlarged to $\Gamma = \Gamma^0 \otimes \mathcal{D}_0^{\mathcal{C}^{op}}$, and we construct $L_\Gamma \mathcal{D}_0^{\mathcal{C}^{op}} = \mathcal{D}_0^{\mathcal{C}^{op}\,\wedge}$, the internal left Bousfield localization of $\mathcal{D}_0^{\mathcal{C}^{op}}$ with respect to Γ. This should provide a notion of category of prestacks, since such a localization in particular provides a localization with respect to Γ^0.

From that point forward, we will introduce a class K_0 of objects in $\mathcal{D}_0^{\mathcal{C}^{op}}\wedge$ and consider the class $\mathcal{K}_0$ of internal K_0-colocal equivalences. We will then construct the internal right Bousfield localization $R_{\mathcal{K}_0}\mathcal{D}_0^{\mathcal{C}^{op}}\wedge$ of $\mathcal{D}_0^{\mathcal{C}^{op}}\wedge$ with respect to $\mathcal{K}_0$, and this will constitute a first step towards defining a generalized notion of stacks on $\mathcal{C}$.

Many of the results we will discuss in this work are stated and proved in the classical sense in [40]. To make comparison with those original statements easier, next to each claim, we will put in brackets the original indexation in [40] along with "mod", indicating that we are stating a modified version thereof.

27.2.1 Preserving equivalences

We would first like functors $F : \mathcal{C}^{op} \to \mathcal{D}_0$ to map equivalences $u : x \to y$ in $\mathcal{C}^{op}$ to equivalences $Fx \to Fy$ in $\mathcal{D}_0$. For this purpose, we introduce the following class:

$$\Gamma^0 = \{h_y \to h_x \mid x \to y \in W^{op}\}.$$

If we were to use the classical Yoneda lemma, to obtain the desired equivalences, it would suffice to show

$$\mathrm{Hom}(h_x, F) \xrightarrow{\simeq} \mathrm{Hom}(h_y, F)$$

which would make F Γ^0-local in the terminology of [40]. However, F is $\mathcal{D}_0$-valued, not Set-valued, so we use the enriched Yoneda lemma [47]. To do so, we start to use internal Homs (whose existence will be warranted below), and hence we want

$$\underline{\mathrm{Hom}}_{\mathcal{D}_0^{\mathcal{C}^{op}}}(h_x, F) \xrightarrow{\simeq} \underline{\mathrm{Hom}}_{\mathcal{D}_0^{\mathcal{C}^{op}}}(h_y, F).$$

We will actually show that

$$\underline{\mathrm{Hom}}_{\mathcal{D}_0^{\mathcal{C}^{op}}}(\tilde{h}_x, \hat{F}) \xrightarrow{\simeq} \underline{\mathrm{Hom}}_{\mathcal{D}_0^{\mathcal{C}^{op}}}(\tilde{h}_y, \hat{F})$$

in the Reedy structure of $(\mathcal{D}_0^{\mathcal{C}^{op}})^{\Delta^{op}}$ ($\tilde{h}$ cofibrant approximation to h, $\hat{F}$ simplicial resolution of F), which will imply that F maps equivalences to equivalences.

As a matter of fact, we will show the more general result:

$$\underline{\mathrm{Hom}}_{\mathcal{D}_0^{\mathcal{C}^{op}}}(\tilde{h}_x \otimes \tilde{G}, \hat{F}) \xrightarrow{\simeq} \underline{\mathrm{Hom}}_{\mathcal{D}_0^{\mathcal{C}^{op}}}(\tilde{h}_y \otimes \tilde{G}, \hat{F})$$

for all $G \in \mathcal{D}_0^{\mathcal{C}^{op}}$, which will imply the above equivalence.

27.2.2 Monoidal structure

27.2.2.1 Definitions

To use internal Homs, we first ask that $\mathcal{D}_0$ be a monoidal model category [42] with internal Hom $\underline{\mathrm{Hom}}_{\mathcal{D}_0}$. For simplicity, every notion of [40] using Hom sets that is generalized using internal Homs will be referred to as a *generalized* or as an *internal concept*. For instance, we would define

$$\mathrm{map}(k_0, F) = \underline{\mathrm{Hom}}_{\mathcal{D}_0^{\mathcal{C}^{\mathrm{op}}}\wedge}(\tilde{k_0}, \hat{F})$$

as an internal right homotopy function complex (abbr. iRhfC), where $\tilde{k_0}$ is a cofibrant approximation to k_0, and $\hat{F}$ is a simplicial resolution of F. The existence of an internal Hom in $\mathcal{D}_0^{\mathcal{C}^{\mathrm{op}}}$ will be discussed below.

For later purposes, we make the following definitions.

Definition 27.2.2.1.1. An *internal right homotopy function complex* (abbr. iRhfC) is an object of $(\mathcal{D}_0^{\mathcal{C}^{\mathrm{op}}})^{\Delta^{\mathrm{op}}}$ of the form $\underline{\mathrm{Hom}}_{\mathcal{D}_0^{\mathcal{C}^{\mathrm{op}}}}(\tilde{X}, \hat{Y})$, where $\tilde{X}$ is a cofibrant approximation to X, and $\hat{Y}$ is a simplicial resolution of Y.

Definition 27.2.2.1.2. An *internal left homotopy function complex* (abbr. iLhfC) is an object of $(\mathcal{D}_0^{\mathcal{C}^{\mathrm{op}}})^{\Delta^{\mathrm{op}}}$ of the form $\underline{\mathrm{Hom}}_{\mathcal{D}_0^{\mathcal{C}^{\mathrm{op}}}}(\tilde{X}, \hat{Y})$, where $\tilde{X}$ is a cosimplicial resolution of X, and $\hat{Y}$ is a fibrant approximation to Y.

Note that $h_x = \mathrm{Hom}_{\mathcal{C}}(-, x)$ is a set-valued functor, not an object of $\mathcal{D}_0^{\mathcal{C}^{\mathrm{op}}}$. Hence we ask that $\mathcal{C}$ be a $\mathcal{D}_0$-enriched category [47]. We can do this since $\mathcal{D}_0$ is a monoidal category.

27.2.2.2 $\mathcal{D}_0^{\mathcal{C}^{\mathrm{op}}}$ is a monoidal model category

We will also need that $\mathcal{D}_0^{\mathcal{C}^{\mathrm{op}}}$ be a monoidal model category. We define the monoidal structure on $\mathcal{D}_0^{\mathcal{C}^{\mathrm{op}}}$ pointwise: if $F, G \in \mathcal{D}_0^{\mathcal{C}^{\mathrm{op}}}$, then for any $x \in \mathcal{C}$, $(F \otimes G)(x) = Fx \otimes Gx$. Endow $\mathcal{D}_0^{\mathcal{C}^{\mathrm{op}}}$ with the injective model structure; equivalences and cofibrations are defined pointwise. This makes the tensor product on $\mathcal{D}_0^{\mathcal{C}^{\mathrm{op}}}$ a Quillen bifunctor. Indeed, let $\alpha : U \to V$ be a cofibration in $\mathcal{D}_0^{\mathcal{C}^{\mathrm{op}}}$, and let $\beta : W \to X$ be a cofibration in $\mathcal{D}_0^{\mathcal{C}^{\mathrm{op}}}$ as well. We need

$$\alpha \square \beta : (V \otimes W) \coprod_{U \otimes W} (U \otimes X) \to V \otimes X$$

to be a cofibration in $\mathcal{D}_0^{\mathcal{C}^{\mathrm{op}}}$, trivial if either of α or β is [42]. Since we take the injective model structure on $\mathcal{D}_0^{\mathcal{C}^{\mathrm{op}}}$, we have to check that pointwise. So let $x \in \mathcal{C}^{\mathrm{op}}$. We are looking at

$$\alpha \square \beta(x) : (Vx \otimes Wx) \coprod_{Ux \otimes Wx} (Ux \otimes Xx) \to Vx \otimes Xx.$$

Now having a cofibration $U \to V$ in $\mathcal{D}_0^{\mathcal{C}^{op}}$ with the injective model structure means we have a cofibration $Ux \to Vx$ in $\mathcal{D}_0$, and a cofibration $W \to X$ means we have a cofibration $Wx \to Xx$ in $\mathcal{D}_0$, and it follows that the above map is a cofibration in $\mathcal{D}_0$ (since it is a monoidal model category), and this for all $x \in \mathcal{C}^{op}$, and hence a cofibration in $\mathcal{D}_0^{\mathcal{C}^{op}}$. Since equivalences are defined pointwise, we also have that $\alpha \Box \beta$ is trivial if either of α or β is. The second condition for being a monoidal model category [42] is that if $Q1 \to 1$ is a cofibrant approximation to the unit 1, then for all cofibrant X, $Q1 \otimes X \to 1 \otimes X$ is a weak equivalence. Here we assume the unit is cofibrant, a condition we will use later. If that is the case, then this condition is automatically satisfied. This makes $\mathcal{D}_0^{\mathcal{C}^{op}}$ a monoidal model category. Observe that having a Quillen bifunctor $\otimes$, if C is cofibrant in $\mathcal{D}_0^{\mathcal{C}^{op}}$, then $C \otimes - : \mathcal{D}_0^{\mathcal{C}^{op}} \to \mathcal{D}_0^{\mathcal{C}^{op}}$ is left Quillen, so preserves cofibrations and trivial cofibrations, a fact that will be very important in the work that follows.

27.2.2.3 Construction of $\underline{\mathrm{Hom}}_{\mathcal{D}_0^{\mathcal{C}^{op}}}$

The internal Hom of $\mathcal{D}_0^{\mathcal{C}^{op}}$ is formally defined as follows: an element $\alpha \in \mathrm{Hom}_{\mathcal{D}_0^{\mathcal{C}^{op}}}(F \otimes G, H)$ is defined pointwise: for $x \in \mathcal{C}$, $\alpha(x) : Fx \otimes Gx \to Hx$ in $\mathcal{D}_0$. Now

$$\mathrm{Hom}_{\mathcal{D}_0}(Fx \otimes Gx, Hx) \cong \mathrm{Hom}_{\mathcal{D}_0}(Fx, \underline{\mathrm{Hom}}_{\mathcal{D}_0}(Gx, Hx))$$

so $\alpha(x)$ would correspond to some $\beta(x) \in \mathrm{Hom}_{\mathcal{D}_0}(Fx, \underline{\mathrm{Hom}}_{\mathcal{D}_0}(Gx, Hx))$. Letting

$$\underline{\mathrm{Hom}}_{\mathcal{D}_0}(Gx, Hx) = \underline{\mathrm{Hom}}_{\mathcal{D}_0^{\mathcal{C}^{op}}}(G, H)(x) \tag{27.2}$$

we have $\underline{\mathrm{Hom}}_{\mathcal{D}_0^{\mathcal{C}^{op}}}(G, H) \in \mathcal{D}_0^{\mathcal{C}^{op}}$, and with this notation,

$$\mathrm{Hom}_{\mathcal{D}_0^{\mathcal{C}^{op}}}(F \otimes G, H) \cong \mathrm{Hom}_{\mathcal{D}_0^{\mathcal{C}^{op}}}(F, \underline{\mathrm{Hom}}_{\mathcal{D}_0^{\mathcal{C}^{op}}}(G, H)).$$

This can be formalized using the language of ends [59]. It suffices to write, still using the identification (27.2),

$$\begin{aligned}
\mathrm{Hom}_{\mathcal{D}_0^{\mathcal{C}^{op}}}(F \otimes G, H) &\cong \int_{x \in \mathcal{C}^{op}} \mathrm{Hom}_{\mathcal{D}_0}(Fx \otimes Gx, Hx) \\
&\cong \int_{x \in \mathcal{C}^{op}} \mathrm{Hom}_{\mathcal{D}_0}(Fx, \underline{\mathrm{Hom}}_{\mathcal{D}_0}(Gx, Hx)) \\
&= \int_{x \in \mathcal{C}^{op}} \mathrm{Hom}_{\mathcal{D}_0}(Fx, \underline{\mathrm{Hom}}_{\mathcal{D}_0^{\mathcal{C}^{op}}}(G, H)(x)) \\
&\cong \mathrm{Hom}_{\mathcal{D}_0^{\mathcal{C}^{op}}}(F, \underline{\mathrm{Hom}}_{\mathcal{D}_0^{\mathcal{C}^{op}}}(G, H)).
\end{aligned}$$

We can make this more precise. From now on, we will assume that $(\mathcal{D}_0, \otimes)$ is also symmetric. Following [27] and working in full generality for later purposes, consider a closed symmetric monoidal category $\mathcal{M}$ (i. e., it has an internal Hom), which presently is $\mathcal{D}_0$ for us. An $\mathcal{M}$-module structure on a category $\mathcal{C}$ is given by an action

$$\begin{aligned}\otimes : \mathcal{C} \times \mathcal{M} &\to \mathcal{C},\\ (X, M) &\mapsto X \otimes M.\end{aligned}$$

This action is closed if we have two functors

$$\begin{aligned}\mathrm{map}_{\mathcal{C}} : \mathcal{C}^{\mathrm{op}} \times \mathcal{C} &\to \mathcal{M},\\ (X, Y) &\mapsto \mathrm{map}_{\mathcal{C}}(X, Y)\end{aligned}$$

and

$$\begin{aligned}\exp : \mathcal{M}^{\mathrm{op}} \times \mathcal{C} &\to \mathcal{C},\\ (K, Y) &\mapsto Y^K\end{aligned}$$

in such a manner that we have, for all $K \in \mathcal{M}$, $X, Y \in \mathcal{C}$, the following natural isomorphisms:

$$\mathrm{Hom}_{\mathcal{C}}(X \otimes K, Y) \cong \mathrm{Hom}_{\mathcal{M}}(K, \mathrm{map}_{\mathcal{C}}(X, Y)) \cong \mathrm{Hom}_{\mathcal{C}}(X, Y^K).$$

Now consider the following functor, where I is an indexing set:

$$\begin{aligned}\mathcal{M}^I \times \mathcal{M} &\to \mathcal{M}^I,\\ (X, K) &\mapsto X \otimes K\end{aligned}$$

defined by $(X \otimes K)_i = X_i \otimes K$. This defines an $\mathcal{M}$-module structure on $\mathcal{M}^I$. It is closed if we use the following definitions: for $X, Y \in \mathcal{M}^I$ and $K \in \mathcal{M}$, define $Y^K = \underline{\mathrm{Hom}}_{\mathcal{M}}(K, Y)$ by $\underline{\mathrm{Hom}}_{\mathcal{M}}(K, Y)_i = \underline{\mathrm{Hom}}_{\mathcal{M}}(K, Y_i)$ and

$$\mathrm{map}_{\mathcal{M}^I}(X, Y) = \int_i \underline{\mathrm{Hom}}_{\mathcal{M}}(X_i, Y_i).$$

In particular, for $\mathcal{M} = \mathcal{D}_0$ and $I = \mathcal{C}^{\mathrm{op}}$, this gives us, for $X, Y \in \mathcal{D}_0^{\mathcal{C}^{\mathrm{op}}}$ and $K \in \mathcal{D}_0$,

$$\mathrm{Hom}_{\mathcal{D}_0^{\mathcal{C}^{\mathrm{op}}}}(X \otimes K, Y) \cong \mathrm{Hom}_{\mathcal{D}_0^{\mathcal{C}^{\mathrm{op}}}}(X, \underline{\mathrm{Hom}}_{\mathcal{D}_0}(K, Y)) \cong \mathrm{Hom}_{\mathcal{D}_0}(K, \mathrm{map}_{\mathcal{D}_0^{\mathcal{C}^{\mathrm{op}}}}(X, Y)).$$

Coming back to the general case, consider

$$\begin{aligned}h_i : I &\to \mathcal{M},\\ j &\mapsto h_i(j) = \coprod_{I(i,j)} 1\end{aligned}$$

where 1 is the unit of $\mathcal{M}$. Define a monoidal structure on $\mathcal{M}^I$ as follows: for $X, Y \in \mathcal{M}^I$, let $(X \otimes Y)_i = X_i \otimes Y_i$, making $(\mathcal{M}^I, \otimes)$ a symmetric monoidal category, which is furthermore closed, with internal Hom given by

$$\begin{aligned}\underline{\mathrm{Hom}}_{\mathcal{M}^I}(X, Y)_i &= \mathrm{map}_{\mathcal{M}^I}(h_i \otimes X, Y)\\ &= \int_{j\in I} \underline{\mathrm{Hom}}_{\mathcal{M}}(h_i(j) \otimes X_j, Y_j).\end{aligned}$$

For $\mathcal{M} = \mathcal{D}_0$ and $I = \mathcal{C}^{\mathrm{op}}$, this gives us the internal Hom in $\mathcal{D}_0^{\mathcal{C}^{\mathrm{op}}}$:

$$\begin{aligned}\underline{\mathrm{Hom}}_{\mathcal{D}_0^{\mathcal{C}^{\mathrm{op}}}}(F, G)(x) &= \mathrm{map}_{\mathcal{D}_0^{\mathcal{C}^{\mathrm{op}}}}(h_x \otimes F, G) &(27.3)\\ &= \int_{y\in\mathcal{C}^{\mathrm{op}}} \underline{\mathrm{Hom}}_{\mathcal{D}_0}(h_x(y) \otimes Fy, Gy)\\ &= \int_{y\in\mathcal{C}^{\mathrm{op}}} \underline{\mathrm{Hom}}_{\mathcal{D}_0}(h_x(y), \underline{\mathrm{Hom}}_{\mathcal{D}_0}(Fy, Gy)) &(27.4)\\ &= \mathrm{map}_{\mathcal{D}_0^{\mathcal{C}^{\mathrm{op}}}}(h_x, \underline{\mathrm{Hom}}_{\mathcal{D}_0}(F-, G-)).\end{aligned}$$

We proceed from (27.4) to get (27.2) as follows. For any W,

$$\begin{aligned}&\mathrm{Hom}_{\mathcal{D}_0}(W, \underline{\mathrm{Hom}}_{\mathcal{D}_0^{\mathcal{C}^{\mathrm{op}}}}(F, G)(x))\\ &\quad= \mathrm{Hom}_{\mathcal{D}_0}(W, \int_{y\in\mathcal{C}^{\mathrm{op}}} \underline{\mathrm{Hom}}_{\mathcal{D}_0}(h_x(y), \underline{\mathrm{Hom}}_{\mathcal{D}_0}(Fy, Gy)))\\ &\quad= \int_{y\in\mathcal{C}^{\mathrm{op}}} \mathrm{Hom}_{\mathcal{D}_0}(W, \underline{\mathrm{Hom}}_{\mathcal{D}_0}(h_x(y), \underline{\mathrm{Hom}}_{\mathcal{D}_0}(Fy, Gy)))\\ &\quad= \int \mathrm{Hom}_{\mathcal{D}_0}(W \otimes h_x(y), \underline{\mathrm{Hom}}_{\mathcal{D}_0}(Fy, Gy))\\ &\quad= \mathrm{Hom}_{\mathcal{D}_0^{\mathcal{C}^{\mathrm{op}}}}(W \otimes h_x, \underline{\mathrm{Hom}}_{\mathcal{D}_0}(F-, G-)).\end{aligned}$$

Now we use the fact that if we define the evaluation functor $Ev_i : \mathcal{M}^I \to \mathcal{M}$ by $Ev_i(X) = X_i$ and $F : \mathcal{M} \to \mathcal{M}^I$ by $F_i(M) = h_i \otimes M$, then we have $F_i \dashv Ev_i$, which reads, for $\mathcal{M} = \mathcal{D}_0$ and $I = \mathcal{C}^{\mathrm{op}}$,

$$\mathrm{Hom}_{\mathcal{D}_0^{\mathcal{C}^{\mathrm{op}}}}(h_x \otimes M, G) \cong \mathrm{Hom}_{\mathcal{D}_0}(M, G(x)).$$

Using this in the previous computation gives

$$\mathrm{Hom}_{\mathcal{D}_0}(W, \underline{\mathrm{Hom}}_{\mathcal{D}_0^{\mathcal{C}^{\mathrm{op}}}}(F, G)(x)) = \mathrm{Hom}_{\mathcal{D}_0}(W, \underline{\mathrm{Hom}}_{\mathcal{D}_0}(Fx, Gx))$$

and this being true for all W, we have

$$\underline{\mathrm{Hom}}_{\mathcal{D}_0^{\mathcal{C}^{\mathrm{op}}}}(F,G)(x) \cong \underline{\mathrm{Hom}}_{\mathcal{D}_0}(Fx,Gx).$$

27.2.2.4 More on tensor products

As part of the construction of $\underline{\mathrm{Hom}}_{\mathcal{D}_0^{\mathcal{C}^{\mathrm{op}}}}$ and for later purposes, we use the following result of [27]: if I is a Reedy category and $\mathcal{M}$ is a cofibrantly generated symmetric monoidal model category with cofibrant unit, then $\mathcal{M}^I$ is a closed symmetric monoidal model category. We use this for $\mathcal{M} = \mathcal{D}_0^{\mathcal{C}^{\mathrm{op}}}\wedge$, cellular, in particular cofibrantly generated, with a symmetric monoidal structure and a cofibrant unit. It follows that if we take $I = \Delta^{\mathrm{op}}$, then $(\mathcal{D}_0^{\mathcal{C}^{\mathrm{op}}}\wedge)^{\Delta^{\mathrm{op}}}$ is a symmetric monoidal model category. In particular, $\otimes$ is a Quillen bifunctor.

Note that one has to keep track of what tensor products are in use, whether they are part of a monoidal structure or a generalization thereof, for example, when we have model structures. For instance,

$$\begin{aligned}\otimes : (\mathcal{D}_0^{\mathcal{C}^{\mathrm{op}}}\wedge)^{\Delta} \times \mathcal{D}_0^{\mathcal{C}^{\mathrm{op}}}\wedge &\to (\mathcal{D}_0^{\mathcal{C}^{\mathrm{op}}}\wedge)^{\Delta},\\ (\tilde{k_0},\tilde{D}) &\mapsto \tilde{k_0}\otimes\tilde{D}\end{aligned}$$

defines a $\mathcal{D}_0^{\mathcal{C}^{\mathrm{op}}}\wedge$-module structure on $(\mathcal{D}_0^{\mathcal{C}^{\mathrm{op}}}\wedge)^{\Delta}$. If $\tilde{D}$ is a cofibrant approximation to D and $\tilde{k_0}$ is a cosimplicial resolution of k_0, then writing

$$\mathrm{Hom}_{\mathcal{D}_0^{\mathcal{C}^{\mathrm{op}}}\wedge}(\tilde{D}, \underline{\mathrm{Hom}}_{\mathcal{D}_0^{\mathcal{C}^{\mathrm{op}}}\wedge}(\tilde{k_0},\hat{X})) \cong \mathrm{Hom}_{\mathcal{D}_0^{\mathcal{C}^{\mathrm{op}}}\wedge}(\tilde{D}\otimes\tilde{k_0},\hat{X})$$

as provided by Theorem 27.3.4.2, the tensor product $\tilde{D}\otimes\tilde{k_0}$ makes sense. Here we have used $\underline{\mathrm{Hom}}_{\mathcal{D}_0^{\mathcal{C}^{\mathrm{op}}}\wedge} = \underline{\mathrm{Hom}}_{\mathcal{D}_0^{\mathcal{C}^{\mathrm{op}}}}$ for the simple reason that the definition of an internal Hom is peculiar to the monoidal structure, not the model structure, so is preserved after localization. This observation, which we will make more precise, presupposes that $\mathcal{D}_0^{\mathcal{C}^{\mathrm{op}}}\wedge$ be a monoidal model category, which we now show.

27.2.2.5 $\mathcal{D}_0^{\mathcal{C}^{\mathrm{op}}}\wedge$ is a monoidal model category

We show that $\mathcal{D}_0^{\mathcal{C}^{\mathrm{op}}}\wedge$ is a monoidal model category. We use the same tensor product as for $\mathcal{D}_0^{\mathcal{C}^{\mathrm{op}}}$, and we need that it be a Quillen bifunctor, that is, it satisfies the pushout-product axiom. First cofibrations in $L_\Gamma\mathcal{D}_0^{\mathcal{C}^{\mathrm{op}}} = \mathcal{D}_0^{\mathcal{C}^{\mathrm{op}}}\wedge$ are those of $\mathcal{D}_0^{\mathcal{C}^{\mathrm{op}}}$ as well. As a monoidal model category, it satisfies the pushout-product axiom, so if $f : U \to V$ and $g : X \to Y$ are cofibrations in $\mathcal{D}_0^{\mathcal{C}^{\mathrm{op}}}\wedge$, then so is $f\Box g$. If in addition either f or g is an iΓ-local equivalence (Definition 27.6.2), then so is $f\Box g$: without loss of generality, let us assume that f is a trivial cofibration in $\mathcal{D}_0^{\mathcal{C}^{\mathrm{op}}}\wedge$. Now i$\Gamma$-local equivalences form an ideal class as shown in Section 27.5, which implies that $U\otimes X \to V\otimes X$ is an iΓ-local equivalence as well, and hence so is $\tilde{U}\otimes\tilde{X} \to \tilde{V}\otimes\tilde{X}$. Since $\otimes$ is a Quillen bifunctor on $\mathcal{D}_0^{\mathcal{C}^{\mathrm{op}}}$, for $\tilde{X}$ a cofibrant approximation to some object X of $\mathcal{D}_0^{\mathcal{C}^{\mathrm{op}}}$, $\tilde{U}\otimes\tilde{X} \to \tilde{V}\otimes\tilde{X}$ is a cofibration in $\mathcal{D}_0^{\mathcal{C}^{\mathrm{op}}}$ if we take $\tilde{U} \to \tilde{V}$ a fibrant cofibrant approximation to $U \to V$ that is a cofibration in $\mathcal{D}_0^{\mathcal{C}^{\mathrm{op}}}$. Hence $\tilde{U} \otimes \tilde{X} \to \tilde{V} \otimes \tilde{X}$ is an iΓ-local equivalence and a cofibration, i. e., a trivial cofibration. Now

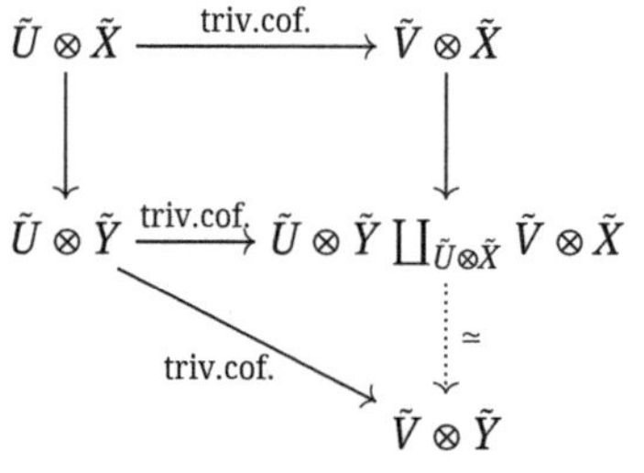

where the bottom horizontal map is a trivial cofibration since those are preserved under pushout, $\tilde{U} \otimes \tilde{Y} \to \tilde{V} \otimes \tilde{Y}$ is a trivial cofibration for the same reason that $\tilde{U} \otimes \tilde{X} \to \tilde{V} \otimes \tilde{X}$ is a trivial cofibration, and hence the dotted arrow is an equivalence by the 2–3 property.

Finally, it suffices to show that $\tilde{U} \otimes \tilde{Y} \coprod_{\tilde{U} \otimes \tilde{X}} \tilde{V} \otimes \tilde{X}$ is a cofibrant approximation to $U \otimes Y \coprod_{U \otimes X} V \otimes X$. It suffices to show that $\tilde{A} \coprod_{\tilde{B}} \tilde{C} \xrightarrow{\simeq} A \coprod_B C$ in the event that $\tilde{B} \to \tilde{C}$ corresponds to $\tilde{U} \otimes \tilde{X} \to \tilde{V} \otimes \tilde{X}$, a trivial cofibration as shown above. Then referring to the commutative diagram

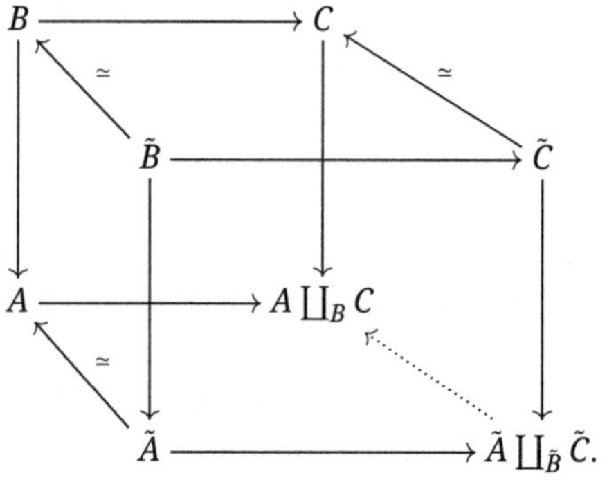

$\tilde{A} \to \tilde{A} \coprod_{\tilde{B}} \tilde{C}$ is a trivial cofibration as a pushout and so a weak equivalence in particular. In the top square, $B \to C$ is an equivalence by the 2–3 property, $C \to A \coprod_B C$ is a cofibration, and $\mathcal{D}_0^{\mathcal{C}^{\mathrm{op}}\wedge}$ is left proper by Theorem 27.6.3, so $A \to A \coprod_B C$ is an equivalence as pushout of an equivalence along a cofibration. In the bottom square, it follows using the 2–3 property that $\tilde{A} \coprod_{\tilde{B}} \tilde{C} \to A \coprod_B C$ is an equivalence.

Then we also need that for any X cofibrant in $L_\Gamma \mathcal{D}_0^{\mathcal{C}^{\mathrm{op}}}$, $Q1 \otimes X \to 1 \otimes X$ is a weak equivalence in $L_\Gamma \mathcal{D}_0^{\mathcal{C}^{\mathrm{op}}}$, i. e., an i$\Gamma$-local equivalence. If X is cofibrant in $\mathcal{D}_0^{\mathcal{C}^{\mathrm{op}}\wedge}$, then it is cofibrant in $\mathcal{D}_0^{\mathcal{C}^{\mathrm{op}}}$ as well. Then since $\mathcal{D}_0^{\mathcal{C}^{\mathrm{op}}}$ is a monoidal model category, $Q1 \otimes X \xrightarrow{\simeq} 1 \otimes X$, but equivalences are also iΓ-local equivalences by Proposition 27.4.1, and hence it is a weak equivalence in $L_\Gamma \mathcal{D}_0^{\mathcal{C}^{\mathrm{op}}}$ as well. We conclude that $L_\Gamma \mathcal{D}_0^{\mathcal{C}^{\mathrm{op}}}$ is a monoidal model category with the same tensor product $\otimes$ as $\mathcal{D}_0^{\mathcal{C}^{\mathrm{op}}}$ and that $\underline{\mathrm{Hom}}_{\mathcal{D}_0^{\mathcal{C}^{\mathrm{op}}\wedge}} = \underline{\mathrm{Hom}}_{\mathcal{D}_0^{\mathcal{C}^{\mathrm{op}}}}$.

27.2.3 Covers

We regard $\mathcal{D}_0$ as the first of a chain of proper cellular symmetric monoidal model categories $\mathcal{D}_i$, the idea being that we are interested in maps $\mathcal{D}_i^{\mathcal{C}^{\text{op}}} \to \mathcal{D}_{i+1}^{\mathcal{C}^{\text{op}}}$ and their localizations. This is especially pertinent in the setting of Segal topos, since internal Homs as defined presently depend very much on the target category. By having a tower of such target categories one thereby pushes further the internal theories we have, producing a tower of internal theories levelwise. Practically, it suffices to determine at which level in this tower one wishes to work. Lower-level theories are in a sense more compact by construction, whereas higher theories are clearly more encompassing.

Denote by $h_i : \mathcal{D}_i \to \mathcal{D}_{i+1}$ a map of model categories, that is, a left Quillen functor. It induces

$$h_i^{\mathcal{C}^{\text{op}}} : \mathcal{D}_i^{\mathcal{C}^{\text{op}}} \to \mathcal{D}_{i+1}^{\mathcal{C}^{\text{op}}}$$

pointwise. It follows from Proposition 11.6.5 of [40] that since our model categories are cellular, in particular, cofibrantly generated, this map is also a left Quillen functor. We refer the reader to Definition 8.5.11 of [40] for defining the left derived functor $\mathbb{L}F$ of a left adjoint F, part of a Quillen adjunction. Recall that $\mathcal{C}$ is $\mathcal{D}_0$-enriched, which means it is also $\mathcal{D}_i$-enriched by composition with the h_i. Define

$$h^{(i)} = h_{i-1} \circ \cdots \circ h_0$$

and

$$h_x^{(i)} = h^{(i-1)} \circ h_x.$$

Let Γ_i be the following class of maps in $\mathcal{D}_i^{\mathcal{C}^{\text{op}}}$:

$$\Gamma_i = \{h_y^{(i)} \otimes G \to h_x^{(i)} \otimes G \mid x \to y \in W(\mathcal{C}^{\text{op}}),\ G \in \mathcal{D}_i^{\mathcal{C}^{\text{op}}}\}.$$

Define

$$\mathbb{L}h_i^{\mathcal{C}^{\text{op}}}\Gamma_i = \{\mathbb{L}h_i^{\mathcal{C}^{\text{op}}}(g) \mid g \in \Gamma_i\}.$$

Then by Theorem 3.3.20 of [40] we have a left Quillen functor

$$h_i^{\mathcal{C}^{\text{op}}} : L_{\Gamma_i}\mathcal{D}_i^{\mathcal{C}^{\text{op}}} \to L_{\mathbb{L}h_i^{\mathcal{C}^{\text{op}}}\Gamma_i}\mathcal{D}_{i+1}^{\mathcal{C}^{\text{op}}}.$$

Observe as an aside that we have $\mathbb{L}h_i^{\mathcal{C}^{\text{op}}}\Gamma_i \subset \Gamma_{i+1}$. Indeed, for $g \in \Gamma_i$,

$$\mathbb{L}h_i^{\mathcal{C}^{\text{op}}}(g) = h_i^{\mathcal{C}^{\text{op}}}(\tilde{C}_i(g)) = h_i \circ \tilde{C}_i(g)$$

where $\tilde{C}_i$ is a cofibrant approximation functor. But we may as well enlarge Γ_i to also contain $\tilde{C}_i(g)$ since we use internal homotopy function complexes, and their image by h_i is therefore in Γ_{i+1} by definition. It follows that the collection of $\mathrm{i}\mathbb{L}h_i^{\mathcal{C}^{\mathrm{op}}}\Gamma_i$-local objects contains $\mathrm{i}\Gamma_{i+1}$-local objects, and, consequently, $\mathrm{i}\mathbb{L}h_i^{\mathcal{C}^{\mathrm{op}}}\Gamma_i$-local equivalences are $\mathrm{i}\Gamma_{i+1}$-local equivalences. We are looking at the following picture:

$$\begin{array}{ccc}
\mathcal{D}_i & \xrightarrow{h_i} & \mathcal{D}_{i+1} \\
\rightsquigarrow & & \rightsquigarrow \\
\mathcal{D}_i^{\mathcal{C}^{\mathrm{op}}} & \xrightarrow{h_i^{\mathcal{C}^{\mathrm{op}}}} & \mathcal{D}_{i+1}^{\mathcal{C}^{\mathrm{op}}} \\
\downarrow L_{\Gamma_i} & & \downarrow L_{\mathbb{L}h_i^{\mathcal{C}^{\mathrm{op}}}\Gamma_i} \\
L_{\Gamma_i}\mathcal{D}_i^{\mathcal{C}^{\mathrm{op}}} & \xrightarrow{h_i^{\mathcal{C}^{\mathrm{op}}}} & L_{\mathbb{L}h_i^{\mathcal{C}^{\mathrm{op}}}\Gamma_i}\mathcal{D}_{i+1}^{\mathcal{C}^{\mathrm{op}}} .
\end{array}$$

If we introduce the right Quillen adjoint $k_{i+1}^{\mathcal{C}^{\mathrm{op}}}$ of $h_i^{\mathcal{C}^{\mathrm{op}}}$, then one can define its right derived form following Definition 8.5.11 of [40], and Theorem 3.3.20 again gives us that we have a Quillen adjunction (here we assume that $L_{\Gamma_i}\mathcal{D}_i^{\mathcal{C}^{\mathrm{op}}}$ is right proper)

$$R_{\mathbb{R}k_{i+1}^{\mathcal{C}^{\mathrm{op}}}\mathcal{K}_{i+1}}L_{\Gamma_i}\mathcal{D}_i^{\mathcal{C}^{\mathrm{op}}} \underset{k_{i+1}^{\mathcal{C}^{\mathrm{op}}}}{\overset{h_i^{\mathcal{C}^{\mathrm{op}}}}{\rightleftarrows}} R_{\mathcal{K}_{i+1}}L_{\mathbb{L}h_i^{\mathcal{C}^{\mathrm{op}}}\Gamma_i}\mathcal{D}_{i+1}^{\mathcal{C}^{\mathrm{op}}}$$

induced from

$$\begin{array}{ccc}
L_{\Gamma_i}\mathcal{D}_i^{\mathcal{C}^{\mathrm{op}}} & \xrightarrow{h_i^{\mathcal{C}^{\mathrm{op}}}} & L_{\mathbb{L}h_i^{\mathcal{C}^{\mathrm{op}}}\Gamma_i}\mathcal{D}_{i+1}^{\mathcal{C}^{\mathrm{op}}} \\
\downarrow R_{\mathbb{R}k_{i+1}^{\mathcal{C}^{\mathrm{op}}}\mathcal{K}_{i+1}} & & \downarrow R_{\mathcal{K}_{i+1}} \\
R_{\mathbb{R}k_{i+1}^{\mathcal{C}^{\mathrm{op}}}\mathcal{K}_{i+1}}L_{\Gamma_i}\mathcal{D}_i^{\mathcal{C}^{\mathrm{op}}} & \xrightarrow[h_i^{\mathcal{C}^{\mathrm{op}}}]{} & R_{\mathcal{K}_{i+1}}L_{\mathbb{L}h_i^{\mathcal{C}^{\mathrm{op}}}\Gamma_i}\mathcal{D}_{i+1}^{\mathcal{C}^{\mathrm{op}}}
\end{array}$$

where K_i is a class of objects of $\mathcal{D}_i^{\mathcal{C}^{\mathrm{op}}\wedge} = L_{\Gamma_i}\mathcal{D}_i^{\mathcal{C}^{\mathrm{op}}}$, and $\mathcal{K}_i = \{iK_i\text{-colocal equivalences}\}$. This presupposes that the left localization of $\mathcal{D}_i^{\mathcal{C}^{\mathrm{op}}}$ is right proper and cellular. That it is cellular is shown when we prove that we have an internal left Bousfield localization, but whether it is right proper is not automatic. This is something we have to assume. In other terms, this construction holds for symmetric monoidal model categories $\mathcal{D}_i^{\mathcal{C}^{\mathrm{op}}}$ whose internal left Bousfield localizations are right proper.

27.3 Foundational results

27.3.1 Elementary results

We list a few results that will be useful in the sequel. First, we have

$$\underline{\mathrm{Hom}}(1, A) \cong A$$

since for any W, $\mathrm{Hom}(W, \underline{\mathrm{Hom}}(1, A)) \cong \mathrm{Hom}(W \otimes 1, A) \cong \mathrm{Hom}(W, A)$. We also have

$$\mathrm{Hom}(1, \underline{\mathrm{Hom}}(A, B)) \cong \mathrm{Hom}(A, B)$$

as it directly follows from the adjunction isomorphism $\mathrm{Hom}(1, \underline{\mathrm{Hom}}(A, B)) \cong \mathrm{Hom}(1 \otimes A, B) \cong \mathrm{Hom}(A, B)$. From this we have the very useful:

Proposition 27.3.1.1. *Internal equivalences* $\underline{\mathrm{Hom}}(\tilde{A}, \hat{B}) \xrightarrow{\simeq} \underline{\mathrm{Hom}}(\tilde{A}, \hat{C})$ *imply classical equivalences* $\mathrm{Hom}(\tilde{A}, \hat{B}) \xrightarrow{\simeq} \mathrm{Hom}(\tilde{A}, \hat{C})$.

Proof. It suffices to write

$$\begin{array}{ccc} \mathrm{Hom}(1, \underline{\mathrm{Hom}}(\tilde{A}, \hat{B})) & \xrightarrow{\simeq} & \mathrm{Hom}(1, \underline{\mathrm{Hom}}(\tilde{A}, \hat{C})) \\ \Vert & & \Vert \\ \mathrm{Hom}(\tilde{A}, \hat{B}) & \underset{\simeq}{\dashrightarrow} & \mathrm{Hom}(\tilde{A}, \hat{C}). \end{array}$$

□

A very important result is the following modified version of Theorem 17.7.7 of [40].

Theorem[17.7.7 mod] 27.3.1.2. *Let* $g : X \to Y$ *be a map in* $\mathcal{D}_0^{\mathcal{C}^{\mathrm{op}}}$. *Then* g *is a weak equivalence if and only if for all* W, $g_* : \underline{\mathrm{Hom}}(\tilde{W}, \hat{X}) \to \underline{\mathrm{Hom}}(\tilde{W}, \hat{Y})$ *is a weak equivalence in* $(\mathcal{D}_0^{\mathcal{C}^{\mathrm{op}}})^{\Delta^{\mathrm{op}}}$ *if and only if for all* Z *in* $\mathcal{D}_0^{\mathcal{C}^{\mathrm{op}}}$, $g^* : \underline{\mathrm{Hom}}(\tilde{Y}, \hat{Z}) \to \underline{\mathrm{Hom}}(\tilde{X}, \hat{Z})$ *is a weak equivalence in* $(\mathcal{D}_0^{\mathcal{C}^{\mathrm{op}}})^{\Delta^{\mathrm{op}}}$.

Proof. Suppose $g : X \to Y$ is an equivalence. Then by the original Theorem 17.7.7 of [40] we have the following top equivalence for any objects W and A:

$$\begin{array}{ccc} \mathrm{Hom}(\widetilde{W \otimes A}, \hat{X}) & \xrightarrow{\simeq} & \mathrm{Hom}(\widetilde{W \otimes A}, \hat{Y}) \\ \Vert & & \Vert \\ \mathrm{Hom}(\tilde{W} \otimes \tilde{A}, \hat{X}) & \xrightarrow{\simeq} & \mathrm{Hom}(\tilde{W} \otimes \tilde{A}, \hat{Y}) \\ \downarrow{\scriptstyle \cong} & & \downarrow{\scriptstyle \cong} \\ \mathrm{Hom}(\tilde{W}, \underline{\mathrm{Hom}}(\tilde{A}, \hat{X})) & \underset{\simeq}{\dashrightarrow} & \mathrm{Hom}(\tilde{W}, \underline{\mathrm{Hom}}(\tilde{A}, \hat{Y})) \end{array}$$

and this being true for any W, it follows that $\underline{\mathrm{Hom}}(\tilde{A}, \hat{X}) \xrightarrow{\simeq} \underline{\mathrm{Hom}}(\tilde{A}, \hat{Y})$. Conversely, if this is true, then $\mathrm{Hom}(\tilde{A}, \hat{X}) \xrightarrow{\simeq} \mathrm{Hom}(\tilde{A}, \hat{Y})$ by Proposition 27.3.1.1 and hence an equivalence $g : X \xrightarrow{\simeq} Y$ by the original Theorem 17.7.7. For the second part of the theorem, we proceed similarly. For any W and Z, we have

$$\begin{array}{ccc}
\mathrm{Hom}(\widetilde{W \otimes Y}, \hat{Z}) & \xrightarrow{\simeq} & \mathrm{Hom}(\widetilde{W \otimes X}, \hat{Z}) \\
\| & & \| \\
\mathrm{Hom}(\tilde{W} \otimes \tilde{Y}, \hat{Z}) & & \mathrm{Hom}(\tilde{W} \otimes \tilde{X}, \hat{Z}) \\
\downarrow{\cong} & & \downarrow{\cong} \\
\mathrm{Hom}(\tilde{W}, \underline{\mathrm{Hom}}(\tilde{Y}, \hat{Z})) & \overset{\simeq}{\dashrightarrow} & \mathrm{Hom}(\tilde{W}, \underline{\mathrm{Hom}}(\tilde{X}, \hat{Z})).
\end{array}$$

The reasoning for showing that g is an equivalence if and only if g^* is an equivalence is identical to the one above for showing that g is an equivalence if and only if g_* is one as well. □

27.3.2 Yoneda

We take the left Bousfield localization of $\mathcal{D}_0^{\mathcal{C}^{\mathrm{op}}}$ with respect to $\Gamma = \Gamma^0 \otimes \mathcal{D}_0^{\mathcal{C}^{\mathrm{op}}}$ with $\Gamma^0 = \{h_y \to h_x \mid x \to y \in W(\mathcal{C}^{\mathrm{op}})\}$. In particular, i$\Gamma$-local $F \in \mathcal{D}_0^{\mathcal{C}^{\mathrm{op}}}$ (see Definition 27.6.1) is therefore iΓ^0-local, that is, it satisfies

$$\underline{\mathrm{Hom}}_{\mathcal{D}_0^{\mathcal{C}^{\mathrm{op}}}}(\tilde{h}_x, \hat{F}) \xrightarrow{\simeq} \underline{\mathrm{Hom}}_{\mathcal{D}_0^{\mathcal{C}^{\mathrm{op}}}}(\tilde{h}_y, \hat{F}).$$

Recall that we initially wanted to have functors from $\mathcal{C}^{\mathrm{op}}$ to $\mathcal{D}_0$ that map equivalences to equivalences. To show that such local objects F as just defined preserve equivalences will the first time necessitate a Yoneda lemma for enriched functors. We use [47] and [59]. Consider a $\mathcal{D}_0$-functor $F : \mathcal{C}^{\mathrm{op}} \to \mathcal{D}_0$. Let $x \in \mathcal{C}$. Then the morphism

$$\phi_y : Fx \to \underline{\mathrm{Hom}}_{\mathcal{D}_0}(h_x(y), Fy)$$

expresses Fx as the end $\int_{y \in \mathcal{C}^{\mathrm{op}}} \underline{\mathrm{Hom}}_{\mathcal{D}_0}(h_x(y), Fy)$. Observe that we could have obtained this result using the $\mathcal{D}_0$-module structure on $\mathcal{D}_0^{\mathcal{C}^{\mathrm{op}}}$. We have

$$\mathrm{map}_{\mathcal{D}_0^{\mathcal{C}^{\mathrm{op}}}}(h_x, F) = \int_{y \in \mathcal{C}^{\mathrm{op}}} \underline{\mathrm{Hom}}_{\mathcal{D}_0}(h_x(y), Fy)$$

but with (27.3), we also have

$$\begin{aligned}
\mathrm{map}_{\mathcal{D}_0^{\mathcal{C}^{\mathrm{op}}}}(h_x, F) &= \mathrm{map}_{\mathcal{D}_0^{\mathcal{C}^{\mathrm{op}}}}(h_x \otimes 1, F) \\
&= \underline{\mathrm{Hom}}_{\mathcal{D}_0^{\mathcal{C}^{\mathrm{op}}}}(1, F)(x)
\end{aligned}$$

$$= F(x)$$

and hence

$$F(x) = \int_{y \in \mathcal{C}^{op}} \underline{\mathrm{Hom}}_{\mathcal{D}_0}(h_x(y), Fy).$$

We improve this result since we still have to connect this to an iRhfC.

Theorem 27.3.2.1.

$$Fx \cong \int_{y \in \mathcal{C}^{op}} \underline{Hom}_{\mathcal{D}_0}(\tilde{h}_x(y), Fy). \tag{27.5}$$

Proof. We adapt the proof found in [47]. Consider any $\mathcal{D}_0$-natural map $\alpha_y^c : X \to \underline{\mathrm{Hom}}(\tilde{h}_x(y), Fy)$ whose adjoint we will represent by the same letter: $\alpha_y^c : \tilde{h}_x(y) \to \underline{\mathrm{Hom}}(X, Fy)$. We have $\phi_y^c : Fx \to \underline{\mathrm{Hom}}(\tilde{h}_x(y), Fy)$, with adjoint again represented by the same letter, $\phi_y^c : \tilde{h}_x(y) \to \underline{\mathrm{Hom}}(Fx, Fy)$. We want to show there is a unique $\eta : X \to Fx$ such that $\alpha_y^c = \phi_y^c \eta$, and that will prove our result. In the adjoint picture, that amounts to showing there exists a unique η such that the triangular diagram below commutes:

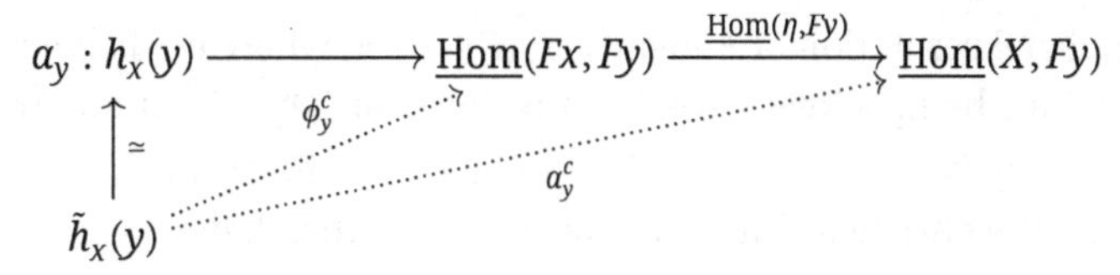

but from the classical proof we know there is such an η, and it is unique, and hence by composition we have our result. □

27.3.3 Localization with respect to Γ produces prestacks

Our aim is to have functors $F : \mathcal{C}^{op} \to \mathcal{D}_0$ that map equivalences to equivalences for obtaining a notion of prestacks. This can be implemented using a Bousfield localization with respect to $\Gamma^0 = \{h_y \to h_x \mid x \to y \in W(\mathcal{C}^{op})\}$. To make this manifest, we needed an enriched Yoneda lemma that we covered in the previous section. With regards to the collection of maps we are using as base for our localization, note that we will localize with respect to $\Gamma = \Gamma^0 \otimes \mathcal{D}_0^{\mathcal{C}^{op}}$ instead, which will in particular imply a localization with respect to Γ^0. Some technical lemmas later will require such an enlargement. For the moment, a heuristic motivation for using Γ goes as follows: Γ^0-local objects recognize morphisms $h_y \to h_x$ as weak equivalences via Hom objects. It is not much of a stretch to ask that Γ^0-local objects F also satisfy $\mathrm{Hom}(\tilde{h}_x \otimes \tilde{C}, \hat{F}) \xrightarrow{\simeq} \mathrm{Hom}(\tilde{h}_y \otimes \tilde{C}, \hat{F})$ for all C as well, and hence we consider Γ-local objects instead.

iΓ-local $F : \mathcal{C}^{\mathrm{op}} \to \mathcal{D}_0$ is iΓ^0-local, which means that for any element of Γ^0, we have

$$\underline{\mathrm{Hom}}_{\mathcal{D}_0^{\mathcal{C}^{\mathrm{op}}}}(\tilde{h}_x, \hat{F}) \xrightarrow{\simeq} \underline{\mathrm{Hom}}_{\mathcal{D}_0^{\mathcal{C}^{\mathrm{op}}}}(\tilde{h}_y, \hat{F}) \tag{27.6}$$

where $\tilde{h}$ is a cofibrant approximation of h, and $\hat{F}$ is a simplicial resolution of F. Recall that for simplicial objects $\hat{X}$, we define the simplicial set $\underline{\mathrm{Hom}}(A, \hat{X})$ as in [40]:

$$\underline{\mathrm{Hom}}(A, \hat{X})_n = \underline{\mathrm{Hom}}(A, \hat{X}_n).$$

We show (27.6) implies $Fx \xrightarrow{\simeq} Fy$ if $x \to y$ is in W for iΓ^0-local F. Since $\underline{\mathrm{Hom}}_{\mathcal{D}_0^{\mathcal{C}^{\mathrm{op}}}}(\tilde{h}_x, \hat{F}) \xrightarrow{\simeq} \underline{\mathrm{Hom}}_{\mathcal{D}_0^{\mathcal{C}^{\mathrm{op}}}}(\tilde{h}_y, \hat{F})$ is an equivalence in the Reedy model structure of $(\mathcal{D}_0^{\mathcal{C}^{\mathrm{op}}})^{\Delta^{\mathrm{op}}}$, this means that for all $[n] \in \Delta$, we have

$$\begin{array}{ccc} \underline{\mathrm{Hom}}_{\mathcal{D}_0^{\mathcal{C}^{\mathrm{op}}}}(\tilde{h}_x, \hat{F})_n & \xrightarrow{\simeq} & \underline{\mathrm{Hom}}_{\mathcal{D}_0^{\mathcal{C}^{\mathrm{op}}}}(\tilde{h}_y, \hat{F})_n \\ \| & & \| \\ \underline{\mathrm{Hom}}_{\mathcal{D}_0^{\mathcal{C}^{\mathrm{op}}}}(\tilde{h}_x, \hat{F}_n) & \overset{\simeq}{\dashrightarrow} & \underline{\mathrm{Hom}}_{\mathcal{D}_0^{\mathcal{C}^{\mathrm{op}}}}(\tilde{h}_y, \hat{F}_n). \end{array} \tag{27.7}$$

$\hat{F}$ being a simplicial resolution of F means $\mathrm{cs}_*(F) \xrightarrow{\simeq} \hat{F}$, where cs_* is the constant simplicial functor. This being a Reedy weak equivalence in $(\mathcal{D}_0^{\mathcal{C}^{\mathrm{op}}})^{\Delta^{\mathrm{op}}}$, it follows that for all $[n] \in \Delta$, we have equivalences in $\mathcal{D}_0^{\mathcal{C}^{\mathrm{op}}}$: $F = (\mathrm{cs}_*(F))_n \xrightarrow{\simeq} \hat{F}_n$. Using this fact, along with (27.7), (27.2), and the enriched Yoneda lemma (27.5), we then have

$$\begin{array}{ccc} \hat{F}_n(x) \cong \int_{t \in \mathcal{C}^{\mathrm{op}}} \underline{\mathrm{Hom}}_{\mathcal{D}_0}(\tilde{h}_x(t), \hat{F}_n(t)) & \xrightarrow{\simeq} & \hat{F}_n(y) \cong \int_{t \in \mathcal{C}^{\mathrm{op}}} \underline{\mathrm{Hom}}_{\mathcal{D}_0}(\tilde{h}_y(t), \hat{F}_n(t)) \\ \simeq \uparrow \mathcal{D}_0 & & \simeq \uparrow \mathcal{D}_0 \\ F(x) & \underset{\simeq}{\dashrightarrow} & F(y) \end{array}$$

where the bottom equivalence follows from the 2–3 property. This shows iΓ-local objects map equivalences to equivalences.

27.3.4 Internal Hom is fibrant

Crucial in all our work is the fact that internal homotopy function complexes are fibrant objects. This is needed to use Theorem 17.7.7 to prove its modified version, Theorem 27.3.1.2. We state this as the following result.

Proposition 27.3.4.1. *Let C be cofibrant in $\mathcal{D}_0^{\mathcal{C}^{\mathrm{op}}}$, and let $\hat{X}$ be a simplicial resolution of $X \in \mathcal{D}_0^{\mathcal{C}^{\mathrm{op}}}$. Then $\underline{\mathrm{Hom}}_{\mathcal{D}_0^{\mathcal{C}^{\mathrm{op}}}}(C, \hat{X})$ is fibrant in the Reedy model structure of $(\mathcal{D}_0^{\mathcal{C}^{\mathrm{op}}})^{\Delta^{\mathrm{op}}}$.*

Proof. $\underline{\mathrm{Hom}}(C,\hat{X})$ fibrant means $\underline{\mathrm{Hom}}(C,\hat{X}) \to *$ is a fibration in $(\mathcal{D}_0^{\mathcal{C}^{\mathrm{op}}})^{\Delta^{\mathrm{op}}}$, where $*$ denotes the terminal object of $(\mathcal{D}_0^{\mathcal{C}^{\mathrm{op}}})^{\Delta^{\mathrm{op}}}$. This is true if for all $[n] \in \Delta$,

$$\underline{\mathrm{Hom}}(C,\hat{X})_n = \underline{\mathrm{Hom}}(C,\hat{X}_n) \to * \times_{M_n *} M_n\underline{\mathrm{Hom}}(C,\hat{X})$$

is a fibration in $\mathcal{D}_0^{\mathcal{C}^{\mathrm{op}}}$, where M_nA is the nth matching object of A (see [40]). Given the definition of the matching object, the object on the right of this map simplifies as $M_n\underline{\mathrm{Hom}}(C,\hat{X})$, which is equal to $\underline{\mathrm{Hom}}(C,M_n\hat{X})$, since the matching object is a limit, and the internal Hom commutes with limits. Thus we seek to show that

$$\underline{\mathrm{Hom}}(C,\hat{X}_n) \to \underline{\mathrm{Hom}}(C,M_n\hat{X})$$

is a fibration in $\mathcal{D}_0^{\mathcal{C}^{\mathrm{op}}}$. Let $D \to E$ be a trivial cofibration in $\mathcal{D}_0^{\mathcal{C}^{\mathrm{op}}}$. We need a lift α in the following commutative diagram:

$$\begin{array}{ccc} D & \longrightarrow & \underline{\mathrm{Hom}}(C,\hat{X}_n) \\ {\scriptstyle \text{triv cof}}\downarrow & \overset{\alpha}{\nearrow} & \downarrow \\ E & \longrightarrow & \underline{\mathrm{Hom}}(C,M_n\hat{X}). \end{array}$$

By adjunction this is equivalent to having a lift in the diagram

$$\begin{array}{ccc} D \otimes C & \longrightarrow & X_n \\ \downarrow & \nearrow & \downarrow \\ E \otimes C & \longrightarrow & M_n\hat{X} \end{array}$$

but C is cofibrant, $\otimes$ is a left Quillen functor, so $D \otimes C \to E \otimes C$ is a trivial cofibration, and $\hat{X}$ being a simplicial resolution, it is fibrant in $(\mathcal{D}_0^{\mathcal{C}^{\mathrm{op}}})^{\Delta^{\mathrm{op}}}$, so we have such a lift, which completes the proof. □

The other internal homotopy function complex we work with is the one we consider after having taken a left Bousfield localization with respect to Γ, and we find ourselves in $\mathcal{D}_0^{\mathcal{C}^{\mathrm{op}}\wedge} = L_\Gamma \mathcal{D}_0^{\mathcal{C}^{\mathrm{op}}}$. The iLhfC of interest is now $\underline{\mathrm{Hom}}_{\mathcal{D}_0^{\mathcal{C}^{\mathrm{op}}\wedge}}(\tilde{k_0},\hat{F})$, where $k_0 \in K_0$, $\tilde{k_0}$ is a cosimplicial resolution of k_0, and $\hat{F}$ is a fibrant approximation to F. As a preliminary result, we prove the following:

Theorem 27.3.4.2. *For a cofibrant approximation $\tilde{D}$ to D in $\mathcal{D}_0^{\mathcal{C}^{\mathrm{op}}\wedge}$, a fibrant approximation $\hat{X}$ to X, and a cosimplicial resolution $\tilde{k_0}$ of k_0, we have*

$$\mathrm{Hom}_{\mathcal{D}_0^{\mathcal{C}^{\mathrm{op}}\wedge}}(\tilde{D},\underline{\mathrm{Hom}}_{\mathcal{D}_0^{\mathcal{C}^{\mathrm{op}}\wedge}}(\tilde{k_0},\hat{X})) \cong \mathrm{Hom}_{\mathcal{D}_0^{\mathcal{C}^{\mathrm{op}}\wedge}}(\tilde{D}\otimes\tilde{k_0},\hat{X}).$$

Proof. Since for a cosimplicial resolution $\tilde{k}_0$, we have $\underline{\mathrm{Hom}}(\tilde{k}_0, A)_n = \underline{\mathrm{Hom}}(\tilde{k}_{0n}, A)$, and for a simplicial resolution $\hat{W}$, we have $\mathrm{Hom}(Y, \hat{W})_n = \mathrm{Hom}(Y, \hat{W}_n)$, it suffices to show the following isomorphism of simplicial sets on components:

$$\begin{aligned}\mathrm{Hom}(\tilde{D}, \underline{\mathrm{Hom}}(\tilde{k}_0, \hat{X}))_n &= \mathrm{Hom}(\tilde{D}, \underline{\mathrm{Hom}}(\tilde{k}_0, \hat{X})_n)\\ &= \mathrm{Hom}(\tilde{D}, \underline{\mathrm{Hom}}(\tilde{k}_{0n}, \hat{X}))\\ &\cong \mathrm{Hom}(\tilde{D} \otimes \tilde{k}_{0n}, \hat{X})\\ &= \mathrm{Hom}((\tilde{D} \otimes \tilde{k}_0)_n, \hat{X})\\ &= \mathrm{Hom}(\tilde{D} \otimes \tilde{k}_0, \hat{X})_n.\end{aligned}$$ □

Proposition 27.3.4.3. *For a cofibrant approximation $\tilde{k}_0$ to k_0, a simplicial resolution $\hat{X}$ of X, and objects of $\mathcal{D}_0^{\mathcal{C}^{\mathrm{op}}}\wedge$, $\underline{\mathrm{Hom}}_{\mathcal{D}_0^{\mathcal{C}^{\mathrm{op}}}\wedge}(\tilde{k}_0, \hat{X})$ is fibrant in the Reedy model structure of $(\mathcal{D}_0^{\mathcal{C}^{\mathrm{op}}}\wedge)^{\Delta^{\mathrm{op}}}$.*

Proof. It follows from the adjunction isomorphism, that the adjoint to

$$\begin{array}{ccc} h_0 & \longrightarrow & \underline{\mathrm{Hom}}(\tilde{k}_0, \hat{X}) \\ \downarrow & & \downarrow \\ l_0 & \longrightarrow & \underline{\mathrm{Hom}}(\tilde{t}_0, \hat{X}) \times_{\underline{\mathrm{Hom}}(\tilde{t}_0, \hat{Y})} \underline{\mathrm{Hom}}(\tilde{k}_0, \hat{Y}) \end{array}$$

is the following diagram:

$$\begin{array}{ccc} \tilde{t}_0 \otimes l_0 \coprod_{\tilde{t}_0 \otimes h_0} \tilde{k}_0 \otimes h_0 & \longrightarrow & \hat{X} \\ \downarrow & & \downarrow \\ \tilde{k}_0 \otimes l_0 & \longrightarrow & \hat{Y}. \end{array}$$

In particular, if L_n denotes the nth-latching object functor, then

$$\begin{array}{ccc} \tilde{C} & \longrightarrow & \underline{\mathrm{Hom}}(\tilde{k}_{0,n}, \hat{X}) \\ \downarrow & & \downarrow \\ \tilde{D} & \longrightarrow & \underline{\mathrm{Hom}}(L_n \tilde{k}_0, \hat{X}) \times_{\underline{\mathrm{Hom}}(L_n \tilde{k}_0, \hat{Y})} \underline{\mathrm{Hom}}(\tilde{k}_{0,n}, \hat{Y}) \end{array}$$

which simplifies to

$$\begin{array}{ccc} \tilde{C} & \longrightarrow & \underline{\mathrm{Hom}}(\tilde{k}_0, \hat{X}) \\ \downarrow & & \downarrow \\ \tilde{D} & \longrightarrow & M_n \underline{\mathrm{Hom}}(\tilde{k}_0, \hat{X}). \end{array}$$

If $\hat{Y} = *$, then we have

$$\begin{array}{ccc} L_n\tilde{k_0} \otimes \tilde{D} \coprod_{L_n\tilde{k_0}\otimes\tilde{C}} \tilde{k_{0,n}} \otimes \tilde{C} & \longrightarrow & \hat{X} \\ \downarrow & & \downarrow \\ \tilde{k_{0,n}} \otimes \tilde{D} & \longrightarrow & \star \end{array}$$

for adjoint. Now the map on the right is a fibration. If the one on the left is a trivial cofibration, then we have a lift, and that would prove our claim. But $\tilde{C} \to \tilde{D}$ is a cofibration, $\otimes$ is a Quillen bifunctor since $(\mathcal{D}_0^{\mathcal{C}^{\mathrm{op}}} {}^\wedge)^{\Delta^{\mathrm{op}}}$ is a symmetric monoidal model category as argued in Section 27.2.2.4. Hence $\tilde{k_0}$ being cofibrant in $(\mathcal{D}_0^{\mathcal{C}^{\mathrm{op}}} {}^\wedge)^{\Delta^{\mathrm{op}}}$ as a cosimplicial resolution, the functor $\tilde{k_0} \otimes - : \mathcal{D}_0^{\mathcal{C}^{\mathrm{op}}} {}^\wedge \to (\mathcal{D}_0^{\mathcal{C}^{\mathrm{op}}} {}^\wedge)^{\Delta^{\mathrm{op}}}$ is a left Quillen functor and so preserves trivial cofibrations, and thus $\tilde{k_0} \otimes \tilde{C} \to \tilde{k_0} \otimes \tilde{D}$ is a trivial cofibration in $(\mathcal{D}_0^{\mathcal{C}^{\mathrm{op}}} {}^\wedge)^{\Delta^{\mathrm{op}}}$, which means exactly that the left vertical map in the above commutative diagram is a trivial cofibration in $\mathcal{D}_0^{\mathcal{C}^{\mathrm{op}}} {}^\wedge$ by the definition of cofibrations in Reedy model categories. This completes our proof. □

27.3.5 Simplicial resolution

In this subsection, we show the following fact, which is implied in the proof of Proposition 27.3.1.2.

Proposition 27.3.5.1. *If $\hat{X}$ is a simplicial resolution of X, then $\underline{\mathrm{Hom}}(A, \hat{X})$ is a simplicial resolution of $\underline{\mathrm{Hom}}(A, X)$.*

Proof. This follows from Proposition 17.4.16 of [40], which states that if $F : \mathcal{M} \rightleftarrows \mathcal{N} : G$ is a Quillen adjunction, X is cofibrant in $\mathcal{M}$, Y is fibrant in $\mathcal{N}$, and $\hat{Y}$ is a simplicial resolution of Y, then $G(\hat{Y})$ is a simplicial resolution of $G(Y)$. We apply this to the case

$$- \otimes A : \mathcal{D}_0^{\mathcal{C}^{\mathrm{op}}} \rightleftarrows \mathcal{D}_0^{\mathcal{C}^{\mathrm{op}}} : \underline{\mathrm{Hom}}(A, -) = G.$$

Let $\hat{Y}$ be a simplicial resolution of Y fibrant. Then $G(\hat{Y}) = \underline{\mathrm{Hom}}(A, \hat{Y})$ is a simplicial resolution of $G(Y) = \underline{\mathrm{Hom}}(A, Y)$. For any Y, take a fibrant replacement of $\underline{\mathrm{Hom}}(A, Y)$

$$\begin{aligned} \underline{\mathrm{Hom}}(A, Y) \xrightarrow{\simeq} R\underline{\mathrm{Hom}}(A, Y) &= RG(Y) \\ &= G(RY) = \underline{\mathrm{Hom}}(A, RY) \end{aligned}$$

so that we have $\underline{\mathrm{Hom}}(A, Y) \xrightarrow{\simeq} \underline{\mathrm{Hom}}(A, RY)$, which implies

$$\mathrm{cs}_* \underline{\mathrm{Hom}}(A, Y) \xrightarrow{\simeq} \mathrm{cs}_* \underline{\mathrm{Hom}}(A, RY) \xrightarrow{\simeq} \underline{\mathrm{Hom}}(A, \hat{Y}).$$

Here $\hat{Y}$ is a simplicial resolution of RY. But $Y \xrightarrow{\simeq} RY$ implies $\mathrm{cs}_* Y \xrightarrow{\simeq} \mathrm{cs}_* RY \xrightarrow{\simeq} \hat{Y}$, so $\hat{Y}$ is also a simplicial resolution of Y. For such a simplicial resolution, $\underline{\mathrm{Hom}}(A, \hat{Y})$ is a simplicial resolution of $\underline{\mathrm{Hom}}(A, Y)$ as shown above. This is independent of the choice of simplicial resolution since all such resolutions are Reedy equivalent by Lemma 16.1.22 of [40]. □

In the same manner, we would show that if $\tilde{Y}$ is a cosimplicial resolution of Y, then $\underline{\mathrm{Hom}}(\tilde{Y}, B)$ is a simplicial resolution of $\underline{\mathrm{Hom}}(Y, B)$.

27.3.6 Equivalence of internal homotopy function complexes

Our internal homotopy function complexes are defined as the homotopy function complexes of [40], save that instead of using Hom sets, we use internal Homs. We want those internal homotopy function complexes to be independent of the choice of resolutions and approximations used in defining them. We first need a couple of definitions, variants of those found in [40].

Definition 27.3.6.1. A *change of iRhfC map*:

$$(\tilde{X}, \hat{Y}, \underline{\mathrm{Hom}}(\tilde{X}, \hat{Y})) \to (\tilde{X'}, \hat{Y'}, \underline{\mathrm{Hom}}(\tilde{X'}, \hat{Y'}))$$

is a triple (f, g, h) formed of a map of cofibrant approximations $f : \tilde{X} \to \tilde{X'}$, a map of simplicial resolutions $g : \hat{Y} \to \hat{Y'}$, and the map of simplicial objects $h : \underline{\mathrm{Hom}}(\tilde{X}, \hat{Y}) \to \underline{\mathrm{Hom}}(\tilde{X'}, \hat{Y'})$ induced by f and g.

Definition 27.3.6.2. For $X, Y \in \mathcal{D}_0^{\mathcal{C}^{\mathrm{op}}}$, we define the *category* $\mathrm{iRhfC}(X, Y)$ to be the category of iRhfCs from X to Y and with changes of iRhfC maps as morphisms.

Theorem 27.3.6.3. *Let $X, Y \in \mathcal{D}_0^{\mathcal{C}^{\mathrm{op}}}$. Then any two iRhfCs from X to Y are connected by an essentially unique zig-zag of changes of iRhfC maps.*

Proof. This follows from Theorem 14.4.5 of [40], which states that if $\mathcal{C}$ is a category, $X, Y \in \mathcal{C}$, and $B\mathcal{C}$ is contractible, then there exists an essentially unique zig-zag from X to Y in $\mathcal{C}$, and from the proposition below, a modified version of Proposition 17.2.10 of [40] to the internal setting. □

Proposition 27.3.6.4. *Let $X, Y \in \mathcal{D}_0^{\mathcal{C}^{\mathrm{op}}}$. Then $B\mathrm{iRhfC}(X, Y)$ is contractible.*

Proof. We have $B\mathrm{iRhfC}(X, Y) = B\mathrm{sRes}(\underline{\mathrm{Hom}}(\tilde{X}, Y))$ by Section 27.3.5, where sRes stands for simplicial resolution, and this latter category is contractible by Proposition 16.1.5 of [40]. □

27.4 The cardinal γ in the proof of Proposition 4.5.1

To prove that we have an internal left Bousfield localization, we use Theorem 11.3.1 of [40], which itself needs Proposition 4.5.1 that we generalize to our setting. The proof of the latter proposition in the classical case uses a cardinal γ. Following Definition 4.5.3 of [40], $\gamma = \xi^{\xi}$, where ξ is the smallest cardinal that is at least as large as any of the following five cardinals. We use the same definitions in the generalized case.

- *Size of the cells of $\mathcal{D}_0^{\mathcal{C}^{\mathrm{op}}}$*: by definition the size of the cells of a cellular model category $\mathcal{M}$ is the smallest cardinal for which Theorem 12.3.1 of [40] holds. This theorem makes no use of a notion of equivalence and can be used as is, hence holds also in the internal setting. Hence we can define the size of the cells of $\mathcal{D}_0^{\mathcal{C}^{\mathrm{op}}}$ following that result.
- *Compactness of the domains of I*: $\mathcal{D}_0^{\mathcal{C}^{\mathrm{op}}}$ is a cellular model category, in particular, cofibrantly generated, and let I denote its set of generating cofibrations. Let η, as in [40], be a cardinal such that the domains of elements of I are η-compact.
- *Cardinal λ in the proof of Theorem 4.3.1*: As in [40], let λ be the cardinal used in the proof of Theorem 4.3.1. This result invokes a set $\widetilde{\Lambda\Gamma}$, originally introduced in Proposition 4.2.5 of [40], which uses equivalences and therefore needs to be stated and proved in the internal setting. To prove it, we invoke the equivalence between Γ-local equivalences and $i\Gamma$-local equivalences (Lemma 27.5.11). In the proof of the original proposition, one result is interesting in its own right, and we prove it in the internal case; it is Proposition 27.4.1. We need it in Section 27.2.2.5.

Proposition[3.1.5 mod] 27.4.1. *Let Γ be a class of maps in $\mathcal{D}_0^{\mathcal{C}^{\mathrm{op}}}$. Then any weak equivalence in $\mathcal{D}_0^{\mathcal{C}^{\mathrm{op}}}$ is also an $i\Gamma$-local equivalence.*

Proof. Let $A \to B$ be a weak equivalence in $\mathcal{D}_0^{\mathcal{C}^{\mathrm{op}}}$. If X is an $i\Gamma$-local object, then we want $\underline{\mathrm{Hom}}(\tilde{B},\hat{X}) \xrightarrow{\simeq} \underline{\mathrm{Hom}}(\tilde{A},\hat{X})$, where $\tilde{A}$ is a cofibrant approximation to A, and $\hat{X}$ is a simplicial resolution to X. For cofibrant C in $\mathcal{D}_0^{\mathcal{C}^{\mathrm{op}}}$, we have:

$$
\begin{array}{ccc}
\mathrm{Hom}(C, \underline{\mathrm{Hom}}(\tilde{B},\hat{X})) & \longrightarrow & \mathrm{Hom}(C, \underline{\mathrm{Hom}}(\tilde{A},\hat{X})) \\
\downarrow{\scriptstyle \cong} & & \downarrow{\scriptstyle \cong} \\
\mathrm{Hom}(C \otimes \tilde{B}, \hat{X}) & \longrightarrow & \mathrm{Hom}(C \otimes \tilde{A}, \hat{X}).
\end{array}
$$

Now we use the fact that $\otimes$ being a left Quillen functor and C being cofibrant, if $\tilde{A} \to \tilde{B}$ is a fibrant–cofibrant approximation to $A \to B$ that is a cofibration, then $C\otimes\tilde{A} \to C\otimes\tilde{B}$ is a trivial cofibration as well; in particular, it is a weak equivalence, so by the original result of [40] it is a Γ-local equivalence, so that the bottom horizontal map above is an equivalence, since $i\Gamma$-local X is also Γ-local by Lemma 27.5.10. It follows from the above commutative diagram and Theorem 17.7.7 of [40] that we have an equivalence $\underline{\mathrm{Hom}}(\tilde{B},\hat{X}) \to \underline{\mathrm{Hom}}(\tilde{A},\hat{X})$ in $(\mathcal{D}_0^{\mathcal{C}^{\mathrm{op}}})^{\Delta^{\mathrm{op}}}$, that is, $A \to B$ is an $i\Gamma$-local equivalence. □

Proposition [4.2.5 mod] 27.4.2. *If I denotes the set of generating cofibrations of $\mathcal{D}_0^{\mathcal{C}^{\mathrm{op}}}$ and Γ is a class of maps in $\mathcal{D}_0^{\mathcal{C}^{\mathrm{op}}}$, then there exists a set $\widetilde{\Lambda\Gamma}$ of relative I-cell complexes* *whose domains are cofibrant such that every element of $\widetilde{\Lambda\Gamma}$ is an $i\Gamma$-local equivalence, and an object W is $i\Gamma$-local if and only if $W \to *$ is $\widetilde{\Lambda\Gamma}$-injective.*

Proof. By Lemmas 27.5.10 and 27.5.11 this is equivalent to the original proposition. □

The cardinal in Theorem 4.3.1 of [40] is $\lambda = \mathrm{succ}(\kappa)$, where κ is a cardinal that, according to Lemma 10.4.6 of [40], is such that the domain of every element of $\widetilde{\Lambda\Gamma}$ is κ-small relative to the subcategory of relative $\widetilde{\Lambda\Gamma}$-cell complexes. This is a classical notion and needs not be generalized.

- *Cardinal κ in Proposition 12.5.3 of* [40]: this proposition is applied to the set $\widetilde{\Lambda\Gamma}$. It mentions a cardinal κ at least as large as four kinds of cardinals, two of which are cardinals given by Proposition 12.5.2 of [40], which makes use of a Hom set. We generalize this proposition as follows:

Proposition [12.5.2 mod] 27.4.0.1. *If X is a cofibrant object of $\mathcal{D}_0^{\mathcal{C}^{\mathrm{op}}}$, then there is a cardinal η such that for a cardinal $\nu \geq 2$ and a cell complex Y of size ν, $\underline{\mathrm{Hom}}(X, Y)$ has cardinality at most ν^{η}.*

Proof. Let C be cofibrant in $\mathcal{D}_0^{\mathcal{C}^{\mathrm{op}}}$. We have $\mathrm{Hom}(C, \underline{\mathrm{Hom}}(X, Y)) \cong \mathrm{Hom}(C \otimes X, Y)$. Since we are in a monoidal model category, $C \otimes X$ is again cofibrant. We apply the original proposition of [40] to cofibrant $C \otimes X$ and Y, which gives $\mathrm{size}(\mathrm{Hom}(C \otimes X, Y)) \leq \nu^{\eta}$. Finally, $\mathrm{size}(\underline{\mathrm{Hom}}(X, Y)) < \mathrm{size}(\mathrm{Hom}(C, \underline{\mathrm{Hom}}(X, Y)) = \mathrm{size}(\mathrm{Hom}(C \otimes X, Y))$, which completes the proof. □

- *Cardinal κ in Proposition 12.5.7 of* [40]: κ is an infinite cardinal at least as large as four types of cardinals, two of which are given by Proposition 27.4.0.1, and one of which is given by Definition 12.5.5 of [40], which invokes a smallness argument and hence does not need to be modified.

27.5 Results needed for an internal left Bousfield localization

The following result is the first one that is needed to prove that we do have an internal left Bousfield localization.

Proposition [3.2.3 mod] 27.5.1. *For a class of maps Γ in $\mathcal{D}_0^{\mathcal{C}^{\mathrm{op}}}$, the class of $i\Gamma$-local equivalences satisfies the 2–3 property.*

Proof. This is just a generalization of Hirschhorn's proof in [40]: let $g : X \to Y$ and $h : Y \to Z$ be maps, and to those apply a functorial cofibrant factorization

$$\begin{array}{ccccc} \tilde{X} & \xrightarrow{\tilde{g}} & \tilde{Y} & \xrightarrow{\tilde{h}} & \tilde{Z} \\ \downarrow & & \downarrow & & \downarrow \\ X & \xrightarrow{g} & Y & \xrightarrow{h} & Z \end{array}$$

where $\tilde{g}$, $\tilde{h}$, and $\tilde{hg}$ are cofibrant approximations to g, h, and hg, respectively. Those exist by Proposition 8.1.23 of [40]. Let W be an iΓ-local object, and let $\hat{W}$ be a simplicial resolution of W. To say, for example, that g is an iΓ-local equivalence would mean that

$$\underline{\mathrm{Hom}}_{\mathcal{D}_0^{\mathcal{C}^{\mathrm{op}}}}(\tilde{Y}, \hat{W}) \xrightarrow{\simeq} \underline{\mathrm{Hom}}_{\mathcal{D}_0^{\mathcal{C}^{\mathrm{op}}}}(\tilde{X}, \hat{W})$$

is an equivalence in $(\mathcal{D}_0^{\mathcal{C}^{\mathrm{op}}})^{\Delta^{\mathrm{op}}}$, where equivalences satisfy the 2–3 property. So if two of $\tilde{g}^* : \underline{\mathrm{Hom}}(\tilde{Y}, \hat{W}) \to \underline{\mathrm{Hom}}(\tilde{X}, \hat{W})$, $\tilde{h}^* : \underline{\mathrm{Hom}}(\tilde{Z}, \hat{W}) \to \underline{\mathrm{Hom}}(\tilde{Y}, \hat{W})$, and $(\tilde{hg})^* : \underline{\mathrm{Hom}}(\tilde{Z}, \hat{W}) \to \underline{\mathrm{Hom}}(\tilde{X}, \hat{W})$ are weak equivalences, then so is the third, which completes the proof. □

In the statement of Theorem 11.3.1 of [40], mention is made of a set J, which exists by Proposition 4.5.1 of the same reference, which we generalize presently. Its proof uses three results of [40], two of which make use of a notion of equivalence and therefore have to be generalized. Their proof needs the following definition, along with two lemmas, Lemmas 27.5.10 and 27.5.11, which we state and prove after those two results.

Definition 27.5.2. A class of maps S in a symmetric monoidal model category $\mathcal{M}$ is said to be an *ideal class* of maps if for all $f : A \to B$ in S and all objects C of $\mathcal{M}$, $C \otimes A \to C \otimes B$ is also in S.

Observe that the set of iΓ-local equivalences and the set of Γ-local equivalences form ideal classes:

Lemma 27.5.3. *The class of* iΓ*-local equivalences forms an ideal class.*

Proof. Let $f : X \to Y$ be an iΓ-local equivalence, and let C be any object. We show that $C \otimes f$ is also an iΓ-local equivalence, that is, if W is an iΓ-local object, then $\underline{\mathrm{Hom}}(\tilde{Y}, \hat{W}) \xrightarrow{\simeq} \underline{\mathrm{Hom}}(\tilde{X}, \hat{W})$ implies $\underline{\mathrm{Hom}}(\tilde{C} \otimes \tilde{Y}, \hat{W}) \xrightarrow{\simeq} \underline{\mathrm{Hom}}(\tilde{C} \otimes \tilde{X}, \hat{W})$. Let Z be any object. We have

$$\begin{array}{ccc}
\mathrm{Hom}(\tilde{Z}, \underline{\mathrm{Hom}}(\tilde{C} \otimes \tilde{Y}, \hat{W})) & & \\
\downarrow \cong & & \\
\mathrm{Hom}(\tilde{Z} \otimes \tilde{C} \otimes \tilde{Y}, \hat{W}) & \xrightarrow{\cong} & \mathrm{Hom}(\tilde{Z} \otimes \tilde{C}, \underline{\mathrm{Hom}}(\tilde{Y}, \hat{W})) \\
 & & \downarrow \cong \\
 & \cong & \mathrm{Hom}(\tilde{Z} \otimes \tilde{C}, \underline{\mathrm{Hom}}(\tilde{X}, \hat{W})) \\
 & & \downarrow \cong \\
 & & \mathrm{Hom}(\tilde{Z} \otimes \tilde{C} \otimes \tilde{X}, \hat{W}) \\
 & & \downarrow \cong \\
 & & \mathrm{Hom}(\tilde{Z}, \underline{\mathrm{Hom}}(\tilde{C} \otimes \tilde{X}, \hat{W})).
\end{array}$$

This for any Z shows by Theorem 17.7.7 of [40] that we have our desired equivalence, and hence $C \otimes f$ is an iΓ-local equivalence. □

The following lemma is the first one that necessitates that we use Γ as a collection of morphisms for localization purposes, and not simply Γ^0.

Lemma 27.5.4. *The class of Γ-local equivalences forms an ideal class.*

Proof. Let $f : X \to Y$ be a Γ-local equivalence, and let C be any object of $\mathcal{D}_0^{\mathcal{C}^{op}}$. We show that $C \otimes f$ is also a Γ-local equivalence, that is, for any Γ-local object W, we have $\mathrm{Hom}(\widetilde{C \otimes Y}, \hat{W}) \xrightarrow{\simeq} \mathrm{Hom}(\widetilde{C \otimes X}, \hat{W})$. Since $\mathrm{Hom}(\widetilde{C \otimes X}, \hat{W}) \cong \mathrm{Hom}(\tilde{X}, \underline{\mathrm{Hom}}(\tilde{C}, \hat{W}))$, it suffices to show that $\underline{\mathrm{Hom}}(\tilde{C}, \hat{W})$ is a Γ-local object for all C and Γ-local W. Consider

$$\begin{array}{ccc}
\mathrm{Hom}(\tilde{h}_x \otimes \tilde{D}, \underline{\mathrm{Hom}}(\tilde{C}, \hat{W})) & \longrightarrow & \mathrm{Hom}(\tilde{h}_y \otimes \tilde{D}, \underline{\mathrm{Hom}}(\tilde{C}, \hat{W})) \\
\cong \downarrow & & \downarrow \cong \\
\mathrm{Hom}(\tilde{h}_x \otimes \tilde{D} \otimes \tilde{C}, \hat{W}) & \longrightarrow & \mathrm{Hom}(\tilde{h}_y \otimes \tilde{D} \otimes \tilde{C}, \hat{W}) \\
\cong \downarrow & & \downarrow \cong \\
\mathrm{Hom}(\tilde{C}, \underline{\mathrm{Hom}}(\tilde{h}_x \otimes \tilde{D}, \hat{W})) & \xrightarrow{\simeq} & \mathrm{Hom}(\tilde{C}, \underline{\mathrm{Hom}}(\tilde{h}_y \otimes \tilde{D}, \hat{W})).
\end{array}$$

Because W is Γ-local, $\underline{\mathrm{Hom}}(\tilde{h}_y \otimes \tilde{D}, \hat{W}) \xrightarrow{\simeq} \underline{\mathrm{Hom}}(\tilde{h}_x \otimes \tilde{D}, \hat{W})$, which gives us the bottom horizontal equivalence above by Theorem 17.7.7 of [40] and hence a top horizontal one as well by commutativity, which shows that $\underline{\mathrm{Hom}}(\tilde{C}, \hat{W})$ is Γ-local. This completes the proof. □

Definition 27.5.5. A class of objects $\mathcal{C}$ in a symmetric monoidal model category $\mathcal{M}$ is said to be an *ideal class of objects* if for any $C \in \mathcal{C}$ and any object X of $\mathcal{M}$, $X \otimes C$ is again in $\mathcal{C}$.

We need the following fact for having an internal right Bousfield localization.

Lemma 27.5.6. *In a right proper cellular model category $\mathcal{M}$, if K is a class of objects in $\mathcal{M}$ and $\mathcal{K}$ is the class of iK-colocal equivalences (Definition 27.8.2), then i$\mathcal{K}$-colocal objects (Definition 27.8.1) form an ideal class.*

Proof. Let W be an i$\mathcal{K}$-colocal object, and let $f : X \to Y$ be an element of $\mathcal{K}$. We have an equivalence of iLhfC's: $\underline{\mathrm{Hom}}(\tilde{W}, \hat{X}) \xrightarrow{\simeq} \underline{\mathrm{Hom}}(\tilde{W}, \hat{Y})$. Now let $D \in \mathcal{M}$. We wish to show that $\underline{\mathrm{Hom}}(\widetilde{D \otimes W}, \hat{X}) \xrightarrow{\simeq} \underline{\mathrm{Hom}}(\widetilde{D \otimes W}, \hat{Y})$. It suffices to consider for all Z cofibrant in $\mathcal{M}$:

$$
\begin{array}{ccc}
\operatorname{Hom}(Z, \underline{\operatorname{Hom}}(\widetilde{D \otimes W}, \hat{X})) & \dashrightarrow & \operatorname{Hom}(Z, \underline{\operatorname{Hom}}(\widetilde{D \otimes W}, \hat{Y})) \\
\Vert & & \Vert \\
\operatorname{Hom}(Z, \underline{\operatorname{Hom}}(\operatorname{diag}\tilde{D} \otimes \tilde{W}, \hat{X})) & & \operatorname{Hom}(Z, \underline{\operatorname{Hom}}(\operatorname{diag}\tilde{D} \otimes \tilde{W}, \hat{Y})) \\
\downarrow \cong & & \downarrow \cong \\
\operatorname{diagHom}(Z \otimes \tilde{D}, \underline{\operatorname{Hom}}(\tilde{W}, \hat{X})) & \xrightarrow[\simeq]{} & \operatorname{diagHom}(Z \otimes \tilde{D}, \underline{\operatorname{Hom}}(\tilde{W}, \hat{Y})).
\end{array}
$$

Having the bottom equivalence by the definition of an $i\mathcal{K}$-colocal object, it follows that the top horizontal map is an equivalence, and hence by Theorem 17.7.7 of [40] we have our desired equivalence, that is, $D \otimes W$ is $i\mathcal{K}$-colocal. □

In the above proof, we used the following:

Lemma 27.5.7.

$$\operatorname{diagHom}(Z \otimes \tilde{D}, \underline{\operatorname{Hom}}(\tilde{W}, \hat{Y})) \cong \operatorname{Hom}(Z, \underline{\operatorname{Hom}}(\operatorname{diag}(\tilde{D} \otimes \tilde{W}), \hat{Y})).$$

Proof. We check this componentwise:

$$
\begin{aligned}
\operatorname{diagHom}(Z \otimes \tilde{D}, \underline{\operatorname{Hom}}(\tilde{W}, \hat{Y}))_n &= \operatorname{Hom}((Z \otimes \tilde{D})_n, (\underline{\operatorname{Hom}}(\tilde{W}, \hat{Y}))_n) \\
&= \operatorname{Hom}(Z \otimes \tilde{D}_n, \underline{\operatorname{Hom}}(\tilde{W}_n, \hat{Y})) \\
&\cong \operatorname{Hom}(Z, \underline{\operatorname{Hom}}(\tilde{D}_n \otimes \tilde{W}_n, \hat{Y})) \\
&= \operatorname{Hom}(Z, \underline{\operatorname{Hom}}(\operatorname{diag}(\tilde{D} \otimes \tilde{W})_n, \hat{Y})) \\
&= \operatorname{Hom}(Z, \underline{\operatorname{Hom}}(\operatorname{diag}(\tilde{D} \otimes \tilde{W}), \hat{Y})_n) \\
&= \operatorname{Hom}(Z, \underline{\operatorname{Hom}}(\operatorname{diag}(\tilde{D} \otimes \tilde{W}), \hat{Y}))_n.
\end{aligned}
$$

□

The first result for an internal localization is the following:

Proposition [4.5.6 mod] 27.5.8. *If Γ is a set of maps in the left proper cellular model category $\mathcal{D}_0^{\mathcal{C}^{\mathrm{op}}}$ and if $p : X \to Y$ has the right lifting property with respect to those inclusions of subcomplexes $i : C \to D$ that are $i\Gamma$-local equivalences and for which the size of D is at most γ, the cardinal described in the previous section, then p has the right lifting property with respect to all inclusions of subcomplexes that are $i\Gamma$-local equivalences.*

Proof. Suppose $p : X \to Y$ is a map that satisfies the conditions of the proposition. Let $i : C \to D$ be an inclusion of cell subcomplexes that is an $i\Gamma$-local equivalence and such that the size of D is at most γ. It is also a Γ-local equivalence by Lemma 27.5.11. Since i is a Γ-local equivalence, p has the right lifting property with respect to all inclusions of subcomplexes that are also Γ-local equivalences and for which the size of D is at most γ. The original Proposition 4.5.6 of [40] can then be used to conclude that p has the right lifting property with respect to all inclusions of subcomplexes that are Γ-local equivalences, which are also $i\Gamma$-local equivalences by Lemma 27.5.11. This completes the proof. □

The second result is the following:

Lemma [4.5.2 mod] 27.5.9. *If Γ is a set of maps in the left proper cellular model category $\mathcal{D}_0^{\mathcal{C}^{\mathrm{op}}}$ and if $p : E \to B$ is a fibration with the right lifting property with respect to all inclusions of cell complexes that are iΓ-local equivalences, then it has the right lifting property with respect to all cofibrations that are iΓ-local equivalences.*

Proof. The reasoning is similar to the previous result; since iΓ-local equivalences are also Γ-local equivalences, the original Lemma 4.5.2 of [40] applies, and p has the right lifting property with respect to all cofibrations that are Γ-local equivalences, which are also iΓ-local equivalences. □

Lemma 27.5.10. *The class of iΓ-local objects of $\mathcal{D}_0^{\mathcal{C}^{\mathrm{op}}}$ coincides with the class of Γ-local objects.*

Proof. Let W be an iΓ-local object in $\mathcal{D}_0^{\mathcal{C}^{\mathrm{op}}}$, and let $A \to B$ be an object of Γ. Then $\underline{\mathrm{Hom}}(\tilde{B}, \hat{W}) \to \underline{\mathrm{Hom}}(\tilde{A}, \hat{W})$ is an equivalence, which implies the equivalence $\mathrm{Hom}(\tilde{B}, \hat{W}) \to \mathrm{Hom}(\tilde{A}, \hat{W})$, which exactly means that W is Γ-local as well. Conversely, let W be Γ-local, and let $A \to B$ in $\Gamma \subset \Gamma - \text{loc.equ's}$. Now those form an ideal class, so for any C, $C \otimes A \to C \otimes B$ is a Γ-local equivalence, and hence we have an equivalence at the bottom of the following commutative diagram:

$$\begin{array}{ccc}
\mathrm{Hom}(\tilde{C}, \underline{\mathrm{Hom}}(\tilde{B}, \hat{W})) & \longrightarrow & \mathrm{Hom}(\tilde{C}, \underline{\mathrm{Hom}}(\tilde{A}, \hat{W})) \\
\cong \downarrow & & \downarrow \cong \\
\mathrm{Hom}(\tilde{C} \otimes \tilde{B}, \hat{W}) & \underset{\simeq}{\dashrightarrow} & \mathrm{Hom}(\tilde{C} \otimes \tilde{A}, \hat{W})
\end{array}$$

and this for any C, so by Theorem 17.7.7 of [40] it follows that $\underline{\mathrm{Hom}}(\tilde{B}, \hat{W}) \xrightarrow{\simeq} \underline{\mathrm{Hom}}(\tilde{A}, \hat{W})$, that is, W is iΓ-local. □

Lemma 27.5.11. *The class of Γ-local equivalences coincides with the class of iΓ-local equivalences.*

Proof. Let $C \to D$ be a Γ-local equivalence. If A is an object of $\mathcal{D}_0^{\mathcal{C}^{\mathrm{op}}}$, then $A \otimes C \to A \otimes D$ is also a Γ-local equivalence since those form an ideal class. This means that for all W iΓ-local and, in particular, Γ-local by the preceding lemma,

$$\begin{array}{ccc}
\mathrm{Hom}(\widetilde{A \otimes D}, \hat{W}) & \xrightarrow{\simeq} & \mathrm{Hom}(\widetilde{A \otimes C}, \hat{W}) \\
\cong \downarrow & & \downarrow \cong \\
\mathrm{Hom}(\tilde{A}, \underline{\mathrm{Hom}}(\tilde{D}, \hat{W})) & \underset{\simeq}{\dashrightarrow} & \mathrm{Hom}(\tilde{A}, \underline{\mathrm{Hom}}(\tilde{C}, \hat{W})).
\end{array}$$

Having the equivalence at the bottom of this diagram implies that we have an equivalence $\underline{\mathrm{Hom}}(\tilde{D}, \hat{W}) \to \underline{\mathrm{Hom}}(\tilde{C}, \hat{W})$ for all iΓ-local objects W by Theorem 17.7.7 of [40], so $C \to D$ is also an iΓ-local equivalence. Conversely, since equivalences of ihfCs imply equivalences of hfCs, it follows that iΓ-local equivalences are also Γ-local equivalences, which completes the proof. □

Proposition [4.5.1 mod] 27.5.12. *In the left proper cellular model category $\mathcal{D}_0^{\mathcal{C}^{\text{op}}}$, if Γ is a set of maps, then there exists a set J_Γ of inclusions of cell complexes such that the class of iJ_Γ-cofibrations equals the class of cofibrations that are iΓ-local equivalences.*

Proof. The proof is almost identical to that of [40], save that it is generalized. We let J_Γ be a set of representatives of isomorphism classes of inclusions of cell complexes that are iΓ-local equivalences of size at most γ. This cardinal is the one described in the previous section. The result follows, as in [40] in the classical case, from Proposition 27.5.8 and Lemma 27.5.9, as well as Corollary 10.5.22 of [40], which we use verbatim since it does not make use of a notion of equivalence. □

27.6 Internal left Bousfield localization

Definition 27.6.1. For a class of maps Γ in $\mathcal{D}_0^{\mathcal{C}^{\text{op}}}$, $W \in \mathcal{D}_0^{\mathcal{C}^{\text{op}}}$ is said to be *iΓ-local* if it is fibrant and for any $f : A \to B$ in Γ, the induced map of iRhfCs $f^* : \underline{\text{Hom}}_{\mathcal{D}_0^{\mathcal{C}^{\text{op}}}}(\tilde{B}, \hat{W}) \to \underline{\text{Hom}}_{\mathcal{D}_0^{\mathcal{C}^{\text{op}}}}(\tilde{A}, \hat{W})$ is a weak equivalence in $(\mathcal{D}_0^{\mathcal{C}^{\text{op}}})^{\Delta^{\text{op}}}$.

Definition 27.6.2. A map $g : X \to Y$ in $\mathcal{D}_0^{\mathcal{C}^{\text{op}}}$ is said to be an *iΓ-local equivalence* if for all iΓ-local objects W, the induced map of iRhfCs $g^* : \underline{\text{Hom}}_{\mathcal{D}_0^{\mathcal{C}^{\text{op}}}}(\tilde{Y}, \hat{W}) \to \underline{\text{Hom}}_{\mathcal{D}_0^{\mathcal{C}^{\text{op}}}}(\tilde{X}, \hat{W})$ is a weak equivalence in $(\mathcal{D}_0^{\mathcal{C}^{\text{op}}})^{\Delta^{\text{op}}}$.

Theorem 27.6.3. *If Γ is a class of maps in the proper cellular model category $\mathcal{D}_0^{\mathcal{C}^{\text{op}}}$, then if we define a class of equivalences on $\underline{\mathcal{D}_0^{\mathcal{C}^{\text{op}}}}$ (the category underlying $\mathcal{D}_0^{\mathcal{C}^{\text{op}}}$) as i$\Gamma$-local equivalences, if we define cofibrations as those of $\mathcal{D}_0^{\mathcal{C}^{\text{op}}}$, and if we define fibrations as maps having the right lifting property with respect to those maps that are cofibrations and iΓ-local equivalences, then this produces a model structure on $\underline{\mathcal{D}_0^{\mathcal{C}^{\text{op}}}}$ that we denote by $L_\Gamma \mathcal{D}_0^{\mathcal{C}^{\text{op}}} = \mathcal{D}_0^{\mathcal{C}^{\text{op}}}\wedge$ and which we call an internal left Bousfield localization of $\mathcal{D}_0^{\mathcal{C}^{\text{op}}}$ along Γ. Further, $\mathcal{D}_0^{\mathcal{C}^{\text{op}}}\wedge$ is a left proper cellular model category.*

Proof. We adapt the classical proof, dependent upon Theorem 11.3.1 of [40], which we use as is. The reader is referred to [40] for its statement. By Proposition 27.5.1 the class of iΓ-local equivalences satisfies the 2–3 property. This is one needed assumption of Theorem 11.3.1. Another assumption about this class we need for Theorem 11.3.1 to hold is that it be closed under retracts. This follows in the classical case from Proposition 3.2.4 of [40], which holds in the internal case as well. Consider now the set J_Γ provided by Proposition 27.5.12. Let I be the set of generating cofibrations of $\mathcal{D}_0^{\mathcal{C}^{\text{op}}}$. By definition, I allows the small object argument, and hence condition 1 of Theorem 11.3.1 is satisfied. Since every element of J_Γ has a cofibrant domain, small relative to the subcategory of cofibrations by Theorem 12.4.3 of [40] as argued in the same paper, hence in particular small relative to J_Γ, this latter satisfies condition 1 of Theorem 11.3.1 as well. Indeed, elements of J_Γ are relative I-cell complexes, which are in I-Cof by Proposition 10.5.10 of [40], and this is the subcategory cof of cofibrations, so $J_\Gamma \subset$ cof. Condition 2 of Theorem 11.3.1 is that J_Γ-cof $\subset I$-Cof $\cap W$ with $W = i\Gamma -$ loc.equ's, and this follows from Proposition 27.5.12: J_Γ-cof $=$ cof $\cap W = I$-Cof $\cap W$. Condition 3 states I-inj $\subset J_\Gamma$-inj $\cap W$. Proposition 27.5.12 implies that J_Γ-cof is a subcategory of cof $= I$-Cof, and hence J_Γ-inf $\supset I$-inj. Finally, I-inj $=$ triv.fibr., in particular, weak equivalences, which are iΓ-local equivalences, i. e., in W, so I-inj $\subset J_\Gamma$-inf $\cap W$. The last condition of Theorem 11.3.1, condition 4a, is satisfied by Proposition 27.5.12, as argued in [40]. This completes the proof that we have a model structure. To show it yields a left proper cellular model category, we follow exactly the proof of [40], since there is no manifest modification going to the general case. □

27.7 Results needed in the proof of Theorem 5.1.1

We will prove a modified version of Theorem 5.1.1 of [40], which necessitates that the model category we are localizing be right proper. For us, that would be $L_\Gamma \mathcal{D}_0^{\mathcal{C}^{op}}$, which is left proper but not necessarily right proper, even if $\mathcal{D}_0^{\mathcal{C}^{op}}$ itself is right proper. For considering covers, we consider only those model categories $L_\Gamma \mathcal{D}_0^{\mathcal{C}^{op}} = \mathcal{D}_0^{\mathcal{C}^{op}\wedge}$ that are right proper as well. Since this may be a strong constraint, we will develop the notion of internal right Bousfield localization not from $\mathcal{D}_0^{\mathcal{C}^{op}\wedge}$, but from a generic right proper cellular model category $\mathcal{M}$.

Proposition[3.2.4 mod] 27.7.1. *If $\mathcal{K}$ is a class of maps in a model category $\mathcal{M}$, then the class of i$\mathcal{K}$-colocal equivalences is closed under retracts.*

Proof. Let $g : X \to Y$ be an i$\mathcal{K}$-colocal equivalence, let $f : V \to W$ be a retract of g, and let $\hat{g} : \hat{X} \to \hat{Y}$ and $\hat{f} : \hat{V} \to \hat{W}$ be fibrant approximations to g and f, respectively, such that $\hat{f}$ is a retract of $\hat{g}$. Let C be an i$\mathcal{K}$-colocal object. We want $\underline{\mathrm{Hom}}(\tilde{C}, \hat{V}) \to \underline{\mathrm{Hom}}(\tilde{C}, \hat{W})$ to also be an equivalence. Let $X \in \mathcal{M}$ with cosimplicial resolution $\tilde{X}$. Then consider

$$\begin{array}{ccc}
\mathrm{Hom}(\widetilde{X \otimes C}, \hat{V}) & \xrightarrow{\simeq} & \mathrm{Hom}(\widetilde{X \otimes C}, \hat{W}) \\
\| & & \| \\
\mathrm{Hom}(\mathrm{diag}\tilde{X} \otimes \tilde{C}, \hat{V}) & & \mathrm{Hom}(\mathrm{diag}\tilde{X} \otimes \tilde{C}, \hat{W}) \\
\downarrow \cong & & \downarrow \cong \\
\mathrm{diag}\mathrm{Hom}(\tilde{X}, \underline{\mathrm{Hom}}(\tilde{C}, \hat{V})) & \dashrightarrow & \mathrm{diag}\mathrm{Hom}(\tilde{X}, \underline{\mathrm{Hom}}(\tilde{C}, \hat{W})).
\end{array}$$

i$\mathcal{K}$-colocal objects form an ideal class, so $X \otimes C$ is again i$\mathcal{K}$-colocal and hence $\mathcal{K}$-colocal by Lemma 27.7.4. Also, an i$\mathcal{K}$-colocal equivalence g is a $\mathcal{K}$-colocal equivalence by Proposition 27.7.5, f is a retract of g and so a $\mathcal{K}$-colocal equivalence by the original Proposition of [40], and hence the top map above is an equivalence. Hence we have an equivalence at the bottom of the above commutative diagram, and Theorem 17.7.7 of [40] allows us to conclude that $\underline{\mathrm{Hom}}(\tilde{C},\hat{V}) \xrightarrow{\simeq} \underline{\mathrm{Hom}}(\tilde{C},\hat{W})$, and hence i$\mathcal{K}$-colocal equivalences are closed under retracts. □

To prove the lifting argument in $R_{\mathcal{K}}\mathcal{M}$, we need to show that a trivial cofibration in $R_{\mathcal{K}}\mathcal{M}$ is also a trivial cofibration in $\mathcal{M}$. The proof of this claim follows from the fact that weak equivalences are also i$\mathcal{K}$-colocal equivalences, which we now prove.

Proposition 27.7.2. *If $\mathcal{K}$ is a class of maps in $\mathcal{M}$, then weak equivalences in $\mathcal{M}$ are also i$\mathcal{K}$-colocal equivalences.*

Proof. This follows readily from Theorem 27.3.1.2 using diagonal objects. □

With this, we can now prove the following:

Lemma [5.3.2 mod] 27.7.3. *Trivial cofibrations in $R_{\mathcal{K}}\mathcal{M}$ are also trivial cofibrations in $\mathcal{M}$.*

Proof. This is verbatim the proof of the original claim in [40], save that we use the internal results of Propositions 27.5.1 and 27.7.1. □

In the proof of the factorization axiom for model categories, we lift the factorization in the classical case to the internal one by invoking Lemma 27.7.5 below, which itself follows from the following:

Lemma 27.7.4. *If $\mathcal{K}$ is a class of maps in $\mathcal{M}$, then i$\mathcal{K}$-colocal objects are also $\mathcal{K}$-colocal objects.*

Proof. This readily follows from Proposition 27.3.1.1; internal equivalences imply classical equivalences. □

Proposition 27.7.5. *If $\mathcal{K}$ is a class of maps in $\mathcal{M}$, then $\mathcal{K}$-colocal equivalences coincide with i$\mathcal{K}$-colocal equivalences.*

Proof. By Proposition 27.3.1.1, i$\mathcal{K}$-colocal equivalences are also $\mathcal{K}$-colocal equivalences. Conversely, if $g : X \to Y$ is a $\mathcal{K}$-colocal equivalence, the for any i$\mathcal{K}$-colocal object W any $C \in \mathcal{M}$, we can write

$$\begin{array}{ccc} \mathrm{Hom}(\widetilde{C \otimes W},\hat{X}) & \xrightarrow{\simeq} & \mathrm{Hom}(\widetilde{C \otimes W},\hat{Y}) \\ \cong\big\downarrow & & \big\downarrow\cong \\ \mathrm{diagHom}(\tilde{C},\underline{\mathrm{Hom}}(\tilde{W},\hat{X})) & \underset{\simeq}{\dashrightarrow} & \mathrm{diagHom}(\tilde{C},\underline{\mathrm{Hom}}(\tilde{W},\hat{Y})). \end{array}$$

Since $i\mathcal{K}$-colocal objects form an ideal class, $C \otimes W$ is also $i\mathcal{K}$-colocal and hence $\mathcal{K}$-colocal by the previous result, so we have the top horizontal equivalence above and hence the one at the bottom as well, and we conclude by using Theorem 17.7.7 of [40]. □

27.8 Internal right Bousfield localization

Definition 27.8.1. For a model category $\mathcal{M}$, a class $\mathcal{K}$ of maps in $\mathcal{M}$, an object W of $\mathcal{M}$ is *i$\mathcal{K}$-colocal* if it is cofibrant and if for any $f : A \to B$ in $\mathcal{K}$, we have an induced equivalence in $(\mathcal{M})^{\Delta^{op}}$ of iLhfCs: $f_* : \underline{\mathrm{Hom}}_{\mathcal{M}}(\tilde{W}, \hat{A}) \to \underline{\mathrm{Hom}}_{\mathcal{M}}(\tilde{W}, \hat{B})$ with a cosimplicial resolution $\tilde{W}$ of W and a fibrant approximation $\hat{A}$ to A.

Definition 27.8.2. A map $g : X \to Y$ in $\mathcal{M}$ is an *i$\mathcal{K}$-colocal equivalence* if for any $\mathcal{K}$-colocal object W, we have an equivalence in $(\mathcal{M})^{\Delta^{op}}$ of iLhfCs: $g_* : \underline{\mathrm{Hom}}_{\mathcal{M}}(\tilde{W}, \hat{X}) \to \underline{\mathrm{Hom}}_{\mathcal{M}}(\tilde{W}, \hat{Y})$.

The proof of Theorem 5.1.1 of [40] makes use of a further notion, which we generalize to our setting:

Definition 27.8.3. For a class K of objects in $\mathcal{M}$, a map $f : A \to B$ in $\mathcal{M}$ is said to be an iK-colocal equivalence if for any object $k \in K$, we have $\underline{\mathrm{Hom}}_{\mathcal{M}}(\tilde{k}, \hat{A}) \xrightarrow{\simeq} \underline{\mathrm{Hom}}_{\mathcal{M}}(\tilde{k}, \hat{B})$.

We now state our theorem, much like Theorem 5.1.1 of [40], which states the existence of an internal right Bousfield localization.

Theorem[5.1.1 mod] 27.8.4. *If $\mathcal{M}$ is a right proper cellular model category, K is a set of objects of $\mathcal{M}$, and $\mathcal{K}$ is the class of iK-colocal equivalences, then the internal right Bousfield localization of $\mathcal{M}$ is a model category structure on $\underline{\mathcal{M}}$ with i$\mathcal{K}$-colocal equivalences as weak equivalences, the same fibrations as those of $\mathcal{M}$, and for cofibrations those maps that have the left lifting property with respect to those fibrations that are also i$\mathcal{K}$-colocal equivalences.*

Proof. The proof, as in [40], consists in making sure that the axioms of a model category are met. We will only focus on those claims that are different from the proof of [40]. The 2–3 property is satisfied because of the dual of Proposition 27.5.1. The retract argument follows from Lemma 7.2.8 of [40], which we use as is, and Proposition 27.7.1, following the same argument as in [40]. The lifting argument involving a cofibration is immediate, since a cofibration in $R_{\mathcal{K}}\mathcal{M}$ has the left lifting property with respect to fibrations that are also i$\mathcal{K}$-local equivalences, i. e., trivial fibrations in $R_{\mathcal{K}}\mathcal{M}$. For the lifting argument involving a trivial cofibration, this follows from Lemma 27.7.3. Finally, for the functorial factorization axiom, we show that any map $g : X \to Y$ can be factored as $X \xrightarrow{p} W \xrightarrow{q} Y$, where p is a trivial cofibration in $R_{\mathcal{K}}\mathcal{M}$, and q is a fibration in $R_{\mathcal{K}}\mathcal{M}$. This follows readily from the classical case by invoking Proposition 27.7.5. The same is true of the factorization where now p is a cofibration, and q is a trivial fibration. This completes the proof. □

28 Supersymmetric derived stacks

28.1 Introduction

Moduli spaces have for a long time been a pervasive object in String Theory. Moduli stacks in particular have been very useful, and with the advent of Derived Algebraic Geometry [51, 76, 79], derived moduli stacks. Geometric Langlands is another avenue of research that has deep connections with Mathematical Physics [5, 18] and makes heavy use of the stacks formalism. Naturally, then it seems appropriate to develop a theory of supersymmetric stacks in its own right. In the present work, we lay the foundations for the theory of such objects, and we define them.

One typically starts with a superspace, a graded vector space. To be more general, one can consider graded modules instead, over a commutative ring k, giving rise to a category of super-k-modules k-sMod. We work in Derived Algebraic Geometry, so we enlarge that to the category of *simplicial graded k-supermodules* sk-sMod$_*$. From that point onward we follow [76] and [79] by showing that sk-sMod$_*$ can be made into a symmetric monoidal model category, with a notion of supercommutativity using the parity function on homogeneous elements that leads to defining sk-sAlg$_*$ = Comm(sk-sMod$_*$), the *category of simplicial graded k-superalgebras*, the opposite category of which is the *category of affine graded k-superschemes* k-D$^-$sAff$_*$, on which we put a bigraded version of the étale topology of [79]. We consider bigraded stacks on this site, functors valued in $S_{*.} = (\mathrm{Set}_\Delta)_{2\times 2}$, a bigraded generalization of the notion of stacks on the ungraded site (k-D$^-$Aff, ét.). The passage from the classical theory as expounded in [76] and [79] to the bigraded setting is made possible precisely because of our choice of working with bigraded objects, along with our choice of tensor product on sk-sMod$_*$.

For each $M \in sk\text{-sMod}_*$, $M = (M_{ip})$, i referred to as the level index, and p as the parity index. For $\sigma = (\sigma^1, \sigma^2)$, $\sigma^i \in M_{i0}$, $\theta = (\theta^1, \theta^2)$, and $\theta^A \in M_{A1}$, we consider maps $\Psi : M \to \mathcal{X}$, $\mathcal{X} = (\mathcal{X}_{ip}) \in \mathrm{Top}_{*.} = (\mathrm{Top})_{2\times 2}$, $\mathcal{X}$ being determined by additional data such as constraint equations, $\Psi = X \oplus \psi$, $X : M_0 \to \mathcal{X}_0$, $\psi : M_1 \to \mathcal{X}_1$, so that X is a function of $\sigma = (\sigma^1, \sigma^2)$, and ψ is a function of $\theta = (\theta^1, \theta^2)$. Instead of using the traditional supersymmetry transformations, say those in [32] for which $M_0 = (\sigma^1, \sigma^2)$, $M_1 = (\theta^1, \theta^2)$, $\delta\sigma^\alpha = \epsilon^T \rho^\alpha \theta$ with anticommuting ϵ, and $\delta\theta^A = \epsilon^A$, we introduce more general transformations that are fully symmetric, both algebraically, but also paritywise, i. e., transformations $\delta = \delta_0 \oplus \delta_1$, $\delta_0 : M_0 \to M_1$ and $\delta_1 : M_1 \to M_0$ such that $\delta_0\sigma^\alpha = \epsilon^T \rho^\alpha \theta$ and $\delta_1\theta^A = \lambda^T \gamma^A \sigma$, for infinitesimal commuting parameters ϵ and λ and matrices ρ and γ. We then show that Ψ transforms via a pullback under δ.

For $A \in sk\text{-sAlg}_*$, define $F(A) = \{\Psi(\sigma, \theta) \mid \sigma \in A_0, \theta \in A_1\}$ endowed with the subspace topology. Those functors F that satisfy hyperdescent on (k-D$^-$sAff$_*$, ét.) are called supersymmetric (derived) stacks.

https://doi.org/10.1515/9783111334073-028

28.2 Simplicial superalgebra

We first remind the reader of various results pertaining to super-algebra that we will later generalize to the simplicial setting. Most of what is presented in the subsection below can be found in [82], and especially in [8], which we modify for our purposes.

28.2.1 Super-algebra

This subsection covers some basic conventions about superalgebra. We fix a commutative ring k. We define a *super-k-module* (also referred to as a k-supermodule) to be a $\mathbb{Z}_2$-graded k-module $M = M_0 \oplus M_1$, endowed with a parity function $|\cdot|$ defined by

$$|x| = \begin{cases} 0, & x \in M_0, \\ 1, & x \in M_1. \end{cases}$$

There is a (distributive) tensor product $\otimes$ on the set of k-supermodules k-sMod, for which the commutativity map σ (also referred to as braiding map), defined on objects by $\sigma_{M,N} : M \otimes N \to N \otimes M$, $M = M_0 \oplus M_1$, $N = N_0 \oplus N_1$, satisfies

$$\sigma_{M,N}(x \otimes y) = (-1)^{|x||y|} y \otimes x$$

on homogeneous elements and is extended by linearity. Morphisms of super-modules are defined to be graded morphisms or, equivalently, morphisms of graded objects of degree zero [58]. In other terms, if $f : M \to N$ is a morphism of super k-modules, then it decomposes into two morphisms $f_0 : M_0 \to N_0$ and $f_1 : M_1 \to N_1$. We denote by $\mathrm{Hom}_k(M, N)$ or simply by $\mathrm{Hom}(M, N)$ the set of graded morphisms from M to N. It is a k-module. Each super-module M has the identity $\mathrm{id}_M = \mathrm{id}_{M_0} \oplus \mathrm{id}_{M_1} \in \mathrm{Hom}(M, M)$, which acts as a left and a right identity for composition. With obvious notations, composition is defined by $f \circ g = f_0 \circ g_0 \oplus f_1 \circ g_1$ and is associative. This makes k-sMod a category.

The identity for the tensor product is k, and we have left and right identity maps

$$\lambda : k \otimes x \xrightarrow{\cong} x,$$
$$\rho : x \otimes k \xrightarrow{\cong} x$$

with $x \in M$, M an object of k-sMod. Here, for instance, $k \otimes (x_0 \oplus x_1) = (k \otimes x_0) \oplus (k \otimes x_1)$, so that we can write $\lambda = \lambda_0 \oplus \lambda_1$ and $\rho = \rho_0 \oplus \rho_1$, where λ_i and ρ_i are the classical left and right identity maps in k-Mod, respectively. We regard elements of k to be of degree zero, so that they preserve parity with the tensor product. The associator is defined by

$$a^{-1}_{M,N,P} : M \otimes (N \otimes P) \xrightarrow{\cong} (M \otimes N) \otimes P$$

and satisfies the pentagon coherence diagram [59]

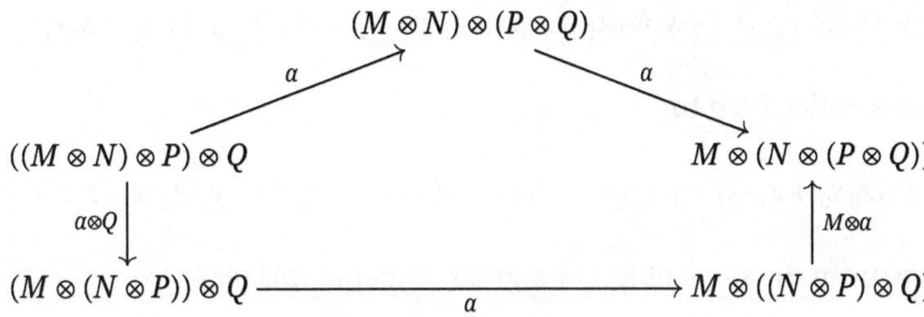

as well as the triangle coherence diagrams

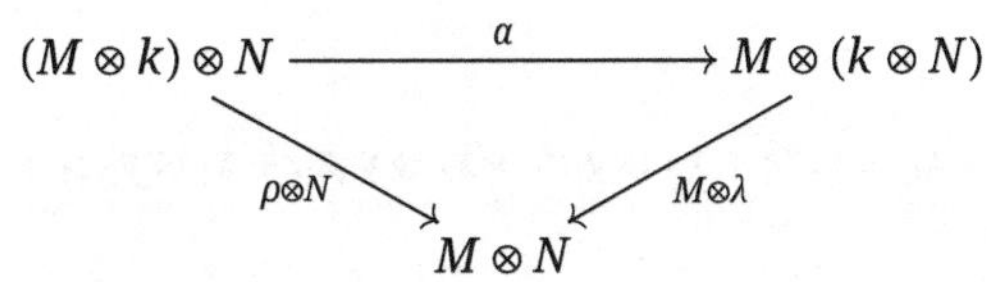

and the bigon relations $\rho_k = \lambda_k$.

The braiding map σ is required to satisfy the hexagon coherence condition

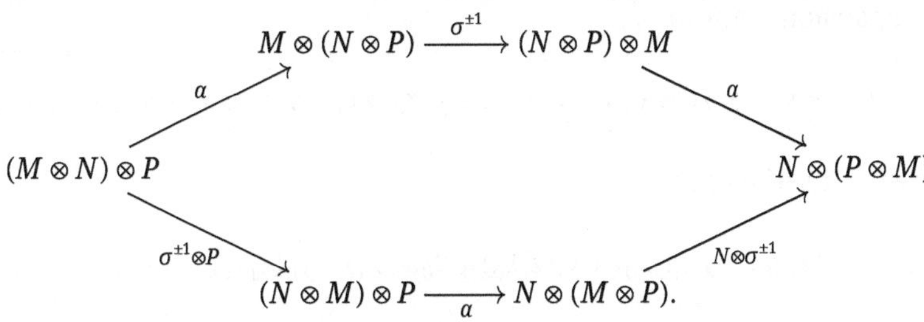

We investigate what does the parity function have to satisfy for this to hold: we start with an element $(x \otimes y) \otimes z$ in $(M \otimes N) \otimes P$, with $x = x_0 \oplus x_1, y = y_0 \oplus y_1$, and $z = z_0 \oplus z_1$. Thus, abbreviating tensor products $x \otimes y$ by xy and direct sums $x \oplus y$ by $x + y$ for simplicity of notation,

$$\begin{aligned}(x \otimes y) \otimes z &= [(x_0 \oplus x_1) \otimes (y_0 \oplus y_1)] \otimes z \\ &= [x_0 y_0 + x_0 y_1 + x_1 y_0 + x_1 y_1] \otimes z\end{aligned}$$

maps to

$$[y_0 x_0 + y_1 x_0 + y_0 x_1 - y_1 x_1] \otimes z$$

under $\sigma \otimes z$, which then maps to

$$y_0(x_0 \otimes z) + y_1(x_0 \otimes z) + y_0(x_1 \otimes z) - y_1(x_1 \otimes z)$$

under a. This expands as

$$y_0(x_0z_0 + x_0z_1) + y_1(x_0z_0 + x_0z_1) + y_0(x_1z_0 + x_1z_1) - y_1(x_1z_0 + x_1z_1)$$

and this maps under $1 \otimes \sigma$ to

$$y_0(z_0x_0 + z_1x_0) + y_1(z_0x_0 + z_1x_0) + y_0(z_0x_1 - z_1x_1) - y_1(z_0x_1 - z_1x_1). \tag{28.1}$$

Now starting from the same object $(x \otimes y) \otimes z$, under α, this maps to

$$(x_0 + x_1) \otimes [y_0z_0 + y_0z_1 + y_1z_0 + y_1z_1]$$

which expands as

$$x_0 \otimes y_0z_0 + x_0 \otimes y_0z_1 + x_0 \otimes y_1z_0 + x_0 \otimes y_1z_1 + x_1 \otimes y_0z_0 + x_1 \otimes y_0z_1 + x_1 \otimes y_1z_0 + x_1 \otimes y_1z_1.$$

Now if we have

$$|x \otimes y| = |x| + |y| \tag{28.2}$$

then this expansion maps to

$$y_0z_0 \otimes x_0 + y_0z_1 \otimes x_0 + y_1z_0 \otimes x_0 + y_1z_1 \otimes x_0 + y_0z_0 \otimes x_1 - y_0z_1 \otimes x_1 - y_1z_0 \otimes x_1 + y_1z_1 \otimes x_1$$

under σ, which itself maps to

$$y_0(z_0x_0) + y_0(z_1x_0) + y_1(z_0x_0) + y_1(z_1x_0) + y_0(z_0x_1) - y_0(z_1x_1) - y_1(z_0x_1) + y_1(z_1x_1)$$

under α, and this is exactly (28.1), showing that the hexagonal diagram does commute with the definition of the parity on tensor products given by (28.2). To see that (28.2) is indeed the correct relation, it suffices to circle the hexagon diagram both ways by starting from the same element $(x \otimes y) \otimes z$ for homogeneous elements x, y, z, as in the following diagram:

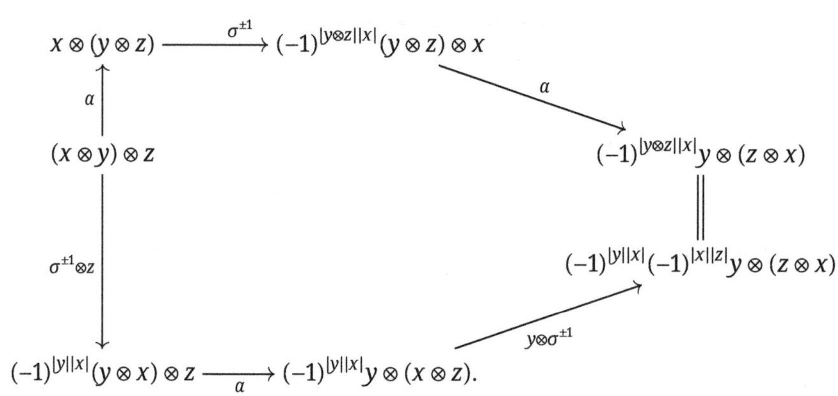

It follows from (28.2) that if the parity function is calculated mod 2, then the tensor product on k-sMod is defined by

$$\begin{aligned}(M \otimes N)_0 &= (M_0 \otimes N_0) \oplus (M_1 \otimes N_1),\\ (M \otimes N)_1 &= (M_0 \otimes N_1) \oplus (M_1 \otimes N_0)\end{aligned}$$

where $M_i \otimes N_j$, $i,j = 0,1$, is the usual tensor product in k-Mod on the underlying k-modules. Finally, the braiding $\sigma = \sigma_0 \oplus \sigma_1$ on k-sMod satisfies $\sigma^2 = \sigma_0^2 \oplus \sigma_1^2 = 1$. Indeed,

$$\begin{aligned}M \otimes N \ni x \otimes y = x_0y_0 + x_1y_1 &\xrightarrow{\sigma_0} y_0x_0 - y_1x_1 \xrightarrow{\sigma_0} x_0y_0 - (-x_1y_1) \in (N \otimes M)_0\\ + x_0y_1 + x_1y_0 &\xrightarrow{\sigma_1} +y_1x_0 + y_0x_1 \xrightarrow{\sigma_1} x_0y_1 + x_1y_0 \in (N \otimes M)_1\end{aligned}$$

where we have omitted tensor products for ease of reading. Further,

$$\begin{aligned}\rho_M &= \rho_{M,0} \oplus \rho_{M,1}\\ &= \oplus_{i=0,1}\lambda_{M,i} \circ (\sigma_{M,k})_i\\ &= \lambda_M \circ \sigma_{M,k}\end{aligned}$$

meaning, taking k of even parity, that the following diagram commutes:

$$\begin{array}{ccc} M \otimes k & \xrightarrow{\sigma_{M,k}} & k \otimes M \\ & \searrow^{\rho_M} \quad \swarrow_{\lambda_M} & \\ & M. & \end{array}$$

At this point, we have shown that $(k\text{-sMod}, \otimes_k, \alpha, \lambda, \rho, k)$ is a symmetric monoidal category.

28.2.2 Diagonal super-algebra

28.2.2.1 Box tensor product ⊠

To develop a theory within an Algebraic Geometry context over graded k-modules, we need commutative monoids whose definition really makes sense only if we take a diagonal version of $(k\text{-sMod}, \otimes_k, \alpha, \lambda, \rho, k)$. One way to achieve this is by considering k-sMod$_*$, the category of graded k-super-modules. We take the grading to be $\mathbb{Z}_2$ as well. An object of k-sMod$_*$ is of the form $M = M_0 \oplus M_1$, $M_i = M_{i0} \oplus M_{i1}$ for $i = 0,1$, where objects of M_{ip} have parity p. Thus one can write $M = \oplus_{i,p=0,1} M_{ip}$, i referring to the graded index, or level, and p referring to parity. We will adopt the shorthand matrix notation $M = (M_{ip})$ to denote such objects. It follows that we will refer to the components M_{ip} as entries of the object $M \in k$-sMod$_*$. In the same manner, morphisms $f : M \to N$ in k-sMod$_*$ are defined entrywise: $f_{ip} : M_{ip} \to N_{ip}$. The tensor product on k-sMod$_*$ is a diagonal tensor product defined by

$$M \boxtimes N = (M_0 \otimes N_0) \oplus (M_1 \otimes N_1)$$

where the tensor products used on the right-hand side are those of k-sMod. Observe that this is performed levelwise and each level is closed under $\otimes$: $M_{ip} \otimes N_{iq} \in k\text{-sMod}_i$, $i = 0, 1$. It follows that $M_i \otimes N_i = (M \boxtimes N)_i$. The advantage of having such a definition of the tensor product on k-sMod$_*$ is that one can use classical module theory with elements that have a parity. We have a braiding

$$\sigma_{M,N} : M \boxtimes N \to N \boxtimes M$$

defined levelwise on supermodules:

$$\sigma_{M,N} : (M_0 \otimes N_0) \oplus (M_1 \otimes N_1) \xrightarrow{\sigma_0 \oplus \sigma_1} (N_0 \otimes M_0) \oplus (N_1 \otimes M_1) = N \boxtimes M$$

with $\sigma_i = \sigma_{i0} \oplus \sigma_{i1} : M_i \otimes N_i \to N_i \otimes M_i$. To be more precise,

$$\sigma_{i0} : (M_{i0} \otimes N_{i0}) \oplus (M_{i1} \otimes N_{i1}) \to (N_{i0} \otimes M_{i0}) \oplus (N_{i1} \otimes M_{i1}),$$
$$\sigma_{i1} : (M_{i0} \otimes N_{i1}) \oplus (M_{i1} \otimes N_{i0}) \to (N_{i1} \otimes M_{i0}) \oplus (N_{i0} \otimes M_{i1}).$$

28.2.2.2 k-sMod$_*$ symmetric monoidal category

Recall that classically morphisms of supermodules $M \to N$ are graded morphisms $f = f_0 \oplus f_1, f_0 : M_0 \to N_0, f_1 : M_1 \to N_1$. In the bigraded case with the box tensor product defined above, morphisms in k-sMod$_*$ are defined levelwise, and on each level, morphisms are morphisms in k-sMod. Thus $f : M \to N$ in k-sMod$_*$ decomposes as $f = f_0 \oplus f_1$ with $f_i : M_i \to N_i$, $i = 0, 1$, and for i fixed, $f_i = f_{i0} \oplus f_{i1}$ with $f_{ip} : M_{ip} \to N_{ip}$, $p = 0, 1$. Thus $f = \oplus_{i,p=0,1} f_{ip}$, which we can represent matricially by $f = (f_{ip})$.

We still denote by $\mathrm{Hom}(M, N)$ the set $\mathrm{Hom}_{k\text{-sMod}_*}(M, N)$. We have $\mathrm{Hom}_{k\text{-sMod}_*} = \mathrm{Hom}_{(k\text{-sMod}_0,\otimes)} \oplus \mathrm{Hom}_{(k\text{-sMod}_1,\otimes)}$. It is a k-module. For any $M \in k$-sMod$_*$, $\mathrm{id}_M = \mathrm{id}_{M_0} \oplus \mathrm{id}_{M_1} \in \mathrm{Hom}(M, M)$. Composition is associative. This makes k-sMod$_*$ a category. We have

$$\lambda = \lambda_0 \oplus \lambda_1 : \ \Delta k \boxtimes x = (k \otimes x_0) \oplus (k \otimes x_1) \xrightarrow{\cong} x,$$
$$\rho = \rho_0 \oplus \rho_1 : \ x \boxtimes \Delta k \xrightarrow{\cong} x,$$
$$\alpha_{M,N,P} : \ (M \boxtimes N) \boxtimes P \xrightarrow{\cong} M \boxtimes (N \boxtimes P).$$

To be more precise,

$$\begin{aligned} \alpha_{M,N,P} &: (M \boxtimes N) \boxtimes P \\ &= \oplus_{i=0,1} (M \boxtimes N)_i \otimes P_i \end{aligned}$$

$$= \oplus_{i=0,1} (M_i \otimes N_i) \otimes P_i \xrightarrow{\oplus_{i=0,1}\alpha_i} \oplus_{i=0,1} M_i \otimes (N_i \otimes P_i) = M \boxtimes (N \boxtimes P)$$

with associators α_0 and α_1 in k-sMod, so we do have

$$\alpha = \alpha_0 \oplus \alpha_1 : (M \boxtimes N) \boxtimes P \xrightarrow{\cong} M \boxtimes (N \boxtimes P).$$

We have a pentagon diagram

$$\begin{array}{ccc}
(M \boxtimes N) \boxtimes (P \boxtimes Q) & & \\
\| & & \\
\oplus_{i=0,1}(M_i \otimes N_i) \otimes (P_i \otimes Q_i) & \overset{\oplus_{i=0,1}\alpha_i}{\searrow} & \\
\uparrow{\scriptstyle \oplus_{i=0,1}\alpha_i} & & \\
\oplus_{i=0,1}((M_i \otimes N_i) \otimes P_i) \otimes Q_i & & M \boxtimes (N \boxtimes (P \boxtimes Q)) \\
\| & & \| \\
((M \boxtimes N) \boxtimes P) \boxtimes Q & & \oplus_{i=0,1} M_i \otimes (N_i \otimes (P_i \otimes Q_i)) \\
\downarrow{\scriptstyle \oplus_{i=0,1}\alpha_i \otimes Q_i} & & \uparrow{\scriptstyle \oplus_{i=0,1} M_i \otimes \alpha_i} \\
\oplus_{i=0,1}(M_i \otimes (N_i \otimes P_i)) \otimes Q_i & \xrightarrow[\oplus_{i=0,1}\alpha_i]{} & \oplus_{i=0,1} M_i \otimes ((N_i \otimes P_i) \otimes Q_i) \\
\| & & \| \\
(M \boxtimes (N \boxtimes P)) \boxtimes Q & & M \boxtimes ((N \boxtimes P) \boxtimes Q)
\end{array}$$

and triangle coherence diagrams

$$\begin{array}{ccc}
(M \boxtimes \Delta k) \boxtimes N & & M \boxtimes (\Delta k \boxtimes N) \\
\| & & \| \\
\oplus_{i=0,1}(M_i \otimes k) \otimes N_i & \xrightarrow{\oplus_{i=0,1}\alpha_i} & \oplus_{i=0,1} M_i \otimes (k \otimes N_i) \\
{\scriptstyle \oplus_{i=0,1}\rho_i \otimes N_i} \searrow & & \swarrow {\scriptstyle \oplus_{i=0,1} M_i \otimes \lambda_i} \\
& \oplus_{i=0,1} M_i \otimes N_i & \\
& \| & \\
& M \boxtimes N &
\end{array}$$

with bigons $\rho_{\Delta k} = \lambda_{\Delta k}$. For the hexagonal coherence condition,

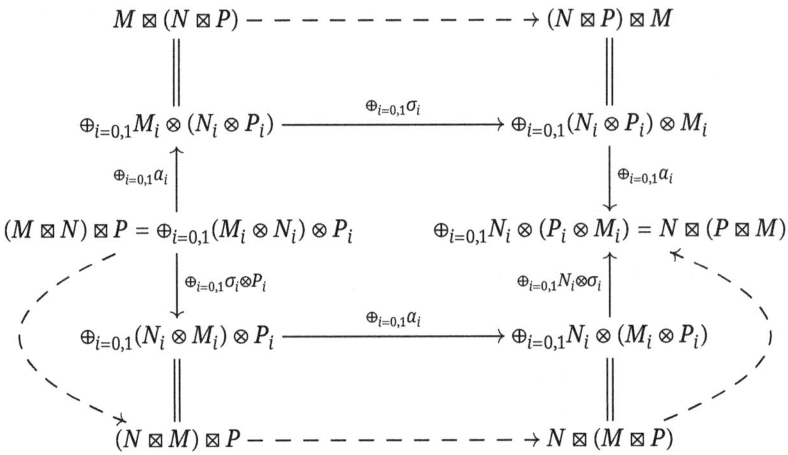

with $\sigma^2 = 1$ and $\lambda_M = \rho_M \circ \sigma_{M,\Delta k}$ for symmetry. We have proved the following:

Theorem 28.2.2.2.1. *(k-sMod$_*$, $\boxtimes, \alpha, \lambda, \rho, \Delta k$) is a symmetric monoidal category.*

28.2.2.3 Commutative monoids in k-sMod$_*$

A monoid in $\mathcal{C} = k$-sMod$_*$ is an object M of $\mathcal{C}$ with an associative binary operation $\mu : M \boxtimes M \to M$, $\mu = \mu_0 \oplus \mu_1$, μ_i the ordinary binary operation in k-sMod$_i$ for $i = 0, 1$. To be more precise,

$$\begin{aligned} \mu_i : M_i \otimes M_i &\to M_i, \\ m_i \otimes m_i' &\mapsto m_i m_i' \end{aligned}$$

where μ_i and $m_i \otimes m_i'$ decompose as

$$(m_{i0} \otimes m_{i0}') \oplus (m_{i1} \otimes m_{i1}') \xrightarrow{\mu_{i0}} m_{i0}m_{i0}' \oplus m_{i1}m_{i1}'$$

and

$$(m_{i0} \otimes m_{i1}') \oplus (m_{i1} \otimes m_{i0}') \xrightarrow{\mu_{i1}} m_{i0}m_{i1}' \oplus m_{i1}m_{i0}'$$

since each μ_{ip} is defined on homogeneous elements. The terms on the right-hand side recombine as $(m_i m_i')_0 \oplus (m_i m_i')_1 = m_i m_i'$.

We also have a unit map $\eta : \Delta k \to M$, $\eta = \eta_0 \oplus \eta_1$, η_i the unit in k-sMod$_i$ for $i = 0, 1$. μ and η satisfy, as in [59], the coherence diagrams

$$\begin{array}{ccccc}
(M \boxtimes M) \boxtimes M & & M \boxtimes (M \boxtimes M) & & M \boxtimes M \\
\| & & \| & & \| \\
\oplus_{i=0,1}(M_i \otimes M_i) \otimes M_i & \xrightarrow{\oplus_{i=0,1}\alpha_i} & \oplus_{i=0,1}M_i \otimes (M_i \otimes M_i) & \xrightarrow{\oplus_{i=0,1}M_i \otimes \mu_i} & \oplus_{i=0,1}M_i \otimes M_i \\
\downarrow{\scriptstyle \oplus_{i=0,1}\mu_i \otimes M_i} & & & & \downarrow{\scriptstyle \oplus_{i=0,1}\mu_i} \\
\oplus_{i=0,1}M_i \otimes M_i = M \boxtimes M & & \xrightarrow[\oplus_{i=0,1}\mu_i]{} & & \oplus_{i=0,1}M_i = M
\end{array}$$

and

$$\begin{array}{ccccc}
\Delta k \boxtimes M & \xrightarrow{\eta \boxtimes M} & M \boxtimes M & \xleftarrow{M \boxtimes \eta} & M \boxtimes \Delta k \\
\| & & \| & & \| \\
\oplus_{i=0,1}k \otimes M_i & \xrightarrow{\oplus_{i=0,1}\eta_i \otimes M_i} & \oplus_{i=0,1}M_i \otimes M_i & \xleftarrow{\oplus_{i=0,1}M_i \otimes \eta_i} & \oplus_{i=0,1}M_i \otimes k \\
& \searrow{\scriptstyle \oplus_{i=0,1}\lambda_i} & \downarrow{\scriptstyle \oplus_{i=0,1}\mu_i} & \swarrow{\scriptstyle \oplus_{i=0,1}\rho_i} & \\
& & \oplus_{i=0,1}M_i = M & &
\end{array}$$

where $f \boxtimes g = \oplus_{i=0,1} f_i \otimes g_i$.

A monoid A in k-sMod$_*$ is said to be *supercommutative*, which we will just refer to as being commutative, if $\mu \circ \sigma = \mu$, i. e., if levelwise, for all x, y homogeneous in A_i for $i = 0, 1$, we have

$$xy = \mu(x \otimes y) = (-1)^{|x||y|} yx.$$

Another way of saying this is $A \in k$-sMod$_*$ is supercommutative if it is so levelwise.

We define the category of graded k-algebras to be the category of commutative monoids (with commutativity defined by σ) in k-sMod$_*$:

$$k\text{-sAlg}_* = \mathrm{Comm}(k\text{-sMod}_*).$$

For $A \in k$-sAlg$_*$, we denote by A-sMod$_*$ the category of elements of k-sMod$_*$ that are A-modules in the classical sense, where the action of A on $M \in k$-sMod$_*$ is defined levelwise. A morphism of A-modules is a morphism $f : M \to N$ of graded supermodules such that $f(am) = af(m)$ for $a \in A$ and $m \in M$. More specifically, am is defined via

$$\begin{aligned} \mu : A \boxtimes M &\to M, \\ a \boxtimes m &\mapsto am \end{aligned}$$

with $A \boxtimes M = (A_0 \otimes M_0) \oplus (A_1 \otimes M_1)$, $\mu = \mu_0 \oplus \mu_1$, as we saw above, so that $am = a_0 m_0 \oplus a_1 m_1$ with $a_i m_i \in M_i$, $i = 0, 1$. Then $f(am) = af(m)$ reads

$$f(a_0 m_0 \oplus a_1 m_1) = f_0(a_0 m_0) \oplus f_1(a_1 m_1)$$

$$
\begin{aligned}
&= a_0 f_0(m_0) \oplus a_1 f_1(m_1)\\
&= a_0 f(m)_0 \oplus a_1 f(m)_1\\
&= af(m)
\end{aligned}
$$

where $f_i \in \mathrm{Mor}\,(A_i\text{-sMod}_i)$ for $i = 0, 1$. We denote by $\mathrm{Hom}_A(M, N)$ the set of morphism of A-modules. There is a tensor product on A-sMod$_*$ defined by

$$M \boxtimes_A N = M \boxtimes_k N / \sim$$

where the equivalence relation $\sim$ is defined by $ma \boxtimes n = m \boxtimes an$ for $a \in A$, $m \in M$, and $n \in N$. However, we will use a simplified product that will allow us to use the work of Toen and Vezzosi; we define

$$
\begin{aligned}
X \tilde{\boxtimes}_A Y = \oplus_{i=0,1} \, [&\mathrm{colim}(X_{i0} \otimes A_{i0} \otimes Y_{i0} \to X_{i0} \otimes Y_{i0})\\
&\oplus \mathrm{colim}(X_{i1} \otimes A_{i1} \otimes Y_{i1} \to X_{i0} \otimes Y_{i1} \vee X_{i1} \otimes Y_{i0})]
\end{aligned}
$$

that is,

$$X \tilde{\boxtimes}_A Y = \oplus_{i,p=0,1} X_{ip} \otimes_{A_{ip}} Y_{ip}$$

which is fully diagonalized.

28.2.3 Simplicial generalization

We now define the simplicial counterparts to all the above definitions. We will not worry about universe considerations, and if needed, they can be transcribed to our case from [76] and [79]. We denote by sk-sMod$_*$ the category of simplicial objects in k-sMod$_*$:

$$sk\text{-sMod}_* = (k\text{-sMod}_*)^{\Delta^{\mathrm{op}}}$$

whose objects will be referred to as *simplicial graded k-supermodules*. We endow this category with the (simplicial) levelwise tensor product $\boxtimes$, that is, for $M, N \in sk$-sMod$_*$, we have $M \boxtimes N = \oplus_{n \geq 0} M_n \boxtimes N_n$, where

$$M_n \boxtimes N_n = \oplus_{i=0,1} M_{n,i} \otimes_k N_{n,i}.$$

We will put a model structure on this category. Following [28], it becomes evident that we will need a notion of model category structure on the category of bigraded simplicial sets. We do this first.

28.2.3.1 Model category of $\mathbb{Z}_2$-bigraded simplicial sets

In this subsection, following [28], we define a model category structure on the category of bigraded simplicial sets. For completeness, we briefly remind the reader of a few results regarding simplicial sets, all of which can be found in [28]. Recall that a map $p : X \to Y$ of simplicial sets is said to be a fibration if for any commutative diagram

$$\begin{array}{ccc} \Lambda^n_k & \longrightarrow & X \\ {\scriptstyle i}\big\downarrow & \nearrow & \big\downarrow{\scriptstyle p} \\ \Delta^n & \longrightarrow & Y \end{array}$$

where as usual $\Delta^n = \mathrm{Hom}_\Delta(-,[n])$, there is a dotted map $\Delta^n \to X$ making the above diagram commute. A fibrant simplicial set, or Kan complex, is a simplicial set X for which $X \to *$ is a fibration, it being the unique map to the final object $*$. For a fibrant simplicial set X and $v \in X_0$, we define the simplicial homotopy group $\pi_n(X,v)$ for $n \geq 1$ to be the set of homotopy classes of maps $\alpha : \Delta^n \to X$ rel $\partial\Delta^n$ that fit in diagrams of the form

$$\begin{array}{ccc} \Delta^n & \xrightarrow{\alpha} & X \\ \big\uparrow & & \big\uparrow{\scriptstyle v} \\ \partial\Delta^n & \longrightarrow & X_0 \end{array}$$

and define $\pi_0(X)$ to be the set of homotopy classes of vertices of X. A cofibration is an inclusion of simplicial sets. An equivalence is defined as follows: $f : X \to Y$ between fibrant simplicial sets is said to be a weak equivalence if for any vertex x of X, $f_* : \pi_n(X,x) \to \pi_n(Y,f(x))$ is an isomorphism for all $n \geq 1$ and $f_* : \pi_0(X) \to \pi_0(Y)$ is a bijection.

We now define all those concepts in the bigraded case. Let $\mathrm{Top}_{*.}$ denote the category of $\mathbb{Z}_2$-bigraded topological spaces. Denote by $S_{*.}$ the category of $\mathbb{Z}_2$-bigraded simplicial sets. We define the realization functor $|\cdot| : S_{*.} \to \mathrm{Top}_{*.}$ entrywise. Introduce the bigraded simplex category $\Delta_{*.} \downarrow X_{*.}$ of a bigraded simplicial set $X_{*.}$, whose objects are bigraded maps, which entry-wise read $\sigma_{ip} : \Delta^n \to X_{ip}$, $i,p = 0,1$.

An arrow of $\Delta_{*.} \downarrow X_{*.}$ consists of commutative diagrams of simplicial maps for $i,p = 0,1$:

$$\begin{array}{ccc} \Delta^n & & \\ & \overset{\sigma_{ip}}{\searrow} & \\ {\scriptstyle \theta_{ip}}\big\downarrow & & X_{ip} \\ & \underset{\tau_{ip}}{\nearrow} & \\ \Delta^m. & & \end{array}$$

We have the following result from [28].

Lemma 28.2.3.1.1. $X_{ip} \cong \operatorname{colim}_{\substack{\Delta^n \to X_{ip} \\ in\ \Delta\downarrow X_{ip}}} \Delta^n \in \mathrm{Set}_\Delta$.

Using the following definition on usual simplicial sets:

$$|X_{ip}| = \operatorname*{colim}_{\substack{\Delta^n \to X_{ip} \\ \text{in } \Delta\downarrow X_{ip}}} |\Delta^n|$$

we define the realization $|X_{*.}|$ of a bigraded simplicial set $X_{*.}$ to be $|X_{*.}| = (|X_{ip}|) \in \mathrm{Top}_{*.}$.

We define the singular functor $S : \mathrm{Top}_{*.} \to S_{*.}$ entrywise. For $T_{*.} \in \mathrm{Top}_{*.}$, $S(T_{*.}) = (S(T_{ip}))$, where $S(T_{ip})$, $i, p = 0, 1$, is the simplicial set given by

$$n \mapsto \mathrm{Hom}_{\mathrm{Top}}(|\Delta^n|, T_{ip})$$

and $|\Delta^n|$ is the standard n-simplex

$$|\Delta^n| = \{(t_0, \dots, t_n) \in \mathbb{R}^{n+1} \mid \sum_{i=0}^{n} t_i = 1;\ t_i \geq 0\}.$$

Proposition 28.2.3.1.2. *For all $X_{*.} \in S_{*.}$ and $Y_{*.} \in \mathrm{Top}_{*.}$, we have*

$$\mathrm{Hom}_{\mathrm{Top}_{*.}}(|X_{*.}|, Y_{*.}) \cong \mathrm{Hom}_{S_{*.}}(X_{*.}, SY_{*.})$$

that is, $|\ | \dashv S$.

Proof. It suffices to write

$$\begin{aligned} \mathrm{Hom}_{\mathrm{Top}_{*.}}(|X_{*.}|, Y_{*.}) &= \oplus_{i,p=0,1} \mathrm{Hom}_{\mathrm{Top}}(|X_{ip}|, Y_{ip}) \\ &\cong \oplus_{i,p=0,1} \mathrm{Hom}_{\mathrm{Set}_\Delta}(X_{ip}, SY_{ip}) \\ &= \mathrm{Hom}_{S_{*.}}(X_{*.}, SY_{*.}). \end{aligned}$$ □

We say that a map $p_{*.} : X_{*.} \to Y_{*.}$ of bigraded simplicial sets is a *fibration* if for $i, q = 0, 1$, the maps $p_{iq} : X_{iq} \to Y_{iq}$ are fibrations, i. e., if it is so entrywise. Thus a fibrant bigraded simplicial set, or bigraded Kan complex, is a bigraded simplicial set $Y_{*.}$ such that $Y_{ip} \to *$ is a fibration for $i, p = 0, 1$. In the same manner, a bigraded continuous map $f_{*.} : T_{*.} \to U_{*.}$ is said to be a bigraded Serre fibration if it is so entrywise. For bigraded simplicial maps $f_{*.}, g_{*.} : K_{*.} \to X_{*.}$, we say there is a *bigraded simplicial homotopy* $f_{*.} \xrightarrow{\simeq} g_{*.}$ if we have a simplicial homotopy $f_{ip} \xrightarrow{\simeq} g_{ip}$ for $i, p = 0, 1$, which we recall means that there is a commutative diagram

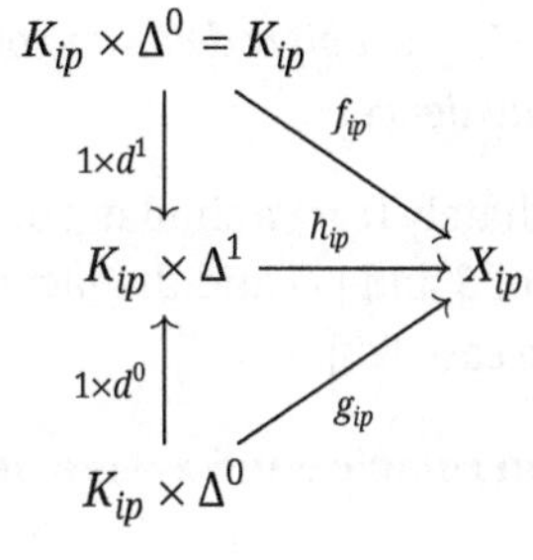

in which case we say $h_{*.} = (h_{ip})$ is a bigraded simplicial homotopy $f_{*.} \to g_{*.}$. If $j_{*.} : L_{*.} \subset K_{*.}$ is a bigraded inclusion such that $f_{*.}|_{L_{*.}} = g_{*.}|_{L_{*.}}$, then we say we have a bigraded simplicial homotopy $f_{*.} \to g_{*.}$ rel $L_{*.}$ if we have a simplicial homotopy $f_{ip} \to g_{ip}$ rel L_{ip} for $i, p = 0, 1$, that is, such that the following diagrams commute:

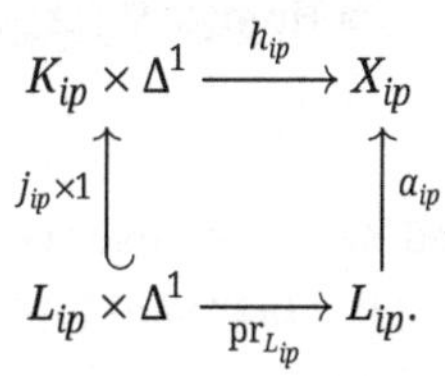

For a fibrant bigraded simplicial set $X_{*.}$ and $v_{*.} = (v_{ip}) \in (X_{*.})_0$, we define $\pi_n(X_{*.}, v_{*.}) = (\pi_n(X_{ip}, v_{ip}))$, where $\pi_n(X_{ip}, v_{ip})$ is the set of homotopy classes of maps $\alpha_{ip} : \Delta^n \to X_{ip}$ rel $\partial\Delta^n$, $i, p = 0, 1$, such that the following diagram is commutative:

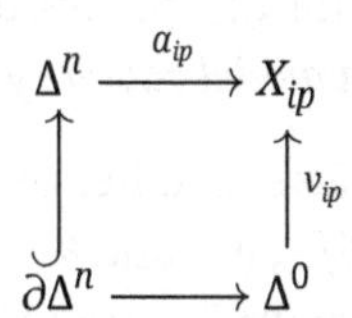

and $\pi_0(X_{*.}) = (\pi_0(X_{ip}))$, $\pi_0(X_{ip})$ the set of path components of X_{ip}, $i, p = 0, 1$. Now a map $f_{*.} : X_{*.} \to Y_{*.}$ between fibrant bigraded simplicial sets is said to be a *weak equivalence* if for all $x_{*.} \in (X_{*.})_0$, the induced map $(f_{*.})_* : \pi_k(X_{*.}, x_{*.}) \to \pi_k(Y_{*.}, f_{*.}(x_{*.}))$ is an isomorphism for all $k \geq 1$, that is, $(f_{ip})_* : \pi_k(X_{ip}, x_{ip}) \xrightarrow{\cong} \pi_k(Y_{ip}, f_{ip}(x_{ip}))$ for $i, p = 0, 1$, and $(f_{*.})_* : \pi_0(X_{*.}) \to \pi_0(Y_{*.})$ is a bijection, that is, $(f_{ip})_* : \pi_0(X_{ip}) \to \pi_0(Y_{ip})$ is a bijection for $i, p = 0, 1$, or in other terms, a weak equivalence is so if it is a weak equivalence entrywise.

Proposition 28.2.3.1.3. *A map $f_{*.} : X_{*.} \to Y_{*.}$ between fibrant bigraded simplicial sets is a trivial fibration if and only if $f_{*.}$ has the right lifting property with respect to all maps $\partial\Delta^n \subset \Delta^n$ for $n \geq 0$ entrywise.*

Proof. The result holds entrywise [28] hence is true for the corresponding bigraded objects by definition. □

Proposition 28.2.3.1.4. *Suppose $X_{*.}$ is a bigraded Kan complex. Then the canonical map $\eta_{X_{*.}} : X_{*.} \to S|X_{*.}|$ is a weak equivalence.*

Proof. The proof follows immediately from a similar proposition of [28] in the ungraded case because of the definitions of S and $|\cdot|$ and the fact that $\eta_{X_{*.}}$ is a weak equivalence if it is so entrywise, which is the case [28]. □

Now if $X_{*.}$ is a bigraded Kan complex and $x_{*.}$ is a vertex of $X_{*.}$, then by the above proposition we have

$$\begin{aligned}\pi_n(X_{*.}, x_{*.}) &\cong \pi_n(S|X_{*.}|, x_{*.}) \\ &= \mathrm{Hom}(\Delta^2\Delta^n/\partial\Delta^n, S|X_{*.}|) \\ &\cong \mathrm{Hom}(\Delta^2|\Delta^n/\partial\Delta^n|, |X_{*.}|) \\ &\cong \mathrm{Hom}(\Delta^2 S^n, |X_{*.}|) \\ &= \pi_n(|X_{*.}|, x_{*.}) \quad n \geq 1\end{aligned}$$

so a map $f_{*.} : X_{*.} \to Y_{*.}$ of bigraded Kan complexes is a weak equivalence if and only if the induced map $|f_{*.}| : |X_{*.}| \to |Y_{*.}|$ is a bigraded topological weak equivalence, which leads us to defining, as in [28], that a map $f_{*.} : X_{*.} \to Y_{*.}$ of bigraded simplicial sets be a *weak equivalence* if the induced map $|f_{*.}| : |X_{*.}| \to |Y_{*.}|$ is a weak equivalence of bigraded spaces. We define a *cofibration* of bigraded simplicial sets to be an entrywise inclusion.

Theorem 28.2.3.1.5. *$S_{*.}$ together with the classes of Kan fibrations, cofibrations, and weak equivalences defined above is a model category.*

Proof. Set_Δ is complete and cocomplete, and hence so is $S_{*.}$. For the 2–3 property, suppose we have $f_{*.} = g_{*.} \circ h_{*.}$. If two of $f_{*.}$, $g_{*.}$, and $h_{*.}$ are weak equivalences, so are their entries, so writing $f_{ip} = g_{ip} \circ h_{ip}$ since the 2–3 property holds at that level for $i, p = 0, 1$, then the third function would be an equivalence for $i, p = 0, 1$, and hence so would be the resulting bigraded function. For the retract property, say $f_{*.}$ is a retract of $g_{*.}$, and $g_{*.}$ is a weak equivalence, fibration or cofibration. The retract breaks up into diagrams for $i, p = 0, 1$ as

$$\begin{array}{ccccc} X & \longrightarrow & Y & \longrightarrow & X \\ \downarrow f_{ip} & & \downarrow g_{ip} & & \downarrow f_{ip} \\ U & \longrightarrow & X & \longrightarrow & U. \end{array}$$

$g_{*.}$ having one of the three properties mentioned above, it is so entrywise, at which level f_{ip} being a retract of g_{ip}, the former map shares the same property for $i, p = 0, 1$, and hence $f_{*.}$ shares that same property $g_{*.}$ had. For the lifting property, suppose we have

$$\begin{array}{ccc} U_{*.} & \longrightarrow & X_{*.} \\ {\scriptstyle i_{*.}}\downarrow & \nearrow & \downarrow{\scriptstyle p_{*.}} \\ V_{*.} & \longrightarrow & Y_{*.} \end{array}$$

with a cofibration $i_{*.}$ and a fibration $p_{*.}$. We show that there is a dotted arrow as shown making the diagram commutative if either $i_{*.}$ or $p_{*.}$ is a weak equivalence. That diagram breaks up into diagrams for $j, q = 0, 1$:

$$\begin{array}{ccc} U_{jq} & \longrightarrow & X_{jq} \\ {\scriptstyle i_{jq}}\downarrow & \nearrow & \downarrow{\scriptstyle p_{jq}} \\ V_{jq} & \longrightarrow & Y_{jq}. \end{array}$$

Suppose for argument's sake that $p_{*.}$ is a weak equivalence. Then it is so entrywise, so in the diagrams above, for $j, q = 0, 1$, there is a dotted arrow, and hence the original bigraded square has a diagonal arrow as claimed. Finally, for the functorial factorization property, let $f_{*.} : X_{*.} \to Y_{*.}$ be a map in $S_{*.}$. We will show half of the property, namely that $f_{*.}$ can factor as $p_{*.} \circ i_{*.}$ with a trivial fibration $p_{*.}$ and a cofibration $i_{*.}$, the other factorization as a trivial cofibration followed by a fibration being proved similarly. Since $f_{*.} = (f_{ip})$ and the factorization property holds entrywise, we can write $f_{jq} = p_{jq} \circ i_{jq}$ with a trivial fibration p_{jq} for $j, q = 0, 1$ and a cofibration i_{jq} for $j, q = 0, 1$. Writing $i = (i_{jq})$ and $p = (p_{jq})$, we have $f_{*.} = p_{*.} \circ i_{*.}$ with a trivial fibration $p_{*.}$ and a cofibration $i_{*.}$. This completes the proof. □

28.2.3.2 Model category structure on sk-sMod$_*$

We adapt Theorem II.4.1 of [28] to our setting. Let $\mathcal{C}_{*.}$ be a $\mathbb{Z}_2$-bigraded category, and let $s\mathcal{C}_{*.}$ be the category of simplicial objects in $\mathcal{C}_{*.}$, $s\mathcal{C}_{*.} = \mathcal{C}_{*.}^{\Delta^{\mathrm{op}}}$. We assume that there is a functor $G_{*.} : s\mathcal{C}_{*.} \to S_{*.}$ with a left adjoint $F_{*.} : S_{*.} \to s\mathcal{C}_{*.}$. We define a morphism $f_{*.} : M_{*.} \to N_{*.}$ in $s\mathcal{C}_{*.}$ to be a weak equivalence (resp., a fibration) if $G_{*}f_{*.}$ is a weak equivalence (resp., a fibration) of bigraded simplicial sets, and $f_{*.}$ is a cofibration if it has the left lifting property with respect to all trivial fibrations in $s\mathcal{C}_{*.}$. Entrywise, that gives us $G_{ip} : s\mathcal{C} \to \mathrm{Set}_\Delta$ with a left adjoint $F_{ip} : \mathrm{Set}_\Delta \to s\mathcal{C}$. Given the model category structure we put on $S_{*.}$, fibrations, cofibrations, and weak equivalences in $s\mathcal{C}_{*.}$ are defined entrywise in $s\mathcal{C}$. We will apply this formalism to the case $\mathcal{C} = k$-Mod, so that $s\mathcal{C} = sk$-Mod and $s\mathcal{C}_{*.} = sk$-sMod$_{*.}$.

Note also that we have a natural map

$$\operatorname*{colim}_I G_{*.}(X_\alpha) \to G_{*.}(\operatorname*{colim}_I X_\alpha)$$

which decomposes into entry maps

$$\operatorname*{colim}_I G_{ip}(X_{a,ip}) \to G_{ip}(\operatorname*{colim}_I X_{a,ip})$$

for $i, p = 0, 1$.

Theorem 28.2.3.2.1. *Suppose $\mathcal{C}$ is bicomplete and $G_{*.} : s\mathcal{C}_{*.} \to S_{*.}$ commutes with filtered colimits. Then with the classes of fibrations, cofibrations, and weak equivalences defined above, along with the assumption that a cofibration with the left lifting property with respect to all fibrations is a weak equivalence, $s\mathcal{C}_{*.}$ is a model category.*

Proof. Since $G_{*.}$ commutes with filtered colimits, it does so entrywise, so the hypotheses of Theorem II.4.1 of [28] are met, and hence $s\mathcal{C}$ is a model category, something we will use to prove that $s\mathcal{C}_{*.}$ itself is a model category. $s\mathcal{C}_{*.}$ is clearly bicomplete. For the 2–3 property, suppose we have a factorization in $s\mathcal{C}_{*.}$:

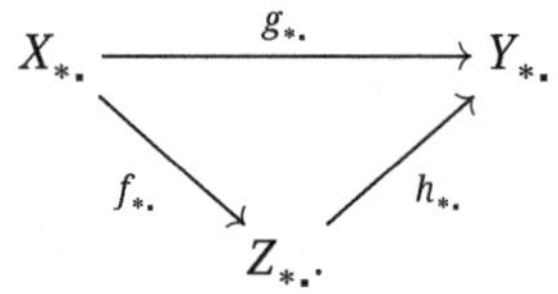

This breaks up into diagrams for $i, p = 0, 1$:

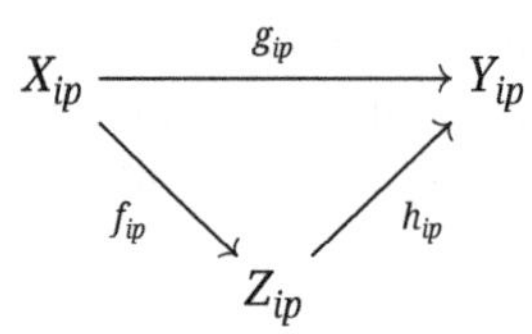

in $s\mathcal{C}$, where the 2–3 property holds since it is a model category. Suppose for illustrative purposes that $g_{*.}$ and $f_{*.}$ are weak equivalences. Then so are g_{ip} and f_{ip} for $i, p = 0, 1$, hence h_{ip} is a weak equivalence for $i, p = 0, 1$, and thus so is $h_{*.} = (h_{ip})$. For the retract property, $f_{*.}$ a retract of $g_{*.}$ is a property defined by a diagram in $s\mathcal{C}_{*.}$, which breaks up into retract diagrams in $s\mathcal{C}$, where the retract property holds, so starting from $g_{*.}$ with a property P, it being a weak equivalence, a fibration, or a cofibration, it is so entrywise, so by the retract property in $s\mathcal{C}$, f_{ip} shares the same property for $i, p = 0, 1$, and those entries recombine into $f_{*.}$ with that same property P $g_{*.}$ had. For the lifting property and the functorial factorization property, the argument is the same: we work entrywise and use the fact that $s\mathcal{C}$ is a model category from [28]. □

Corollary 28.2.3.2.2. *With the above notions of fibrations, cofibrations, and weak equivalences, sk-sMod$_*$ is a model category.*

Proof. Let $\mathcal{C} = k$-Mod, so that $s\mathcal{C} = sk$-Mod and $s\mathcal{C}_{*.} = sk$-sMod$_*$. We know $\mathcal{C}$ is bicomplete. Start from the forgetful functor $G : k\text{-Mod} \to \text{Set}$ with a left adjoint F, which we both prolong to the simplicial case to get maps, which we will again denote by G and

F: G : sk-Mod $\to$ Set$_\Delta$ and its left adjoint F, by defining $G(X)_n = G(X_n)$. G preserving filtered colimits, so will its bigraded generalization $G_{*.}$: sk-sMod$_*$ $\to S_{*.}$, which has a bigraded generalization of F for left adjoint, denoted $F_{*.}$. At this point, we just use the previous theorem. □

28.2.3.3 Bigraded simplicial categories

About notations, if $\mathcal{C}$ is a bigraded category, then we write $\underline{\mathcal{C}}$ for its ungraded counterpart. For instance, if $\mathcal{C} = sk$-sMod$_*$, then $\underline{\mathcal{C}} = sk$-Mod, which generically refers to either sk-Mod$_0$ or sk-Mod$_1$. Also, for ease of reading, we will just write M for $M_{*.} \in k$-sMod$_*$. We also define the enhanced Hom set Hom$^+$ as follows: $(\mathrm{Hom}^+(M,N))_0 = \mathrm{Hom}(M,N)$, whereas $(\mathrm{Hom}^+(M,N))_1$ is the set of parity-reversing morphisms from M to N.

Definition 28.2.3.3.1. A bigraded category $\mathcal{C}$ is a *bigraded simplicial category*, following the ungraded definition in [28], if there is a mapping space functor

$$\underline{\mathrm{Hom}}_{\mathcal{C}}(-,-) : \mathcal{C}^{\mathrm{op}} \times \mathcal{C} \to S_{*.}$$

such that for all $M, N \in \mathrm{Ob}\,(\mathcal{C})$:
1. $\underline{\mathrm{Hom}}_{\mathcal{C}}(M,N)_{0\times 0} = \mathrm{Hom}^+_{\mathcal{C}}(M,N)$
2. $\underline{\mathrm{Hom}}_{\mathcal{C}}(M,-) : \mathcal{C} \to S_{*.}$ has a left-adjoint $M \boxtimes - : S_{*.} \to \mathcal{C}$. Adjointness means that

$$\mathrm{Hom}_{\mathcal{C}}(M \boxtimes K, N) \cong \mathrm{Hom}_{S_{*.}}(K, \underline{\mathrm{Hom}}_{\mathcal{C}}(M,N))$$

   with associativity $M \boxtimes (K \times L) \cong (M \boxtimes K) \boxtimes L$.
3. $- \boxtimes K : \mathcal{C} \to \mathcal{C}$ has a right adjoint $\exp\_(K) = \underline{\mathrm{hom}}_{\mathcal{C}}(K,-) : \mathcal{C} \to \mathcal{C}$, i. e.,

$$\mathrm{Hom}_{\mathcal{C}}(M \boxtimes K, N) \cong \mathrm{Hom}_{\mathcal{C}}(M, N^K).$$

Theorem 28.2.3.3.2. $s\underline{\mathcal{C}} = sk$-Mod *being a simplicial category* [79], *it follows that* $s\mathcal{C} = sk$-sMod$_*$ *becomes a bigraded simplicial category with*

$$\begin{aligned}\underline{\mathrm{Hom}}_{s\mathcal{C}}(M,N)_{n\times n} &= \mathrm{Hom}_{s\mathcal{C}}(M \boxtimes \Delta^2\Delta^n, N)\\ &= (\mathrm{Hom}_{s\underline{\mathcal{C}}}(M_{ip}, N_{ip})_n).\end{aligned}$$

Proof. Here, as in [76],

$$- \otimes - : sk\text{-Mod} \times \mathrm{Set}_\Delta \to sk\text{-Mod}$$

is defined by

$$(M \otimes K)_n = \coprod_{k \in K_n} M_n.$$

For $\psi : [m] \to [n]$ in Δ, we have an induced map $\psi^* : (M \otimes K)_m \to (M \otimes K)_n$ that comes from

$$\coprod_{k\in K_m} M_m \to \coprod_{k\in K_m} M_n \to \coprod_{k\in K_n} M_n.$$

We determine $\mathrm{Hom}_{sC}(M \boxtimes K, N)$. First, sk-Mod being a simplicial category, if $M, N \in sk$-Mod and $K \in \mathrm{Set}_\Delta$, then we have $\mathrm{Hom}(M \otimes K, N) \cong \mathrm{Hom}(K, \underline{\mathrm{Hom}}_{sk\text{-Mod}}(M, N))$. If now $M, N \in sk$-sMod, $K \in S_{\bullet} = (\mathrm{Set}_\Delta)_0 \oplus (\mathrm{Set}_\Delta)_1$:

$$\begin{aligned}
\mathrm{Hom}(M \otimes K, N) &= \mathrm{Hom}((M \otimes K)_0 \oplus (M \otimes K)_1, N_0 \oplus N_1) \\
&= \mathrm{Hom}((M_0 \otimes K_0) \oplus (M_1 \otimes K_1), N_0) \\
&\quad \oplus \mathrm{Hom}((M_0 \otimes K_1) \oplus (M_1 \otimes K_0), N_1) \\
&= \mathrm{Hom}(M_0 \otimes K_0, N_0) \oplus \mathrm{Hom}(M_1 \otimes K_1, N_0) \\
&\quad \oplus \mathrm{Hom}(M_0 \otimes K_1, N_1) \oplus \mathrm{Hom}(M_1 \otimes K_0, N_1) \\
&\cong \mathrm{Hom}_{\mathrm{Set}_\Delta}(K_0, \underline{\mathrm{Hom}}_{sk\text{-Mod}}(M_0, N_0)) \oplus \mathrm{Hom}(K_1, \underline{\mathrm{Hom}}(M_1, N_0)) \\
&\quad \oplus \mathrm{Hom}(K_1, \underline{\mathrm{Hom}}(M_0, N_1)) \oplus \mathrm{Hom}(K_0, \underline{\mathrm{Hom}}(M_1, N_1)) \\
&= \mathrm{Hom}(K_0, \underline{\mathrm{Hom}}(M_0, N_0) \oplus \underline{\mathrm{Hom}}(M_1, N_1)) \\
&\quad \oplus \mathrm{Hom}(K_1, \underline{\mathrm{Hom}}(M_1, N_0) \oplus \underline{\mathrm{Hom}}(M_0, N_1)) \\
&= \mathrm{Hom}_{S_{\bullet}}(K, \underline{\mathrm{Hom}}_{sk\text{-sMod}}(M, N))
\end{aligned}$$

if we define

$$\begin{aligned}
(\underline{\mathrm{Hom}}_{sk\text{-sMod}}(M, N))_0 &= \underline{\mathrm{Hom}}(M_0, N_0) \oplus \underline{\mathrm{Hom}}(M_1, N_1), \\
(\underline{\mathrm{Hom}}_{sk\text{-sMod}}(M, N))_1 &= \underline{\mathrm{Hom}}(M_0, N_1) \oplus \underline{\mathrm{Hom}}(M_1, N_0).
\end{aligned}$$

Now for $M, N \in sk\text{-sMod}_* = sC$ and $K \in S_{*\bullet}$, we have

$$\begin{aligned}
\mathrm{Hom}_{sC}(M \boxtimes K, N) &= \oplus_i \mathrm{Hom}_{sk\text{-sMod}}(M_i \otimes K_i, N_i) \\
&\cong \oplus_i \mathrm{Hom}_{S_{\bullet}}(K_i, \underline{\mathrm{Hom}}_{sk\text{-sMod}}(M_i, N_i)) \\
&= \mathrm{Hom}_{S_{*\bullet}}(K, \underline{\mathrm{Hom}}_{sk\text{-sMod}_*}(M, N))
\end{aligned} \tag{28.3}$$

if we define

$$\underline{\mathrm{Hom}}_{sC}(M, N) = \oplus_i \underline{\mathrm{Hom}}_{sk\text{-sMod}}(M_i, N_i).$$

Thus (28.3) shows that $M \boxtimes - \dashv \underline{\mathrm{Hom}}_{sC}(M, -)$. We also have

$$\begin{aligned}
\underline{\mathrm{Hom}}_{sC}(M, N)_{0\times 0} &= \mathrm{Hom}_{S_{*\bullet}}(\Delta^2\Delta^0, \underline{\mathrm{Hom}}_{sC}(M, N)) \\
&= \oplus_i \mathrm{Hom}_{S_{\bullet}}(\Delta\Delta^0, \underline{\mathrm{Hom}}_{sk\text{-sMod}}(M_i, N_i))
\end{aligned}$$

$$
\begin{aligned}
&= \oplus_i\{\mathrm{Hom}_{\mathrm{Set}_\Delta}(\Delta^0, \oplus_p \underline{\mathrm{Hom}}_{sk\text{-Mod}}(M_{ip}, N_{ip})) \\
&\quad \oplus \mathrm{Hom}(\Delta^0, \oplus_{p\neq q} \underline{\mathrm{Hom}}_{sk\text{-Mod}}(M_{ip}, N_{iq}))\} \\
&= \oplus_i\{\oplus_p \mathrm{Hom}_{\mathrm{Set}_\Delta}(\Delta^0, \underline{\mathrm{Hom}}_{sk\text{-Mod}}(M_{ip}, N_{ip})) \\
&\quad \oplus \oplus_{p\neq q} \mathrm{Hom}(\Delta^0, \underline{\mathrm{Hom}}_{sk\text{-Mod}}(M_{ip}, N_{iq}))\} \\
&= \oplus_i\{\oplus_p (\underline{\mathrm{Hom}}_{sk\text{-Mod}}(M_{ip}, N_{ip}))_0 \oplus \oplus_{p\neq q} (\underline{\mathrm{Hom}}(M_{ip}, N_{iq}))_0\} \\
&= \oplus_i\{\oplus_p \mathrm{Hom}_{s\underline{C}}(M_{ip}, N_{ip}) \oplus \oplus_{p\neq q} \mathrm{Hom}_{s\underline{C}}(M_{ip}, N_{iq})\} \\
&= \oplus_i\{\mathrm{Hom}^+_{sk\text{-sMod}}(M_i, N_i)\} \\
&= \mathrm{Hom}^+_{sC}(M, N).
\end{aligned}
$$

From (28.3) again, we have

$$
\begin{aligned}
\underline{\mathrm{Hom}}_{sC}(M, N)_{n\times n} &= \mathrm{Hom}_{S_{*.}}(\Delta^2\Delta^n, \underline{\mathrm{Hom}}_{sk\text{-sMod}_*}(M, N)) \\
&\cong \mathrm{Hom}_{sC}(M \boxtimes \Delta^2\Delta_n, N)
\end{aligned}
$$

as claimed. For the associativity, we use the associativity of $\otimes$ in the ungraded case:

$$
\begin{aligned}
(M \boxtimes K) \boxtimes L &= \oplus_i (M_i \otimes K_i) \otimes L_i \\
&= \oplus_i \oplus_{p,q,r} (M_{ip} \otimes K_{iq}) \otimes L_{ir} \\
&\cong \oplus_i \oplus_{p,q,r} M_{ip} \otimes (K_{iq} \times L_{ir}) \\
&= \oplus_i (\oplus_p M_{ip}) \times (\oplus_{q,r} K_{iq} \times L_{ir}) \\
&= \oplus_i M_i \otimes (K_i \times L_i) \\
&= \oplus_i M_i \otimes (K \times L)_i \\
&= M \boxtimes (K \times L).
\end{aligned}
$$

For the exponent map,

$$
\begin{aligned}
\mathrm{Hom}_{sC}(M \boxtimes K, N) &= \oplus_i \mathrm{Hom}_{sk\text{-sMod}}(M_i \otimes K_i, N_i) \\
&= \oplus_i\{\mathrm{Hom}_{sk\text{-Mod}}(\oplus_p M_{ip} \otimes K_{ip}, N_{i0}) \\
&\quad \oplus \mathrm{Hom}(\oplus_{p\neq q} M_{ip} \otimes K_{iq}, N_{i1})\} \\
&= \oplus_i\{\oplus_p \mathrm{Hom}_{s\underline{C}}(M_{ip} \otimes K_{ip}, N_{i0}) \\
&\quad \oplus \oplus_{p\neq q} \mathrm{Hom}_{s\underline{C}}(M_{ip} \otimes K_{iq}, N_{i1})\} \\
&\cong \oplus_i\{\oplus_p \mathrm{Hom}_{s\underline{C}}(M_{ip}, N_{i0}^{K_{ip}}) \oplus \oplus_{p\neq q} \mathrm{Hom}_{s\underline{C}}(M_{ip}, N_{i1}^{K_{iq}})\} \\
&= \oplus_i\{\mathrm{Hom}_{s\underline{C}}(M_{i0}, N_{i0}^{K_{i0}}) \oplus \mathrm{Hom}_{s\underline{C}}(M_{i1}, N_{i0}^{K_{i1}}) \\
&\quad \oplus \mathrm{Hom}_{s\underline{C}}(M_{i0}, N_{i1}^{K_{i1}}) \oplus \mathrm{Hom}_{s\underline{C}}(M_{i1}, N_{i1}^{K_{i0}})\} \\
&= \oplus_i\{\mathrm{Hom}_{s\underline{C}}(M_{i0}, (N_i^{K_i})_0) \oplus \mathrm{Hom}_{s\underline{C}}(M_{i1}, (N_i^{K_i})_1)\}
\end{aligned}
$$

if we define

$$(N_i^{K_i})_0 = N_{i0}^{K_{i0}} \oplus N_{i1}^{K_{i1}},$$
$$(N_i^{K_i})_1 = N_{i0}^{K_{i1}} \oplus N_{i1}^{K_{i0}}.$$

From there

$$\begin{aligned}\mathrm{Hom}_{s\mathcal{C}}(M \boxtimes K, N) &= \oplus_i \mathrm{Hom}_{sk\text{-sMod}}(M_i, N_i^{K_i}) \\ &= \mathrm{Hom}_{s\mathcal{C}}(M, N^K)\end{aligned}$$

if we define $N^K = \oplus_i N_i^{K_i}$. This shows that $- \boxtimes K \dashv \exp\_(K)$. This completes the proof. □

28.2.3.4 Bigraded simplicial model categories

We axiomatize the definition of bigraded simplicial model category as done in [28] in the ungraded case. We first need to define pullbacks in $S_{*,\cdot}$. It being a bigraded category, pullbacks are defined entrywise. In what follows, $\mathcal{C} = sk\text{-sMod}_*$. We also use the abbreviation $\underline{\mathrm{Hom}}_{\mathcal{C}}(X, Y)_{ip} = XY_{ip}$ for $i, p = 0, 1$. Consider the Cartesian square

$$\begin{array}{ccc} \Gamma = \underline{\mathrm{Hom}}_{\mathcal{C}}(A, X) \times_{\underline{\mathrm{Hom}}_{\mathcal{C}}(A,Y)} \underline{\mathrm{Hom}}_{\mathcal{C}}(B, Y) & \longrightarrow & \underline{\mathrm{Hom}}_{\mathcal{C}}(B, Y) = (BY_{ip}) \\ \downarrow & & \downarrow \\ \underline{\mathrm{Hom}}_{\mathcal{C}}(A, X) = (AX_{ip}) & \longrightarrow & \underline{\mathrm{Hom}}_{\mathcal{C}}(A, Y) = (AY_{ip}). \end{array}$$

It breaks up into individual Cartesian squares:

$$\begin{array}{ccc} \Gamma_{ip} & \longrightarrow & BY_{ip} \\ \downarrow & & \downarrow \\ AX_{ip} & \longrightarrow & AY_{ip} \end{array}$$

for $i, p = 0, 1$, giving $\Gamma = \oplus_{i,p} AX_{ip} \times_{AY_{ip}} BY_{ip} = (AX_{ip} \times_{AY_{ip}} BY_{ip})$.

Graded simplicial model category axiom – grsModCat 28.2.3.4.1. Let $\mathcal{C}$ be a bigraded model category and a bigraded simplicial category. Suppose $j : A \to B$ is a cofibration and $q : X \to Y$ is a fibration. Then

$$\underline{\mathrm{Hom}}_{\mathcal{C}}(B, X) \xrightarrow{(j^*, q_*)} \underline{\mathrm{Hom}}_{\mathcal{C}}(A, X) \times_{\underline{\mathrm{Hom}}_{\mathcal{C}}(A,Y)} \underline{\mathrm{Hom}}_{\mathcal{C}}(B, Y)$$

is a fibration in $S_{*,\cdot}$, which is trivial if either of j or q is.

Definition 28.2.3.4.2. A category satisfying the axiom grsModCat is called a *bigraded simplicial model category.*

This definition follows exactly the definition of such categories in the ungraded case as laid out in [28]. We prove a preliminary result found in the same reference, which will be instrumental in proving that sk-sMod$_*$ is a bigraded simplicial model category.

Proposition 28.2.3.4.3. *Let C be a bigraded model category and a bigraded simplicial category, let $i : K \to L$ be a cofibration in $S_{*,\cdot}$, and let $q : X \to Y$ be a fibration in C. Then the axiom grsModCat is equivalent to*

$$\underline{\mathrm{hom}}_C(L,X) \to \underline{\mathrm{hom}}_C(K,X) \times_{\underline{\mathrm{hom}}_C(K,Y)} \underline{\mathrm{hom}}_C(L,Y)$$

being a fibration, trivial if either of i or q is. Here we have denoted $N^K = \underline{\mathrm{hom}}_C(K,N)$ for ease of reading.

Proof. It suffices to work entrywise. We have a fibration

$$\underline{\mathrm{Hom}}_C(B,X)_{ip} \to \underline{\mathrm{Hom}}_C(A,X)_{ip} \times_{\underline{\mathrm{Hom}}_C(A,Y)_{ip}} \underline{\mathrm{Hom}}_C(B,Y)_{ip}.$$

For $p = 0$, we have a fibration

$$\begin{array}{c} \underline{\mathrm{Hom}}(B_{i0},X_{i0}) \oplus \underline{\mathrm{Hom}}(B_{i1},X_{i1}) \\ \downarrow \\ \oplus_p \underline{\mathrm{Hom}}(A_{ip},X_{ip}) \times_{\oplus_p \underline{\mathrm{Hom}}(A_{ip},Y_{ip})} \oplus_p \underline{\mathrm{Hom}}(B_{ip},Y_{ip}) \\ \| \\ \oplus_p \underline{\mathrm{Hom}}(A_{ip},X_{ip}) \times_{\underline{\mathrm{Hom}}(A_{ip},Y_{ip})} \underline{\mathrm{Hom}}(B_{ip},Y_{ip}) \end{array}$$

which decomposes into

$$\underline{\mathrm{Hom}}(B_{ip},X_{ip}) \xrightarrow{(j_{ip}^*,(q_{ip})_*)} \underline{\mathrm{Hom}}(A_{ip},X_{ip}) \times_{\underline{\mathrm{Hom}}(A_{ip},Y_{ip})} \underline{\mathrm{Hom}}(B_{ip},Y_{ip})$$

for $p = 0, 1$, a fibration in Set$_\Delta$, trivial if either of j or q is and, in particular, if j_{ip} or q_{ip} is. Using the ungraded counterpart of the proposition from [28], this is equivalent to

$$\underline{\mathrm{hom}}_{\underline{C}}(L_{ip},X_{ip}) \to \underline{\mathrm{hom}}_{\underline{C}}(K_{ip},X_{ip}) \times_{\underline{\mathrm{hom}}_{\underline{C}}(K_{ip},Y_{ip})} \underline{\mathrm{hom}}_{\underline{C}}(L_{ip},Y_{ip}) \tag{28.4}$$

being a fibration, trivial if either $i_{ip} : K_{ip} \to L_{ip}$ or $q_{ip} : X_{ip} \to Y_{ip}$ is for $p = 0, 1$.

For $p = 1$, we have

$$\begin{array}{c}\underline{\mathrm{Hom}}(B_{i0}, X_{i1}) \oplus \underline{\mathrm{Hom}}(B_{i1}, X_{i0}) \\ \downarrow \\ \oplus_{p\neq q}\underline{\mathrm{Hom}}(A_{ip}, X_{iq}) \times_{\oplus_{p\neq q}\underline{\mathrm{Hom}}(A_{ip}, Y_{iq})} \underline{\mathrm{Hom}}(B_{ip}, Y_{iq}) \\ \downarrow \\ \oplus_{p\neq q}\underline{\mathrm{Hom}}(A_{ip}, X_{iq}) \times_{\underline{\mathrm{Hom}}(A_{ip}, Y_{iq})} \underline{\mathrm{Hom}}(B_{ip}, Y_{iq})\end{array}$$

which decomposes as

$$\underline{\mathrm{Hom}}(B_{i0}, X_{i1}) \xrightarrow{(j_{i0}^*, (q_{i1})_*)} \underline{\mathrm{Hom}}(A_{i0}, X_{i1}) \times_{\underline{\mathrm{Hom}}(A_{i0}, Y_{i1})} \underline{\mathrm{Hom}}(B_{i0}, Y_{i1})$$

and

$$\underline{\mathrm{Hom}}(B_{i1}, X_{i0}) \xrightarrow{(j_{i1}^*, (q_{i0})_*)} \underline{\mathrm{Hom}}(A_{i1}, X_{i0}) \times_{\underline{\mathrm{Hom}}(A_{i1}, Y_{i0})} \underline{\mathrm{Hom}}(B_{i1}, Y_{i0})$$

both fibrations, trivial if j_{ik} or q_{il} is trivial with $k \neq l$. Again, by the ungraded counterpart of this result from [28], this is equivalent to

$$\underline{\mathrm{hom}}_{\mathcal{C}}(L_{ik}, X_{il}) \to \underline{\mathrm{hom}}_{\mathcal{C}}(K_{ik}, X_{il}) \times_{\underline{\mathrm{hom}}_{\mathcal{C}}(K_{ik}, Y_{il})} \underline{\mathrm{hom}}_{\mathcal{C}}(L_{ik}, Y_{il}) \tag{28.5}$$

a fibration, trivial if $i_{ik} : K_{ik} \to L_{ik}$ or $q_{il} : X_{il} \to Y_{il}$ is trivial for $k \neq l$. Recombining (28.4) and (28.5) yields

$$\underline{\mathrm{hom}}_{\mathcal{C}}(L, X) \to \underline{\mathrm{hom}}_{\mathcal{C}}(K, X) \times_{\underline{\mathrm{hom}}_{\mathcal{C}}(K, Y)} \underline{\mathrm{hom}}_{\mathcal{C}}(L, Y)$$

a fibration, trivial if $i : K \to L$ or $q : X \to Y$ is trivial. □

Theorem 28.2.3.4.4. sk-sMod$_*$ *is a bigraded simplicial model category.*

Proof. We use the functor $G : \mathcal{C} = sk\text{-sMod}_* \to S_{*.}$ used to put a model structure on sk-sMod$_*$. We wish to show that sk-sMod$_*$ satisfies the axiom grsModCat, which we just showed is equivalent to saying in particular that the map

$$\underline{\mathrm{hom}}_{\mathcal{C}}(L, X) \to \underline{\mathrm{hom}}_{\mathcal{C}}(K, X) \times_{\underline{\mathrm{hom}}_{\mathcal{C}}(K, Y)} \underline{\mathrm{hom}}_{\mathcal{C}}(L, Y)$$

is a fibration for a cofibration $i : K \to L$ in $S_{*.}$ and a fibration $q : X \to Y$ in $\mathcal{C}$ (trivial if either of i or q is), which means that $G(X^L) \to G(X^K \times_{Y^K} Y^L)$ is a fibration in $S_{*.}$ (trivial if either of i or q is). Since G is a right adjoint, it commutes with finite limits, so this is equivalent to showing that

$$G(X^L) \to G(X^K) \times_{G(Y^K)} G(Y^L)$$

is a fibration, trivial if either of i or q is.

Now $F \dashv G : S_{*.} \to \mathcal{C}$ satisfies, for $L, K \in S_{*.}$,

$$\begin{aligned}
F(L \boxtimes K) &= F(\oplus_i L_i \otimes K_i) \\
&= \oplus_i F_i(L_i \otimes K_i) \\
&= \oplus_i F_i(\oplus_{p,q} L_{ip} \otimes K_{iq}) \\
&= \oplus_i [F_{i0}((L_{i0} \otimes K_{i0}) \oplus (L_{i1} \otimes K_{i1})) \\
&\qquad \oplus F_{i1}((L_{i0} \otimes K_{i1}) \oplus (L_{i1} \otimes K_{i0}))] \\
&= \oplus_i (F_{i0}(L_{i0} \otimes K_{i0}) \oplus F_{i0}(L_{i1} \otimes K_{i1}) \\
&\qquad \oplus F_{i1}(L_{i0} \otimes K_{i1}) \oplus F_{i1}(L_{i1} \otimes K_{i0})) \\
&\cong \oplus_i ((F_{i0} L_{i0} \otimes K_{i0}) \oplus (F_{i1} L_{i1} \otimes K_{i1}) \\
&\qquad \oplus (F_{i0} L_{i0} \otimes K_{i1}) \oplus (F_{i1} L_{i1} \otimes K_{i0})) \\
&= \oplus_i ([(F_i L_i)_0 \otimes K_{i0}] \oplus [(F_i L_i)_1 \otimes K_{i1}] \\
&\qquad \oplus [(F_i L_i)_0 \otimes K_{i1}] \oplus [(F_i L_i)_1 \otimes K_{i0}]) \\
&= \oplus_i \oplus_{p,q} (FL)_{ip} \otimes K_{iq} \\
&= FL \boxtimes K
\end{aligned} \tag{28.6}$$

where we have used the fact that $F_{ip} \dashv G_{ip}$, $i, p = 0, 1$, so it commutes with colimits as a left adjoint. Note that we used (for $p = q + r$)

$$F_{ip}(L_{iq} \otimes K_{ir}) \cong F_{iq} L_{iq} \otimes K_{ir}$$

throughout. We need the following lemma.

Lemma 28.2.3.4.5. *If for any $K, L \in S_{*.}$ there is a natural isomorphism $F(L \boxtimes K) \cong F(L) \boxtimes K$, then for any $X \in \mathcal{C} = sk\text{-sMod}_*$,*

$$G\underline{\hom}_{\mathcal{C}}(K, X) \cong \underline{\hom}_{S_{*.}}(K, G(X)).$$

Proof. It suffices to write

$$\begin{aligned}
\mathrm{Hom}_{S_{*.}}(L, G\underline{\hom}_{\mathcal{C}}(K, X)) &= \oplus_{i,p} \mathrm{Hom}_S(L_{ip}, G_{ip} \underline{\hom}_{\mathcal{C}}(K, X)_{ip}) \\
&= \oplus_i [\mathrm{Hom}_S(L_{i0}, G_{i0}(X_{i0}^{K_{i0}} \oplus X_{i1}^{K_{i1}})) \\
&\qquad \oplus \mathrm{Hom}_S(L_{i1}, G_{i1}(X_{i0}^{K_{i1}} \oplus X_{i1}^{K_{i0}}))] \\
&= \oplus_i [\mathrm{Hom}_S(L_{i0}, G_{i0} X_{i0}^{K_{i0}}) \oplus \mathrm{Hom}_S(L_{i0}, G_{i0} X_{i1}^{K_{i1}}) \\
&\qquad \oplus \mathrm{Hom}_S(L_{i1}, G_{i1} X_{i0}^{K_{i1}}) \oplus \mathrm{Hom}_S(L_{i1}, G_{i1} X_{i1}^{K_{i0}})] \\
&\cong \oplus_i [\mathrm{Hom}_{\underline{\mathcal{C}}}(F_{i0} L_{i0}, X_{i0}^{K_{i0}}) \oplus \mathrm{Hom}_{\underline{\mathcal{C}}}(F_{i0} L_{i0}, X_{i1}^{K_{i1}}) \\
&\qquad \oplus \mathrm{Hom}_{\underline{\mathcal{C}}}(F_{i1} L_{i1}, X_{i0}^{K_{i1}}) \oplus \mathrm{Hom}_{\underline{\mathcal{C}}}(F_{i1} L_{i1}, X_{i1}^{K_{i0}})] \\
&\cong \oplus_i [\mathrm{Hom}_{\underline{\mathcal{C}}}(F_{i0} L_{i0} \otimes K_{i0}, X_{i0}) \oplus \mathrm{Hom}_{\underline{\mathcal{C}}}(F_{i0} L_{i0} \otimes K_{i1}, X_{i1})
\end{aligned}$$

$$
\begin{aligned}
&\quad \oplus \operatorname{Hom}_{\underline{\mathcal{C}}}(F_{i1}L_{i1} \otimes K_{i1}, X_{i0}) \oplus \operatorname{Hom}_{\underline{\mathcal{C}}}(F_{i1}L_{i1} \otimes K_{i0}, X_{i1})] \\
&= \oplus_i [\operatorname{Hom}_{\underline{\mathcal{C}}}((FL \boxtimes K)_{i0}, X_{i0}) \oplus \operatorname{Hom}_{\underline{\mathcal{C}}}((FL \boxtimes K)_{i1}, X_{i1})] \\
&= \oplus_i \operatorname{Hom}_{sk\text{-sMod}}((FL \boxtimes K)_i, X_i) \\
&= \operatorname{Hom}_{\mathcal{C}}(FL \boxtimes K, X) \\
&\cong \operatorname{Hom}_{\mathcal{C}}(F(L \boxtimes K), X) \\
&\cong \operatorname{Hom}_{\mathcal{C}}(L \boxtimes K, GX) \\
&= \oplus_i [\operatorname{Hom}_{\underline{\mathcal{C}}}(L_{i0} \otimes K_{i0}, (GX)_{i0}) \oplus \operatorname{Hom}_{\underline{\mathcal{C}}}(L_{i1} \otimes K_{i1}, (GX)_{i0}) \\
&\quad \oplus \operatorname{Hom}_{\underline{\mathcal{C}}}(L_{i0} \otimes K_{i1}, (GX)_{i1}) \oplus \operatorname{Hom}_{\underline{\mathcal{C}}}(L_{i1} \otimes K_{i0}, (GX)_{i1})] \\
&\cong \oplus_i [\operatorname{Hom}_S(L_{i0}, \underline{\operatorname{hom}}_{\mathrm{Set}_\Delta}(K_{i0}, (GX)_{i0})) \\
&\quad \oplus \operatorname{Hom}_S(L_{i1}, \underline{\operatorname{hom}}_{\mathrm{Set}_\Delta}(K_{i1}, (GX)_{i0})) \\
&\quad \oplus \operatorname{Hom}_S(L_{i0}, \underline{\operatorname{hom}}_{\mathrm{Set}_\Delta}(K_{i1}, (GX)_{i1})) \\
&\quad \oplus \operatorname{Hom}_S(L_{i1}, \underline{\operatorname{hom}}_{\mathrm{Set}_\Delta}(K_{i0}, (GX)_{i1}))] \\
&= \oplus_i \operatorname{Hom}_{S_\cdot}(L_i, \underline{\operatorname{hom}}_{S_{*\cdot}}(K, GX)_i) \\
&= \operatorname{Hom}_{S_{*\cdot}}(L, \underline{\operatorname{hom}}_{S_{*\cdot}}(K, GX))
\end{aligned}
$$

with $S = \mathrm{Set}_\Delta$. □

From this we have

$$G(X^L) \to G(X^K) \times_{G(Y^K)} G(Y^L)$$

that is,

$$G\underline{\operatorname{hom}}_{\mathcal{C}}(L,X) \to G\underline{\operatorname{hom}}_{\mathcal{C}}(K,X) \times_{G\underline{\operatorname{hom}}_{\mathcal{C}}(K,Y)} G\underline{\operatorname{hom}}_{\mathcal{C}}(L,Y)$$

is equivalent to

$$\underline{\operatorname{hom}}_{S_{*\cdot}}(L, GX) \to \underline{\operatorname{hom}}_{S_{*\cdot}}(K, GX) \times_{\underline{\operatorname{hom}}_{S_{*\cdot}}(K,GY)} \underline{\operatorname{hom}}_{S_{*\cdot}}(L, GY).$$

Now $q : X \to Y$ a fibration in $\mathcal{C}$ means $G(q) : GX \to GY$ is a fibration in $S_{*\cdot}$, and since $S_{*\cdot}$ is a bigraded simplicial model category, it satisfies the axiom grsModCat, so the above map is a fibration, trivial if either of q or i is, that is, sk-sMod$_*$ is a bigraded simplicial model category. □

28.2.3.5 Internal hom

From [8] and [82] we know that k-sMod has an internal hom $\underline{\mathcal{H}om}_{k\text{-sMod}}$. This comes from the fact that the tensor product $\otimes_k$ on k-sMod involves terms of mixed parity. In those references the internal hom is defined as follows:

$$\underline{\mathcal{H}om}_{k\text{-sMod}}(M,N)_0 = \mathrm{Hom}_{k\text{-sMod}}(M,N) = \mathrm{Hom}_k(M,N)$$

whereas $\underline{\mathcal{H}om}_{k\text{-sMod}}(M,N)_1$ is the set of morphisms $\phi : M \to N$ that reverse parity, i. e., those morphisms $M_0 \to N_1$ and $M_1 \to N_0$. We can see this is indeed the correct definition by starting from the formal definition of the internal hom:

$$\mathrm{Hom}_k(M \otimes_k N, P) \cong \mathrm{Hom}_k(M, \underline{\mathcal{H}om}_{k\text{-sMod}}(N,P)).$$

Expanding the left-hand side in full, we have

$$\begin{aligned}
&\mathrm{Hom}(M_0 \otimes N_0, P_0) \oplus \mathrm{Hom}(M_1 \otimes N_1, P_0) \\
&\qquad \oplus \mathrm{Hom}(M_0 \otimes N_1, P_1) \oplus \mathrm{Hom}(M_1 \otimes N_0, P_1) \\
&\cong \mathrm{Hom}(M_0, \mathrm{Hom}_k(N_0, P_0)) \oplus \mathrm{Hom}(M_1, \mathrm{Hom}_k(N_1, P_0)) \\
&\qquad \oplus \mathrm{Hom}(M_0, \mathrm{Hom}_k(N_1, P_1)) \oplus \mathrm{Hom}(M_1, \mathrm{Hom}_k(N_0, P_1))
\end{aligned}$$

where we used the fact that k-Mod is a closed monoidal category in the sense of [42] with internal hom Hom_k. It follows as claimed above that

$$\begin{aligned}
\underline{\mathcal{H}om}(N,P)_0 &= \mathrm{Hom}(N_0, P_0) \oplus \mathrm{Hom}(N_1, P_1), \\
\underline{\mathcal{H}om}(N,P)_1 &= \mathrm{Hom}(N_1, P_0) \oplus \mathrm{Hom}(N_0, P_1).
\end{aligned}$$

Note that this shows that what we previously called the enhanced hom Hom^+ is actually the internal hom: $\mathrm{Hom}^+ = \underline{\mathcal{H}om}$. Henceforward we will use the notation $\underline{\mathcal{H}om}$.

For the tensor product in k-sMod_*,

$$\begin{aligned}
\mathrm{Hom}(M \boxtimes N, P) &= \oplus_{i=0,1} \mathrm{Hom}(M_i \otimes N_i, P_i) \\
&\cong \oplus_{i=0,1} \mathrm{Hom}(M_i, \underline{\mathcal{H}om}_{k\text{-sMod}}(N_i, P_i)) \\
&= \mathrm{Hom}(M, \underline{\mathcal{H}om}_{k\text{-sMod}_*}(N,P))
\end{aligned}$$

if $\underline{\mathcal{H}om}_{sk\text{-sMod}_*}(N,P) = \oplus_{i=0,1} \underline{\mathcal{H}om}_{sk\text{-sMod}}(N_i, P_i)$. We prolong this to the simplicial case levelwise in the simplicial index:

$$\begin{array}{ccc}
\mathrm{Hom}_{k\text{-sMod}_*}(M_n \boxtimes N_n, P_n) & \xrightarrow{\cong} & \mathrm{Hom}(M_n, \underline{\mathcal{H}om}_{k\text{-sMod}_*}(N_n, P_n)) \\
\Big\downarrow & & \Big\downarrow \\
\mathrm{Hom}_{sk\text{-sMod}_*}(M \boxtimes N, P) & \underset{\cong}{\dashrightarrow} & \mathrm{Hom}(M, \underline{\mathcal{H}om}_{sk\text{-sMod}_*}(N,P)).
\end{array}$$

At this point, $\mathcal{C} = sk$-sMod_*, the category of simplicial graded k-supermodules, is endowed with the levelwise tensor product $\boxtimes$ for which we have an internal hom $\underline{\mathcal{H}om}_{sk\text{-sMod}_*}$ and is endowed with the "usual" model structure whereby weak equivalences and fibrations are defined on the underlying bigraded simplicial sets. We have also proved that $\mathcal{C}$ is a bigraded simplicial model category with

$$\underline{\mathrm{Hom}_{\mathcal{C}}}(M,N)_{n\times n} \cong \mathrm{Hom}_{\mathcal{C}}(M \boxtimes \Delta^2\Delta^n, N).$$

Let

$$\mathrm{Comm}(\mathcal{C}) = sk\text{-sAlg}_*$$

where again commutativity is defined on superalgebras by $ab = (-1)^{|a||b|}ba$ on homogeneous elements.

Now for $A \in sk\text{-sAlg}_*$, denote by A-sMod$_*$ the category of objects of sk-sMod$_*$ that are A-modules with $(A\text{-sMod}_*)_n = A_n\text{-sMod}$. A morphism f of A-supermodules is a simplicial morphism of supermodules $f : M \to N$ such that $f(am) = af(m)$ for $a \in A, m \in M$, $M \in A\text{-sMod}_*$. To be more specific, f has components $f_n : M_n \to N_n$ in $\mathrm{Hom}_{A_n}(M_n, N_n)$, and

$$\begin{aligned} f(am) &= \oplus_n f_n(a_n m_n) \\ &= \oplus_n a_n f_n(m_n) \\ &= af(m). \end{aligned}$$

There is a tensor product $M \tilde{\boxtimes}_A N$ defined levelwise:

$$(M\tilde{\boxtimes}_A N)_n = M_n \tilde{\boxtimes}_{A_n} N_n.$$

28.3 Prehomotopical algebraic context

The notion of homotopical algebraic context is introduced in [79]. The reader is referred to that reference for a full definition. We call it *pre-homotopical* for the simple reason that we neither use $\mathcal{C}_0$, nor do we need $\mathcal{A}$, a subcategory of good objects, or, equivalently, we just work with a symmetric monoidal model category $\mathcal{C}$ satisfying only the first four assumptions of [79], which we will adapt to our bigraded setting.

28.3.1 sk-sMod$_*$ symmetric monoidal model category

We use the definition of symmetric monoidal model category as presented in [42]. We already have a monoidal structure $(\boxtimes_k, \alpha, \lambda, \rho, \Delta k)$ on k-sMod$_*$, which we prolong to a monoidal structure on sk-sMod$_*$ levelwise. For that tensor product, we have an internal hom $\underline{\mathcal{H}om}_{sk\text{-sMod}_*}$. We have half of an adjunction of two variables:

$$\mathrm{Hom}_r = \underline{\mathcal{H}om}_{sk\text{-sMod}_*}$$

with

$$\mathrm{Hom}_{sk\text{-sMod}_*}(M \boxtimes N, P) \xrightarrow[\cong]{\phi_r} \mathrm{Hom}_{sk\text{-sMod}_*}(M, \underline{\mathcal{H}om}_{sk\text{-sMod}_*}(N, P)).$$

We have the braiding $\sigma : M \boxtimes N \xrightarrow{\cong} N \boxtimes M$ in k-sMod$_*$, which we prolonged to sk-sMod$_*$. Now

$$\begin{array}{ccc} \mathrm{Hom}_{sk\text{-sMod}_*}(M \boxtimes N, P) & & \\ \cong \big\downarrow \sigma^* & \overset{\phi_l = \phi_r \circ \sigma^*}{\searrow} & \\ \mathrm{Hom}_{sk\text{-sMod}_*}(N \boxtimes M, P) & \xrightarrow[\cong]{\phi_r} & \mathrm{Hom}_{sk\text{-sMod}_*}(N, \underline{\mathcal{H}om}_{sk\text{-sMod}_*}(M, P)) \end{array}$$

so $(\boxtimes, \underline{\mathcal{H}om}_{sk\text{-sMod}_*}, \underline{\mathcal{H}om}_{sk\text{-sMod}_*}, \phi_r, \phi_r \circ \sigma^*)$ is an adjunction of two variables, and hence we have a closed monoidal structure on sk-sMod$_*$ making it into a closed monoidal category, as defined in [42].

We now show that $\boxtimes$ is a Quillen bifunctor in $\mathcal{C} = sk$-sMod$_*$. Let $f : U \to V$ be a cofibration in $\mathcal{C}$, and let $g : W \to X$ be a cofibration in $\mathcal{C}$ as well. This means they are cofibrations on the underlying bigraded simplicial sets, so entrywise cofibrations of simplicial sets, i. e., inclusions. More generally, maps in sk-sMod$_*$ are fibrations, cofibrations, or weak equivalences if and only if they are respectively fibrations, cofibrations, or weak equivalences in sk-Mod or, equivalently, if they are so entrywise.

We need

$$f \square g : (V \boxtimes W) \coprod_{U \boxtimes W} (U \boxtimes X) \to V \otimes X$$

a cofibration in $\mathcal{C}$, trivial if either of f or g is.

$f : U \to V$ and $g : W \to X$ being cofibrations in sk-sMod$_*$ means we have entrywise cofibrations in sk-Mod $f_{ip} : U_{ip} \to V_{ip}$ and $g_{ip} : W_{ip} \to X_{ip}$ for $i, p = 0, 1$ with $U = (U_{ip})$, $V = (V_{ip})$, $W = (W_{ip})$, and $X = (X_{ip})$. Consider the coproduct $(V \boxtimes W) \coprod_{U \boxtimes W} (U \boxtimes X) = \oplus_{i=0,1} (V_i \otimes W_i) \coprod_{U_i \otimes W_i} (U_i \otimes X_i)$, which we simply denote by $\coprod$, computed in sk-sMod$_*$. Levelwise, it reads

$$\begin{array}{ccc} U_i \otimes W_i & \longrightarrow & U_i \otimes X_i \\ \big\downarrow & & \big\downarrow \\ V_i \otimes W_i & \longrightarrow & \coprod_i \end{array}$$

further decomposing into paritywise coproducts:

$$\begin{array}{ccc} (U_{i0} \otimes W_{i0}) \oplus (U_{i1} \otimes W_{i1}) & \longrightarrow & (U_{i0} \otimes X_{i0}) \oplus (U_{i1} \otimes X_{i1}) \\ \big\downarrow & & \big\downarrow \\ (V_{i0} \otimes W_{i0}) \oplus (V_{i1} \otimes W_{i1}) & \longrightarrow & \coprod_{i0} \end{array}$$

with a similar coproduct defining $\coprod_{i1}$ in such a manner that $\coprod = (\coprod_{ip})$. Note that the maps in this diagram are induced by f_{ip} and g_{ip}. Thus the diagram defining $\coprod_{i0}$ further decomposes into two coproduct diagrams

$$\begin{array}{ccc} U_{ip} \otimes W_{ip} & \longrightarrow & U_{ip} \otimes X_{ip} \\ \downarrow & & \downarrow \\ V_{ip} \otimes W_{ip} & \longrightarrow & C_{ipp} \end{array}$$

so that $\coprod_{i0} = C_{i00} \oplus C_{i11}$. One would obtain a similar decomposition for $\coprod_{i1}$. Working levelwise, we want

$$V_i \otimes W_i \coprod_{U_i \otimes W_i} U_i \otimes X_i \to V_i \otimes X_i \tag{28.7}$$

to be a cofibration. But we know that $f_{ip} : U_{ip} \to V_{ip}$ and $g_{iq} : W_{iq} \to X_{iq}$ are cofibrations, so $C_{ipq} = V_{ip} \otimes W_{iq} \coprod_{U_{ip} \otimes W_{iq}} U_{ip} \otimes X_{iq} \to V_{ip} \otimes X_{iq}$ is a cofibration for $p, q = 0, 1$, $\otimes$ being a Quillen bifunctor on sk-Mod, and all such maps recombine into (28.7), which is therefore a cofibration. Now if either of f or g is trivial, then so are its entries. For instance, if this is true for f, then the entries f_{ip} are trivial, making $U_{ip} \otimes W_{iq} \coprod_{U_{ip} \otimes W_{iq}} U_{ip} \otimes X_{iq} \to V_{ip} \otimes X_{iq}$ trivial for $i, p, q = 0, 1$, and those maps recombine into $f \square g$ trivial. This shows that $\boxtimes$ is a Quillen bifunctor.

Since k is the unit for k-Mod and Δk is the unit in k-sMod$_*$, if c_* is the constant simplicial functor, then $c_*(k) = k_*$ is the unit for sk-Mod, and $c_*(\Delta k) = \Delta k_*$ is the unit for sk-sMod$_*$, a constant bigraded simplicial object. If we call Q the cofibrant replacement functor in sk-sMod$_*$, entrywise that can be obtained from the cofibrant replacement functor on sk-Mod. We want

$$Q\Delta k_* \boxtimes X \to \Delta k_* \boxtimes X \tag{28.8}$$

to be a weak equivalence for any cofibrant object X in sk-sMod$_*$. That means the entries of X are cofibrant in sk-Mod as well, and this latter being a symmetric monoidal category [76, 79], we have $Qk_* \otimes X_{ip} \to k_* \otimes X_{ip}$ cofibrations for $i, p = 0, 1$, all of which recombine into (28.8), a cofibration, and this for all cofibrant X in sk-sMod$_*$. Finally, $\boxtimes_k$ being symmetric in k-sMod$_*$, so is its prolongation to sk-sMod$_*$. Thus we have proved the following:

Theorem 28.3.1.0.1. sk-sMod$_*$ *is a symmetric monoidal model category.*

28.3.2 Assumption 1

This is referred to as Assumption 1.1.0.1 in [79]. A homotopical algebraic context has six assumptions; we use the first four, and that is why we call our theory as being based on a pre-homotopical context. In what follows, we use the fact that sk-Mod satisfies our four assumptions [68].

Assumption 1 28.3.2.1. sk-sMod$_*$ is proper and pointed, and for any $X, Y \in sk$-sMod$_*$, the natural morphism

$$QX \coprod QY \to X \coprod Y \to RX \times RY \tag{28.9}$$

is an equivalence. Moreover, Ho (sk-sMod$_*$) is an additive category.

Proof. sk-Mod is proper and pointed, and hence so is sk-sMod$_*$. Letting $X, Y \in sk$-sMod$_*$, we show that the map $QX \coprod QY \to X \coprod Y \to RX \times RY$ is a natural equivalence. This composition decomposes levelwise into

$$QX_i \coprod QY_i \to X_i \coprod Y_i \to RX_i \times RY_i$$

which itself decomposes as follows:

$$\oplus_{p,q=0,1} QX_{ip} \coprod QY_{iq} \to \oplus_{p,q} X_{ip} \coprod Y_{iq} \to \oplus_{p,q} RX_{ip} \times RY_{iq}$$

and thus entrywise, for fixed p, q,

$$QX_{ip} \coprod QY_{ip} \to X_{ip} \coprod Y_{iq} \to RX_{ip} \times RY_{iq}.$$

This however is an equivalence since sk-Mod is assumed to satisfy this assumption. All such maps recombine into (28.9), which is therefore an equivalence. Hence the assumption holds in sk-sMod$_*$. Finally, Ho (sk-sMod$_*$) is an additive category since this is true for Ho (sk-Mod). □

28.3.3 Assumption 2

Assumption 2 28.3.3.1. Let $C = sk$-sMod$_*$ and $A \in \mathrm{Comm}(C) = sk$-sAlg$_*$. Define a morphism in A-sMod$_*$ to be a fibration or an equivalence if it is so on the underlying objects of C. With this notion, A-sMod$_*$ is a combinatorial proper model category, on which a tensor product $X \tilde{\boxtimes}_A Y$ is defined as $\oplus_{i,p} X_{ip} \otimes_{A_{ip}} Y_{ip}$. With the monoidal structure defined by $-\tilde{\boxtimes}_A-$, A-sMod$_*$ is a symmetric monoidal model category.

Proof. We first show properness, that is, left properness and right properness. Starting with left properness, we must show that any pushout of a weak equivalence along a cofibration is a weak equivalence. Consider any such pushout in A-sMod$_*$

$$\begin{array}{ccc} X & \longrightarrow & U \\ {\scriptstyle\simeq}\downarrow & & \downarrow \\ Y & \underset{\text{cofibr.}}{\longrightarrow} & V. \end{array}$$

Those maps being defined on the underlying objects in sk-sMod$_*$ we have that same diagram in sk-sMod$_*$, which is proper, so $U \to V$ is a weak equivalence in sk-sMod$_*$, and hence so it is in A-sMod$_*$. We show right properness similarly.

Regarding being combinatorial, we first show that A-sMod$_*$ is cofibrantly generated. Since we will use the notations of [42], where that concept is discussed, we remind the reader the following facts: if M is a model category, then to say that M is cofibrantly generated means that there are small sets I and J (again not worrying about universe considerations) of morphisms in M and a small cardinal α such that:

1. Domains and codomains of maps in I and J are α-small.
2. The class of fibrations is J-inj.
3. The class of trivial fibrations is I-inj.

See [42] for the relation between I and J. Since A-Mod is cofibrantly generated for $A \in sk$-CAlg, it follows that A_{ip}-Mod$_{iq}$, $i, p, q = 0, 1$, is cofibrantly generated for $A = (A_{ip}) \in sk$-sAlg$_*$ with small sets I_{ipq} and J_{ipq} such that domains and codomains of maps in those sets are α_{ipq}-small for some α_{ipq}. Then

$$A\text{-sMod}_* = (\oplus_{p,q=0,1} A_{0p}\text{-Mod}_{0q}) \oplus (\oplus_{r,s=0,1} A_{1r}\text{-Mod}_{1s})$$

where each summand is of the form A-Mod for some cofibrantly generated $A \in sk$-CAlg. Then define $I = (I_{ipq})$ and $J = (J_{ipq})$ with $\alpha = \max\{\alpha_{ipq}\}$, I-inj the fibrations in A-sMod$_*$, and J-inj the trivial fibrations in A-sMod$_*$, thereby cofibrantly generated. Finally, recall from [76], for example, that a category $\mathcal{C}$ is locally presentable if there exists a small set of α-small objects $\mathcal{C}_0 \subset \mathcal{C}$ for some cardinal α such that any object in $\mathcal{C}$ is an α-filtered colimit of objects in $\mathcal{C}_0$. It is clear that a bigraded category is locally presentable if it is so entrywise. Then a combinatorial model category is a cofibrantly generated model category whose underlying category is locally presentable. Since for $A \in sk$-CAlg, A-Mod is combinatorial and, in particular, locally presentable, so is A-sMod$_*$ for $A \in sk$-sAlg$_*$, and being cofibrantly generated as well, it is therefore combinatorial.

We now show that A-sMod$_*$ is a symmetric monoidal model category. We first prove that $-\tilde{\boxtimes}_A-$ gives a monoidal structure on A-sMod$_*$. A-Mod has an internal hom $\underline{\mathrm{Hom}}_A$ for $A \in sk$-CAlg:

$$\mathrm{Hom}(M \otimes_A N, P) \cong \mathrm{Hom}(M, \underline{\mathrm{Hom}}_A(N, P)).$$

Now for $A \in sk$-sAlg$_*$ and $M, N, P \in A$-sMod$_*$, we have

$$\mathrm{Hom}(M \tilde{\boxtimes}_A N, P) = \mathrm{Hom}(\oplus_{i,p} M_{ip} \otimes_{A_{ip}} N_{ip}, P_{ip})$$

$$= \oplus_{i,p}\mathrm{Hom}(M_{ip} \otimes_{A_{ip}} N_{ip}, P_{ip})$$
$$\cong \mathrm{Hom}(M_{ip}, \underline{\mathrm{Hom}}_{A_{ip}}(N_{ip}, P_{ip}))$$
$$= \mathrm{Hom}(M, \underline{\mathrm{Hom}}_A(N, P))$$

so there is an internal hom $\underline{\mathrm{Hom}}_A$ in A-sMod$_*$ with

$$\underline{\mathrm{Hom}}_A(N, P) = (\underline{\mathrm{Hom}}_{A_{ip}}(N_{ip}, P_{ip})).$$

So far we have half of an adjunction of two variables:

$$\mathrm{Hom}_{A\text{-sMod}_*}(M\tilde{\boxtimes}_A N, P) \xrightarrow[\cong]{\phi_r} \mathrm{Hom}_{A\text{-sMod}_*}(M, \underline{\mathrm{Hom}}_A(N, P)).$$

Recall that we have the braiding $\sigma : M \boxtimes N \to N \boxtimes M$ in sk-sMod$_*$ defined entrywise, so accordingly we also have a braiding operator $\sigma_A : M\tilde{\boxtimes}_A N \to N\tilde{\boxtimes}_A M$. We have

$$\begin{array}{ccc} \mathrm{Hom}_{A\text{-sMod}_*}(M\tilde{\boxtimes}_A N, P) & & \\ \cong \downarrow \sigma_A^* & \overset{\phi_l = \phi_r \circ \sigma_A^*}{\searrow} & \\ \mathrm{Hom}_{A\text{-sMod}_*}(N\tilde{\boxtimes}_A M, P) & \xrightarrow[\cong]{\phi_r} & \mathrm{Hom}_{A\text{-sMod}_*}(N, \underline{\mathrm{Hom}}_A(M, P)) \end{array} \tag{28.10}$$

so $(\tilde{\boxtimes}_A, \underline{\mathrm{Hom}}_A, \underline{\mathrm{Hom}}_A, \phi_r, \phi_r \circ \sigma_A^*)$ is an adjunction of two variables, and hence we have a closed monoidal structure on A-sMod$_*$ making it a closed monoidal category.

We now show that $\tilde{\boxtimes}_A$ is a Quillen bifunctor. Write $\mathcal{C} = A$-sMod$_*$, let $f : U \to V$ and $\mathcal{C}, g : W \to X$ be cofibrations in $\mathcal{C}$. We need

$$f\square g : (V\tilde{\boxtimes}_A W) \coprod_{U\tilde{\boxtimes}_A W} (U\tilde{\boxtimes}_A X) \to V\tilde{\boxtimes}_A X$$

to be a cofibration in $\mathcal{C}$, trivial if either of f or g is. This decomposes entrywise into

$$(V_{ip} \otimes_{A_{ip}} W_{ip}) \coprod_{U_{ip}\otimes_{A_{ip}} W_{ip}} (U_{ip} \otimes_{A_{ip}} X_{ip}) \to V_{ip} \otimes_{A_{ip}} X_{ip}$$

for $i, p = 0, 1$ to be a cofibration in $\mathcal{C}$, trivial if either of f or g is. This follows from the same computation done when showing that $\boxtimes$ was a Quillen bifunctor on sk-sMod$_*$.

Now A being the unit in A-sMod$_*$ for $\tilde{\boxtimes}_A$ and $QA \to A$ a cofibrant replacement, for any cofibrant object X, we show that $QA\tilde{\boxtimes}_A X \to A\tilde{\boxtimes}_A X$ is a weak equivalence. This follows from the definition of $\tilde{\boxtimes}_A$ and the fact that for $A \in sk$-CAlg and X cofibrant in sk-Mod, $QA \otimes_A X \to A \otimes_A X$ is a weak equivalence. Finally, since $M\tilde{\boxtimes}_A N = \oplus_{i,p} M_{ip} \otimes_{A_{ip}} N_{ip}$ and $- \otimes_A -$ defines a symmetric monoidal structure on A-Mod, for $A \in sk$-CAlg, it follows that $-\tilde{\boxtimes}_A-$ defines a symmetric monoidal structure on A-sMod$_*$ as well for $A \in sk$-sAlg$_*$. This completes the proof. □

28.3.4 Assumption 3

Assumption 3 28.3.4.1. For $A \in \mathrm{Comm}(\mathcal{C}) = sk\text{-sAlg}_*$ and for any M fibrant in A-sMod$_*$, the functor $-\tilde{\boxtimes}_A M : A\text{-sMod}_* \to A\text{-sMod}_*$ preserves equivalences.

Proof. For M fibrant in A-sMod$_*$, writing 1 for the terminal object of A-sMod$_*$, $M \to 1$ being a fibration, we have that $M_{ip} \to 1_{ip}$ is a fibration in sk-Mod for $i, p = 0, 1$, i. e., M_{ip} fibrant in A_{ip}-Mod for $i, p = 0, 1$. Let f be an equivalence in A-sMod$_*$, and $f = (f_{ip})$, f_{ip} a weak equivalence in A_{ip}-Mod for $i, p = 0, 1$. We know that for $A \in sk$-CAlg and $M \in A$-Mod, $- \otimes_A M$ preserves weak equivalences, so having M_{ip} fibrant and f_{ip} an equivalence, $f_{ip} \otimes_{A_{ip}} M_{ip}$ is a weak equivalence for $i, p = 0, 1$, and hence so is $f \tilde{\boxtimes}_A M$ for $M \in A$-sMod$_*$. □

28.3.5 Assumption 4

Assumption 4 28.3.5.1. For $A \in \mathrm{Comm}(\mathcal{C}) = sk\text{-sAlg}_*$, there exist categories A-Comm$(\mathcal{C})$ and A-Comm$_{nu}(\mathcal{C})$ (nonunital) whose morphisms are fibrations and equivalences if they are so on the underlying objects of $\mathcal{C}$. This makes those categories combinatorial proper model categories. If B is cofibrant in A-Comm$(\mathcal{C})$, then the functor $B \tilde{\boxtimes}_A - : A\text{-sMod}_* \to B\text{-sMod}_*$ preserves equivalences.

Proof. Given how fibrations and equivalences are defined in those categories, proving they are combinatorial proper model categories is done exactly the same way we proved those statements for A-sMod$_*$. Proving that $B \tilde{\boxtimes}_A -$ preserves equivalences for B cofibrant is done in the same manner we proved that $- \tilde{\boxtimes}_A C$ preserves equivalences for C fibrant in A-sMod$_*$. □

28.4 Bigraded derived algebraic stacks

28.4.1 Finitely presented morphisms

Let $\mathcal{C} = sk\text{-sMod}_*$, so that $\mathrm{Comm}(\mathcal{C}) = sk\text{-sAlg}_* = M$, our model category of interest. We will need M to be proper, but weak equivalences are defined on the underlying objects of sk-sMod$_*$, and it being proper, so is M. We fix a bigraded cofibrant resolution functor $(\Gamma : M \to M^{\Delta}, \iota)$, $\Gamma = (\Gamma_{ip})$ [76] with $\Gamma_{ip} : \underline{M} \to \underline{M}^{\Delta}$ cofibrant resolution functors in sk-Mod for $i, p = 0, 1$, where as usual the underlined object corresponds to an ungraded counterpart. Here $M = \mathrm{Comm}(sk\text{-sMod}_*)$, so $\underline{M} = \mathrm{Comm}(\underline{sk\text{-sMod}_*}) = \mathrm{Comm}(sk\text{-Mod}) = sk\text{-CAlg}$. We have $c^* = (c^*)_{2X2}$ with weak equivalences $\Gamma_{ip}(A) \xrightarrow{\iota_{ip}(A)} c^* A$ for $i, p = 0, 1$, with c^* the constant cosimplicial object functor. Since we consider bigraded morphisms in M, we have

$$\begin{aligned}\mathrm{Map}_M(A,B) &= \mathrm{Hom}_M(\Gamma(A),B)\\ &= \oplus_{i=0,1}\mathrm{Hom}_{sk\text{-sAlg}}(\Gamma_i(A_i),B_i)\\ &= \oplus_{i,p}\mathrm{Hom}_{\underline{M}}(\Gamma_{ip}A_{ip},B_{ip})\\ &= \oplus_{i,p}\mathrm{Map}_{\underline{M},ip}(A_{ip},B_{ip})\end{aligned}$$

so we define $\mathrm{Map}_{\underline{M}} = \mathrm{Map}_{\underline{M},0} \oplus \mathrm{Map}_{\underline{M},1}$ so that we have

$$\mathrm{Map}_M = (\mathrm{Map}_{\underline{M},ip}(-,-)) = (\mathrm{Hom}_{\underline{M}}(\Gamma_{ip}-,-)).$$

Now let $A \to B$ be a morphism in $M = sk\text{-sAlg}_*$. Consider any filtered diagram of objects under A and $\{C_i\}_{i\in I} \in A/M$. We have a natural morphism

$$\begin{aligned}&\operatorname*{hocolim}_{i\in I} \mathrm{Map}_{A/M}(B,C_i)\\ &\quad= \Big(\operatorname*{hocolim}_{i\in I} \mathrm{Map}_{A_{jp}/\underline{M}}(B_{jp},C_{i,jp})\Big)\\ &\quad\xrightarrow{\cong} \Big(\mathrm{Map}_{A_{jp}/\underline{M}}(B_{jp},\operatorname*{hocolim}_{i\in I} C_{i,jp})\Big)\\ &\quad= \mathrm{Map}_{A/M}(B,\operatorname*{hocolim}_{i\in I} C_i)\end{aligned}$$

and those objects are defined in Ho $(S_{*.})$. This shows a morphism in M is finitely presented if and only if it is so in $\underline{M} = sk\text{-CAlg}$.

28.4.2 Bigraded derivations and cotangent complexes

For $\mathcal{C} = sk\text{-sMod}_*$, $A \in \mathrm{Comm}(\mathcal{C})$, and $M \in A\text{-Comm}(\mathcal{C})$, we define a commutative monoid $A \oplus M$ with $A \coprod M$ as underlying object, just as in [79], with multiplication

$$(A\oplus M)\boxtimes(A\oplus M) \xrightarrow{\pi} A\oplus M$$

defined by

$$\begin{aligned}&(A\textstyle\coprod M)\boxtimes(A\coprod M)\\ &\quad= A\boxtimes A\textstyle\coprod A\boxtimes M\coprod M\boxtimes A\coprod M\boxtimes M\\ &\quad\longrightarrow A\textstyle\coprod M\end{aligned}$$

where

$$\begin{aligned}\mu\textstyle\coprod * &: A\boxtimes A \xrightarrow{\mu} A \to A\textstyle\coprod M,\\ *\textstyle\coprod \rho &: A\boxtimes M \xrightarrow{\rho} M \to A\textstyle\coprod M,\\ &\quad M\boxtimes M \to M \to A\textstyle\coprod M.\end{aligned}$$

Here μ and ρ have entries as defined in [79], and in the last line, we use the fact that M is a commutative monoid in A-sMod$_*$. To show $A \oplus M$ is commutative, it suffices to write

$$A \coprod M = \oplus_{i=0,1} A_i \coprod M_i = \oplus_{i=0,1} \oplus_p A_{ip} \coprod M_{ip}.$$

Then

$$\begin{aligned}(A \coprod M) \boxtimes (A \coprod M) &= \oplus_{i=0,1}(A_i \coprod M_i) \otimes (A_i \coprod M_i)\\ &= \oplus_{i=0,1}(\oplus_p A_{ip} \coprod M_{ip}) \otimes (\oplus_q A_{iq} \coprod M_{iq})\\ &= \oplus_{i=0,1} \oplus_{p,q} (A_{ip} \coprod M_{ip}) \otimes (A_{iq} \coprod M_{iq}).\end{aligned}$$

Now $A \in k$-sMod$_*$ is supercommutative if $\mu \circ \sigma = \mu$, that is, for any homogeneous elements x, y in A, $xy = \mu(x \otimes y) = (-1)^{|x||y|} yx$. Presently, we have π on

$$\begin{aligned}(A \oplus M) \boxtimes (A \oplus M) &= \oplus_{i=0,1} \oplus_{p,q} (A_{ip} \otimes A_{iq}) \coprod (A_{ip} \otimes M_{iq})\\ &\quad \coprod (M_{ip} \otimes A_{iq}) \coprod (M_{ip} \otimes M_{iq})\end{aligned}$$

and on each entry, π is supercommutative, so $A \oplus M$ is supercommutative.

Now for a morphism $A \to B$ in $\mathrm{Comm}(\mathcal{C})$ and $M \in B$-$\mathrm{Comm}(\mathcal{C})$, the morphism $\mathrm{id} \coprod * : B \oplus M \to B$ is a morphism in A-$\mathrm{Comm}(\mathcal{C})$, and hence we can regard $B \oplus M \in A$-$\mathrm{Comm}(\mathcal{C})/B$.

Definition 28.4.2.1. For a morphism $A \to B$ in $\mathrm{Comm}(\mathcal{C})$, $M \in B$-$\mathrm{Comm}(\mathcal{C})$, and $\mathcal{C} = sk$-sMod$_*$, we define, as in [79], the *set of A-derivations $B \to M$* as

$$\mathbb{D}\mathrm{er}_A(B, M) = \mathrm{Map}_{A\text{-}\mathrm{Comm}(\mathcal{C})/B}(B, B \oplus M) \in \mathrm{Ho}\,(S_{*,\cdot})$$

where $B \to B \oplus M$ is a section of $\mathrm{id} \coprod *$.

We have the following result, much as in [79].

Proposition 28.4.2.2. *For any morphism $A \to B$ in sk-sAlg$_*$, there is an object $\mathbb{L}_{B/A} \in B$-sMod$_*$ and $d \in \pi_0(\mathbb{D}\mathrm{er}_A(B, \mathbb{L}_{B/A}))$ such that for any $M = (M_{ip})$, $M_{ip} \in B_{ip}$-Mod, i. e., $M \in B$-sMod$_*$, the natural induced morphism*

$$d^* : \mathrm{Map}_{B\text{-sMod}_*}(\mathbb{L}_{B/A}, M) \to \mathbb{D}\mathrm{er}_A(B, M)$$

is an isomorphism in $\mathrm{Ho}\,(S_{*,\cdot})$.

Note that for $\psi \in \mathrm{Map}_{B\text{-sMod}_*}(\mathbb{L}_{B/A}, M)$, the element $d^*\psi$ is constructed as follows:

$$\begin{array}{ccccc} & & \Gamma\mathbb{L}_{B/A} & \xrightarrow{\psi} & M \\ & & \downarrow & & \downarrow \\ B & \xrightarrow{d} & B \oplus \Gamma\mathbb{L}_{B/A} & \xrightarrow{B\oplus\psi} & B \oplus M. \end{array}$$

Proof. A morphism $A \to B \in \mathrm{Comm}(\mathcal{C})$ gives rise to entry maps $A_{ip} \to B_{ip} \in \mathrm{Comm}(\underline{\mathcal{C}})$ with $\underline{\mathcal{C}} = sk\text{-Mod}$ for $i, p = 0, 1$, so we are guaranteed by [79] that there is some $\mathbb{L}_{B_{ip}/A_{ip}} \in B_{ip}\text{-Mod}$ and $d_{ip} \in \pi_0(\mathbb{D}\mathrm{er}_{A_{ip}}(B_{ip}, \mathbb{L}_{B_{ip}/A_{ip}}))$ such that for any $M_{ip} \in B_{ip}\text{-Mod}$,

$$d_{ip}^* : \mathrm{Map}_{B_{ip}\text{-Mod}}(\mathbb{L}_{B_{ip}/A_{ip}}, M_{ip}) \to \mathbb{D}\mathrm{er}_{A_{ip}}(B_{ip}, M_{ip})$$

is an isomorphism in Ho (Set_Δ). We view $\mathbb{L}_{B/A} = (\mathbb{L}_{B_{ip}/A_{ip}})$ as an object of $(B_{ip}\text{-Mod}_{ip}) \subset B\text{-sMod}_*$, and the same is true of $M = (M_{ip})$. Then it suffices to write

$$\begin{aligned} \mathrm{Map}_{B\text{-sMod}_*}(\mathbb{L}_{B/A}, M) &= \oplus_{i,p} \mathrm{Map}_{B_{ip}\text{-Mod}}(\mathbb{L}_{B_{ip}/A_{ip}}, M_{ip}) \\ &\xrightarrow{(d_{ip}^*)} \oplus_{i,p} \mathrm{Map}_{A_{ip}-\mathrm{Comm}(\mathcal{C})/B_{ip}}(B_{ip}, B_{ip} \oplus M_{ip}) \\ &= \mathrm{Map}_{A\text{-Comm}(\mathcal{C})/B}(B, B \oplus M) \\ &= \mathbb{D}\mathrm{er}_A(B, M) \end{aligned}$$

with $d = (d_{ip}) \in \pi_0(\mathbb{D}\mathrm{er}_A(B, \mathbb{L}_{B/A}))$. □

Definition 28.4.2.3. Let $A \to B$ be a morphism in $\mathrm{Comm}(sk\text{-sMod}_*)$. As in [79], the B-module $\mathbb{L}_{B/A} \in$ Ho $(B\text{-sMod}_*)$ is called the *cotangent complex of B over A*. If $A = 1 = \Delta k_*$, then write $\mathbb{L}_B = \mathbb{L}_{B/1}$, which we will call the cotangent complex of B.

28.4.3 Étale morphisms

We have shown that $A\text{-sMod}_*$ is a symmetric monoidal model category with the tensor product $\tilde{\boxtimes}_A$. This means its homotopy category Ho $(A\text{-sMod}_*)$ has a natural symmetric monoidal structure $\tilde{\boxtimes}_A^{\mathbb{L}}$.

Definition 28.4.3.1. Let $f : A \to B$ be a morphism in $\mathrm{Comm}(sk\text{-sMod}_*) = sk\text{-sAlg}_*$. The morphism f is said to be *formally étale* if

$$\mathbb{L}_A \tilde{\boxtimes}_A B \xrightarrow{\cong} \mathbb{L}_B$$

or entrywise

$$\mathbb{L}_{A_{ip}} \otimes_{A_{ip}} B_{ip} \xrightarrow{\cong} \mathbb{L}_{B_{ip}}$$

for $i, p = 0, 1$, i. e., f is formally étale if it is so entrywise.

Note that this is a well-defined morphism: if $p = 0$, then the parity of $\mathbb{L}_{A_{ip}} \otimes_{A_{ip}} B_{ip}$ is $p+p+p = 0$, so we do have $\mathbb{L}_{A_{i0}} \otimes_{A_{i0}} B_{i0} \to \mathbb{L}_{B_{i0}}$, and if $p = 1$, then the parity of $\mathbb{L}_{A_{i1}} \otimes_{A_{i1}} B_{i1}$ is 3 mod 2, which is 1, so we also have a well-defined map $\mathbb{L}_{A_{i1}} \otimes_{A_{i1}} B_{i1} \to \mathbb{L}_{B_{i1}}$.

Definition 28.4.3.2. We define a morphism of commutative monoids in sk-sMod$_*$ to be *étale* if it is finitely presented and formally étale.

This is the same definition as initially presented in [79]. Following the same reference, we will use the same terminology for morphisms in Ho (Comm($\mathcal{C}$)) and for the corresponding morphisms of representable stacks. We typically write $\mathcal{A}\mathrm{ff}_{\mathcal{C}} = (\mathrm{Comm}(\mathcal{C}))^{\mathrm{op}}$, but if $\mathcal{C} = sk\text{-sMod}_*$ and $\mathrm{Comm}(\mathcal{C}) = sk\text{-sAlg}_*$, then $(sk\text{-sAlg}_*)^{\mathrm{op}}$ is denoted, as in [79], k-D$^-$sAff$_*$.

If $\mathcal{C} = sk\text{-sAlg}_* \subset sk\text{-sMod}_*$ and $A \in \mathcal{C}$, then we define the underlying bigraded space of A as

$$\begin{aligned} |A| &= \mathrm{Map}_{\mathcal{C}}(\Delta k_*, A) \\ &= (\mathrm{Map}_{\underline{\mathcal{C}}}(k_*, A_{ip}) \\ &= (|A_{ip}|) \in S_*. \end{aligned}$$

so that following [79], we define

$$\begin{aligned} \pi_k(A) &= \pi_k(|A|, *) \\ &= (\pi_k(|A_{ip}|, *_{ip})) \\ &= (\pi_k A_{ip}). \end{aligned}$$

Write $\mathrm{Spec}\, A \in k\text{-D}^-\mathrm{sAff}_*$ if $A \in sk\text{-sAlg}_*$. Following [79], we define families $\{\mathrm{Spec}\, A_i \to \mathrm{Spec}\, A\}_{i \in I}$ to be *étale covering families* if and only if there is $J \subset I$ such that for all $i \in I$,

$$\pi_n A \tilde{\boxtimes}_{\pi_0 A} \pi_0 A_i \xrightarrow{\cong} \pi_n A_i \tag{28.11}$$

and

$$\coprod_{j \in J} \mathrm{Spec}\, \pi_0 A_j \to \mathrm{Spec}\, \pi_0 A$$

is étale and surjective, where $\mathrm{Spec}\, A \to \mathrm{Spec}\, B$ is étale if $B \to A$ is an étale morphism in sk-sAlg$_*$. Note that (28.11) reads as follows entrywise:

$$\pi_n A_{jp} \otimes_{\pi_0 A_{jp}} \pi_0 A_{i,jp} \xrightarrow{\cong} \pi_n A_{i,jp}.$$

Now if $A \in k\text{-sAlg}_*$, then

$$\mathrm{Spec}\, A = \mathrm{Spec}\, \oplus_{i,p} A_{ip}$$

$$= (\oplus_{i,p} A_{ip})^{\mathrm{op}}$$
$$= \oplus_{i,p} A_{ip}^{\mathrm{op}}$$
$$= \oplus_{i,p} \mathrm{Spec}\, A_{ip}$$

along with $\pi_0 A = (\pi_0 A_{ip})$, which gives, for $A \in sk\text{-sAlg}_*$,

$$\mathrm{Spec}\, \pi_0 A = \mathrm{Spec}\, \oplus_{i,p} \pi_0 A_{ip} = \oplus_{i,p} \mathrm{Spec}\, \pi_0 A_{ip}.$$

It follows that

$$\coprod_{j \in J} \mathrm{Spec}\, \pi_0 A_j = \coprod_{j \in J} \oplus_{l,p} \mathrm{Spec}\, \pi_0 A_{j,lp}$$
$$= \oplus_{l,p} \coprod_{j \in J} \mathrm{Spec}\, \pi_0 A_{j,lp} \overset{\text{ét.}}{\twoheadrightarrow} \oplus_{l,p} \mathrm{Spec}\, \pi_0 A_{lp}$$

so that entrywise

$$\coprod_{j \in J} \mathrm{Spec}\, \pi_0 A_{j,lp} \twoheadrightarrow \mathrm{Spec}\, \pi_0 A_{lp}.$$

However, finitely presented morphisms in $sk\text{-sAlg}_*$ are so entrywise, and given the definition of formally étale morphisms and the definition of $\tilde{\boxtimes}_A$, formally étale maps are so entrywise as we argued. It follows that étale morphisms in $sk\text{-sMod}_*$ are so entrywise, so the above morphism is entrywise étale. We conclude that a family of coverings in $k\text{-D}^-\mathrm{sAff}_*$ is an étale covering if and only if it is so entrywise in $\mathrm{k\text{-}D^-Aff}$.

28.4.4 Descent on supermodules

We now define the descent condition on supermodules, which is just an adaptation to our setting of the work done in [79]. We define a cosimplicial object A_* in $\mathrm{Comm}(\mathcal{C})$, $\mathcal{C} = k\text{-sMod}_*$, to be an object of $\mathrm{Comm}(\mathcal{C})^{\Delta}$ that with $[n]$ associates $A_n = (A_{n,ip})$. A cosimplicial A_*-module M_* is given by a graded A_n-supermodule M_n for all $n \in \Delta$, $M_n = (M_{n,ip})$, and for any morphism $u : [n] \to [m]$ in Δ of a morphism of graded A_n-supermodules $\alpha_u : M_n \to M_m$ satisfying the usual covariance condition. In the same manner, a morphism of cosimplicial graded A_*-supermodules $f : M_* \to N_*$ is given by a collection of morphisms $f_n : M_n \to N_n$ for all $n \in \Delta$ commuting with the α. This defines a category $\mathrm{cs}A_*\text{-sMod}_*$ of cosimplicial graded A_*-supermodules. This is a combinatorial proper model category where equivalences (resp., fibrations) are morphisms $f : M_* \to N_*$ such that each $f_n : M_n \to N_n$ is an equivalence (resp., a fibration) in $A_n\text{-sMod}_*$, a levelwise projective model structure. It will also be convenient sometimes, for $A \in k\text{-sAlg}_*$, to regard B_* a cosimplicial commutative graded A-superalgebra as a cosimplicial commutative monoid with a coaugmentation morphism $A \to B_*$, where we regard A as a constant cosimplicial object. This latter defines a category csA-sMod$_*$ of cosimplicial graded cc(A)-supermodules, which is really cs(A-sMod$_*$), along with its levelwise projective model structure.

For a cosimplicial graded A-supermodule M_*, we define a cosimplicial graded B_*-supermodule $B_*\tilde{\boxtimes}_A M_*$ by $(B_*\tilde{\boxtimes}_A M_*)_n = B_n\tilde{\boxtimes}_A M_n$, with morphisms between different degrees given by the ones on B_* and M_*. This gives a functor

$$B_*\tilde{\boxtimes}_A - : csA\text{-sMod}_* \to csB_*\text{-sMod}_*$$

with a right adjoint

$$csA\text{-sMod}_* \leftarrow csB_*\text{-sMod}_* : f$$

sending a graded B_*-supermodule M_* to its underlying graded A-supermodule, clearly a Quillen adjunction. As in [79], we have an additional adjunction

$$\text{cst} : \text{Ho}\,(A\text{-sMod}_*) \rightleftarrows \text{Ho}\,(csA\text{-sMod}_*) : \text{holim} = \mathbb{R}\lim$$

and by composition

$$B_*\tilde{\boxtimes}_A^{\mathbb{L}} - = \mathbb{L}(B_*\tilde{\boxtimes}_A -) \circ \text{cst} : \text{Ho}\,(A\text{-sMod}_*) \rightleftarrows \text{Ho}\,(csB_*\text{-sMod}) : \text{holim} \circ \mathbb{R}f = \int .$$

Definition 28.4.4.1. Let B_* be a cosimplicial commutative graded supermonoid, and let M_* be a cosimplicial graded B_* supermodule. We say M_* is *homotopy Cartesian* if for all $u : [n] \to [m]$ in Δ, the map induced by $\alpha_u : M_n \to M_m$, $M_n \boxtimes^{\mathbb{L}}_{B_n} B_m \to M_m$ is an isomorphism in Ho (B_m-sMod$_*$).

Now if $A \in \text{Comm}(\mathcal{C})$, $\mathcal{C} = k$-sMod$_*$, and B_* is cosimplicial commutative algebra over A, seen as a cosimplicial commutative graded super monoid, then we say the augmentation $A \to B_*$ satisfies the *descent condition* if in the adjunction

$$B_*\tilde{\boxtimes}_A^{\mathbb{L}} - : \text{Ho}\,(A\text{-sMod}_*) \rightleftarrows \text{Ho}\,(csB_*\text{-sMod}_*) : \int .$$

$B_*\tilde{\boxtimes}_A^{\mathbb{L}}-$ is fully faithful and gives an equivalence between Ho (A-sMod$_*$) and the full subcategory of Ho (csB_*-sMod$_*$) spanned by homotopy Cartesian objects.

28.4.5 Étale model topology

In this subsection, we prove that étale covering families define a model topology on k-D$^-$sAff$_*$, which further satisfies an assumption that we will refer to as the “Cover” assumption, the bigraded generalization of Assumption 1.3.2.2 of [79].

28.4.5.1 Model topology

Recall that a family of morphisms $\{\operatorname{Spec} A_i \to \operatorname{Spec} A\}_{i\in I}$ in k-D$^-$sAff$_*$ is an étale covering family if and only if there is a finite subset $J \subset I$ such that for all $i \in I$, $\pi_* A \tilde{\boxtimes}_{\pi_0 A} \pi_0 A_i \xrightarrow{\cong} \pi_* A_i$, and $\coprod_{j\in J} \operatorname{Spec} \pi_0 A_j \to \operatorname{Spec} \pi_0 A$ is étale and surjective. Recall that being étale means finitely presented and formally étale, where $A \to B$ is formally étale if $\mathbb{L}_A \tilde{\boxtimes}_A B \xrightarrow{\cong} \mathbb{L}_B$, and finitely presented means that for any family $\{C_i\}$, $\operatorname{hocolim}_{i\in I} \operatorname{Map}_{A/M}(B, C_i) \xrightarrow{\cong} \operatorname{Map}_{A/M}(B, \operatorname{hocolim} C_i)$. Recall also that a family is a covering étale family if and only if it is so entrywise. We will use the fact, proven in [79], that étale covering families define a model topology in k-D$^-$Aff $=$ $(sk\text{-CAlg})^{\mathrm{op}}$. We refer the reader to [76] for the notion of model topology. There are three conditions to be met. Let $M = k$-D$^-$sAff$_*$, and let $\operatorname{Spec} A \in M$. Suppose we have an isomorphism $\operatorname{Spec} B \to \operatorname{Spec} A$. Entry-wise, this reads $\operatorname{Spec} B_{ip} \to \operatorname{Spec} A_{ip}$ isomorphism for $i, p = 0, 1$ in k-D$^-$Aff, which gives $\{\operatorname{Spec} B_{ip} \to \operatorname{Spec} A_{ip}\} \in \operatorname{cov}_{\text{ét}}(\operatorname{Spec} A_{ip})$ for $i, p = 0, 1$. But since $\operatorname{cov}_{\text{ét}}(X) = (\operatorname{cov}_{\text{ét}}(X_{ip}))$, it follows that condition 1 is met. For the second point, let $\{\operatorname{Spec} A_i \to \operatorname{Spec} A\}_{i\in I} \in \operatorname{cov}_{\text{ét}}(\operatorname{Spec} A)$. This means $\{\operatorname{Spec} A_{i,kp} \to \operatorname{Spec} A_{kp}\}_{i\in I} \in \operatorname{cov}_{\text{ét}}(\operatorname{Spec} A_{kp})$ for $k, p = 0, 1$. Let $\{\operatorname{Spec} B_{ij} \to \operatorname{Spec} A_i\}_{j\in J} \in \operatorname{cov}_{\text{ét}}(\operatorname{Spec} A_i)$ for $i \in I$, and again that means $\{\operatorname{Spec} B_{ij,kp} \to \operatorname{Spec} A_{i,kp}\}_{j\in J} \in \operatorname{cov}_{\text{ét}}(\operatorname{Spec} A_{i,kp})$ for $k, p = 0, 1$. Then $\{\operatorname{Spec} B_{ij,kp} \to \operatorname{Spec} A_{kp}\}_{i\in I, j\in J} \in \operatorname{cov}_{\text{ét}}(\operatorname{Spec} A_{kp})$ for $k, p = 0, 1$, i. e., $\{\operatorname{Spec} B_{ij} \to \operatorname{Spec} A\}_{i\in I, j\in J} \in \operatorname{cov}_{\text{ét}}(\operatorname{Spec} A)$. This is the second condition. Third point, for any $\{\operatorname{Spec} A_i \to \operatorname{Spec} A\}_{i\in I} \in \operatorname{cov}_{\text{ét}}(\operatorname{Spec} A)$, that is, $\{\operatorname{Spec} A_{i,jp} \to \operatorname{Spec} A_{jp}\}_{i\in I} \in \operatorname{cov}_{\text{ét}}(\operatorname{Spec} A_{jp})$ for $j, p = 0, 1$, and for any morphism $\operatorname{Spec} B \to \operatorname{Spec} A$, which entry-wise reads $\operatorname{Spec} B_{jp} \to \operatorname{Spec} A_{jp}$, we have $\{\operatorname{Spec} A_{i,jp} \times^h_{\operatorname{Spec} A_{jp}} \operatorname{Spec} B_{jp} \to \operatorname{Spec} B_{jp}\} \in \operatorname{cov}_{\text{ét}}(\operatorname{Spec} B_{jp})$ for $j, p = 0, 1$, which recombine into $\{\operatorname{Spec} A_i \times^h_{\operatorname{Spec} A} \operatorname{Spec} B \to \operatorname{Spec} B\} \in \operatorname{cov}_{\text{ét}}(\operatorname{Spec} B)$. Hence étale covering families define a model topology on k-D$^-$sAff$_*$.

Now part of showing the cover assumption is satisfied hinges on the fact that $(s\mathrm{Aff}_{\mathcal{C}})_*$ is a simplicial model category on which we can put a Reedy model structure.

28.4.5.2 Reedy model structure on $s\mathrm{Aff}_{\mathcal{C}}$

$s\mathrm{Aff}_{\mathcal{C}} = \mathcal{A}\mathrm{ff}_{\mathcal{C}}^{\Delta^{\mathrm{op}}}$ with $\mathcal{A}\mathrm{ff}_{\mathcal{C}} = k$-D$^-$sAff$_* = (sk\text{-sAlg}_*)^{\mathrm{op}}$ is a simplicial category with $\mathcal{C} = sk$-sMod$_*$, thus it is cotensored over Set_Δ, and so for $X_* \in s\mathrm{Aff}_{\mathcal{C}}$, $K \in \mathrm{Set}_\Delta$, X_*^K is well defined, and its zeroth part will just be denoted X^K. In particular, we have

$$X_*^K = (\oplus_{i,p} X_{*,ip})^K = \oplus_{i,p} X_{*,ip}^K.$$

Indeed, for $X_*, Y_* \in s\mathrm{Aff}_{\mathcal{C}}$ and $K \in \mathrm{Set}_\Delta$,

$$\begin{aligned}\operatorname{Hom}(Y_* \otimes K, X_*) &= \operatorname{Hom}((\oplus_{i,p} Y_{*,ip}) \otimes K, (\oplus_{i,p} X_{*,ip}))\\ &= \oplus_{i,p} \operatorname{Hom}_{s\mathrm{Aff}_{\underline{\mathcal{C}}}}(Y_{*,ip} \otimes K, X_{*,ip})\\ &\cong \oplus_{i,p} \operatorname{Hom}(Y_{*,ip}, X_{*,ip}^K)\end{aligned}$$

since $s\mathrm{Aff}_{\underline{\mathcal{C}}}$ is a simplicial category. This further recombines into

$$\mathrm{Hom}(Y_*, \oplus_{i,p} X^K_{*,ip}) = \mathrm{Hom}(Y_*, X^K_*).$$

We also have

$$\begin{aligned} X^K &= (\oplus_{i,p} X^K_{*,ip})_0 \\ &= \oplus_{i,p} (X^K_{*,ip})_0 \\ &= \oplus_{i,p} X^K_{ip} \\ &= (X^K_{ip}). \end{aligned}$$

Thus the simplicial structure on $s\mathrm{Aff}_C$ is an entrywise generalization of that on $s\mathrm{Aff}_{\underline{C}}$.

One could also show that it is a bigraded simplicial category, and this is proved formally in exactly the same manner that we showed sk-sMod$_*$ is a bigraded simplicial category. Then for the bigraded simplicial structure, we have $X^K_* = \oplus_{i,p,q} K^{K_{iq}}_{*,ip}$. If $K = \Delta^n \times \Delta^n$ with Δ^n regarded as being of even parity, then we have

$$K^{\Delta\Delta^n}_* = \oplus_{i,p} K^{\Delta^n}_{*,ip} = (K_*)^{\Delta^n}$$

for the simplicial structure. For $Y \in \mathcal{A}\mathit{ff}_C \hookrightarrow s\mathrm{Aff}_C$ regarded as a constant simplicial object via the constant simplicial functor $\mathrm{cs}_*(Y)$ with $\mathrm{cs}_*(Y)_n = Y$ for all n, when we write Y^K, we really mean $(\mathrm{cs}_*(Y))^K$.

We now put a Reedy structure on $s\mathrm{Aff}_C$. Equivalences $X_* \to Y_*$ are such that $X_n \to Y_n$ are equivalences in $\mathcal{A}\mathit{ff}_C$ for all n, and fibrations $X_* \to Y_*$ are those maps such that for all n,

$$\begin{array}{ccc} X^{\Delta^n} \cong X_n & \longrightarrow & X^{\partial\Delta^n} \times_{Y^{\partial\Delta^n}} Y^{\Delta^n} \\ \| & & \| \\ (X^{\Delta^n}_{ip}) & & (X^{\partial\Delta^n}_{ip} \times_{Y^{\partial\Delta^n}_{ip}} Y^{\Delta^n}_{ip}) \end{array}$$

is a fibration in $\mathcal{A}\mathit{ff}_C$. Entrywise this decomposes into $X^{\Delta^n}_{ip} \to X^{\partial\Delta^n}_{ip} \times_{Y^{\partial\Delta^n}_{ip}} Y^{\Delta^n}_{ip}$, a fibration in $\mathrm{Aff}_{\underline{C}}$ for $i, p = 0, 1$, i. e., the Reedy model structure on $s\mathrm{Aff}_C$ is an entrywise Reedy model structure on $s\mathrm{Aff}_{\underline{C}}$. Observe that for $K \in \mathrm{Set}_\Delta$, the functor

$$\begin{aligned} s\mathrm{Aff}_C &\to \mathcal{A}\mathit{ff}_C \\ X_* &\mapsto (X^K_*)_0 = X^K \end{aligned}$$

is a right Quillen functor for the Reedy model structure on $s\mathrm{Aff}_C$. Indeed, it decomposes into $X_* = (X_{*,ip}) \mapsto (X^K_{ip})$, right Quillen in the Reedy model structure on $s\mathrm{Aff}_{\underline{C}}$ entrywise, recombining into a right Quillen functor in $s\mathrm{Aff}_C$. Its right derived functor is given by

$$
\begin{aligned}
\mathrm{Ho}\,(s\mathrm{Aff}_{\mathcal{C}}) &\to \mathrm{Ho}\,(\mathcal{A}\mathrm{ff}_{\mathcal{C}}),\\
X_* &\mapsto \mathbb{R}X^k\\
&= \mathbb{R}(X_{ip}^K)\\
&= (\mathbb{R}X_{ip}^K)\\
&= (X_{ip}^{\mathbb{R}K})\\
&= X^{\mathbb{R}K}.
\end{aligned}
$$

28.4.5.3 The Cover assumption

This assumption, which mirrors Assumption 1.3.2.2 of [79], is just the bigraded version of the latter, so the reader is referred to that reference for the full statement. Essentially, we assume that the étale topology on $\mathrm{Ho}\,(k\text{-}\mathrm{D}^{-}\mathrm{sAff}_*)$ is quasi-compact, projections onto coproducts form étale-covering families, both of which are fairly immediate to verify, and thirdly if $X_* \to Y$ is an augmented simplicial object in $\mathcal{A}\mathrm{ff}_{\mathcal{C}}$ corresponding to a coaugmented cosimplicial object $A \to B_*$ in $\mathrm{Comm}(\mathcal{C})$ and if we assume that for all n, the morphism

$$X_n \to X^{\mathbb{R}\partial\Delta^n} \times^h_{Y^{\mathbb{R}\partial\Delta^n}} Y \tag{28.12}$$

by itself forms an étale covering family in $\mathcal{A}\mathrm{ff}_{\mathcal{C}}$, then $A \to B_*$ satisfies descent. We discuss this third point.

We have

$$X^{\mathbb{R}\partial\Delta^n} \times^h_{Y^{\mathbb{R}\partial\Delta^n}} Y = (X_{ip}^{\mathbb{R}\partial\Delta^n} \times^h_{Y_{ip}^{\mathbb{R}\partial\Delta^n}} Y_{ip}).$$

With this the map, (28.12) entrywise reads

$$X_{n,ip} \to X_{ip}^{\mathbb{R}\partial\Delta^n} \times^h_{Y_{ip}^{\mathbb{R}\partial\Delta^n}} Y_{ip}$$

for $i, p = 0, 1$. Now since étale covering families are so entrywise, those maps correspond to étale covering families in $\mathrm{Aff}_{\underline{\mathcal{C}}}$, so the maps $A_{ip} \to B_{*,ip}$ satisfy descent by the original ungraded cover assumption of [79].

We show this implies that they can be recombined into $A \to B_*$ that also satisfies descent. This will imply that our bigraded cover assumption is just a diagonalization of the original cover assumption of [79].

Recall that to satisfy descent means

$$B_* \tilde{\boxtimes}_A^{\mathbb{L}} - : \mathrm{Ho}\,(A\text{-}\mathrm{sMod}_*) \to \mathrm{Ho}\,(\mathrm{cs}B_*\text{-}\mathrm{sMod}_*)$$

is fully faithful and establishes an equivalence of $\mathrm{Ho}\,(A\text{-}\mathrm{sMod}_*)$ with the subcategory of homotopy Cartesian objects in $\mathrm{Ho}\,(\mathrm{cs}B_*\text{-}\mathrm{sMod}_*)$. That the entries $A_{ip} \to B_{*,ip}$ satisfy descent then means that $B_{*,ip} \otimes^{\mathbb{L}}_{A_{ip}} - : \mathrm{Ho}(A_{ip}\text{-Mod}) \xrightarrow{\mathrm{ff}} \mathrm{Ho}(\mathrm{cs}B_{*,ip}\text{-Mod})$ for $i,p = 0,1$. Those maps recombine into $B_* \tilde{\boxtimes}^{\mathbb{L}}_A - : \mathrm{Ho}(A\text{-sMod}_*) \to \mathrm{Ho}(\mathrm{cs}B_*\text{-sMod}_*)$ fully faithful. Regarding homotopy Cartesian objects, those are objects M_* in $\mathrm{Ho}(\mathrm{cs}B_*\text{-Mod}_*)$ such that $M_n \tilde{\boxtimes}^{\mathbb{L}}_{B_n} B_m \to M_m$ is an isomorphism in $\mathrm{Ho}(B_m\text{-sMod}_*)$. We quickly check that $B_* \tilde{\boxtimes}^{\mathbb{L}}_A -$ maps objects of $\mathrm{Ho}(A\text{-sMod}_*)$ into Cartesian objects of $\mathrm{Ho}(\mathrm{cs}B_*\text{-sMod}_*)$. Let $C_* \in \mathrm{Ho}(A\text{-sMod}_*)$. We know from [79] in the ungraded case that entrywise

$$(B_{n,ip} \otimes^{\mathbb{L}}_{A_{ip}} C_{n,ip}) \otimes^{\mathbb{L}}_{B_{n,ip}} B_{m,ip} \cong B_{m,ip} \otimes^{\mathbb{L}}_{A_{ip}} C_{m,ip}$$

for $i,p = 0,1$, since $B_* \otimes^{\mathbb{L}}_A -$ maps objects of $\mathrm{Ho}(A\text{-Mod})$ into homotopy Cartesian objects of $\mathrm{Ho}(\mathrm{cs}B_*\text{-Mod})$. Now we have

$$\begin{aligned}(B_n \tilde{\boxtimes}^{\mathbb{L}}_A C_n) \tilde{\boxtimes}^{\mathbb{L}}_{B_n} B_m &= \oplus_{i,p} (B_n \tilde{\boxtimes}^{\mathbb{L}}_A C_n)_{ip} \otimes^{\mathbb{L}}_{B_{n,ip}} B_{m,ip} \\ &= \oplus_{i,p} (B_{n,ip} \otimes^{\mathbb{L}}_{A_{ip}} C_{n,ip}) \otimes^{\mathbb{L}}_{B_{n,ip}} B_{m,ip} \\ &\cong \oplus_{i,p} B_{m,ip} \otimes^{\mathbb{L}}_{A_{ip}} C_{m,ip} = B_m \tilde{\boxtimes}^{\mathbb{L}}_A C_m\end{aligned}$$

and hence $B_* \tilde{\boxtimes}^{\mathbb{L}}_A -$ maps objects of $\mathrm{Ho}(A\text{-sMod}_*)$ into homotopy Cartesian objects. The essential surjectivity is dealt with entrywise. Thus the third point is satisfied.

To conclude this subsection on the model topology, the étale topology does define a model topology on $k\text{-D}^-\mathrm{sAff}_*$ and makes $(k\text{-D}^-\mathrm{sAff}_*, \text{ét.})$ into a model site. Further, this topology satisfies our technical cover assumption, bigraded generalization of the cover assumption as originally introduced in [79].

28.4.6 Bigraded simplicial presheaves

Let M be a small bigraded model category (the choice of universes $\mathbb{U} \in \mathbb{V}$ being implied, M $\mathbb{V}$-small, $\mathbb{U}$-cofibrantly generated...), $W(M)$ its class of equivalences. For us, $M = k\text{-D}^-\mathrm{sAff}_*$. Define

$$\mathrm{sPr}_{\mathbb{Z}_2^2}(M) = S_*^{M^{\mathrm{op}}}.$$

the category of bigraded simplicial presheaves on $M = (M_{ip})$. Equivalently,

$$\mathrm{sPr}_{\mathbb{Z}_2^2}(M) = \oplus_{i,p} \mathrm{sPr}(\underline{M}).$$

We put the entrywise projective model structure on $\mathrm{sPr}_{\mathbb{Z}_2^2}(M)$, where equivalences and fibrations are defined objectwise.

Our aim at this point is to define a model category of prestacks $M^\wedge$ obtained as a bigraded left Bousfield localization of $\mathrm{sPr}_{\mathbb{Z}_2^2}(M)$ along $\{h_u \mid u \in W(M)\}$, where $h : M \to \mathrm{Pr}_{\mathbb{Z}_2}(M) \hookrightarrow \mathrm{sPr}_{\mathbb{Z}_2^2}(M)$ is the bigraded Yoneda embedding. In other terms, we localize each individual $\mathrm{sPr}(\underline{M})$, i. e., we do an entrywise Bousfield localization. Then $\mathrm{Ho}\,(M^\wedge)$ is identified with the subcategory of $\mathrm{Ho}\,(\mathrm{sPr}_{\mathbb{Z}_2^2}(M))$ that consists of presheaves preserving weak equivalences entrywise. $M^\wedge$ is a bigraded simplicial model category since $\mathrm{sPr}(\underline{M})^\wedge = \mathrm{sPr}(\text{k-D}^-\mathrm{Aff})^\wedge$ is a simplicial model category from [76], and we denote by $\mathbb{R}\underline{\mathrm{Hom}} \in \mathrm{Ho}\,(S_{*,})$ its derived hom. From there we define a model category $M^{\sim,\text{ét.}}$ of stacks on $(M, \text{ét.})$, left bigraded Bousfield localization of $M^\wedge$ along homotopy étale-hypercovers. We obtain a pair of adjoint Quillen functors

$$\mathrm{id} : M^\wedge \rightleftarrows M^{\sim,\text{ét.}} : \mathrm{id}$$

giving rise to

$$a = \mathbb{L}\mathrm{id} : \mathrm{Ho}\,(M^\wedge) \rightleftarrows \mathrm{Ho}\,(M^{\sim,\text{ét.}}) : \mathbb{R}\mathrm{id} = j.$$

We finally define a *bigraded stack* on $(k\text{-D}^-\mathrm{sAff}_*, \text{ét.})$ as an object of $\mathrm{sPr}_{\mathbb{Z}_2^2}(M)$ whose image in $\mathrm{Ho}\,(M^\wedge)$ is in the essential image of the functor j, exactly in the same way that derived stacks were defined in [79].

28.4.7 Bigraded left Bousfield localization

The reference we use for Bousfield localizations is [40], which we follow closely since it is just a matter of adapting Hirschhorn's definition to the bigraded case. Let $M = (M_{ip})$ be a $\mathbb{Z}_2$-graded model category, such as $k\text{-D}^-\mathrm{sAff}_*$. Let $\mathcal{C}$ be a class of maps in M, $\mathcal{C} = (\mathcal{C}_{ip})$. An object W of M is $\mathcal{C}$-local if it is fibrant and for any $f : A \to B$ in $\mathcal{C}$, or entrywise $f_{ip} : A_{ip} \to B_{ip}$ in $\underline{\mathcal{C}}$ for $i, p = 0, 1$, the induced map of mapping spaces is a weak equivalence

$$f^* : \mathrm{Map}(B, W) \xrightarrow{\simeq} \mathrm{Map}(A, W)$$

or entrywise

$$f_{ip}^* : \mathrm{Map}(B_{ip}, W_{ip}) \xrightarrow{\simeq} \mathrm{Map}(A_{ip}, W_{ip}).$$

In other terms, $\mathcal{C}$-local objects are entrywise $\underline{\mathcal{C}}$-local objects. We define a map $g : X \to Y$ in M to be a $\mathcal{C}$-local equivalence if for any $\mathcal{C}$-local object W, the induced map of mapping spaces

$$g^* : \mathrm{Map}(Y, W) \to \mathrm{Map}(X, W)$$

is a weak equivalence, or, equivalently, entrywise

$$g_{ip}^* : \mathrm{Map}(Y_{ip}, W_{ip}) \xrightarrow{\simeq} \mathrm{Map}(X_{ip}, W_{ip})$$

for $i, p = 0, 1$, so again $\mathcal{C}$-local equivalences are entrywise $\underline{\mathcal{C}}$-local equivalences.

Finally, we define the *bigraded left Bousfield localization* of M with respect to $\mathcal{C}$ as a model category structure $\mathrm{L}_{\mathcal{C}}M$ on the underlying bigraded category of M such that the class of weak equivalences is the class of $\mathcal{C}$-local equivalences, cofibrations are those of M, and fibrations have the right lifting property with respect to those cofibrations that are also $\mathcal{C}$-local equivalences. If $A \to B$ is such a cofibration and $X \to Y$ is a fibration, then a diagram such as

$$\begin{array}{ccc} A & \longrightarrow & X \\ \downarrow & \nearrow & \downarrow \\ B & \longrightarrow & Y \end{array}$$

breaks up into diagrams

$$\begin{array}{ccc} A_{ip} & \longrightarrow & X_{ip} \\ \downarrow & \nearrow & \downarrow \\ B_{ip} & \longrightarrow & Y_{ip} \end{array}$$

for $i, p = 0, 1$ with a cofibration $A_{ip} \to B_{ip}$ that is also a $\underline{\mathcal{C}}$-local equivalence. This shows fibrations are also entrywise fibrations, and hence

$$\mathrm{L}_{\mathcal{C}}M = (L_{\mathcal{C}_{ip}}M_{ip}).$$

28.4.8 Model category of prestacks $M^{\wedge}$

We first define restricted diagrams following [76]. For a small category M (relative to $\mathbb{V}$), such as $k\text{-}\mathrm{D}^{-}\mathrm{sAff}_{*,}$, given a bigraded simplicial model category $S_{*.}$, cofibrantly generated, we consider the *category* $S_{*.}^{M}$ *of bigraded simplicial functors* $M \to S_{*.}$ with its entrywise projective model structure. For any $x \in M$, we have an induced map

$$\begin{aligned} \iota_x^* : S_{*.}^M &\to S_{*.}, \\ F &\mapsto F(x) \end{aligned}$$

with a left adjoint $(\iota_x)_! : S_{*.} \to S_{*.}^M$ that is a left Quillen functor. Let I be a set of generating cofibrations in $S_{*.}$, let $f : A \to B$ be any morphism in I, and let $u : x \to y$ be any morphism in M. Consider the natural morphism in $S_{*.}^M$

$$f \Box u : (\iota_y)_! A \coprod_{(\iota_x)_! A} (\iota_x)_! B \to (\iota_y)_! B$$

with $(\iota_x)_{!jp} : S_{jp} \to S_{jp}^{M_{jp}}$, left adjoint to

$$(\iota_x^*)_{jp} : F_{jp} \mapsto (F(x))_{jp} = F_{jp}(x_{jp})$$

that is, $(\iota_x^*)_{jp} = \iota_{x_{jp}}^*$, so that $(\iota_x)_{!jp} = (\iota_{x_{jp}})_!$. Thus

$$(\iota_y)_! A = ((\iota_{y_{jp}})_! A_{jp})$$

which means $f\square u = (f_{jp}\square u_{jp})$, so that entrywise

$$f_{lp}\square u_{lp} : (\iota_{y_{lp}})_! A_{lp} \coprod_{(\iota_{x_{lp}})_! A_{lp}} (\iota_{x_{lp}})_! B_{lp} \to (\iota_{y_{lp}})_! B_{lp}.$$

Definition 28.4.8.1. The model category of restricted diagrams from M to $S_{*.}$, denoted $M^{\mathrm{op},\wedge}$, is defined as the left Bousfield localization of $S_{*.}^M$ along the set of morphisms of the form $f\square u$ for $f \in I$ and weak equivalences u in M.

If we denote by W the set of weak equivalences, then this gives

$$L_{I\square W} S_{*.}^M = (L_{I_{jp}\square W_{jp}} S_{jp}^{M_{jp}})$$

or, in other terms,

$$M^{\mathrm{op},\wedge} = (M_{jp}^{\mathrm{op},\wedge}).$$

With this notion, we define $M^\wedge = L_{I\square W} S_{*.}^{M^{\mathrm{op}}}$ to be the model category of prestacks on M.

28.4.9 Hyperdescent

There is a slew of pseudo-representable objects we have to define before we get the definition of hyperdescent. Those objects were initially defined in [79] in the ungraded case, and we briefly reproduce their bigraded definitions here for convenience.

We say $F \in M^\wedge$ is pseudo-representable if it is a small disjoint union of representable presheaves:

$$F \simeq \coprod_{u\in I} h_u = (\coprod_{u_{lp}\in I_{lp}} h_{u_{lp}}) = (F_{lp})$$

that is, it is entrywise pseudo-representable.

A pseudo-fibration is a morphism of pseudo-representable objects represented by a fibration in M. This breaks up into

$$\oplus_{l,p=0,1} \prod_{u_{lp}\in I_{lp}} \coprod_{v_{lp}\in J_{lp}} \mathrm{Hom}(h_{u_{lp}}, h_{v_{lp}}) \to \oplus_{l,p=0,1} \coprod_{v_{lp}\in J_{lp}} \mathrm{Hom}(h_{u_{lp}}, h_{v_{lp}})$$
$$\simeq \oplus_{l,p} \coprod_{v_{lp}\in J_{lp}} \mathrm{Hom}_{M_{lp}}(u_{lp}, v_{lp})$$

with entries that are pseudo-fibrations, so pseudo-fibrations are so entrywise in $\underline{M}$.

To avoid repetition, the reader is referred to [76] for the definition of pseudo-covering. It is clearly entrywise a pseudo-covering.

For pseudo-representable hypercovers, let x be a fibrant object of M. Then a pseudo-representable hypercover is an object $F_* \to h_x$ in $sM^\wedge/h_x$ such that for any $n \geq 0$, the induced morphism

$$F_n \to F^{\partial\Delta^n} \times_{h_x^{\partial\Delta^n}} h_x^{\Delta^n}$$

is a pseudo-fibration and a pseudo-covering of pseudo-representable objects. We write $F_* = (F_{*,ip})$ with $F_{*,ip} \in s\underline{M}^\wedge/h_{x_{ip}}$ with $x = (x_{ip})$ in M, so that

$$\begin{array}{ccc} \oplus_{i,p}F_{n,ip} & \longrightarrow & \left(\oplus_{i,p} F_{ip}^{\partial\Delta^n}\right) \times_{(\oplus_{i,p} h_{x_{ip}}^{\partial\Delta^n})} \left(\oplus_{i,p} h_{x_{ip}}^{\Delta^n}\right) \\ & \searrow & \| \\ & & \oplus_{i,p}F_{ip}^{\partial\Delta^n} \times_{h_{x_{ip}}^{\partial\Delta^n}} h_{x_{ip}}^{\Delta^n} \end{array}$$

showing that pseudo-representable hypercovers are defined entrywise.

Note that $M^\wedge = L_{I\square W}S_{*.}^{M^{\mathrm{op}}}$ is naturally tensored and cotensored over $S_{*.}$ with external products and exponentials defined objectwise. This makes $M^\wedge$ a bigraded simplicial model category with bigraded simplicial hom $\underline{\mathrm{Hom}}$ and derived bigraded simplicial hom, which we will denote $\mathbb{R}_\omega\underline{\mathrm{Hom}}$ as in [79].

Regarding Yoneda, fix a cofibrant resolution functor $(\Gamma : M \to M^\Delta, \iota)$ with $\Gamma = (\Gamma_{ip})$ and $\iota = (\iota_{ip})$, that is, for all $x \in M$, $\Gamma(x) = (\Gamma_{ip}(x_{ip}))$ is a cosimplicial object in M, cofibrant for the Reedy model structure on M^Δ, together with a natural equivalence $\iota(x) : \Gamma(x) \to c^*(x)$, a bigraded constant cosimplicial object in M at x. We have

$$\begin{aligned} \underline{h} = (\underline{h}_{ip}) &: M \to \mathrm{sPr}_{\mathbb{Z}_2^2}(M), \\ \underline{h}_x = (\underline{h}_{x_{ip}}) &: M^{\mathrm{op}} \to S_{*.}, \\ & y \mapsto \mathrm{Hom}_M(\Gamma(y), x). \end{aligned}$$

Since $\underline{h}_{ip}$ preserves fibrant objects and equivalences between them, so does $\underline{h}$, and so $\mathbb{R}\underline{h} : \mathrm{Ho}\,(M) \to \mathrm{Ho}\,(M^\wedge)$ is well-defined. Further,

$$\mathbb{R}_\omega\underline{\mathrm{Hom}}_{\underline{M}^\wedge}(\underline{h}_{x_{ip}}, F_{ip}) \cong F_{ip}(x_{ip})$$

for $i, p = 0, 1$, leading to

$$\mathbb{R}_\omega\underline{\mathrm{Hom}}_{\underline{M}^\wedge}(\underline{h}_x, F) \cong F(x).$$

This follows from the classical enriched Yoneda isomorphism $\mathbb{R}_\omega\underline{\mathrm{Hom}}(h_x, F) \cong F(x)$ and the fact that $\mathrm{Ho}\,(h) \cong \mathbb{R}\underline{h}$ in $\mathrm{Ho}\,(M^\wedge)$ entrywise [76].

Finally, $F \in M^\wedge$ is said to have hyperdescent if for any fibrant object $x \in M$ and any pseudo-representable hypercover $H_* \to h_x$ with realization $|H_*| = (|H_{*,ip}|)$, the induced morphism

$$F(x) \simeq \mathbb{R}_\omega \underline{\mathrm{Hom}}(h_x, F) \to \mathbb{R}_\omega \underline{\mathrm{Hom}}(|H_.|, F)$$

is an isomorphism in $\mathrm{Ho}(S_{*.})$. A stack on $(M, \text{ét.})$ is a prestack $F \in M^\wedge$ that satisfies ét.-hyperdescent. We denote by $M^{\sim,\text{ét.}}$ the model category of stacks on M. Another way to say this is by considering $H_\beta(x)$ for x fibrant in M the set of representatives of the set of isomorphism classes of objects $F_* \to h_x$ in $sM^\wedge/h_x$ (see [79] for details), those morphisms that are pseudo-representable hypercovers with a bound on the cardinality of each F_n. We have

$$H_\beta(x) = (H_{\beta_{ip}}(x_{ip})).$$

Then we can equivalently say that $M^{\sim,\text{ét.}}$ is the left Bousfield localization of $M^\wedge$ with respect to morphisms in H_β.

28.4.10 Local equivalences

Another way to define stacks is as follows. First, consider the presheaf of connected components of RF

$$\begin{aligned} \pi_0^{\mathrm{pr}}(F) : M^{\mathrm{op}} &\to \mathrm{Set}_{*.}, \\ x &\mapsto \pi_0(RF(x)) = (\pi_0(R_{ip}F_{ip}(x_{ip}))) \end{aligned}$$

leading to

$$\begin{aligned} \pi_0^{\mathrm{eq}} : M^\wedge &\to \mathrm{Pr}_{\mathbb{Z}_2^2}(M), \\ F &\mapsto \pi_0^{\mathrm{pr}}(RF) \end{aligned}$$

factoring as

$$\begin{aligned} \pi_0^{\mathrm{eq}} : M^\wedge &\to \mathrm{Pr}_{\mathbb{Z}_2^2}(\mathrm{Ho}(M)), \\ F &\mapsto \pi_0^{\mathrm{eq}}(F) \end{aligned}$$

which further factors as $\pi_0^{\mathrm{eq}} : \mathrm{Ho}(M^\wedge) \to \mathrm{Pr}_{\mathbb{Z}_2^2}(\mathrm{Ho}(M))$. We also have an evaluation functor

$$\begin{aligned} j_x^* = (j_{x_{ip}}^*) : M^\wedge &\to (M/x)^\wedge, \\ F &\mapsto F(x) = (F_{ip}(x_{ip})) \end{aligned}$$

with a left adjoint

$$j_{x!} = (j_{x_{ip}!}) : (M/x)^{\wedge} = ((\underline{M}/x_{ip})^{\wedge}) \to M^{\wedge}$$

so once right derived,

$$\mathbb{R}j_x^* : \mathrm{Ho}\,(M^{\wedge}) \to \mathrm{Ho}\,((M/x)^{\wedge}).$$

Now for $F \in M^{\wedge}$ and $x \in M$ fibrant, $s \in \pi_0^{\mathrm{eq}}(F)(x)$ can be represented by a morphism $s : h_x \to F$ in $\mathrm{Ho}\,(M^{\wedge})$, giving by pullback $\mathbb{R}j_x^* s : \mathbb{R}j_x^*(h_x) \to \mathbb{R}j_x^* F$. Since we have a point $* \to \mathbb{R}j_x^* h_x$, by composition we get a global point $* \to \mathbb{R}j_x^* F$. All of this can be found in [76]. We now define, much as in this reference, for any $n > 0$, the sheaf

$$\pi_n(F, s) = \pi_0\Big(\mathbb{R}j_x^*(F)^{\mathbb{R}\Delta^n} \times^h_{\mathbb{R}j_x^*(F)^{\mathbb{R}\partial\Delta^n}} *\Big)$$

which decomposes as follows:

$$\begin{aligned}
\pi_n(F, s) &= \pi_0\Bigg(\Big(\oplus_{i,p}\,\mathbb{R}j_{x_{ip}}^*(F_{ip})^{\mathbb{R}\Delta^n}\Big) \times^h_{\oplus_{i,p}\mathbb{R}j_{x_{ip}}^*(F_{ip})^{\mathbb{R}\partial\Delta^n}} \oplus_{i,p} *_{ip}\Bigg) \\
&= \pi_0(\oplus_{i,p}\mathbb{R}j_{x_{ip}}^*(F_{ip})^{\mathbb{R}\Delta^n} \times^h_{\mathbb{R}j_{x_{ip}}^*(F_{ip})^{\mathbb{R}\partial\Delta^n}} *_{ip}) \\
&= \oplus_{i,p}\pi_0(\mathbb{R}j_{x_{ip}}^*(F_{ip})^{\mathbb{R}\Delta^n} \times^h_{\mathbb{R}j_{x_{ip}}^*(F_{ip})^{\mathbb{R}\partial\Delta^n}} *_{ip}) \\
&= \oplus_{i,p}\pi_n(F_{ip}, s_{ip})
\end{aligned}$$

where $s \in \pi_0^{\mathrm{eq}}(F)(x) = (\pi_0^{\mathrm{eq}}(F_{ip})(x_{ip}))$, so we write $s = (s_{ip})$.

We now define local equivalences exactly as in [76]. Those are also referred to as π_*-equivalences. A morphism $f : F \to G$ in $M^{\wedge}$ is such an equivalence if the induced morphism $\pi_0(F) \to \pi_0(G)$ is an isomorphism on $\mathrm{Ho}\,(M)$, i. e., if it is entrywise so, and additionally if for any fibrant x in M, any $s \in \pi_0^{\mathrm{eq}}(F)(x)$, and $n > 0$, we have a bijection of sheaves on $\mathrm{Ho}\,(M/x)$: $\pi_n(F, s) \to \pi_n(G, f(s))$, and those are entrywise isomorphisms as well.

In other terms, local equivalences are entrywise local equivalences.

Finally, given M, and for us this is really k-$\mathrm{D}^-\mathrm{sAff}_*$ we have in mind with the étale topology, there is a closed model structure on $\mathrm{sPr}_{\mathbb{Z}_2^2}(M)$ called the entrywise local projective model structure for which equivalences are local equivalences and cofibrations are cofibrations in $M^{\wedge}$ for its projective model structure. It turns out that $\mathrm{sPr}_{\mathbb{Z}_2^2}(M)$ with this model structure is $M^{\sim,\text{ét.}}$. The proof of the existence of such a model structure is done entrywise, and at that level, this is proved in [76]. Now as pointed out above, we can also construct $M^{\sim,\text{ét.}}$ as the left Bousfield localization of $M^{\wedge}$ with respect to $H = \{|F_*| \to h_x \mid x \in M^f, F_* \in H_\beta(x)\}$. Maps in H break down as $|F_{*,ip}| \to h_{x_{ip}}$ so that it is clear we have $H = (H_{ip})$, and hence

$$\begin{aligned} M^{\sim,\text{ét.}} &= L_H M^{\wedge} \\ &= (L_{H_{ip}} M^{\wedge}_{ip}) \\ &= (M^{\sim,\text{ét.}}_{ip}). \end{aligned}$$

28.4.11 Truncations

Define $X \in S_{*.}$ to be n-truncated if it is entrywise n-truncated, and $x \in \mathrm{Ho}\,(M_{*.})$ is said to be n-truncated if it is so entrywise. Then define $\pi_{\leq n}$-equivalences as in [76]. There is a model structure on $\mathrm{sPr}_{\mathbb{Z}_2^2}(M)$, called the n-truncated bigraded local projective model structure, for which equivalences are bigraded $\pi_{\leq n}$-equivalences and cofibrations are cofibrations for the bigraded projective model structure on $M^{\wedge}$. That we have such a model structure follows directly for the same result in the ungraded case proved in [76], since we can work entrywise. This model category is denoted $M^{\sim,\text{ét.}}_{\leq n}$ and can be seen as a substitute for working with compactified theories in Theoretical Physics. We also have as a corollary that this model category can be obtained as the left Bousfield localization of $M^{\sim,\text{ét.}}$ with respect to $\{\partial\Delta^i \otimes \underline{h}_x \to \Delta^i \otimes \underline{h}_x, i > n, x \in M\}$.

28.5 Supersymmetric stacks

In what follows, we will make use of the strictification theorem, which can be found in [41], [76], and [74]: for a category C with a subcategory S of morphisms, a cofibrantly generated model category M, the localization $M^{C,S}$ of M^C with respect to restricted diagrams, we have

$$L(M^{C,S}) \simeq \mathbb{R}\underline{\mathrm{Hom}}(L(C,S), LM)$$

where L stands for Segal localization. In particular, if T is an S-site [76], then we just have $L(M^T) \cong \mathbb{R}\underline{\mathrm{Hom}}(T, LM)$. If $M = S_{*.}$ with $L(S_{*.}) = \mathrm{Top}_{*.}$, then we have $L(S^T_{*.}) \cong \mathbb{R}\underline{\mathrm{Hom}}(T, \mathrm{Top}_{*.})$. For us, $T = L(sk\text{-sAlg}_*)^{\mathrm{op}} = L(k\text{-D}^-\mathrm{sAff}_*)$ will be denoted $\mathrm{d}k\text{-sAff}_*$. We define by

$$\widehat{\mathrm{d}k\text{-sAff}_*} = \mathbb{R}\underline{\mathrm{Hom}}(\mathrm{d}k\text{-sAff}^{\mathrm{op}}_*, \mathrm{Top}_{*.})$$

the Segal localization of $\mathrm{sPr}_{\mathbb{Z}_2^2}(k\text{-D}^-\mathrm{sAff}_*)$. Then we define the category of stacks $\mathrm{d}k\text{-sAff}^{\sim,\text{ét.}}_*$ as a left exact localization of $\widehat{\mathrm{d}k\text{-sAff}_*}$, which can equivalently be obtained as the Segal localization of $k\text{-D}^-\mathrm{sAff}^{\sim,\text{ét.}}_*$, itself the left Bousfield localization of $\mathrm{sPr}_{\mathbb{Z}_2^2}(k\text{-D}^-\mathrm{sAff}_*)$, regarded as a simplicial category for strictification and then as a bigraded simplicial category. Diagrammatically,

$$\begin{array}{ccc} k\text{-D}^{-}\text{sAff}_*^{\wedge} & \xrightarrow{L} & \mathbb{R}\underline{\text{Hom}}(L(k\text{-D}^{-}\text{sAff}_*)^{\text{op}}, \text{Top}_{*.}) = \widehat{\text{d}k\text{-sAff}_*} \\ {\scriptstyle L_{\text{Bous}}}\downarrow & & \downarrow{\scriptstyle \text{LexLoc}} \\ k\text{-D}^{-}\text{sAff}_*^{\sim,\text{ét.}} & \xrightarrow[L]{} & \text{d}k\text{-sAff}_*^{\sim,\text{ét.}}. \end{array}$$

The stacks we have defined are objects of $k\text{-D}^{-}\text{sAff}_*^{\sim,\text{ét.}}$, and after Segal localization, they become objects of the category $\text{d}k\text{-sAff}_*^{\sim,\text{ét.}}$, where they become functors $F : L(sk\text{-sAlg}_*) \to \text{Top}_{*.}$. Now for every $A \in sk\text{-sAlg}_*$, $\mathcal{X} \in \text{Top}_{*.}$ determined by constraint equations such as equations of motion, $\Psi : A \to \mathcal{X}$ a map satisfying those equations, define $F(A) = \{\Psi(\sigma_i, \theta_i) \mid \sigma_i \in A_0, \theta_i \in A_1, i = 0, 1\}$ with the induced topology. Suppose F thus defined is a stack. If in addition we have a notion of supersymmetric transformation on each $A \in sk\text{-sAlg}_*$ under which Ψ is well-behaved, then the resulting functor F is called a *supersymmetric stack*.

We first go over simple supersymmetry transformations using a standard reference such as [32]. This will motivate our modifications, which will come afterwards.

28.5.1 Ordinary supersymmetry

Everything in this section is from [32], and the reader is referred to that reference for more details. Suppose we have, for $M \in sk\text{-sMod}$, two elements σ^a of $M_0 \in sk\text{-Mod}$, $a = 1, 2$, and two elements θ^A of M_1, $A = 1, 2$. Those are seen as vector components. We also introduce matrices

$$\rho^0 = \begin{pmatrix} 0 & -i \\ i & 0 \end{pmatrix}$$

and

$$\rho^1 = \begin{pmatrix} 0 & i \\ i & 0 \end{pmatrix}.$$

We write $\bar{\theta} = \theta^T \rho^0$. Suppose elements of M are variables. We can define the following derivatives:

$$Q_A = \frac{\partial}{\partial \bar{\theta}^A} + i(\rho^a \theta)_A \partial_a$$

defining a so-called supersymmetry transformation

$$\begin{aligned} \delta\theta^A &= [\bar{\epsilon} Q, \theta^A] = \epsilon^A, \\ \delta\sigma^a &= [\bar{\epsilon} Q, \sigma^a] = i\bar{\epsilon}\rho^a\theta \end{aligned}$$

where ϵ is an odd parameter. Define a field on A valued in Top, with a dimensional index μ:

$$Y^\mu(\sigma,\theta) = X^\mu(\sigma) + \bar{\theta}\psi^\mu(\sigma) + \frac{1}{2}\bar{\theta}\theta B^\mu(\sigma).$$

Then defining an ansatz for the supersymmetry transformation of Y under δ as in

$$\delta Y^\mu(\sigma,\theta) = \delta X^\mu(\sigma) + \bar{\theta}\delta\psi^\mu(\sigma) + \frac{1}{2}\bar{\theta}\theta\delta B^\mu(\sigma)$$

one finds

$$\begin{aligned}
\delta X^\mu &= \bar{\epsilon}\psi^\mu \\
\delta\psi^\mu &= -i\rho^\alpha\epsilon\partial_\alpha X^\mu + B^\mu\epsilon \\
\delta B^\mu &= -i\bar{\epsilon}\rho^\alpha\partial_\alpha\psi^\mu.
\end{aligned}$$

28.5.2 Supersymmetric maps

For our purposes, we need to have Y valued in a bigraded topological space. Note that in the definition of Y^μ in the preceding section, there is no odd part. Additionally, the supersymmetry transformation of θ produces ϵ, another anticommuting variable, though if we want symmetry in our formalism, then we would like to obtain an even element. Finally, the supersymmetry transformations on σ and θ are not symmetric themselves, so this is something we might want to enforce. Finally, δ on each of X^μ, ψ^μ, and B^μ in the previous section is defined via an ansatz, so those transformations are really specific to each field and do not correspond to a single algebraic operation δ, something we would like to have for a more functorial treatment of supersymmetry transformations.

We introduce $\mathcal{X} = (\mathcal{X}_{ip}) \in \mathrm{Top}_{*,*}$, $\Psi : M \to \mathcal{X}$, $M \in k\text{-sMod}_*$, with entries $\Psi_{ip} : M_{ip} \to \mathcal{X}_{ip}$ with $i,p = 0,1$. We have seen that k-sMod$_*$ has an internal Hom, its odd parts being made of morphisms in k-sMod$_*$ that change parity. Let δ be such a morphism.

We consider those δ that provide supersymmetric transformations in the sense that they are elements of $\underline{\mathcal{H}om}(M,M)_1$, and their definition displays some symmetry. For that purpose, we are led to introducing even parameters ϵ and λ and matrices ρ and γ such that

$$\begin{aligned}
\delta_0\sigma^\alpha &= \epsilon^T\rho^\alpha\theta, \\
\delta_1\theta^A &= \lambda^T\gamma^A\sigma
\end{aligned}$$

with $\alpha, A = 0,1$. Thus $\delta_0 : M_0 \to M_1$ and $\delta_1 : M_1 \to M_0$. The induced transformation on Ψ is given by

$$\delta_0^*\Psi_1 : M_0 \to \mathcal{X}_1,$$

$$\delta_1^* \Psi_0 : M_1 \to \mathcal{X}_0$$

or $\delta = \delta_0 \oplus \delta_1 \in \underline{\mathcal{H}\text{om}}(M, M)_1$. Thus

$$\begin{aligned} \delta^* \Psi(\sigma, \theta) &= \Psi(\delta(\sigma, \theta)) \\ &= \Psi(\delta_0 \sigma \oplus \delta_1 \theta) \\ &= \Psi_0(\delta_1 \theta) \oplus \Psi_1(\delta_0 \sigma)) \\ &= \delta_1^* \Psi_0 \oplus \delta_0^* \Psi_1 \end{aligned}$$

that is, $\delta^* = \delta_0^* \oplus \delta_1^*$. We now generalize this construction to the case $M \in sk\text{-sMod}_*$. This means we consider objects $\sigma_n, \theta_n, \gamma_n, \rho_n, \lambda_n, \epsilon_n$, and δ_n. We have

$$\begin{array}{ccc} [n-1] & \xrightarrow{M} & M_{n-1} \\ {\scriptstyle d^i}\downarrow & & \uparrow{\scriptstyle M(d^i)=d_i} \\ [n] & \xrightarrow[M]{} & M_n \end{array}$$

where $d_i \sigma_n^a = (d_i \sigma^a)[n-1] = d^{i*} \sigma^a [n-1] = \sigma^a \circ d^i [n-1]$. δ is a natural transformation of simplicial objects, breaking into $\delta_0 : M_{*0} \to M_{*1}$ and $\delta_1 : M_{*1} \to M_{*0}$ with

$$\begin{array}{ccc} M_{n-1,0} & \xrightarrow{\delta_{n-1,0}} & M_{n-1,1} \\ {\scriptstyle d_i}\uparrow & & \uparrow{\scriptstyle d_i} \\ M_{n,0} & \xrightarrow[\delta_{n,0}]{} & M_{n,1} \end{array}$$

as well as

$$\begin{array}{ccc} M_{n-1,0} & \xrightarrow{\delta_{n-1,0}} & M_{n-1,1} \\ {\scriptstyle s_j}\downarrow & & \downarrow{\scriptstyle s_j} \\ M_{n,0} & \xrightarrow[\delta_{n,0}]{} & M_{n,1} \end{array}$$

with the degeneracy map s_j defined by

$$\begin{array}{ccc} [n-1] & \xrightarrow{M} & M_{n-1} \\ {\scriptstyle s^j}\uparrow & & \downarrow{\scriptstyle M(s^j)=s_j} \\ [n] & \xrightarrow[M]{} & M_n \end{array}$$

where d^i and s^j are the usual connecting maps on Δ [28]. We have similar commutative diagrams for δ_{*1}. In all those, we have

$$\delta_{n,0}\sigma_n^{\alpha} = \epsilon_n^T \rho_n^{\alpha} \theta_n,$$
$$\delta_{n,1}\theta_n^A = \lambda_n^T \gamma_n^A \sigma_n$$

with fixed $\epsilon, \lambda, \rho, \gamma$, independently of the coordinates σ and θ, and

$$\begin{aligned}\delta_n^* \Psi(\sigma_n, \theta_n) &= (\delta_{n,1}^* \Psi_0)(\theta_n) \oplus (\delta_{n,0}^* \Psi_1)(\sigma_n)\\ &= \Psi_0(\delta_{n,1}\theta_n) \oplus \Psi_1(\delta_{n,0}\sigma_n)\\ &= \Psi_0(\lambda_n^T \gamma_n^A \sigma_n) \oplus \Psi_1(\epsilon_n^T \rho_n^{\alpha} \theta_n)\end{aligned}$$

the collection of which defines

$$\delta^* F(A) = \{\delta^* \Psi(\sigma, \theta) \mid \sigma \in A_0, \theta \in A_1\}$$

thereby defining

$$\begin{aligned}\delta^* : \mathrm{Top}_{*.} &\to \mathrm{Top}_{*.},\\ F(A) &\mapsto \delta^* F(A)\end{aligned}$$

objectwise presentation of

$$\begin{aligned}\delta^* : \widehat{\mathrm{d}k\text{-sAff}}_* &\to \widehat{\mathrm{d}k\text{-sAff}}_*,\\ F &\mapsto \delta^* F\end{aligned}$$

with $(\delta^* F)(A) = \delta^* F(A)$. If δ^* commutes with the left exact localization $\widehat{\mathrm{d}k\text{-sAff}}_* \to \mathrm{d}k\text{-sAff}_*^{\sim,\text{ét}}$, then F is referred to as a supersymmetric stack. By doing so, we preserve the physical origin of such a functor: to say that it is supersymmetric is in reference to the supersymmetry transformation in the base, that is, the context is supersymmetric insofar as we have supersymmetry transformations, not that F itself transforms in a symmetric manner under such transformations.

29 Taking stock – Physical Mathematics

29.1 Introduction

This chapter has two aims, the first one being to develop a notion of dynamics in towers of Segal topoi, any Segal topos in the tower being seen as embedded in the one that follows, and the second aim being an application of this formalism to Physics and to strings in particular, all of this being developed in the spirit of Physical Mathematics in the sense Greg Moore gives it [64, 65, 4].

In [23], we focused on the Segal category of geometric morphisms $\mathbb{R}\underline{\mathrm{Hom}}^*_{\mathrm{SeT}}(\mathcal{X}, \mathcal{X})$, with $\mathcal{X} = \mathrm{dSt}(k) = L(\mathrm{D}^- k\text{-}\mathrm{Aff}^{\sim,\mathrm{ffqc}})$ [74] for studying higher Galois. Recall that $\mathrm{dSt}(k) = \mathrm{d}k\text{-}\mathrm{Aff}^{\sim,\mathrm{ffqc}}$ is the localization of $\widehat{\mathrm{d}k\text{-}\mathrm{Aff}} = \mathbb{R}\underline{\mathrm{Hom}}(\mathrm{d}k\text{-}\mathrm{Aff}^{\mathrm{op}}, \mathrm{Top})$ with respect to ffqc hypercovers, where $\mathrm{Top} = L\mathrm{Set}_\Delta$ [74, 77]. Such a Segal category is a Segal groupoid however, and with a view towards applying the present formalism to Physics, where the notion of entropy is fundamental, it is therefore natural to just focus on $\mathbb{R}\underline{\mathrm{Hom}}(\mathcal{X}, \mathcal{X})$ instead. We will work in full generality and take a base Segal category built from a generic (simplicial) symmetric monoidal model category $\mathcal{C}$ satisfying the conditions of a homotopical algebraic context as enunciated in [79]. Under these conditions, $\mathrm{d}\mathcal{A}\mathrm{ff}_{\mathcal{C}} = L\mathrm{Comm}(\mathcal{C})^{\mathrm{op}}$ is our base Segal category on which we put a Segal topology τ, giving rise to a Segal category of (derived) stacks $\mathrm{d}\mathcal{A}\mathrm{ff}_{\mathcal{C}}^{\sim,\tau}$, a Segal topos. For applications to Physical Mathematics, this is really the Segal topos we have in mind, but for a formal development of the underlying theory, we will work with a generic Segal topos $\mathcal{X}$.

Identifying $\mathcal{X}$ with $\mathrm{id}_{\mathcal{X}} \in \mathbb{R}\underline{\mathrm{Hom}}(\mathcal{X}, \mathcal{X})$, we regard objects of $\mathbb{R}\underline{\mathrm{Hom}}(\mathcal{X}, \mathcal{X})$ other than the identity as providing incremental changes in our system $\mathcal{X}$. Indeed, an instantaneous snapshot of our system is given by providing all objects of $\mathcal{X}$ and the morphisms between them. For instance, a morphism $f : F \to G$ in $\mathcal{X}$ relates F and G together. Now given $\zeta \neq \mathrm{id}$ in $\mathbb{R}\underline{\mathrm{Hom}}(\mathcal{X}, \mathcal{X})$, $\zeta F \to \zeta G$ provides a modification to such a relation for all such morphisms f and for all objects F and G of $\mathcal{X}$. It is because such functors ζ provide an overall change in our system $\mathcal{X}$ that we regard them as deformations of our system. For that reason, we will refer to $\mathbb{R}\underline{\mathrm{Hom}}(\mathcal{X}, \mathcal{X})$ as the space of deformations of our system $\mathcal{X}$.

For fixed $F \in \mathcal{X}$, if one initially obtains a shift $F \mapsto \zeta F$ upon applying ζ, using ζ again, one obtains a sequence of incremental changes in the dynamics of F as in $F \mapsto \zeta F \mapsto \zeta^2 F \mapsto \cdots$. The limit $\zeta^\infty F = \lim \zeta^n F$ of this sequence in $\mathcal{X} \in \mathrm{SeT}$ (the Segal category of Segal topos) provides a discrete flow of F in the direction of ζ. A (smooth) flow is obtained if ζ is an infinitesimal deformation. To characterize such deformations, we will develop a notion of quantum deformation, defined as any functor $\zeta : \mathcal{X} \to \mathcal{X}$ that is minimal in the way it modifies objects of $\mathcal{X}$, something that we will define formally in Section 29.3.4. Now there is no reason for keeping ζ throughout. Thus we consider how such functors ζ evolve, and if we do this functorially, then we are led to considering functors $\Psi : \mathbb{R}\underline{\mathrm{Hom}}(\mathcal{X}, \mathcal{X}) \to \mathbb{R}\underline{\mathrm{Hom}}(\mathcal{X}, \mathcal{X})$. Iteratively applying Ψ to ζ, we obtain a sequence

https://doi.org/10.1515/9783111334073-029

$$\zeta \to \Psi\zeta \to \Psi^2\zeta \to \cdots$$

with a limit $\Psi^\infty\zeta = \lim \Psi^n\zeta$, which once applied to $F \in \mathcal{X}$ provides a discrete flow

$$F \to \zeta F \to \Psi\zeta F \to \Psi^2\zeta F \to \cdots$$

for F, which becomes a smooth, physical flow if both ζ and Ψ are quantum deformations. Iterating this construct, there is no reason that Ψ always be used to modify ζ, so we should consider how Ψ itself evolves. Adopting the notation $\mathbb{R}\underline{\mathrm{Hom}}(\mathcal{X}, \mathcal{X}) = \delta\mathcal{X}$, we have $\Psi \in \delta^2\mathcal{X}$, and we are therefore looking for an overarching object $\Gamma \in \delta^3\mathcal{X}$ to repeat this construction. Formalizing this, if $\psi^{(-,m)} \in \delta^m\mathcal{X}$ and $\psi^{(n+1,m)} = \psi^{(n,m+1)}\psi^{(n,m)}$, then at the level of $\mathcal{X}$, we get a discrete flow

$$F \xrightarrow{\psi^{(0,1)}} \psi^{(0,1)}F \xrightarrow{\psi^{(1,1)}} \psi^{(1,1)}\psi^{(0,1)}F \xrightarrow{\psi^{(2,1)}} \psi^{(2,1)}\psi^{(1,1)}\psi^{(0,1)}F \to \cdots$$

where the objects $\psi^{(n,1)}$ of $\delta\mathcal{X}$ fit into a sequence

$$\psi^{(0,1)} \xrightarrow{\psi^{(0,2)}} \psi^{(0,2)}\psi^{(0,1)} = \psi^{(1,1)} \xrightarrow{\psi^{(1,2)}} \psi^{(1,2)}\psi^{(1,1)} = \psi^{(2,1)} \to \cdots$$

and the objects $\psi^{(n,2)}$ of $\delta^2\mathcal{X}$ themselves fit into a sequence:

$$\psi^{(0,2)} \xrightarrow{\psi^{(0,3)}} \psi^{(0,3)}\psi^{(0,2)} = \psi^{(1,2)} \xrightarrow{\psi^{(1,3)}} \psi^{(1,3)}\psi^{(1,2)} = \psi^{(2,2)} \to \cdots$$

ad infinitum. We argue that $\lim_n \delta^n\mathcal{X}$ contains all the possible higher dynamics of objects of $\mathcal{X}$, which exists by virtue of [41], or even [67]. Further, the flow of some $F \in \mathcal{X}$ can be made smooth by taking all the higher deformations $\psi^{(n,m)}$ involved to be quantum in nature for all $m \geq 1$ and $n \geq 0$.

For simplicity and for illustrative purposes, we will focus on the case where we have one single deformation $\zeta = \psi^{(0,1)} \in \delta\mathcal{X}$ and one $\Psi = \psi^{(0,2)} \in \delta^2\mathcal{X}$, and we consider a (local) flow of ζ induced by Ψ. We are led to introducing the notion of cusp at (Ψ, ζ), which we will define by way of a tangent object using as impetus the derivation of the cotangent complex the way it was developed in [79]. To be precise, $\mathrm{Cusp}[\Psi, \zeta] = \{\Phi : \delta\mathcal{X} \to \delta\mathcal{X} \mid \mathbb{T}_{\Psi,\zeta} \cong \mathbb{T}_{\Phi,\zeta}\}$, where $\mathbb{T}_{\Psi,\zeta}$ is the tangent object of Ψ at ζ, defined as the dual of the cotangent object $\mathbb{L}_{\Psi,\zeta}$, a crude first approximation thereof being provided by $\mathrm{Map}(\Psi\zeta, \Psi\rho \circ \Psi\zeta)$ for some $\rho \in \delta\mathcal{X}$. We did not yet specify the exact nature of such objects, the reason being that we will give two presentations of this formalism, one being Set-based and the other one being simplicial in nature, and accordingly, depending on the context, one will talk about a (co)tangent set or a (co)tangent simplex.

Finally, we apply this formalism to strings by focusing on $\mathcal{X} = d\mathcal{A}\mathrm{ff}_{\mathcal{C}}^{\sim,\tau}$, which we introduced in the second paragraph above, and by defining a string between $a \in A$ and $a' \in A'$ in $\mathrm{Comm}(\mathcal{C})$ as an equivalence relation defined by $a \overset{\gamma}{\sim} a'$ if $A \cong A'$ in $L\mathrm{Comm}(\mathcal{C})$. It then becomes apparent that in algebraic geometry we obtain an algebraic version of string theory. From this perspective, stacks $F : \mathrm{Comm}(\mathcal{C}) \to \mathrm{Set}_\Delta$ map algebras $A \in \mathrm{Comm}(\mathcal{C})$, viewed as natural laws, to simplicial sets FA, and $a \overset{\gamma}{\sim} a'$ in A maps to $Fa \overset{F\gamma}{\sim} Fa'$, a string in FA. We regard γs as formal, algebraic strings with realization $F\gamma$ in an ambient space FA. All such images FA are in a sense representations of a law A and are related to one another by functoriality of F. Since we expect physical phenomena to be derived from the dynamics of strings, to exhibit natural laws in a coherent fashion, we are therefore led to viewing such functors F as modeling physical phenomena, some of which in particular will be physical fields.

In Section 29.2, we fix notations about our base symmetric monoidal model category $\mathcal{C}$ and construct the Segal category of stacks $\mathrm{d}\mathcal{A}\mathrm{ff}_{\mathcal{C}}^{\sim,\tau}$. In Section 29.3, we start with a Segal topos $\mathcal{X}$, which we consider as our system, and consider deformations on it, morphisms of deformations, and accompanying flows. The iterative picture is presented and discussed. We also provide an analysis of what would make a deformation quantum in nature for having a working definition of flows. In Section 29.4, we focus on local flows. To do this, we have to introduce the concepts of generalized categories and of tangent and cotangent objects. Finally, we apply this formalism to the study of strings in the last section, which we regard as equivalences between objects of commutative monoids of $\mathcal{C}$, which was fixed at the onset. In particular, we explain why the present setting is a natural place to have strings; not only does it explain that strings are not seen in nature, but it also suggests that a functorial treatment may be a good place to develop M-theory.

Notations: we denote by Set_Δ the category of simplicial sets, by SeCat the category of Segal categories, and by SeT the Segal category of Segal topos. Often, we will abbreviate $\mathbb{R}\underline{\mathrm{Hom}}(\mathcal{X}, \mathcal{X})$ by just $\delta\mathcal{X}$.

29.2 Building up the base Segal topos

As mentioned in the introduction, in lieu of working with $\mathrm{d}k\text{-}\mathrm{Aff} = Lsk\text{-}\mathrm{CAlg}^{\mathrm{op}}$ as we would in derived algebraic geometry, we abstract this formalism and start with a base (simplicial) symmetric monoidal category $\mathcal{C}$ endowed with a model category structure compatible with its monoidal structure. This makes $\mathcal{C}$ a symmetric monoidal model category. We ask that it be combinatorial for localization purposes, that it be pointed, that $\mathrm{Ho}(\mathcal{C})$ be additive, and that $\mathrm{Comm}(\mathcal{C})$ as well as modules over its objects have a model category structure. Those conditions are covered in [79], and collectively they define a homotopical algebra (HA) context. To be precise, asking that $\mathcal{C}$ be pointed and combinatorial and that $\mathrm{Ho}(\mathcal{C})$ be additive, this corresponds to Assumption 1.1.0.1 of [79]. For later purposes, if $A \in \mathrm{Comm}(\mathcal{C})$, then we want A-Mod to be combinatorial, proper, with its induced tensor product $- \otimes_A -$ making it a symmetric monoidal model category. This is Assumption 1.1.0.2 of the same reference. Assumption 1.1.0.3 ensures that $- \otimes_A M$ preserves equivalences. Assumption 1.1.0.4 ensures that A-$\mathrm{Comm}(\mathcal{C})$ as well as $(A\text{-}\mathrm{Comm}(\mathcal{C}))_{\mathrm{nu}}$ (nu stands for nonunital) are combinatorial, proper, and if B is cofibrant in A-$\mathrm{Comm}(\mathcal{C})$, then $B \otimes_A -$ preserves equivalences. Now less obvious is the following assumption, which we will adopt for infinitesimal lifting properties: if we fix a subcategory $\mathcal{C}_0$ of $\mathcal{C}$, Assumption 1.1.0.6 asks that $1 \in \mathcal{C}_0$, that $\mathcal{C}_0$ be stable by equivalences and homotopy colimits, and that $\mathrm{Ho}(\mathcal{C}_0) \subset \mathrm{Ho}(\mathcal{C})$ be stable by the derived tensor product.

All this being defined, and using the notations and definitions of [79], our base model category $\mathrm{Comm}(\mathcal{C})^{\mathrm{op}}$ is denoted $\mathcal{A}\mathrm{ff}_{\mathcal{C}}$, whose objects are referred to as affine objects (derived affine objects if $\mathcal{C}$ is simplicial). If $A \in \mathrm{Comm}(\mathcal{C})$, then the corresponding object in $\mathcal{A}\mathrm{ff}_{\mathcal{C}}$ is denoted $\mathrm{Spec}\,A$. The category of (derived) prestacks on $\mathcal{A}\mathrm{ff}_{\mathcal{C}}$ is denoted $\mathcal{A}\mathrm{ff}_{\mathcal{C}}^{\wedge}$, and it is the left Bousfield localization of the category of simplicial presheaves $\mathrm{sPr}(\mathcal{A}\mathrm{ff}_{\mathcal{C}})$ with respect to equivalences, that is, $\mathcal{A}\mathrm{ff}_{\mathcal{C}}^{\wedge} = L_{\mathrm{Bous}}\mathrm{sPr}(\mathcal{A}\mathrm{ff}_{\mathcal{C}})$. Its objects are functors $F : \mathcal{A}\mathrm{ff}_{\mathcal{C}}^{\mathrm{op}} \to \mathrm{Set}_{\Delta}$ that preserve equivalences. One denotes by $\mathbb{R}\underline{\mathrm{Hom}}$ the derived simplicial Hom of the simplicial model category $\mathcal{A}\mathrm{ff}_{\mathcal{C}}^{\wedge}$. If we fix a model topology τ on $\mathcal{A}\mathrm{ff}_{\mathcal{C}}$, then one obtains a model category of stacks $\mathcal{A}\mathrm{ff}_{\mathcal{C}}^{\sim,\tau}$ on $(\mathcal{A}\mathrm{ff}_{\mathcal{C}}, \tau)$, whose homotopy category $\mathrm{St}(\mathcal{C}, \tau) = \mathrm{Ho}(\mathcal{A}\mathrm{ff}_{\mathcal{C}}^{\sim,\tau})$ is the category of (derived) stacks on $\mathcal{A}\mathrm{ff}_{\mathcal{C}}$ for the topology τ. Its objects are functors $F : \mathcal{A}\mathrm{ff}_{\mathcal{C}}^{\mathrm{op}} \to \mathrm{Set}_{\Delta}$ that preserve equivalences and satisfy the descent property for homotopy τ-hypercovers. One assumption we need at this point is Assumption 1.3.2.2 of [79], which asks that τ on $\mathrm{Ho}(\mathcal{A}\mathrm{ff}_{\mathcal{C}})$ be quasi-compact, plus two other technical requirements that the reader can find in [79], both of which being needed for the functorial behavior of representable stacks under equivalences, as well as for the convenient characterization of $\mathcal{A}\mathrm{ff}_{\mathcal{C}}^{\sim,\tau}$ as the left Bousfield localization of $\mathcal{A}\mathrm{ff}_{\mathcal{C}}^{\wedge}$. This construction of $\mathcal{A}\mathrm{ff}_{\mathcal{C}}^{\sim,\tau}$, starting from a model category $\mathcal{A}\mathrm{ff}_{\mathcal{C}}$, can equivalently be done starting from a simplicial site (T, τ) instead, with $T = L\mathcal{A}\mathrm{ff}_{\mathcal{C}}$, giving rise to a model category of stacks $\mathrm{sPr}_{\tau}(T)$.

All that was done so far does not yet involve Segal categories. The transition is being made by regarding the simplicial category $\mathrm{d}\mathcal{A}\mathrm{ff}_{\mathcal{C}} = L(\mathrm{Comm}(\mathcal{C})^{\mathrm{op}})$ as an object of $\mathrm{Ho}(\mathrm{SeCat})$, whose objects are generically referred to as (derived) affine schemes, following standard usage [74]. From [77] one defines $\mathrm{d}\mathcal{A}\mathrm{ff}_{\mathcal{C}}^{\wedge} = \mathbb{R}\underline{\mathrm{Hom}}(\mathrm{d}\mathcal{A}\mathrm{ff}_{\mathcal{C}}^{\mathrm{op}}, \mathrm{Top})$ – where $\mathrm{Top} = L\mathrm{Set}_{\Delta}$ – referred to as the Segal category of (derived) prestacks on $\mathrm{d}\mathcal{A}\mathrm{ff}_{\mathcal{C}}$. Let τ be a Segal topology on $\mathrm{d}\mathcal{A}\mathrm{ff}_{\mathcal{C}}$, that is, a Grothendieck topology on $\mathrm{Ho}(\mathrm{d}\mathcal{A}\mathrm{ff}_{\mathcal{C}})$. Then the Segal category of (derived) stacks $\mathrm{d}\mathcal{A}\mathrm{ff}_{\mathcal{C}}^{\sim,\tau}$ on $(\mathrm{d}\mathcal{A}\mathrm{ff}_{\mathcal{C}}, \tau)$ is a full subcategory of $\mathrm{d}\mathcal{A}\mathrm{ff}_{\mathcal{C}}^{\wedge}$ (a Segal topos as well), which one can construct as a left exact localization of $\mathrm{d}\mathcal{A}\mathrm{ff}_{\mathcal{C}}^{\wedge}$ or as a left Bousfield localization thereof with respect to τ-hypercovers.

The connection with our initial construction starting from a simplicial site (T, τ) is given by

$$L(\mathrm{sPr}_{\tau}(\mathrm{d}\mathcal{A}\mathrm{ff}_{\mathcal{C}})) \cong \mathrm{d}\mathcal{A}\mathrm{ff}_{\mathcal{C}}^{\sim,\tau}$$

where L is a simplicial localization of model categories, and we used the facts that $\mathrm{d}\mathcal{A}\mathrm{ff}_{\mathcal{C}} = L(\mathrm{Comm}(\mathcal{C})^{\mathrm{op}})$ is a simplicial category and that a topology on the simplicial category $\mathrm{d}\mathcal{A}\mathrm{ff}_{\mathcal{C}}$ also corresponds to a Segal topology τ on $\mathrm{d}\mathcal{A}\mathrm{ff}_{\mathcal{C}}$ viewed as a Segal category, since simplicial sites and Segal sites are defined similarly [77, 79].

29.3 Dynamics

We regard $\mathcal{X} = \mathrm{d}\mathcal{A}\mathrm{ff}_{\mathcal{C}}^{\sim,\tau}$ as modeling natural phenomena for an appropriately chosen fundamental (simplicial) symmetric monoidal model category $\mathcal{C}$ and a topology τ on $\mathrm{d}\mathcal{A}\mathrm{ff}_{\mathcal{C}}$, both providing a homotopical algebraic geometry context (see [79] for details). In this picture, natural laws are modeled by objects of $\mathrm{Comm}(\mathcal{C})$, and a given phenomenon in this picture is represented by a stack $F \in \mathcal{X}$ valued in Set_{Δ}, since $\mathcal{X}$ can be modeled by functors $F : \mathrm{Comm}(\mathcal{C}) \to \mathrm{Set}_{\Delta}$. The problem is that this is very much inert; for $A \in \mathrm{Comm}(\mathcal{C})$, $F(A) \in \mathrm{Set}_{\Delta}$ presents how F responds to a law A. For another $B \in \mathrm{Comm}(\mathcal{C})$, assuming that we have a morphism $f : A \to B$ between those laws, by functoriality we have a map $Ff : F(A) \to F(B)$, thereby displaying some coherence in the representations of F, but there is no manifest dynamics. In other terms, $F \in \mathcal{X}$ provides one coherent manifestation of physical laws, and morphisms in $\mathcal{X}$ relate different manifestations together, without providing any sense in which those stacks are deformed to produce dynamics.

In what follows, $\mathcal{X} \in \mathrm{SeT}$. We identify $\mathcal{X}$ with $\mathrm{id}_{\mathcal{X}} \in \mathbb{R}\underline{\mathrm{Hom}}(\mathcal{X}, \mathcal{X})$, the identity therefore providing a snapshot of inter-relations between objects of $\mathcal{X}$ as argued above, so that other nonidentity morphisms $\psi : \mathcal{X} \to \mathcal{X}$ in $\mathbb{R}\underline{\mathrm{Hom}}(\mathcal{X}, \mathcal{X})$ can consequently be interpreted as providing an overall shift in the objects F of $\mathcal{X}$, a first step towards defining a dynamics in $\mathcal{X}$. Yet any such morphism ψ is an instantaneous deformation: $\psi : \mathcal{X} \to \mathcal{X}$ maps any $F \in \mathcal{X}$ to ψF. For a smoother representation, since this is just incremental, we seek a flow generated by such functors ψ. The most basic flow consists in having $\psi \in \mathbb{R}\underline{\mathrm{Hom}}(\mathcal{X}, \mathcal{X})$ fixed and on repeatedly applying it to obtain a sequence of the form

$$F \xrightarrow{\psi} \psi F \xrightarrow{\psi} \psi^2 F \xrightarrow{\psi} \cdots.$$

More generally, one can consider various objects $\psi = \psi_0, \psi_1, \ldots$ of $\mathbb{R}\underline{\mathrm{Hom}}(\mathcal{X}, \mathcal{X})$ being used to construct a flow, as in

$$F \xrightarrow{\psi_0} \psi_0 F \xrightarrow{\psi_1} \psi_1\psi_0 F \xrightarrow{\psi_2} \psi_2\psi_1\psi_0 F \xrightarrow{\cdots} .$$

Such a flow can be provided by a functor $\Psi : \mathbb{R}\underline{\mathrm{Hom}}(\mathcal{X}, \mathcal{X}) \to \mathbb{R}\underline{\mathrm{Hom}}(\mathcal{X}, \mathcal{X})$ for which $\Psi\psi_i = \psi_{i+1}$. Those are the first stages of the iterative construction presented in the introduction. In the present section, we motivate referring to objects of $\mathbb{R}\underline{\mathrm{Hom}}(\mathcal{X}, \mathcal{X})$ as deformations, we construct flows, and we study the iterative picture.

29.3.1 States and operators

29.3.1.1 General formalism

For simplicity, we write $\delta\mathcal{X}$ for $\mathbb{R}\underline{\mathrm{Hom}}(\mathcal{X}, \mathcal{X})$. We informally refer to $\mathcal{X}$ as a *system*, and we define a *deformation* of a system $\mathcal{X}$ as being an object of $\delta\mathcal{X}$, that is, a functor of

Segal categories $\mathcal{X} \to \mathcal{X}$. Indeed, $\psi \in \delta\mathcal{X}$ maps all $F's \in \mathcal{X}$ to $\psi F \in \mathcal{X}$ and therefore corresponds in this picture to an overall change in the system $\mathcal{X}$, whence the definition of deformation. We can formalize this as follows. Amongst objects $\psi \in \delta\mathcal{X}$, there are those functors that change the intrinsic nature of objects $F \in \mathcal{X}$ and those that merely cast them in another state without modifying them otherwise. We are interested in the latter case of transformation. If ψF is F shifted to a certain state, in the physical sense of the word, this implies that F itself was initially in a given state. This means we are not modifying the intrinsic object F, but the state it finds itself in. To make this manifest, we introduce the underlying object $\underline{F}$ associated with F prior to being put in a certain state, it being understood that the latter is built from $\underline{F}$ and a certain physical state. We introduce a symmetric monoidal category of (physical) states Σ, whose objects are denoted by $|-\rangle$, which we assume to form a complete orthonormal set with product $m : \langle p| \otimes |q\rangle \to \langle p|q\rangle = \delta_{pq}$. We regard $\underline{F}$ as an object of the same type as F and an operator as well, and if the letter $\mathcal{O}$ designates operators, then we denote by $\mathcal{X} \wedge \mathcal{O}$ the category of underlying objects of $\mathcal{X}$ that are also operators. $\underline{F}$ is an operator insofar as its eigenvalues on Σ are objects of $\mathcal{X}$ as in

$$\underline{F}|n\rangle = F_n|n\rangle$$

where $F_n \in \mathcal{X}$. In particular, we will designate by $|0\rangle$ the state of Σ that corresponds to F in such a manner that $\underline{F}|0\rangle = F|0\rangle$. With these notations,

$$F = \langle 0|\underline{F}|0\rangle$$

appears as the expectation value of $\underline{F}$. More precisely, we first have operator-states evaluation maps

$$\begin{aligned} \mathrm{ev} : \mathcal{O} \otimes \Sigma &\to \Sigma, \\ \Gamma \otimes |p\rangle &\mapsto \Gamma|p\rangle = \sum c_n|n\rangle \end{aligned}$$

for certain coefficients c_n, which one can generalize to

$$\begin{aligned} \mathrm{ev} : (\mathcal{X} \wedge \mathcal{O}) \otimes \Sigma &\to \mathcal{X} \otimes \Sigma, \\ \underline{F} \otimes |p\rangle &\mapsto \underline{F}|p\rangle = F_p|p\rangle. \end{aligned} \tag{29.1}$$

To be precise, if one defines $\pi : \mathcal{X} \to \Sigma$, $F_p \mapsto |p\rangle$, then denoting by $\mathcal{X} \otimes^{\Delta} \Sigma$ the image of the composite $\mathcal{X} \xrightarrow{\Delta} \mathcal{X} \otimes \mathcal{X} \xrightarrow{\mathrm{id}\otimes\pi} \mathcal{X} \otimes \Sigma$, the evaluation map in (29.1) is valued in $\mathcal{X} \otimes^{\Delta} \Sigma$. Focusing now on the product in Σ, $m : \Sigma \otimes \Sigma \to \mathbb{R}$, it has an obvious generalization

$$\begin{aligned} m : \Sigma \otimes (\mathcal{X} \otimes^{\Delta} \Sigma) &\to \mathcal{X}, \\ \langle p| \otimes \sum F_n|n\rangle &\mapsto \sum F_n\langle p|n\rangle = F_p. \end{aligned} \tag{29.2}$$

Combining (29.1) and (29.2), we get

$$\begin{aligned} \Sigma \otimes (\mathcal{X} \wedge \mathcal{O} \otimes \Sigma) &\xrightarrow{1\otimes \mathrm{ev}} \Sigma \otimes (\mathcal{X} \otimes^{\Delta} \Sigma) \xrightarrow{m} \mathcal{X}, \\ \langle 0| \otimes (\underline{F} \otimes |0\rangle) &\mapsto \langle 0| \otimes F|0\rangle \longmapsto \langle 0|F|0\rangle = F. \end{aligned}$$

Alternatively, one may want a state-like representation of $\underline{F}$ for the purpose of doing such evaluations. If $\underline{F}$ is fully defined by its eigenvalues F_n on states $|n\rangle$, then it suffices to consider the isomorphism

$$\begin{aligned} \mathcal{X} \wedge \mathcal{O} &\xrightarrow{\cong} \mathcal{X} \otimes^{\Delta} \Sigma, \\ \underline{F} &\mapsto |\underline{F}\rangle = \sum |n\rangle\langle n|\underline{F}|n\rangle = \sum F_n |n\rangle. \end{aligned}$$

With these notations,

$$\langle 0|\underline{F}\rangle = \langle 0| \sum |n\rangle\langle n|\underline{F}|n\rangle = \langle 0|\underline{F}|0\rangle = F.$$

29.3.1.2 Objects of $\mathcal{X}$ as functors

Since $\mathcal{X}$ is a functor category in practical applications, we focus on this case presently. We want to make sure the functorial calculus at the level of $\underline{F} \in \mathcal{X} \wedge \mathcal{O}$ projects down to the same calculus for objects of $\mathcal{X}$ in any state. Consider a map between objects of $\mathcal{X} \wedge \mathcal{O}$: $\underline{\tau} : \underline{F} \Rightarrow \underline{G}$, defined levelwise by $\underline{\tau}_n : F_n \Rightarrow G_n$ by virtue of the isomorphism $\mathcal{X} \wedge \mathcal{O} \cong \mathcal{X} \otimes^{\Delta} \Sigma$. Precisely, in the same manner that we had $\underline{F}|n\rangle = F_n|n\rangle$, which led to the representation $|\underline{F}\rangle = \sum F_n|n\rangle$, presently we have $\underline{\tau}|n\rangle = \underline{\tau}_n|n\rangle$, so we have a state-like representation of $\underline{\tau}$ given by $|\underline{\tau}\rangle = \sum \underline{\tau}_n|n\rangle : \sum F_n|n\rangle \Rightarrow \sum G_n|n\rangle$. Projecting to any state $|n\rangle$, we have $\langle n|\underline{\tau}\rangle : F_n = \langle n|\underline{F}\rangle \Rightarrow \langle n|\underline{G}\rangle = G_n$, so that $\langle n|\underline{\tau}\rangle = \underline{\tau}_n$. Already at this point, it is clear that if $\underline{\tau}$ is a natural transformation, then so is $\underline{\tau}_0 : F \Rightarrow G$, but we nevertheless work this out in full for illustrative purposes. Using the fact that $\underline{\tau}$ is a natural transformation, since objects of $\mathcal{X}$ are regarded as functors, for $f : A \to B$ in the base category of $\mathcal{X}$, we have a commutative square in $\mathrm{Set}_{\Delta} \wedge \mathcal{O}$ if objects of $\mathcal{X}$ are valued in Set_{Δ}:

$$\begin{array}{ccc} \underline{F}A & \xrightarrow{\underline{\tau}_A} & \underline{G}A \\ {\scriptstyle \underline{F}f}\downarrow & & \downarrow{\scriptstyle \underline{G}f} \\ \underline{F}B & \xrightarrow[\underline{\tau}_B]{} & \underline{G}B. \end{array}$$

If $|\underline{F}\rangle = \sum F_n|n\rangle$, then $|\underline{F}\rangle A = \sum F_n A|n\rangle$ in $\mathrm{Set}_{\Delta} \otimes^{\Delta} \Sigma$, and likewise $|\underline{\tau}\rangle_A = \sum (\underline{\tau}_n)_A|n\rangle$. Now because we have an isomorphism

$$\begin{aligned} \mathcal{X} \wedge \mathcal{O} &\xrightarrow{\cong} \mathcal{X} \otimes^{\Delta} \Sigma, \\ \underline{F} &\mapsto |\underline{F}\rangle \end{aligned}$$

being defined pointwise at objects A of the base category of $\mathcal{X}$, we have pointwise isomorphisms

$$\begin{aligned} \mathrm{Set}_\Delta \wedge \mathcal{O} &\xrightarrow{\cong} \mathrm{Set}_\Delta \otimes^\Delta \Sigma, \\ \underline{F}A &\mapsto |\underline{F}\rangle A. \end{aligned}$$

The state-like representation of the above CD in $\mathrm{Set}_\Delta \wedge \mathcal{O}$ expressed in $\mathrm{Set}_\Delta \otimes^\Delta \Sigma$ using this isomorphism now reads

$$\begin{array}{ccc} |\underline{F}\rangle A & \xrightarrow{|\underline{\tau}\rangle_A} & |\underline{G}\rangle A \\ {\scriptstyle |\underline{F}\rangle f}\downarrow & & \downarrow{\scriptstyle |\underline{G}\rangle f} \\ |\underline{F}\rangle B & \xrightarrow[|\underline{\tau}\rangle_B]{} & |\underline{G}\rangle B \end{array}$$

which upon projection to a state $|0\rangle$ produces

$$\begin{array}{ccc} \langle 0|\underline{F}\rangle A & \xrightarrow{\langle 0|\underline{\tau}\rangle_A} & \langle 0|\underline{G}\rangle A \\ {\scriptstyle \langle 0|\underline{F}\rangle f}\downarrow & & \downarrow{\scriptstyle \langle 0|\underline{G}\rangle f} \\ \langle 0|\underline{F}\rangle B & \xrightarrow[\langle 0|\underline{\tau}\rangle_B]{} & \langle 0|\underline{G}\rangle B \end{array}$$

and with $F = \langle 0|\underline{F}\rangle$, $Ff = \langle 0|\underline{F}\rangle f$, and $\tau = \langle 0|\underline{\tau}\rangle$, this reads

$$\begin{array}{ccc} FA & \xrightarrow{\tau_A} & GA \\ {\scriptstyle Ff}\downarrow & & \downarrow{\scriptstyle Gf} \\ FB & \xrightarrow[\tau_B]{} & GB \end{array}$$

and hence $\tau : F \Rightarrow G$.

29.3.1.3 Maps on physical states

We argued above that one should focus on those functors ψ for which the map $F \to \psi F$ is effectively just changing the state the essential intrinsic functor $\underline{F}$ initially finds itself in another state. Thus we seek a map of physical states since this is primarily what we are interested in. We consider only those changes of states that are functors $\underline{\psi} : \Sigma \to \Sigma$. Let $\underline{\psi}$ be such a functor. We consider the following defining CD for ψ:

$$\begin{array}{ccc}
\Sigma \otimes (\mathcal{X} \wedge \mathcal{O}) & \xrightarrow{\underline{\psi}\otimes \mathrm{id}} & \Sigma \otimes (\mathcal{X} \wedge \mathcal{O}) \\
{\scriptstyle\cong}\downarrow & & \downarrow{\scriptstyle\cong} \\
\Sigma \otimes (\mathcal{X} \otimes^{\Delta} \Sigma) & \xrightarrow{\underline{\psi}\otimes \mathrm{id}} & \Sigma \otimes (\mathcal{X} \otimes^{\Delta} \Sigma) \\
{\scriptstyle m}\downarrow & & \downarrow{\scriptstyle m} \\
\mathcal{X} & \xrightarrow[\psi]{} & \mathcal{X}
\end{array}$$

with $\psi = m(\underline{\psi} \otimes \mathrm{id})$. To fix ideas, let $\underline{\psi} : |0\rangle \to |1\rangle$ in $\mathrm{End}(\Sigma)$, and take $\underline{F}$ such that $\underline{F}|0\rangle = F|0\rangle$. Then we are looking at

$$\begin{array}{ccc}
\langle 0| \otimes \underline{F} & \xrightarrow{\underline{\psi}\otimes \mathrm{id}_{\underline{F}}} & \langle 1| \otimes \underline{F} \\
{\scriptstyle\cong}\downarrow & & \downarrow{\scriptstyle\cong} \\
\langle 0| \otimes |\underline{F}\rangle & \xrightarrow{\underline{\psi}\otimes \mathrm{id}_{|\underline{F}\rangle}} & \langle 1| \otimes |\underline{F}\rangle \\
{\scriptstyle m}\downarrow & & \downarrow{\scriptstyle m} \\
\langle 0|\underline{F}\rangle = F & \xrightarrow[\psi]{} & \langle 1|\underline{F}\rangle = \langle 1| \sum F_n |n\rangle = \langle 1|F_1|1\rangle = F_1 = \psi F.
\end{array}$$

That ψ is a functor is due in part to the fact that $\underline{\psi}$ itself is a functor on Σ. Consider

$$\begin{array}{ccc}
\langle p| & \xrightarrow{\underline{\psi}} & \langle p'| \\
{\scriptstyle \underline{u}}\downarrow & & \downarrow{\scriptstyle \underline{\psi}\underline{u}} \\
\langle q| & \xrightarrow[\underline{\psi}]{} & \langle q'|.
\end{array} \tag{29.3}$$

Evaluated at some $\underline{F} \cong |\underline{F}\rangle$, this yields

$$\begin{array}{ccc}
\langle p|\underline{F}\rangle & \xrightarrow{m(\underline{\psi}\otimes \mathrm{id}_{|\underline{F}\rangle})} & \langle p'|\underline{F}\rangle \\
{\scriptstyle m(\underline{u}\otimes \mathrm{id}_{|\underline{F}\rangle})}\downarrow & & \downarrow{\scriptstyle m(\underline{\psi}\underline{u}\otimes \mathrm{id}_{|\underline{F}\rangle})} \\
\langle q|\underline{F}\rangle & \xrightarrow[m(\underline{\psi}\otimes \mathrm{id}_{|\underline{F}\rangle})]{} & \langle q'|\underline{F}\rangle
\end{array} \tag{29.4}$$

which reads

$$\begin{array}{ccc} F_p & \xrightarrow{\psi} & \psi F_p \\ {\scriptstyle u}\downarrow & & \downarrow{\scriptstyle \psi u} \\ F_q & \xrightarrow[\psi]{} & \psi F_q \end{array} \tag{29.5}$$

with $u = m(\underline{u} \otimes \mathrm{id}_{|\underline{F}\rangle})$ and $\psi u = m(\underline{\psi u} \otimes \mathrm{id}_{|\underline{F}\rangle})$. Now note that the notation $F = \langle 0|\underline{F}\rangle$ is bifunctorial. A natural transformation $F \Rightarrow G$ in $\mathcal{X}$ can originate from a change of states, as we just considered, or from a change in the underlying functor $\underline{F}$ of F, something we call a *mutation*, $\underline{\tau} : \underline{F} \Rightarrow \underline{G}$. Thus we also consider the following CD:

$$\begin{array}{ccc} \langle 0| \otimes |\underline{F}\rangle & \xrightarrow{\underline{\psi}\otimes \mathrm{id}_{|\underline{F}\rangle}} & \langle 1| \otimes |\underline{F}\rangle \\ {\scriptstyle \mathrm{id}_{\langle 0|}\otimes|\underline{\tau}\rangle = \langle 0|\otimes|\underline{\tau}\rangle}\downarrow & & \downarrow{\scriptstyle \mathrm{id}_{\langle 1|}\otimes|\underline{\tau}\rangle = \langle 1|\otimes|\underline{\tau}\rangle} \\ \langle 0| \otimes |\underline{G}\rangle & \xrightarrow[\underline{\psi}\otimes \mathrm{id}_{|\underline{G}\rangle}]{} & \langle 1| \otimes |\underline{G}\rangle \end{array} \tag{29.6}$$

which upon multiplication yields

$$\begin{array}{ccc} F = \langle 0|\underline{F}\rangle & \xrightarrow{\psi = m(\underline{\psi}\otimes \mathrm{id}_{|\underline{F}\rangle})} & \langle 1|\underline{F}\rangle = \psi F \\ {\scriptstyle \tau = \langle 0|\underline{\tau}\rangle}\downarrow & & \downarrow{\scriptstyle \langle 1|\underline{\tau}\rangle = \psi\tau} \\ G = \langle 0|\underline{G}\rangle & \xrightarrow[\psi = m(\underline{\psi}\otimes \mathrm{id}_{|\underline{G}\rangle})]{} & \langle 1|\underline{G}\rangle = \psi G \end{array} \tag{29.7}$$

which simplifies as

$$\begin{array}{ccc} F & \xrightarrow{\psi} & \psi F \\ {\scriptstyle \tau}\downarrow & & \downarrow{\scriptstyle \psi\tau} \\ G & \xrightarrow[\psi]{} & \psi G. \end{array} \tag{29.8}$$

Hence (29.5) and (29.8) show that ψ is a functor in $\mathbb{R}\underline{\mathrm{Hom}}(\mathcal{X}, \mathcal{X})$.

We pointed out that $F = \langle 0|\underline{F}\rangle$ is bifunctorial. Note that this really follows from m being bifunctorial on $\Sigma \otimes (\mathcal{X} \otimes^{\Delta} \Sigma)$ as $F = \langle 0|\underline{F}\rangle = m(\langle 0| \otimes |\underline{F}\rangle)$. Using this line of argument, right tensoring (29.3) by $|\underline{F}\rangle$ in $\Sigma \otimes (\mathcal{X} \otimes^{\Delta} \Sigma)$ and using the fact that m is functorial in its first argument, one obtains (29.4). Then using the bifunctoriality of m again, one goes from (29.6) to (29.7). Thus ψ being a functor follows from the bifunctoriality of m and the fact that $\underline{\psi}$ is a functor.

29.3.2 Iterative shifts

So far applying ψ to some $F \in \mathcal{X}$ just produces a shift in F. Dynamics comes up when one considers a sequence of such shifts. For simplicity, we initially focus on dynamics generated by a single $\psi \in \delta\mathcal{X}$. Consider a sequence of states in Σ generated by $\underline{\psi}$ from a physical state $|0\rangle$ as in

$$|0\rangle \xrightarrow{\underline{\psi}} \underline{\psi}|0\rangle = |1\rangle \xrightarrow{\underline{\psi}} \underline{\psi}|1\rangle = \underline{\psi}^2|0\rangle \xrightarrow{\underline{\psi}} \cdots$$

which once tensored with $|\underline{F}\rangle$ and then evaluated produces

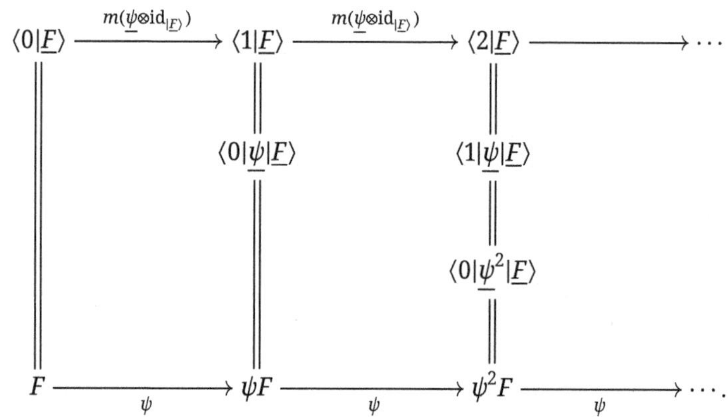

As argued before, the next step consists in considering deformations that vary via a functor $\Psi : \delta\mathcal{X} \to \delta\mathcal{X}$, thereby producing a sequence in $\delta\mathcal{X}$, and take the limit thereof. Let us denote by ϵ_0 an initial deformation in $\delta\mathcal{X}$. We consider

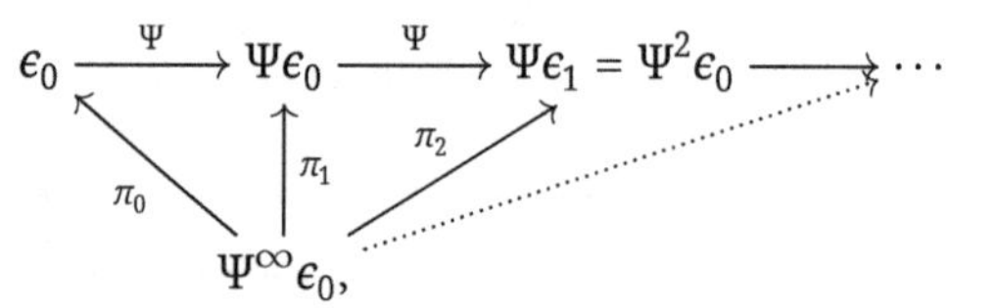

which exists since $\mathcal{X}$ being a Segal topos, so is $\delta\mathcal{X} = \mathbb{R}\underline{\mathrm{Hom}}(\mathcal{X}, \mathcal{X})$, and hence it has all small limits. One defines a *flow* of our system from an initial deformation ϵ_0 and induced by some $\Psi \in \mathbb{R}\underline{\mathrm{Hom}}(\delta\mathcal{X}, \delta\mathcal{X})$ given by $\Psi^\infty\epsilon_0 = \lim_n \Psi^n\epsilon_0$. This is defined pointwise. Thus for some object $F \in \mathcal{X}$, a flow in the dynamics of F starting from an initial deformation ϵ_0 in the direction of Ψ is given by an augmented diagram

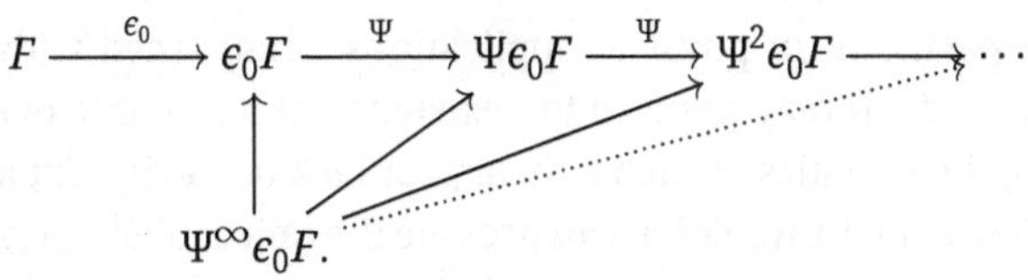

To summarize the current picture, our system $\mathcal{X}$ is identified with $\mathrm{id}_{\mathcal{X}} \in \mathbb{R}\underline{\mathrm{Hom}}(\mathcal{X}, \mathcal{X}) := \delta\mathcal{X}$, a Segal category of deformations of our system, which is enhanced to $\mathbb{R}\underline{\mathrm{Hom}}(\delta\mathcal{X}, \delta\mathcal{X}) := \delta^2\mathcal{X}$, and objects thereof are the functors Ψ discussed above, which effectively move our system from one deformation to another. Iterative compositions of a given Ψ on an initial deformation ultimately produce a flow of our system in the limit. The repetitiveness of using $\psi \in \delta\mathcal{X}$ above was fixed by introducing $\Psi \in \delta^2\mathcal{X}$. Now however that problem is repeating itself for Ψ. Thus iterating this overall construct and letting $\mathbb{R}\underline{\mathrm{Hom}}(\delta^n\mathcal{X}, \delta^n\mathcal{X}) := \delta^{n+1}\mathcal{X}$ represent higher deformations, we then claim the ∞-Segal topos $\lim_{n\infty} \delta^n\mathcal{X}$ – the limit taken in the 2-Segal category of Segal categories – contains all possible higher dynamics of our initial system $\mathcal{X}$ and consequently corresponds to a complete description thereof, making $\mathcal{X}$ a closed system in the physical sense of the word. To fix ideas, the first stages of this construction go as follows. If $\psi^{(-,m)} \in \delta^m\mathcal{X}$ and $\psi^{(n+1,m)} = \psi^{(n,m+1)}\psi^{(n,m)}$, then we get a discrete flow at the level of $\mathcal{X}$:

$$F \xrightarrow{\psi^{(0,1)}} \psi^{(0,1)}F \xrightarrow{\psi^{(1,1)}} \psi^{(1,1)}\psi^{(0,1)}F \xrightarrow{\psi^{(2,1)}} \psi^{(2,1)}\psi^{(1,1)}\psi^{(0,1)}F \to \cdots$$

where the objects $\psi^{(n,1)}$ of $\delta\mathcal{X}$ fit into a sequence

$$\psi^{(0,1)} \xrightarrow{\psi^{(0,2)}} \psi^{(0,2)}\psi^{(0,1)} = \psi^{(1,1)} \xrightarrow{\psi^{(1,2)}} \psi^{(1,2)}\psi^{(1,1)} = \psi^{(2,1)} \to \cdots$$

and the objects $\psi^{(n,2)}$ of $\delta^2\mathcal{X}$ themselves fit into a sequence

$$\psi^{(0,2)} \xrightarrow{\psi^{(0,3)}} \psi^{(0,3)}\psi^{(0,2)} = \psi^{(1,2)} \xrightarrow{\psi^{(1,3)}} \psi^{(1,3)}\psi^{(1,2)} = \psi^{(2,2)} \to \cdots$$

and higher deformations are constructed similarly.

29.3.3 Remarks on higher deformations

To come back to the iterative step $\delta^{n+1}\mathcal{X} = \mathbb{R}\underline{\mathrm{Hom}}(\delta^n\mathcal{X}, \delta^n\mathcal{X})$, an object of $\delta^{n+1}\mathcal{X}$ is a functor of Segal categories $\delta^n\mathcal{X} \to \delta^n\mathcal{X}$ that provides a global shift in the objects of $\delta^n\mathcal{X}$. The point we want to make is that $\Psi_{n+1} \in (\delta^{n+1}\mathcal{X})_0$ is an object of $\delta^{n+1}\mathcal{X}$, yet from the perspective of $\delta^n\mathcal{X}$, for all $\zeta \in (\delta^n\mathcal{X})_0, \Psi_{n+1} : \zeta \mapsto \Psi_{n+1}\zeta$; in other terms, its action can be quite complex. Essentially, then simple objects in a higher category can have complex realizations in a lower category, or viewed differently, complex operations in a given categorical context can be repackaged into something far simpler in a higher category.

This is especially pertinent in physical applications when faced with local behaviors at the level of $\delta^n\mathcal{X}$, which may point to the existence of a possible overarching theory without providing the specifics of such a theory for lack of clarity. Expanding our views at the level of $\delta^n\mathcal{X}$ to something global can provide the object of $\delta^{n+1}\mathcal{X}$, if one can extract it, that explains the overall picture and local behaviors in particular.

Another interesting observation is the following: if we place ourselves at the level of $\delta^n\mathcal{X}$ and consider an object ζ of $\delta^n\mathcal{X}$, and $\Psi \in \mathbb{R}\underline{\mathrm{Hom}}(\delta^n\mathcal{X}, \delta^n\mathcal{X}) = \delta^{n+1}\mathcal{X}$ is a higher deformation of our system, from the perspective of $\delta^n\mathcal{X}$, ζ and $\Psi\zeta$ correspond to a "before" and an "after", and hence a de facto notion of time, whereas at the level of $\delta^{n+1}\mathcal{X}$ already, objects Ψ thereof are defined pointwise, so Ψ is defined by $\zeta \xrightarrow{\Psi} \Psi\zeta$, and hence regard past (or present) and future events in $\delta^n\mathcal{X}$ as being simultaneous.

29.3.4 Quantum deformations

Note that we have defined a generic flow as being defined from higher deformations $\psi^{(n,m)} \in \delta^m\mathcal{X}$ for $m \geq 1$ and $n \geq 0$. In practice, one should consider only those initial deformations that are fundamental, such as quantum deformations. In the present section, we define such fundamental, quantum deformations in our formalism. Since this is fairly applied, we will work with $\mathcal{X} = \mathrm{d}\mathcal{A}\mathrm{ff}_C^{\sim,\tau}$ for definiteness. Let $\psi \in \delta\mathcal{X}$. For ψ to be a nontrivial quantum deformation, we first ask that there are $F \in \mathcal{X}$ such that $\psi F \neq F$ and $A \in \mathrm{Comm}(C)$ such that $\psi FA \neq FA$. We define the *differential matrix* of ψ by

$$\Delta\psi = \left(d_{\mathrm{Set}_\Delta}(\psi FA, FA)\right)_{\substack{F\in\mathcal{X}\\ A\in\mathrm{Comm}(C)}}$$

and for simplicity, we write $(\Delta\psi)_{F,A} = d_{\mathrm{Set}_\Delta}(\psi FA, FA)$. Since the geometric realization of a simplicial set is a CW-complex, for $X, Y \in \mathrm{Set}_\Delta$, we define

$$d_{\mathrm{Set}_\Delta}(X, Y) = d_{\mathrm{CW}}(|X|, |Y|)$$

where we consider homotopy types of CW-complexes determined by their homotopy classes of attaching maps, and thus we can define a distance function d_{CW} on CW-complexes, denoted $d_{\mathrm{CW}}(C, C')$ for two CW-complexes C and C', by the number of different homotopy classes of attaching maps C and C' have. This number may be unbounded, but this is inconsequential for defining quantum deformations whose differential matrices have finite entries. Denote by $\delta(C_n, C'_n)$ the number of different homotopy classes of attaching maps on the n-skeleta C_n and C'_n of C and C', respectively, to get C_{n+1} and C'_{n+1}, and let

$$\delta(C_{\leq n}, C'_{\leq n}) = \sum_{0\leq p\leq n} \delta(C_p, C'_p)$$

where $C_{\leq n} = \bigcup_{0\leq p\leq n} C_p$. Using this, we let

$$d_{\mathrm{CW}}(C, C') = \sum_{n \geq 0} \delta(C_n, C'_n) = \lim_{n\infty} \delta(C_{\leq n}, C'_{\leq n}).$$

This is clearly symmetric and reflexive, and the only situation where we would run into difficulties while showing $\delta(C_n, C''_n) \leq \delta(C_n, C'_n) + \delta(C'_n, C''_n)$ is when the left-hand side is 1, and the right-hand side is 0 + 0, which cannot occur since homotopy is an equivalence relation and hence transitive. It follows that d_{CW} is a well-defined distance on CW-complexes. Having defined this, it follows that $d_{\mathrm{Set}_\Delta}(X, Y) = d_{\mathrm{CW}}(|X|, |Y|)$ is a well-defined distance function on simplicial sets in a homotopical context. Hence $(\Delta\psi)_{F,A} = d_{\mathrm{Set}_\Delta}(\psi FA, FA)$ is well-defined. We now define *quantum deformations* to be those deformations ψ that make their differential matrix minimal. To be precise, ψ is a quantum deformation if there is no other deformation η such that

$$(\Delta\eta)_{F,A} < (\Delta\psi)_{F,A}$$

for some $F \in \mathcal{X}$ and $A \in \mathrm{Comm}(\mathcal{C})$, all other entries of $\Delta\eta$ being higher than or equal to those of $\Delta\psi$.

We can iterate this for higher deformations. Consider $\Psi \in \delta^2\mathcal{X}$. It is a nontrivial (quantum) deformation if there exists $\eta \in \delta\mathcal{X}$ such that $\Psi\eta \neq \eta$, which holds if there exists $F \in \mathcal{X}$ such that $\Psi\eta F \neq \eta F$, and again for objects F that are functors, this is true if there exists $A \in \mathrm{Comm}(\mathcal{C})$ such that $\Psi\eta FA \neq \eta FA$ in Set_Δ. We are led to considering the differential matrix of Ψ

$$\Delta\Psi = \left(d_{\mathrm{Set}_\Delta}(\Psi\eta FA, \eta FA)\right)_{\eta,F,A}$$

and deformations $\Psi \in \delta^2\mathcal{X}$ that are quantum in nature are those that make this cubic matrix minimal. More generally, physical, smooth flows are built from higher deformations $\psi^{(n,m)} \in \delta^m\mathcal{X}$ that must all be quantum in nature, each having a minimal differential matrix of size $m + 1$.

29.4 Cusps

A natural question is, given a deformation ϵ in $\delta\mathcal{X}$, and $\Psi : \delta\mathcal{X} \to \delta\mathcal{X}$, what are the other such functors Ξ that yield a same local flow as Ψ immediately at ϵ. We claim that those satisfy $T_{\Xi,\epsilon} \cong T_{\Psi,\epsilon}$ for a same deformation ϵ in $\delta\mathcal{X}$, an isomorphism of tangent spaces. Since one of the things we want to achieve is the ability to clarify the functorial dynamics at the level of a Segal category $\mathcal{X}$ such as $\mathrm{d}\mathcal{A}\mathrm{ff}_{\mathcal{C}}^{\sim,\tau}$, we focus on this particular example presently, and for our base monoidal category, we take a symmetric monoidal model category $\mathcal{C}$ that we choose to be simplicial, that is, we place ourselves within the realm of derived algebraic geometry. The correct notion of tangent space in derived algebraic geometry being that of a tangent complex, we define $T_{\Psi,\epsilon} := \mathbb{T}_{\Psi,\epsilon}$ as the proper adaptation in our context of the tangent complex of Ψ at ϵ. Then we define the *cusp* of

Ψ at ϵ by

$$\mathrm{Cusp}[\Psi,\epsilon] = \{\Xi : \delta\mathcal{X} \to \delta\mathcal{X} \mid \mathbb{T}_{\Xi,\epsilon} \cong \mathbb{T}_{\Psi,\epsilon}\}$$

which gives those functors $\Xi : \delta\mathcal{X} \to \delta\mathcal{X}$ that have a same flow as Ψ at ϵ. Having thus defined cusps, note that this defines equivalence classes of local flows in the space of deformations; from any given deformation ϵ, functors $\Psi : \delta\mathcal{X} \to \delta\mathcal{X}$ can be grouped into equivalence classes $[\Psi] := \mathrm{Cusp}[\Psi,\epsilon]$.

29.4.1 Formal construction of the tangent complex

We remind the reader of the formal construction of a tangent complex as covered in [79]. We will not need this construction in our work, but it is nevertheless important to understand how the tangent complex comes about to justify our present definition of the tangent object $\mathbb{T}_{\Psi,x}$. For a stack $F \in \mathcal{A}\mathrm{ff}_{\mathcal{C}}^{\sim,\tau}$ and an A-point $x : \mathbb{R}\underline{\mathrm{Spec}}\,A \to F$ of F, $\mathbb{T}_{F,x}$ is an element of the homotopy category of spectra $\mathrm{Ho}\,(\mathrm{Sp}(A\text{-Mod}))$. Here $A \in \mathcal{A}$, $\mathcal{A} \subset \mathrm{Comm}(\mathcal{C})$ having $\mathcal{C}_0$-good objects for elements, where according to Definition 1.1.0.10 of [79], $A \in \mathrm{Comm}(\mathcal{C})$ is good with respect to $\mathcal{C}_0$ if the functor $\mathrm{Ho}\,(A\text{-Mod}^{\mathrm{op}}) \to \mathrm{Ho}\,((A\text{-Mod}_0^{\mathrm{op}})^\wedge)$ is fully faithful, where $A\text{-Mod}_0 = A\text{-Mod} \cap \mathcal{C}_0$, and $\mathcal{C}_0$ is a full subcategory of $\mathcal{C}$ satisfying Assumption 1.1.0.6 of [79] (see reference for more details on why it is needed). Here we have used the notation $\mathbb{R}\underline{\mathrm{Spec}}\,A = \mathbb{R}\underline{h}_{\mathrm{Spec}\,A}$ for $\mathbb{R}\underline{h} : \mathrm{Ho}\,(\mathcal{A}\mathrm{ff}_{\mathcal{C}}) \to \mathrm{St}(\mathcal{C},\tau)$, the right derived functor of the model Yoneda embedding. In this definition, for a cofibrant resolution functor Γ_* on the model category $\mathcal{A}\mathrm{ff}_{\mathcal{C}}$, $\underline{h}_X(Y) = \mathrm{Hom}_{\mathcal{A}\mathrm{ff}_{\mathcal{C}}}(\Gamma_*(Y), X)$.

With regards to spectra, one uses the fact that $\mathcal{C}$ being pointed, one has a suspension functor $S : \mathrm{Ho}\,(\mathcal{C}) \to \mathrm{Ho}\,(\mathcal{C})$, $x \mapsto S(x) = * \coprod_x^{\mathbb{L}} *$. Fix once and for all, as done in [79], an object $S_{\mathcal{C}}^1$ in $\mathcal{C}$, a cofibrant model for $S(1) \in \mathrm{Ho}\,(\mathcal{C})$, and write $S_A^1 = S_{\mathcal{C}}^1 \otimes A \in A\text{-Mod}$. That allows one to consider transition morphisms $\sigma_n : S_A^1 \otimes_A M_n \to M_{n+1}$. Following the same argument as in [79], one gets a category of spectra $\mathrm{Sp}(A\text{-Mod})$. Put the stable model structure on $\mathrm{Sp}(A\text{-Mod})$ by taking the left Bousfield localization of this category, using as local objects stable modules $M_* \in \mathrm{Sp}(A\text{-Mod})$, objects such that the induced morphisms $M_n \to \mathbb{R}\underline{\mathrm{Hom}}_A(S_A^1, M_{n+1})$ are isomorphisms in $\mathrm{Ho}\,(A\text{-Mod})$, where $\underline{\mathrm{Hom}}_A(S_A^1, -)$ is the right adjoint of $S_A^1 \otimes_A -$. Further, $\mathrm{Ho}\,(\mathrm{Sp}(A\text{-Mod}))$ inherits from $\mathrm{Ho}\,(A\text{-Mod})$ a symmetric monoidal structure $- \otimes_A^{\mathbb{L}} -$ making it a closed symmetric monoidal category with an internal hom $\mathbb{R}\underline{\mathrm{Hom}}_A^{\mathrm{Sp}}(M_*, N_*) \in \mathrm{Ho}\,(\mathrm{Sp}(A\text{-Mod}))$ for two stable A-modules M_* and N_*. This allows one to define

$$\mathbb{T}_{F,x} = \mathbb{R}\underline{\mathrm{Hom}}_A^{\mathrm{Sp}}(\mathbb{L}_{F,x}, A)$$

for $\mathbb{L}_{F,x} \in \mathrm{Ho}\,(\mathrm{Sp}(A\text{-Mod}))$, the cotangent complex of F at x. The latter is defined by

$$\mathbb{D}\mathrm{er}_F(X,-) \cong \mathbb{R}\underline{h}_s^{\mathbb{L}_{F,x}} \in \mathrm{Ho}\,((A\text{-Mod}_0^{\mathrm{op}})^\wedge)$$

with $\mathbb{D}\mathrm{er}_F(X,M) = \mathrm{Map}(X[M],F) \in \mathrm{Ho}\,(\mathrm{Set}_\Delta)$, $X[M] = \mathbb{R}\underline{\mathrm{Spec}}\,(A \oplus M)$, $A \oplus M$ the square zero extension of A by $M \in A$-Mod. This defining isomorphism also makes use of a functor $\underline{h}_s : \mathrm{Sp}(A\text{-Mod})^{\mathrm{op}} \to (A\text{-Mod}_0^{\mathrm{op}})^\wedge$ with $\underline{h}_s^{M_*} : A\text{-Mod}_0 \to \mathrm{Set}_\Delta$ mapping N to $\mathrm{Hom}(M_*, \Gamma_*(S_A(N)))$, Γ_* a simplicial resolution functor on Sp(A-Mod), and the functor $S_A : \mathrm{Ho}\,(A\text{-Mod}) \to \mathrm{Ho}\,(\mathrm{Sp}(A\text{-Mod}))$ is defined by $S_A(M)_n = (S_A^1)^{\otimes_A^n} \otimes_A M$. One has that for all $A \in \mathrm{Comm}(\mathcal{C})$, $\mathbb{R}\underline{h}_s : \mathrm{Ho}\,(\mathrm{Sp}(A\text{-Mod}))^{\mathrm{op}} \to \mathrm{Ho}\,((A\text{-Mod}_0^{\mathrm{op}})^\wedge)$ is well-defined as the right derived functor of $\underline{h}_s$. All of this work can be found in [79].

Perhaps, we can pause at this point and note a first departure on our end from this standard formalism. Stacks are valued in simplicial sets, but our functors, to which we apply this construction, are valued in Segal categories. Indeed, $\Psi : \delta\mathcal{X} \to \delta\mathcal{X}$ for us plays the role of F above. Thus the definition of $\mathbb{D}\mathrm{er}_F(X,-)$ will have to be modified in our situation (see [79] for the discussion about this functor having a lift to a functor A-Mod $\to \mathrm{Set}_\Delta$). Additionally, Map is the mapping complex for a model category, which here in particular is $\mathcal{A}\mathrm{ff}_{\mathcal{C}}^{\sim,\tau}$, and in our case, we work with $\mathbb{R}\underline{\mathrm{Hom}}(\delta\mathcal{X},\delta\mathcal{X})$, for which we do not have an obvious mapping complex. We will remedy this below.

29.4.2 Derivations

Letting $\delta^2\mathcal{X} = \mathbb{R}\underline{\mathrm{Hom}}(\delta\mathcal{X},\delta\mathcal{X}) \in \mathrm{Ho}\,(\mathrm{SeCat})$ and supposing for a moment that for $A \in \delta\mathcal{X}$, we have a functor $\mathbb{R}\underline{\mathrm{Spec}}\,A \in \delta^2\mathcal{X}$, a natural transformation $x : X = \mathbb{R}\underline{\mathrm{Spec}}\,A \to \Psi$ in $\delta^2\mathcal{X}$ providing an A-point of Ψ would be an element of $(\delta^2\mathcal{X})_{(\mathbb{R}\underline{\mathrm{Spec}}\,A,\Psi)} \in \mathrm{Ho}\,(\mathrm{Set}_\Delta)$. However, our base category $\delta\mathcal{X}$ is no symmetric monoidal category, so we have no obvious notion of linear algebra over it, and thus a priori one cannot talk about A-modules, let alone square zero extensions $A \oplus M$. Note however that we are looking for deformations of A, and this can be implemented with infinitesimal compositions MA for some functor $M \in \delta\mathcal{X}$. We therefore define

$$\mathbb{D}\mathrm{er}_\Psi(X,M) = \mathrm{Map}(\mathbb{R}\underline{\mathrm{Spec}}\,(MA),\Psi).$$

We first have to formally define $\mathbb{R}\underline{\mathrm{Spec}}\,A$, which will lead us to defining certain mapping objects in generalized categories. Having done that, the definition of the analog in our context of the cotangent and tangent complexes will fall out of our formalism.

29.4.3 Mapping spaces

Presently, we discuss $\mathbb{R}\underline{\mathrm{Spec}}\,A \in \delta^2\mathcal{X}$ for $A \in \delta\mathcal{X}$, and what entails for morphisms of $\delta^2\mathcal{X}$ that have it for domain. For $B \in \delta\mathcal{X}$, we define

$$\mathbb{R}\underline{\mathrm{Spec}}\,A(B) = \mathrm{Nat}(A,B)$$

to be the collection of natural transformations $\Lambda_{i,B} : A \Rightarrow B$. To make sense of a morphism $\Omega_B : \mathbb{R}\underline{\mathrm{Spec}}\,A(B) \to \Psi(B)$, we therefore have to look at Ψ differently. $\Psi(B) \in \delta\mathcal{X} = \mathbb{R}\underline{\mathrm{Hom}}(\mathcal{X}, \mathcal{X})$ being a functor from $\mathcal{X}$ to itself, for $f : F \to G$ in $\mathcal{X}$, we have a commutative diagram

$$\begin{array}{ccc} F & \longrightarrow & \Psi(B)F \\ {\scriptstyle f}\downarrow & & \downarrow{\scriptstyle \Psi(B)f} \\ G & \longrightarrow & \Psi(B)G \end{array}$$

which we can regard as defining a natural transformation $\Gamma_B : \mathrm{id} \Rightarrow \Psi(B)$:

$$\begin{array}{ccc} \mathrm{id}(F) & \xrightarrow{\Gamma_{B,F}} & \Psi(B)F \\ {\scriptstyle \mathrm{id}(f)}\downarrow & & \downarrow{\scriptstyle \Psi(B)f} \\ \mathrm{id}(G) & \xrightarrow[\Gamma_{B,G}]{} & \Psi(B)G \end{array}$$

in a natural way; equivalently, $\Psi : \delta\mathcal{X} \to \delta\mathcal{X}$ can be viewed as a natural transformation-valued functor $\underline{\mathrm{id}} \Rightarrow \Psi$, where $\underline{\mathrm{id}}$ is identically the identity functor on $\delta\mathcal{X}$. In this manner, a morphism $\Lambda_{i,B} \to \Psi(B)$ is regarded as a 2-morphism $\Omega_{i,B} : \Lambda_{i,B} \to \Gamma_B$ in $\delta\mathcal{X}$ with $(\Lambda_{i,B} : A \Rightarrow B) \to (\Gamma_B : \mathrm{id} \Rightarrow \Psi(B))$ as presentation, and the collection of all such maps gives us $\{\mathrm{Nat}(A, B) = \mathbb{R}\underline{\mathrm{Spec}}\,A(B) \to (\mathrm{id} \Rightarrow \Psi(B))\}$. We define $\mathrm{Map}(\mathbb{R}\underline{\mathrm{Spec}}\,A, \Psi)$ pointwise in this manner. Observe that this gives us a notion of an A-multipoint of Ψ.

29.4.4 Generalized categories and generalized functors

One pressing issue at this point is making sense of the collection of morphisms $\mathbb{D}\mathrm{er}_\Psi(X, M)(B) = \mathrm{Map}(\mathbb{R}\underline{\mathrm{Spec}}\,MA(B), \Psi(B))$ mentioned above. We start by defining generalized categories, and then we define the corresponding notion of generalized functors.

In an ordinary category $\mathcal{C}$, an object c of $\mathcal{C}$ is an element of the set $\mathrm{Ob}\,(\mathcal{C})$ of objects of $\mathcal{C}$. We generalize this to considering the case where we have *amalgams* of objects $c = (c_1, \ldots, c_n) \in \mathrm{Ob}\,(\mathcal{C})^n$. We are led to introducing the generalized category $\mathcal{C}^{\mathrm{gen}}$ induced by $\mathcal{C}$ as the graded object $\mathcal{C}^{\mathrm{gen}} = \oplus_{n\geq 1}\mathcal{C}_n^{\mathrm{gen}}$ with $\mathrm{Ob}\,(\mathcal{C}_n^{\mathrm{gen}}) \cong (\mathrm{Ob}\,\mathcal{C})^n$. The appropriate notion of morphism in this case is defined as follows: given objects $c = (c_1, \ldots, c_n) \in \mathcal{C}_n^{\mathrm{gen}}$ and $d = (d_1, \ldots, d_m) \in \mathcal{C}_m^{\mathrm{gen}}$, a morphism f from c to d is given by a matrix of morphisms: $f = (f_{ij} : c_i \to d_j) \in \mathrm{Mor}\,(\mathcal{C})^{n\times m}$. More generally, denoting such a morphism by f_{nm}, a generic morphism in $\mathcal{C}^{\mathrm{gen}} = \oplus\mathcal{C}_n^{\mathrm{gen}}$ is of the form $f = \oplus f_{nm}$. Coming back to the special case where $f = (f_{ij}) : c \to d$ above, to account for the fact that for i and j fixed, there may be several morphisms between c_i and d_j, we denote such morphisms by $f_{ij}^\lambda : c_i \to d_j$ with $\lambda \in A_{ij} := A[c_i, d_j]$ an indexing set for such morphisms, whose elements are referred to as *flavors*. In the present formalism, we regard $f_{ij}^{A_{ij}} = (f_{ij}^{\lambda})_{\lambda \in A_{ij}}$ as a *quiver* [20] between c_i and d_j. Write

$$f^A = (f_{ij}^{A_{ij}}) = ((f_{ij}^{\lambda})_{\lambda \in A_{ij}})_{\substack{1 \le i \le n \\ 1 \le j \le m}} := (f_{ij}^{\lambda}) \in \mathrm{Mor}\,(\mathcal{C}_n^{\mathrm{gen}}, \mathcal{C}_m^{\mathrm{gen}})$$

with $A = \times A_{ij}$, and we refer to f^A as a (higher-dimensional) *cube* in reference to the fact that we have three indices i, j, and λ (and later multiflavors, as we will soon see). By abuse of notation we will just write $f = (f_{ij}^{\lambda})$, A being implied. Immediately, for simplicity, we focus on morphisms that are not made of quivers. For another object $e = (e_1, \dots, e_p)$ and $g : d \to e$ given by $g = (g_{jk} : d_j \to e_k)$, the composition of g and f is defined by $g \circ f = (g_{jk} \circ f_{ij} : c_i \to e_k)$. It is clearly associative, with unit $\mathrm{id}_c = (\mathrm{id}_{c_i})$. It is worth noting that with this definition, $(g \circ f)_{ik} : c_i \to e_k$ has a flavor j, keeping track of what object d_j is $(g \circ f)_{ik}$ going through, and we write $(g \circ f)_{ik}^{j}$ for such a morphism. We now add flavors to the picture: let $g = (g_{jk}^{\mu} : d_j \to e_k)$ and $f = (f_{ij}^{\lambda} : c_i \to d_j)$. The composite $g \circ f$ will now have flavors j and, for fixed j, flavors $\lambda \in A[c_i, d_j]$ and $\mu \in A[d_j, e_k]$.

For fixed j, we therefore have two flavors λ and μ. If one is fixed, then the other one acts as a flavor relative to the first one as follows:

$$(g \circ f)_{ik}^{j,\lambda} = ((g \circ f)_{ik}^{j,\lambda,\mu})_{\mu \in A[d_j, e_k]} := ((g \circ f)_{ik}^{j,\lambda,\mu})$$

so that

$$\begin{aligned} g \circ f &= ((g \circ f)_{ik}) \\ &= (((g \circ f)_{ik}^{j})) \\ &= (\cdots (g \circ f)_{ik}^{j,\lambda} \cdots) \\ &= (\cdots (g \circ f)_{ik}^{j,\lambda,\mu} \cdots) \\ &= ((g \circ f)_{ik}^{j,\lambda,\mu}) \end{aligned}$$

where at each level the new index is a flavor relative to the indices of the one line prior. One could have fixed μ and obtained λ as a flavor relative to μ for the same end result entrywise. The fact that we have towers of flavors justifies calling such objects higher-dimensional cubes. Note however that with this notion of generalized category, one cannot define inverse morphisms without running into pathological behavior. The only morphisms that are really invertible are those "horizontal" morphisms $f = (f_{ii})$ for which f_{ii} is invertible for all i. More generally still, a *generalized category* is a graded object $\mathcal{C} = \oplus \mathcal{C}_n$ with $\mathcal{C}_n \cong \prod_{1 \le i \le n} \mathcal{C}_{(i)}$, the $\mathcal{C}_{(i)}$ being ordinary categories, and morphisms in $\mathcal{C}$ are defined levelwise from objects $c = (c_1, \dots, c_n) \in \mathcal{C}_n$ to $d = (d_1, \dots, d_m) \in \mathcal{C}_m$ by $f = (f_{ij})$ with $f_{ij} \in \mathrm{Hom}(\mathcal{C}_{(i)}, \mathcal{C}_{(j)})$. Finally, note that if $\mathcal{C}$ is a generalized category, so is $\mathrm{Mor}\,\mathcal{C}$, and hence we can apply this formalism to define higher morphisms in a generalized setting.

For completeness, we define the notion of *generalized functors*, which are functors between generalized categories. The simplest case is the following: any ordinary functor $F : \mathcal{C} \to \mathcal{D}$ between ordinary categories gives rise in a natural way to a generalized functor $F^{\text{gen}} : \mathcal{C}^{\text{gen}} \to \mathcal{D}^{\text{gen}}$ defined by $F^{\text{gen}} = \oplus_{n\geq 1} F_n^{\text{gen}}$ with $F_n^{\text{gen}} : \mathcal{C}_n \to \mathcal{D}_n$. By abuse of notation we will just denote F_n^{gen} by F. Then if $c = (c_1, \dots, c_n) \in \mathcal{C}_n^{\text{gen}}$, then $Fc = (Fc_1, \dots, Fc_n)$. If $d = (d_1, \dots, d_m)$ is another object of $\mathcal{C}^{\text{gen}}$ and $f : c \to d$ is a morphism given by $f = (f_{ij} : c_i \to d_j)$, then $Ff = (F(f_{ij}) : Fc_i \to Fd_j)$, that is, if $f \in \text{Hom}(\mathcal{C}_n^{\text{gen}}, \mathcal{C}_m^{\text{gen}})$, then $Ff \in \text{Hom}(\mathcal{D}_n^{\text{gen}}, \mathcal{D}_m^{\text{gen}})$.

If $e = (e_1, \dots, e_p)$ is another object of $\mathcal{C}^{\text{gen}}$ and $g : d \to e$ is a morphism given by $g = (g_{jk} : d_j \to e_k)$, then

$$\begin{aligned} Fg \circ Ff &= (F(g_{jk}) \circ F(f_{ij})) \\ &= (F(g_{jk} \circ f_{ij}) : Fc_i \to Fe_k) \\ &= (F((g \circ f)^j_{ik})) \\ &= F((g \circ f)^j_{ik}) \\ &= F(g \circ f). \end{aligned}$$

We also have

$$F(\text{id}_c) = F((\text{id}_{c_i})) = (F(\text{id}_{c_i})) = (\text{id}_{Fc_i}) = \text{id}_{Fc}.$$

Presently, we have not used the fact that the entries of f and g could have been quivers, but adding in flavors is just a tedious exercise, yielding the same results as above. Functors between generic generalized categories can be defined in an obvious way from the above work.

An equivalent definition of generalized categories consists in using sets to represent amalgams and morphisms: $c = \{c_1, \dots, c_n\}$, and $f : c \to d$ is given by $f = \{f_{ij} : c_i \to d_j\}$. If $c_i \in \text{Ob}(\mathcal{C})$, $\mathcal{C}$ an ordinary category, then c is an object of $\mathcal{C}^{\text{gen}}$, and with $f_{ij} \in \mathcal{C}_1$, we have $f \in \mathcal{C}_1^{\text{gen}}$, but note that in both instances, we are dealing with sets, while in the cubic picture, for $c = (c_1, \dots, c_n)$ and $d = (d_1, \dots, d_m)$, $f = (f_{ij}) \in \mathcal{C}_1^{\text{gen}}$ is an object of $(\mathcal{C}_1)^{n\times m}$, and if we have multiple flavors $\lambda_{k_{p+1}} \in A_{ij,\lambda_{k_1},\dots,\lambda_{k_p}}$, then $f = (f_{ij}^{\lambda_{k_1},\dots,\lambda_{k_r}}) \in (\mathcal{C}_1)^N$, where $N = \sum_{ij} |A_{ij}| \times \cdots \times |A_{ij,\lambda_{k_1},\dots,\lambda_{k_{r-1}}}|$. To make a distinction between concepts appearing in both the Set-based formalism of generalized categories and its cubic counterpart, we will use a superscript σ in the latter case. We have initially given a cubical approach to generalized categories since it is new, and we have given a short Set-based description since it is most natural. In what follows, we will use the Set-based approach for the same reason, and we will later present the cubical approach, which we refer to as the simplicial case for reasons that will then become obvious.

To come back to our problem of formally constructing $\mathbb{D}\text{er}_\Psi(X, M)$, in the Set-based setting, one can write $\Lambda_B^M = \{\Lambda_{i,B}^M : MA \Rightarrow B\}$, i being regarded as a flavor here, thereby casting $\delta\mathcal{X}$ as a generalized Segal category, albeit with just quivers on single objects, and

$\Lambda_B^M = \mathrm{Nat}(MA, B)$ as a generalized morphism in $\delta\mathcal{X}^{\mathrm{gen}}$. Observe that if $\Lambda_{i,B}^M \in \delta\mathcal{X}_{(MA,B)} \in \mathrm{Ho}\,(\mathrm{Set}_\Delta)$, then in a generalized context, $\Lambda_B^M \in (\delta\mathcal{X}^{\mathrm{gen}})_{(MA,B)}$ is an object of Set. We also have $\Gamma_B : \mathrm{id} \Rightarrow \Psi(B)$ as another morphism in $\delta\mathcal{X}$. The various 2-morphisms between Λ_B^M and Γ_B are given entrywise by some $\Omega_{i,B}^{M,\lambda} : \Lambda_{i,B}^M \to \Gamma_B \in \delta\mathcal{X}_{(MA,\mathrm{id},\Psi(B))} \in \mathrm{Ho}\,(\mathrm{Set}_\Delta)$ – as we will show in the next subsection – for a flavor λ in some indexing set L_i. A collection of such maps gives a (quiver) morphism $\Omega_{i,B}^{M,L_i} = \{\Omega_{i,B}^{M,\lambda} \mid \lambda \in L_i\}$, and such morphisms themselves can be collected into a morphism $\Omega_B^{M,L} = \{\Omega_{i,B}^{M,L_i} : \Lambda_{i,B}^M \to \Gamma_B\}$ with an object $L = \times_i L_i$ of $(\delta\mathcal{X}^{\mathrm{gen}})_{(MA,\mathrm{id},\Psi(B))}$. Note that this gives us a generalized morphism $\Omega_B^{M,L} : \Lambda_B^M \to \Gamma_B$ if each index i is accounted for in such a collection. Henceforth we will only consider such *"full"* collections. Using the fact that

$$\Lambda_B^M = \{\Lambda_{i,B}^M : MA \to B\} = \mathrm{Map}(MA, B) = \mathbb{R}\underline{\mathrm{Spec}}\,(MA)(B)$$

it follows that one can define

$$\begin{aligned}\Omega_B^M &= \{\Omega_B^{M,L} : \Lambda_B^M \to \Gamma_B\} \\ &= \mathrm{Map}(\mathbb{R}\underline{\mathrm{Spec}}\,(MA)(B), \Psi(B)) \\ &= \mathbb{D}\mathrm{er}_\Psi(X, M)(B).\end{aligned}$$

29.4.5 Cotangent set

Following the above philosophy of regarding $\delta A = MA$ as a small deformation of $A \in \delta\mathcal{X}$, one can naively expect the analog in our situation of the cotangent complex of Ψ at A to be an object of the form $\mathrm{Map}(\Psi A, \Psi(MA))$. This will help us see that the abstract definition of the cotangent object $\mathbb{L}_{\Psi,X}$ as we will define it does what we want it to do. For now, observe that $\Psi : \delta\mathcal{X} \to \delta\mathcal{X}$ is a functor on $\delta\mathcal{X}$, whose objects are deformations, which are really functors. Being functors from $\mathcal{X}$ to $\mathcal{X}$, they can be composed. If A and B are two such functors, then one can define their composite BA. For consistency, we ask that Ψ be a morphism on $(\delta\mathcal{X}, \circ, \mathrm{id})$, so that $\Psi(BA) = \Psi(B) \circ \Psi(A)$. Thus we are looking at $\mathrm{Nat}(\Psi A, \Psi M \Psi A)$. We regard ΨM as providing an infinitesimal deformation of ΨA in the M direction.

As a first step towards defining a cotangent object, we now justify $\Omega_{i,B}^{M,\lambda} : \Lambda_{i,B}^M \to \Gamma_B$ being valued in $\delta\mathcal{X}_{(MA,\mathrm{id},\Psi(B))} \in \mathrm{Ho}\,(\mathrm{Set}_\Delta)$, as initially mentioned in the preceding section. To put things in perspective, recall that $\mathbb{D}\mathrm{er}_\Psi(X, M)(B) = \mathrm{Map}(\mathbb{R}\underline{\mathrm{Spec}}\,(MA)(B), \Psi(B))$ is a set of full morphisms $\Omega_B^{M,L}$ of the form $\{\Omega_{i,B}^{M,L_i} : \Lambda_{i,B}^M \to \Gamma_B\}$ with $L = \times_i L_i$, and where $\Lambda_{i,B}^M : MA \Rightarrow B$ and $\Gamma_B : \mathrm{id} \Rightarrow \Psi(B)$. The quivers $\Omega_{i,B}^{M,L_i}$ are collections of maps $\Omega_{i,B}^{M,\lambda}$ between natural transformations, given by commutative diagrams

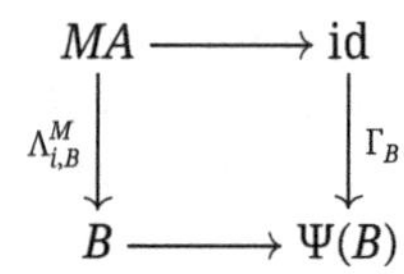

which because of the commutativity can be represented as

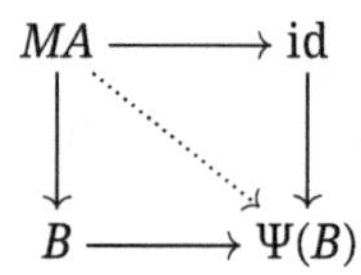

that is, $\Omega^{M,\lambda}_{i,B} : MA \to \mathrm{id} \to \Psi(B)$ or, equivalently, $\Omega^{M,\lambda}_{i,B} : MA \to B \to \Psi(B)$, and thus $\Omega^{M,\lambda}_{i,B} \in \delta\mathcal{X}_{(MA,\mathrm{id},\Psi B)}$ or $\Omega^{M,\lambda}_{i,B} \in \delta\mathcal{X}_{(MA,B,\Psi B)}$, depending on which factorization one is looking at. Since $\delta\mathcal{X}$ is a Segal category, we therefore have $\Omega^{M,\lambda}_{i,B} \in \mathrm{Ho}\,(\mathrm{Set}_\Delta)$. Note however that $\Omega^{M,L_i}_{B} = \{\Omega^{M,\lambda}_{i,B} : \Lambda^M_{i,B} \to \Gamma_B \mid \lambda \in L_i\} \in (\delta\mathcal{X})^{\mathrm{gen}}_{(MA,\mathrm{id},\Psi(B))}$ is already a set.

It follows from the above analysis that each of the maps $\Omega^{M,\lambda}_{i,B}$ making up $\mathrm{Map}(\underline{\mathbb{R}\mathrm{Spec}}\,(MA)(B), \Psi(B))$ corresponds to an inner commutative square in the diagram below for a given deformation $\delta : A \to MA$:

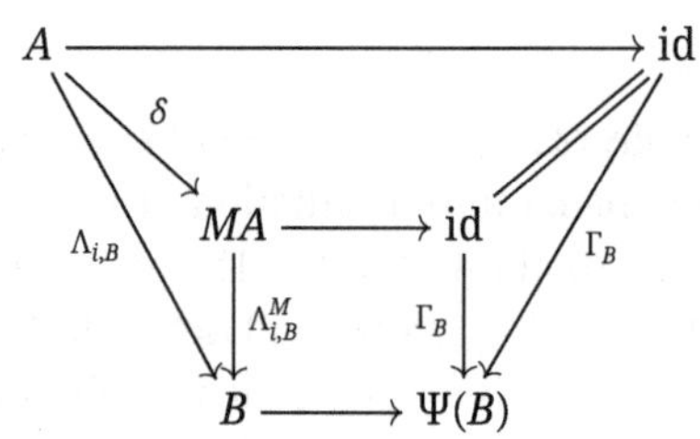

thereby presenting the natural transformations $\Lambda^M_{i,B}$ as a factorization of $\Lambda_{i,B}$. By the functoriality of Ψ one has

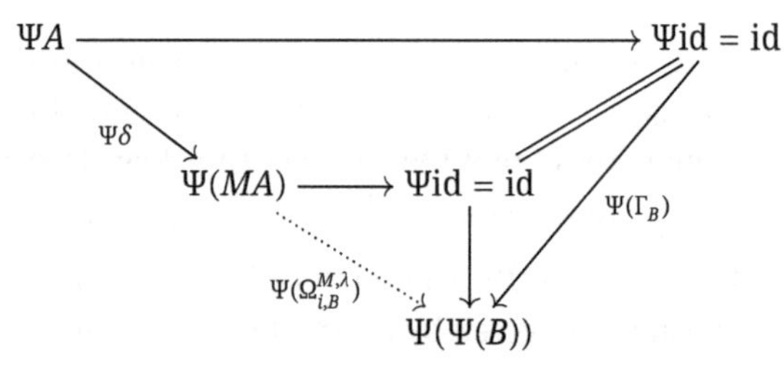

where we used $\Omega^{M,\lambda}_{i,B} : MA \to \mathrm{id} \to \Psi(B)$ from the previous paragraph, and $\Psi(\mathrm{id}) = \mathrm{id}$ since Ψ is a morphism on $(\delta\mathcal{X}, \circ, \mathrm{id})$. This is best presented in the form

$$\begin{array}{ccc} \Psi A & \longrightarrow & \mathrm{id} \\ {\scriptstyle \Psi\delta}\big\downarrow & & \big\downarrow{\scriptstyle \Psi(\Gamma_B)} \\ \Psi(MA) & \xrightarrow[\Psi(\Omega_{i,b}^{M,\lambda})]{} & \Psi^2(B) \end{array}$$

which is a map $(\Psi A \Rightarrow \Psi(MA)) \to (\mathrm{id} \Rightarrow \Psi^2(B))$, the first part of which would be a contribution to our naive definition of a cotangent set, and one would hope the converse holds, that is, from the above diagram one retrieves the maps $\Omega_{i,B}^{M,\lambda}$ and hence $\mathbb{D}\mathrm{er}_\Psi(X, M)(B)$. Precisely, because that is not necessarily the case however, we do not yet have an isomorphism of the form $\mathbb{D}\mathrm{er}_F(X, M) \cong \mathbb{R}\underline{h}_s^{\mathbb{L}_{F,x}}$ as in [79]. To circumvent this difficulty, we define the *cotangent set* of Ψ at $x : X \equiv \mathbb{R}\underline{\mathrm{Spec}}\, A \to \Psi$ in the direction of M as the object $\mathbb{L}_{\Psi,x}M$ such that

$$\mathbb{D}\mathrm{er}_\Psi(X, M)(B) \cong \mathrm{Map}(\mathbb{L}_{\Psi,x}M, \{\mathrm{id} \Rightarrow \Psi^2(B)\}) \tag{29.9}$$

an isomorphism of sets, or $\mathbb{D}\mathrm{er}_\Psi(X, M) \cong \mathrm{Map}(\mathbb{L}_{\Psi,x}M, \Psi^2)$, an isomorphism of functors $\delta\mathcal{X} \to \mathrm{Set}$, defined pointwise as in (29.9). This being defined, the *cotanget set functor* of Ψ at x is given by

$$\begin{aligned} \mathbb{L}_{\Psi,x} : \delta\mathcal{X} &\to \mathrm{Set}, \\ M &\mapsto \mathbb{L}_{\Psi,x}M. \end{aligned}$$

We define the *tangent set functor* of Ψ at x by $\mathbb{T}_{\Psi,x} = \mathrm{Map}(\mathbb{L}_{\Psi,x}, A)$, defined pointwise by $\mathbb{T}_{\Psi,x}M = \mathrm{Map}(\mathbb{L}_{\Psi,x}M, \{A = A\}) \in \mathrm{Set}$. The latter expression gives the *tangent set* of Ψ at x in the direction M. This being defined, we finally have $\mathrm{Cusp}[\Psi, \epsilon] = \mathbb{T}_{\Psi,\epsilon}$, where $\epsilon = A$, and we abused notation by writing $\mathbb{T}_{\Psi,\epsilon}$ for $\mathbb{T}_{\Psi,x}$, where $x : \mathbb{R}\underline{\mathrm{Spec}}\,(\epsilon) \to \Psi$.

29.4.6 Alternate simplicial formalism

We defined $\mathbb{D}\mathrm{er}_\Psi(X, M) = \mathrm{Map}(\mathbb{R}\underline{\mathrm{Spec}}\, MA, \Psi)$, a Set-valued functor, because we considered collections of maps $\Omega_B^{M,L} : \Lambda_B^M \to \Psi(B)$ for all $B \in \delta\mathcal{X}$. We now consider the cubical coverage of this formalism. Starting with the natural transformations $\Lambda_{i,B}^M : MA \Rightarrow B$, we consider the maximal cubic quiver $\Lambda_B^{\sigma,M} = (\Lambda_{i,B}^M)$. Define

$$\mathbb{R}\underline{\mathrm{Spec}}^{\,\sigma} MA(B) = \mathrm{maxMap}^{\mathrm{gen}}(MA, B)$$

where $\mathrm{maxMap}^{\mathrm{gen}}(MA, B)$ indicates that we only consider the maximal quiver from MA to B, and gen is in reference to the fact that we work with generalized categories. With this definition, it then follows that

$$\mathbb{R}\underline{\mathrm{Spec}}^{\,\sigma} MA(B) = \Lambda_B^{\sigma,M}.$$

If now instead of considering select few morphisms $\Lambda_{i,B}^{M} \to \Gamma_B$ to populate a quiver, we take a maximal quiver $\Omega_{i,B}^{\sigma,M} = (\Lambda_{i,B}^{M} \to \Gamma_B)$ for each i, and if further we consider the full cube of all such maximal quivers, then we get an object $\Omega_B^{\sigma,M} \in (\delta\mathcal{X})^{\mathrm{gen}}_{(MA,\mathrm{id},\Psi(B))} \in \mathrm{Ho}\,(\mathrm{Set}_\Delta)^{\mathrm{gen}}$. We can then define

$$\mathbb{D}\mathrm{er}_\Psi^\sigma(X,M)(B) = \mathrm{maxMap}^{\mathrm{gen}}(\mathbb{R}\underline{\mathrm{Spec}}^{\sigma} MA(B), \Gamma_B) = \Omega_B^{\sigma,M}$$

to be the maximal generalized morphism from $\mathbb{R}\underline{\mathrm{Spec}}^{\sigma} MA(B)$ to Γ_B, an object of $(\delta\mathcal{X})^{\mathrm{gen}}_{(MA,\mathrm{id},\Psi(B))} \in \mathrm{Ho}\,(\mathrm{Set}_\Delta)^{\mathrm{gen}}$. Alternatively, note that for categories $\mathcal{C}$ and $\mathcal{D}$, $\mathrm{Fun}(\mathcal{C}, \mathcal{D}^{\mathrm{gen}}) \subset \mathrm{Fun}(\mathcal{C},\mathcal{D})^{\mathrm{gen}}$, that is,

$$\mathbb{R}\underline{\mathrm{Hom}}(\delta\mathcal{X}, (\delta\mathcal{X})^{\mathrm{gen}}) \subset \mathbb{R}\underline{\mathrm{Hom}}(\delta\mathcal{X}, \delta\mathcal{X})^{\mathrm{gen}} = \delta^2\mathcal{X}^{\mathrm{gen}}. \tag{29.10}$$

This implies $\mathbb{R}\underline{\mathrm{Spec}}^{\sigma}(MA) \in \delta^2\mathcal{X}^{\mathrm{gen}}$ and $\Psi \in \delta^2\mathcal{X}^{\mathrm{gen}}$, from which it follows that $\mathbb{D}\mathrm{er}_\Psi^\sigma(X,M) = \mathrm{maxMap}^{\mathrm{gen}}(\mathbb{R}\underline{\mathrm{Spec}}^{\sigma}(MA), \Psi) \in (\delta^2\mathcal{X})^{\mathrm{gen}}_{(\mathbb{R}\underline{\mathrm{Spec}}^{\sigma}(MA),\Psi)} \in \mathrm{Ho}\,(\mathrm{Set}_\Delta)^{\mathrm{gen}}$.

We define the *cotangent simplex* of Ψ at $x : X \equiv \mathbb{R}\underline{\mathrm{Spec}}^{\sigma} A \to \Psi$ in the direction of M as the object $\mathbb{L}_{\Psi,x}^\sigma M$ such that

$$\mathbb{D}\mathrm{er}_\Psi^\sigma(X,M)(B) \cong \mathrm{maxMap}^{\mathrm{gen}}(\mathbb{L}_{\Psi,x}^\sigma M, (\mathrm{id} \Rightarrow \Psi^2(B))) \tag{29.11}$$

an isomorphism in $\mathrm{Ho}\,(\mathrm{Set}_\Delta)^{\mathrm{gen}}$, or $\mathbb{D}\mathrm{er}_\Psi^\sigma(X,M) \cong \mathrm{maxMap}^{\mathrm{gen}}(\mathbb{L}_{\Psi,x}^\sigma M, \Psi^2)$, an isomorphism of functors $\delta\mathcal{X} \to \mathrm{Ho}\,(\mathrm{Set}_\Delta)^{\mathrm{gen}}$ defined pointwise as in (29.11). To be precise, by the very definition of $\mathbb{L}_{\Psi,x}^\sigma$ we have $\mathbb{L}_{\Psi,x}^\sigma M \in (\delta\mathcal{X})_1^{\mathrm{gen}}$, so that $\mathrm{maxMap}^{\mathrm{gen}}(\mathbb{L}_{\Psi,x}^\sigma M, (\mathrm{id} \Rightarrow \Psi^2(B))) \in (\delta\mathcal{X})_2^{\mathrm{gen}}$, and hence $\mathrm{maxMap}^{\mathrm{gen}}(\mathbb{L}_{\Psi,x}^\sigma M, \Psi^2) \in \delta^2\mathcal{X}^{\mathrm{gen}}$ by (29.10), so that the isomorphism $\mathbb{D}\mathrm{er}_\Psi^\sigma(X,M) \cong \mathrm{maxMap}^{\mathrm{gen}}(\mathbb{L}_{\Psi,x}^\sigma M, \Psi^2)$ takes place in $\delta^2\mathcal{X}^{\mathrm{gen}}$.

Finally, we define the *tangent simplex* of Ψ at x by

$$\mathbb{T}_{\Psi,x}^\sigma = \mathrm{maxMap}^{\mathrm{gen}}(\mathbb{L}_{\Psi,x}^\sigma, A)$$

defined pointwise by $\mathbb{T}_{\Psi,x}^\sigma M = \mathrm{maxMap}^{\mathrm{gen}}(\mathbb{L}_{\Psi,x}^\sigma M, (A = A)) \in \mathrm{Ho}\,(\mathrm{Set}_\Delta)^{\mathrm{gen}}$. The latter expression gives the tangent simplex of Ψ at x in the direction M. This being defined, we finally have $\mathrm{Cusp}^\sigma[\Psi, \epsilon] = \mathbb{T}_{\Psi,\epsilon}^\sigma$, where $\epsilon = A$, and we abused notation by writing $\mathbb{T}_{\Psi,\epsilon}^\sigma$ for $\mathbb{T}_{\Psi,x}^\sigma$ with $x : \mathbb{R}\underline{\mathrm{Spec}}^{\sigma}(\epsilon) \to \Psi$.

29.4.7 Connections between cubic and Set-based formalisms

We can recover $\mathbb{D}\mathrm{er}_\Psi(X,M)(B)$ for all $B \in \delta\mathcal{X}$ from its cubical counterpart by a truncation process: recall that if L_i is an indexing set of flavors of maps $\Omega_{i,B}^{M,\lambda} : \Lambda_{i,B}^M \to \Gamma_B$, then one can define $\Omega_{i,B}^{\sigma,M,L_i} = (\Omega_{i,B}^{M,\lambda})_{\lambda \in L_i}$. We have an obvious projection

$$\pi_{L_i} : \Omega_{i,B}^{\sigma,M} = (\Omega_{i,B}^{M,\lambda})_{\forall\lambda} \mapsto (\Omega_{i,B}^{M,\lambda})_{\lambda\in L_i} = \Omega_{i,B}^{\sigma,M,L_i}$$

which we can generalize to

$$\pi_L \Omega_B^{\sigma,M} = (\pi_{L_i} \Omega_{i,B}^{\sigma,M}) = (\Omega_{i,B}^{\sigma,M,L_i}) = \Omega_B^{\sigma,M,L}$$

as well as an iterative *deconstruction map* simply defined by

$$\begin{aligned} \dagger(\Omega_{i,B}^{\sigma,M,L_i}) &= \{\dagger\Omega_{i,B}^{\sigma,M,L_i}\} \\ &= \{\{\Omega_{i,B}^{M,\lambda} \mid \lambda \in L_i\}\} \\ &= \{\Omega_{i,B}^{M,L_i}\} \\ &= \Omega_B^{M,L} \end{aligned}$$

from which we have

$$\begin{aligned} \dagger\pi\mathbb{D}\mathrm{er}_\Psi^\sigma(X,M)(B) &= \dagger\pi\Omega_B^{\sigma,M} \\ &:= \dagger\{\pi_L \Omega_L^{\sigma,M}\}_L \\ &= \{\dagger(\Omega_{i,B}^{\sigma,M,L_i})\} \\ &= \{\Omega_B^{M,L}\} \\ &= \mathbb{D}\mathrm{er}_\Psi(X,M)(B). \end{aligned}$$

Vice versa, by definition $\mathbb{D}\mathrm{er}_\Psi^\sigma(X,M)$ is constructed from $\mathbb{D}\mathrm{er}_\Psi(X,M)$.

Using this truncation process, we now show that those two notions of cusps we have defined, one being Set-based, and the other one being simplicial, carry the same information. The same composite $\dagger\pi$ we used above also allows one to go from $\mathrm{maxMap}^{\mathrm{gen}}(\mathbb{L}_{\Psi,x}^\sigma M, (\mathrm{id} \Rightarrow \Psi^2(B)))$ to $\mathrm{Map}(\mathbb{L}_{\Psi,x} M, (\mathrm{id} \Rightarrow \Psi^2(B)))$:

$$\begin{array}{ccc} \mathbb{D}\mathrm{er}_\Psi^\sigma(X,M)(B) & \xrightarrow{\cong} & \mathrm{maxMap}^{\mathrm{gen}}(\mathbb{L}_{\Psi,x}^\sigma M, (\mathrm{id} \Rightarrow \Psi^2(B))) \\ \downarrow{\scriptstyle \dagger\pi} & & \vdots\,{\scriptstyle \dagger\pi} \\ \mathbb{D}\mathrm{er}_\Psi(X,M)(B) & \xrightarrow[\cong]{} & \mathrm{Map}(\mathbb{L}_{\Psi,x} M, \{\mathrm{id} \Rightarrow \Psi^2(B)\}) \end{array}$$

from which it follows that

$$\dagger\pi\mathbb{L}_{\Psi,x}^\sigma M = \mathbb{L}_{\Psi,x} M.$$

Now

$$\begin{aligned} \dagger\pi\mathbb{T}_{\Psi,x}^\sigma M &= \dagger\pi\mathrm{maxMap}^{\mathrm{gen}}(\mathbb{L}_{\Psi,x}^\sigma M, (A = A)) \\ &= \mathrm{Map}(\dagger\pi\mathbb{L}_{\Psi,x}^\sigma M, \{A = A\}) \end{aligned}$$

$$= \mathrm{Map}(\mathbb{L}_{\Psi,x}M, \{A = A\})$$
$$= \mathbb{T}_{\Psi,x}M.$$

29.5 Strings

Recall that objects of $\mathrm{d}\mathcal{A}\mathrm{ff}_{\mathcal{C}}^{\sim,\tau}$ are stacks, which are modeled by functors $F : \mathrm{Comm}(\mathcal{C}) \to \mathrm{Set}_\Delta$. For $A \in \mathrm{Comm}(\mathcal{C})$, $F(A)$ can be viewed as the manifestation, or realization, of A via F. All those manifestations $F(A)$ of a phenomenon F come together coherently by functoriality of F. In other terms, such a functor F should therefore correspond to a physical system of some sort, since such systems vary in accordance to some laws, and this in a coherent fashion. Vice versa, typical solutions to physical equations reflect certain laws, so assuming that they depend functorially on them, if D is a differential operator and Ψ is a solution to the equation $D\Psi = 0$, then one can identify $\Psi \equiv F : \mathrm{Comm}(\mathcal{C}) \to \mathrm{Set}_\Delta$.

Stacks are really objects of $\mathbb{R}\underline{\mathrm{Hom}}(\mathrm{d}\mathcal{A}\mathrm{ff}_{\mathcal{C}}^{\mathrm{op}}, \mathrm{Top})$ with $\mathrm{d}\mathcal{A}\mathrm{ff}_{\mathcal{C}} = L(\mathrm{Comm}(\mathcal{C})^{\mathrm{op}})$, a Segal category, where L denotes the Dwyer–Kan simplicial localization. Recall how this is performed [15]: if $\mathcal{C}$ is an ordinary category, then $L\mathcal{C} = F_*\mathcal{C}[F_*W^{-1}]$, where $F_k\mathcal{C} = F^{k+1}\mathcal{C}$, and $F\mathcal{C}$ is the free category on $\mathcal{C}$ with generators Fc for each nonidentity morphism c of $\mathcal{C}$. Thus after a simplicial localization, equivalences have been sequentially contracted.

This will prove essential in our applications to Physical Mathematics, in which we regard strings as modeling equivalences in $\mathrm{Comm}(\mathcal{C})$, which are therefore no longer apparent in $\mathrm{d}\mathcal{A}\mathrm{ff}_{\mathcal{C}}$, and since we will regard $\mathrm{d}\mathcal{A}\mathrm{ff}_{\mathcal{C}}^{\sim,\tau}$, as done already in [23, 25, 26], to model physical phenomena, this provides a reason for not being able to observe strings in nature.

Some physicists will argue that strings are fundamental objects. Already in Homotopy Type theory, we see that this is not far-fetched. Indeed, one can regard sets as discrete homotopy types, thereby presenting homotopy types as fundamental objects. Those provide identification of elements through paths, which we regard as strings, in contrast to working with point-based Physics.

Practically, for us, working with strings corresponds to working with model categories, and after localization, we are left with point-based Physics, which corresponds to the observable universe modeled by stacks $F : \mathrm{d}\mathcal{A}\mathrm{ff}_{\mathcal{C}}^{\mathrm{op}} \to \mathrm{Top}$.

To be precise, if $\mathcal{C}$ is a simplicial, symmetric monoidal model category, assuming that we have a model category structure on $\mathrm{Comm}(\mathcal{C})$ as well in accordance with the assumptions put on a homotopical algebraic context, equivalences in $\mathrm{Comm}(\mathcal{C})$ are defined on underlying simplicial objects, so $A \simeq A'$ in $\mathrm{Comm}(\mathcal{C})$ means $\underline{A} \simeq \underline{A'}$ in Set_Δ. Those objects are therefore simplicially isomorphic in $\mathrm{d}\mathcal{A}\mathrm{ff}_{\mathcal{C}} = L(\mathrm{Comm}(\mathcal{C})^{\mathrm{op}})$, which means $\underline{A}_0 \cong \underline{A'}_0$, so we have an isomorphism between the objects of A and A'. Define $a \in A_0$ and $a' \in A'_0$ in $\mathrm{Comm}(\mathcal{C})$ to be equivalent: $a \sim a'$ if they are identified under the above isomorphism. We regard such an equivalence as a *string* between a and a', and we will sometimes write $a \overset{\gamma}{\sim} a'$. One important note at this point: those paths are formal, and there is no parameterization σ along such a string, the way there would be in string theory. Also, there is no concept of propagation of such a string and hence no time parameterization τ. In other terms, we have an algebraicization of physical string theory.

There are open and closed strings. Algebraically, an open string is given by $a \sim b$ with $a \neq b$.

A closed string is given by $a \sim a$.

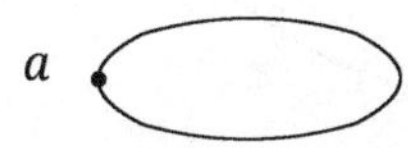

In particular, every object of $A \in \mathrm{Comm}(\mathcal{C})$ gives rise to its own closed string in a trivial way. Strings can split and join. For open strings, a join reads $a \sim b \wedge b \sim c \Rightarrow a \sim c$,

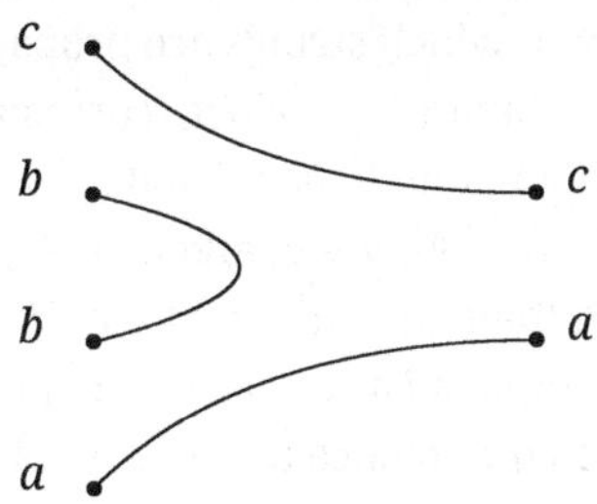

and a splitting reads $a \sim c \Rightarrow a \sim b \wedge b \sim c$ whenever $a \sim b$ or $b \sim c$:

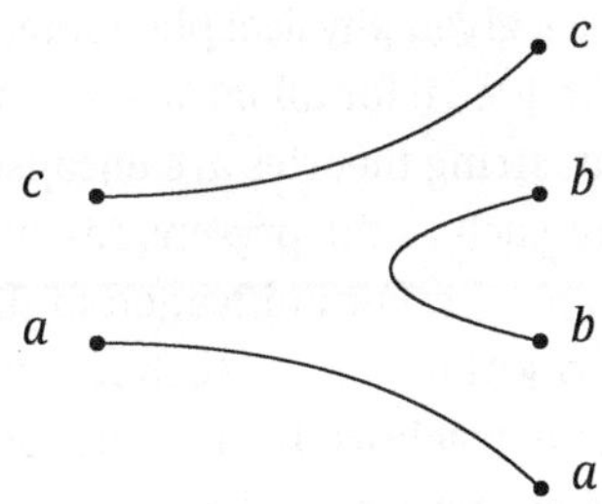

For closed strings, this reads $(a \sim a) \wedge (a \sim a) \Rightarrow a \sim a$, a fold map

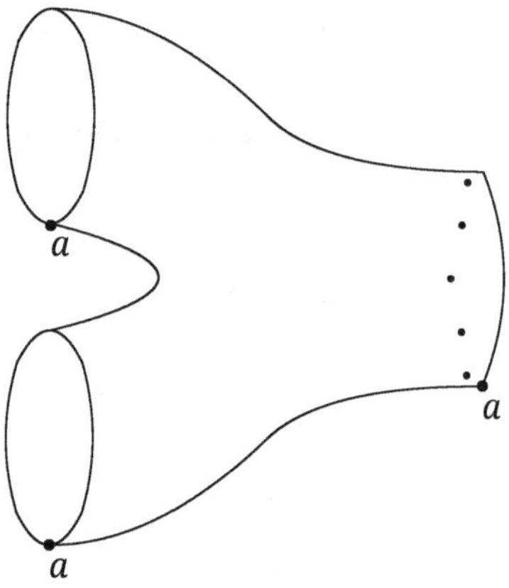

and $a \sim a \Rightarrow (a \sim a) \wedge (a \sim a)$, a diagonal map of sorts, with the reverse diagram. What is referred to as the world sheet in Physics for us corresponds to working in $\mathrm{Comm}(\mathcal{C})$. A string in physical space-time is given by $X^{\mu}(\sigma, \tau)$, where μ is an index that keeps track of the dimension of space-time in which strings are propagating [32]. We argued that a stack modeled by a functor $F : \mathrm{Comm}(\mathcal{C}) \to \mathrm{Set}_{\Delta}$ corresponds to a physical field that satisfies some equations of motion, and hence F and X^{μ} are indistinguishable. If A and A' are two objects of $\mathrm{Comm}(\mathcal{C})$, $A \simeq A'$, $a \in A$ and $a' \in A'$, identified after localization, then we have a string between them: $a \sim a'$, and therefore a map between $Fa \in (FA)_0$ and $Fa' \in (FA')_0$. But $a \cong a'$ implies $Fa \cong Fa'$, and hence $Fa \sim Fa'$ if we are willing to extend to Set_{Δ} that notion of equivalence at the level of objects we had in $\mathrm{Comm}(\mathcal{C})$. Then $Fa \overset{F\gamma}{\sim} Fa'$ plays the role of $X^{\mu}(\sigma, \tau)$, a string in Set_{Δ}.

Now one has to be mindful of the fact that $F\gamma$ is a string, where $F \in \mathrm{d}\mathcal{A}\mathrm{ff}_{\mathcal{C}}^{\sim,\tau}$, for γ in $A \in \mathrm{Comm}(\mathcal{C})$. F corresponds to a given physical phenomenon and hence contains in its image all possible strings $F\gamma$ for $\gamma \in A$, for all laws $A \in \mathrm{Comm}(\mathcal{C})$. It would be interesting to investigate how different string theories are encapsulated by different laws and whether a functorial treatment such as the present one would shed light on M-theory, since a stack F corresponds to a coherent unification of different string theories from that perspective. Indeed, one way of looking at such functors F is that they provide coherent manifestations of various algebras. If those algebras are regarded as physical laws, then one should regard F as one aspect of a given string theory.

If however we focus on those algebras, each of which corresponds to a given string theory in the categorical sense, then F provides a unification of different string theories in the physical sense.

In contrast to string theory, which has its own equations of motion, F for us is just static and tells us how a given phenomenon depends on physical laws $A \in \mathrm{Comm}(\mathcal{C})$. Thus $\mathcal{X} = \mathrm{d}\mathcal{A}\mathrm{ff}_{\mathcal{C}}^{\sim,\tau}$ offers a snapshot of a system at any moment. Dynamics, as argued earlier, corresponds to flows in $\mathbb{R}\underline{\mathrm{Hom}}(\mathcal{X}, \mathcal{X})$. For simplicity, consider a flow built from a deformation ϵ of our system, an object of $\delta\mathcal{X}$, and some $\Psi \in \delta^2\mathcal{X}$. $\Psi\epsilon$ is just another deformation of our system, and it is only $\Psi^\infty\epsilon$ that offers a flow of our system. In other terms, the dynamics of F in the present setting, and consequently that of the strings it is depicting, is given by $\Psi^\infty\epsilon(F)$, which is formally defined by the following diagram, where we let $\epsilon = \epsilon_0$ and $\Psi\epsilon_i = \epsilon_{i+1}$:

$$\begin{array}{ccccccc} F & \xrightarrow{\epsilon_0} & \epsilon_0 F & \overset{\Psi}{\longmapsto} & \epsilon_1 F & \overset{\Psi}{\longmapsto} & \cdots \\ & & \nwarrow{\scriptstyle \pi_0} & & \uparrow{\scriptstyle \pi_1} & \nearrow & \\ & & & & \Psi^\infty\epsilon F. & & \end{array}$$

To come back to strings, suppose $A \simeq A'$ in $\mathrm{Comm}(\mathcal{C})$, so that $a \in A$ and $a' \in A'$, being identified under the simplicial isomorphism $A \cong A'$ in $L\mathrm{Comm}(\mathcal{C})$, are equivalent: $a \overset{\gamma}{\sim} a'$. If $F \in \mathrm{d}\mathcal{A}\mathrm{ff}_{\mathcal{C}}^{\sim,\tau}$, then we therefore have $Fa \overset{F\gamma}{\sim} Fa'$ in Set_Δ. Now let us consider a deformation $\epsilon = \zeta$ of $\mathbb{R}\underline{\mathrm{Hom}}(\mathcal{X}, \mathcal{X})$. In the same manner that a model for F is given by a functor $F : \mathrm{Comm}(\mathcal{C}) \to \mathrm{Set}_\Delta$, a model for ζ is a functor $\zeta : \mathrm{Mor}\,(\mathrm{Comm}(\mathcal{C}), \mathrm{Set}_\Delta) \to \mathrm{Mor}\,(\mathrm{Comm}(\mathcal{C}), \mathrm{Set}_\Delta)$ that maps F to ζF and hence $F\gamma$ to $\zeta F\gamma$. Thus a deformation of our system does move strings around. Now for $\Psi \in \delta^2\mathcal{X}$, $\Psi\zeta$ is some other deformation ϕ. Concretely, we have the following picture:

$$\begin{array}{ccccc} F\gamma : Fa \sim Fa' & \overset{\zeta}{\dashrightarrow} & \zeta Fa \sim \zeta Fa' & \overset{\phi}{\dashrightarrow} & \phi\zeta Fa \sim \phi\zeta Fa' \\ {\scriptstyle L}\downarrow & & \uparrow{\scriptstyle L^{-1}} & & \uparrow{\scriptstyle L^{-1}} \\ Fa \cong Fa' & \xrightarrow[\zeta]{} & \zeta Fa \cong \zeta Fa' & \xrightarrow[\phi]{} & \phi\zeta Fa \cong \phi\zeta Fa'. \end{array}$$

Regarding Ψ, the fashion in which it picks the deformation ϕ from a given deformation ζ of the system really depends on how it interprets natural laws, which are therefore inherently built into Ψ. Another interesting question is what mathematical formalism would allow us to make sense of such a concept.

Bibliography

[1] J. Adamek and J. Rosicky, *Locally Presentable and Accessible Categories*, London Math. Soc. Lecture Note Series, **189**, Cambridge University Press, 1994.
[2] G. Arone and M. Ching, *Goodwillie Calculus*, arXiv:1902.00803 [math.AT].
[3] M. Artin, *Versal Deformations and Algebraic Stacks*, Invent. Math. **27** (1974), 165–189.
[4] I. Bah, D. Freed, G. Moore, N. Nekrasov, S. Razamat, and S. Schafer-Nameki, *Snowmass Whitepaper: Physical Mathematics 2021*, arXiv:2203.05078 [hep-th].
[5] A. Balasubramanian and J. Teschner, *Supersymmetric Field Theories and Geometric Langlands: the Other Side of the Coin*, arXiv:1702.06499 [hep-th].
[6] C. Barwick, *On Left and Right Model Categories and Left and Right Bousfield Localizations*, Homol. Homotopy Appl. **12** (2010), no. 2, 245–320.
[7] O. Caramello, *Theories, Sites, Toposes*, Oxford University Press, 2018.
[8] C. Carmeli, L. Caston and R. Fioresi, *Mathematical Foundations of Supersymmetry*, EMS Series of Lectures in Mathematics, EMS, Zurich, 2011.
[9] D. C. Cisinski and F. Deglise, *Triangulated Categories of Mixed Motives*, Springer Monographs in Mathematics, Springer, 2019.
[10] B. Conrad and A. J. de Jong, *Approximation of Versal Deformations*, J. Algebra **225** (2002), no. 2, 489–515.
[11] J. M. Cordier, *Sur la Notion de Diagramme Homotopiquement Coherent*, Cah. Topol. Géom. Différ. Catég. **23** (1982), no. 1, 93–112.
[12] P. Deligne, *Le Groupe Fondamental de la Droite Projective moins Trois Points, Galois Groups over* $\mathbb{Q}$, Math. Sci. Res. Inst. Publ., **16**, Springer-Verlag, New York, 1989.
[13] P. Deligne, *Categories Tannakiennes, Grothendieck Festschrift Vol. II*, Progress in Math., **87**, Birkhauser, Boston, 1990.
[14] E. J. Dubuc and C. S. de la Vega, *On the Galois Theory of Grothendieck*, arxiv:math/0009145v1 [math.CT].
[15] W. G. Dwyer and D. M. Kan, *Simplicial Localizations of Categories*, J. Pure Appl. Algebra, **17** (1980), 267–284.
[16] W. G. Dwyer and D. M. Kan, *Calculating Simplicial Localizations*, J. Pure Appl. Algebra **18** (1980), 17–35.
[17] W. G. Dwyer and D. M. Kan, *Function Complexes in Homotopical Algebra*, Topology **19** (1980), 427–440.
[18] C. Elliott and P. Yoo, *Geometric Langlands Twists of* $N = 4$ *Gauge Theory From Derived Algebraic Geometry*, arXiv:1507.03048v3 [math-ph].
[19] A. D. Elmendorf, I. Kerz, M. A. Mandell, and J. P. May, *Rings, Modules and Algebras in Stable Homotopy Theory*, Mathematical Surveys and Monographs, **47**, AMS, Providence, RI, 1997.
[20] P. Gabriel, *Unzerlegbare Darstellunger. I*, Manuscr. Math. **6** (1972), 71–103.
[21] P. Gabriel and M. Zisman, *Calculus of Fractions and Homotopy Theory*, Ergebn. der Math., **35**, Springer-Verlag, Berlin, 1967.
[22] M. Gallauer, *Une Introduction to Six-Functor Formalisms*, arXiv:2112.10456 [math.AG].
[23] R. Gauthier, *Higher Galois for Segal Topos and Natural Phenomena*, arXiv:1607.00940 [math.AG].
[24] R. Gauthier, ∞*-Links,* ∞*-4-Manifolds and Kirby Categories*, arXiv:1309.7514 [math.GT].
[25] R. Gauthier, ∞*-Topoi and Natural Phenomena: Generation*, arXiv:1802.02284 [math.AG].
[26] R. Gauthier, *Segal Topoi and Natural Phenomena: Universality of Physical Laws*, arXiv:1803.09611 [math.CT].
[27] M. Ghazel and F. Kadhi, *Reedy Diagrams in Symmetric Monoidal Model Categories*, arXiv:1609.09623v1 [math.AT].
[28] P. G. Goerss and J. F. Jardine, *Simplicial Homotopy Theory*, Modern Birkhauser Classics, Birkhauser Verlag AG, 2009.
[29] Goodwillie, *The Differential Calculus of Homotopy Functors*, Proceeding of the International Congress of Mathematicians, Vol. I, II (Kyoto, 1990), 621–630, Math. Soc. Japan, Tokyo, 1991.
[30] Goodwillie, *Calc II: Analytic Functors*, K-Theory **5** (1991/1992), no. 4, 295–332.

https://doi.org/10.1515/9783111334073-030

[31] Goodwillie, *Calc III: Taylor Series*, Geom. Topol. **7** (2003), 645–711.
[32] M. B. Green, J. H. Schwarz and E. Witten, *Superstring Theory, Vol. 1*, Cambridge Monographs on Mathematical Physics, Cambridge University Press, 1987.
[33] A. Grothendieck, *SGA 1 (1960-61)*, Springer Lecture Notes, **224**, 1971.
[34] A. Grothendieck, *Recoltes et Semailles*, unpublished manuscript, available at web.ma.utexas.edu/users/slaoui/notes/recoltes_et_semailles.pdf.
[35] A. Grothendieck, *Pursuing Stacks*, unpublished manuscript.
[36] J. J. Gutierrez and C. Roitzheim, *Bousfield Localisations along Quillen Bifunctors*, Appl. Categ. Struct., **25** (2017), no. 6, 1113–1136.
[37] M. Hakim, *Topos Anneles et Schemas Relatifs*, Ergebnisse der Mathematik und ihrer Grenzgebiete, **64**, Springer-Verlag, Berlin–New York, 1972.
[38] J. Hall and D. Rydh, *Addendum: étale Devissage, Descent and Pushouts of Stacks*, J. Algebra **498** (2018), 398–412.
[39] R. Hartshorne, *Algebraic Geometry*, Graduate Texts in Mathematics, Springer-Verlag, New York, 1977.
[40] P. S. Hirschhorn, *Model Categories and their Localizations*, Math. Surveys and Monographs Series, **99**, AMS, Providence, 2003.
[41] A. Hirschowitz and C. Simpson, *Descente pour les n-champs*, arxiv:math/9807049 [math.AG].
[42] M. Hovey, *Model Categories*, Mathematical Surveys and Monographs, **63**, Amer. Math. Soc., Providence, 1998.
[43] M. Hovey, B. Shipley and J. Smith, *Symmetric Spectra*, J. Am. Math. Soc. **13** (2000), no. 1, 149–208.
[44] J. F. Jardine, *Simplicial Presheaves*, J. Pure Appl. Algebra **47** (1987), no. 1, 35–87.
[45] A. Joyal and M. Tierney, *Notes on Simplicial Homotopy Theory*, https://mat.uab.cat/~kock/crm/hocat/advanced-course/Quadern47.pdf.
[46] M. Kapranov, *DG-Schemes in Algebraic Geometry*, MSRI, Como, 1990, available at msri.org/workshops/138/schedules/710.
[47] Kelly, *Basic Concepts of Enriched Category Theory*, London Math. Soc. Lecture Note Series, **64**, Cambridge University Press, New York, 1982.
[48] A. Khan, *Motivic Homotopy Theory in Derived Algebraic Geometry*, Ph.D. thesis, Universitat Duisburg-Essen, 2016.
[49] M. Kontsevich, *Enumeration of Rational Curves via Torus Action*, The Moduli Space of Curves (Texel Island, 1994), 335–368, Progr. Math., **129**, Birkhauser, Boston, 1995.
[50] M. Kontsevich and Y. Soibelman, *Homological Mirror Symmetry and Torus Fibrations*, arxiv:math/0011041v2 [math.SG].
[51] J. Lurie, *Derived Algebraic Geometry*, thesis, MIT, 2004.
[52] J. Lurie, *Higher Topos Theory*, Annals of Mathematics Studies, Princeton University Press, 2009.
[53] J. Lurie, *Higher Algebra*, available at www.math.ias.edu/~lurie/.
[54] J. Lurie, *Spectral Algebraic Geometry*, available at www.math.ias.edu/~lurie/.
[55] J. Lurie, *Representability Theorems*, available at www.math.ias.edu/~lurie/.
[56] J. Lurie, *On ∞-Topoi*, arXiv:math/0306109v2 [math.CT].
[57] J. Lurie, *Structured Spaces*, available at www.math.ias.edu/~lurie/.
[58] S. Mac Lane, *Homology*, Classics in Mathematics, Springer, Berlin, 1995.
[59] S. Mac Lane, *Categories for the Working Mathematician*, 2nd Ed., Springer Verlag, New York, 2010.
[60] S. Mac Lane and I. Moerdijk, *Sheaves in Geometry and Logic*, Universitext, Springer-Verlag, New York (1992).
[61] P. May, *A Concise course in Algebraic Topology*, Chicago Lectures in Mathematics, University of Chicago Press, Chicago, IL, 1999.
[62] J. Milne, *Motives – Grothendieck's Dream*, Open Problems and Surveys of Contemporary Mathematics, 325–342, Surv. Mod. Math., **6**, Int. Press, Somerville, MA, 2013.
[63] J. Milne, *Lectures on Etale Cohomology*, available at jmilne.org/math/CourseNotes.
[64] G. Moore, *Some Comments on Physical Mathematics*, available at physics.rutgers.edu/~gmoore/.

[65] G. Moore, *Physical Mathematics and the Future*, available at physics.rutgers.edu/~gmoore/.
[66] R. Pellissier, *Categories Enrichies Faibles*, arXiv:math/0308246v1 [math.AT].
[67] E. Riehl and D. Verity, *On the Construction of Limits and Colimits in ∞-Categories*, arXiv:1808.09835 [math.CT].
[68] S. Schwede and B. Shipley, *Stable Model Categories are Categories of Modules*, Topology, **42** (2003), 103–153.
[69] J. P. Serre, *Algebre Locale. Multiplicites*, Lecture Notes in Mathematics, **11**, Springer-Verlag, Berlin–New York, 1965.
[70] B. Shipley, *A Convenient Model Category for Commutative Ring Spectra*, Homotopy Theory: Relations with Algebraic Geometry, Group Cohomology and Algebraic K-Theory, 473–483, Contemp. Math., **346**, Amer. Math. Soc., Providence, RI, 2004.
[71] C. Simpson, *A Giraud-type Characterization of the Simplicial Categories Associated to Closed Model Categories as ∞-Pretopoi*, arXiv:9903167 [math.AT].
[72] M. Spitsweck, *Operads, Algebras and Modules in Model Categories and Motives*, Ph.D. thesis, Mathematisches Institut, Friedrich-Wilhelms-Universitat Bonn (2001), bonndoc.ulb.uni-bonn.de/xmlui/handle/20.500.11811/1732.
[73] B. Toen, *Derived Algebraic Geometry*, EMS Surv. Math. Sci. **1** (2014), 153–240.
[74] B. Toen, *Higher and Derived Stacks: a Global Overview*, arXiv:0604504 [math.AG].
[75] B. Toen, *Homotopical and Higher Categorical Structures in Algebraic Geometry*, arXiv:math/03122262 [math.AG].
[76] B. Toen and G. Vezzosi, *Homotopical Algebraic Geometry I: Topos Theory*, Adv. Math. **193** (2005), no. 2, 257–372.
[77] B. Toen and G. Vezzosi, *Segal Topoi and Stacks over Segal Categories*, arxiv:math/0212330v2 [math.AG].
[78] B. Toen and G. Vezzosi, *Algebraic Geometry Over Model Categories: a General Approach to Derived Algebraic Geometry*, arXiv:0110109 [math.AG].
[79] B. Toen and G. Vezzosi, *Homotopical Algebraic Geometry II: Geometric Stacks and Applications*, arXiv:0404373 [math.AG].
[80] B. Toen and G. Vezzosi, *Brave New Algebraic Geometry and Global Derived Moduli Spaces of Ring Spectra*, arXiv:math/0309145v2 [math.AT].
[81] B. Toen and G. Vezzosi, *From HAG to DAG: Derived Moduli Stacks*, arXiv:0210407 [math.AG].
[82] V. S. Varadarajan, *Supersymmetry for Mathematicians: An Introduction*, Courant Lecture Notes in Mathematics **11**, AMS, Rhode Island, 2004.
[83] A. Vistoli, *Grothendieck Topologies, Fibered Categories and Descent Theory*, FGA, 1–104, Math. Surveys Monogr., **123**, Amer. Math. Soc., Providence, RI, 2005.
[84] D. White, *Monoidal Bousfield Localizations and Algebras Over Operads*, arXiv:1404.5197v2 [math.AT].
[85] D. White and D. Yau, *Right Bousfield Localization and Operadic Algebras*, arXiv:1512.07570v2 [math.AT].

Index

https://doi.org/10.1515/9783111334073-031

Notations

https://doi.org/10.1515/9783111334073-032

De Gruyter Expositions in Mathematics

Volume 74
Walter Roth
Integral Representation. Choquet Theory for Linear Operators on Function Spaces, 2023
ISBN 978-3-11-131450-1, e-ISBN 978-3-11-131547-8

Volume 59
Friedrich Haslinger
The d-bar Neumann Problem and Schrödinger Operators, 2nd Edition, 2023
ISBN 978-3-11-118290-2, e-ISBN 978-3-11-118292-6

Volume 69/3
Mariusz Urbański, Mario Roy, Sara Munday
Non-Invertible Dynamical Systems. Volume 3: Analytic Endomorphisms of the Riemann Sphere, 2023
ISBN 978-3-11-076984-5, e-ISBN 978-3-11-076987-6

Volume 73
Sergei S. Akbarov
Stereotype Spaces and Algebras, 2022
ISBN 978-3-11-078086-4, e-ISBN 978-3-11-078091-8

Volume 72
Ali Baklouti
Deformation Theory of Discontinuous Groups, 2022
ISBN 978-3-11-076529-8, e-ISBN 978-3-11-076530-4

Volume 71
Victor P. Maslov, Yurievich Oleg Shvedov†
The Canonical Operator in Multiparticle Problems and Quantum Field Theory, 2022
ISBN 978-3-11-076238-9, e-ISBN 978-3-11-076270-9

Volume 70
Aron Simis, Zaqueu Ramos
Graded Algebras in Algebraic Geometry, 2022
ISBN 978-3-11-063754-0, e-ISBN 978-3-11-064069-4

Bei Fragen zur Produktsicherheit wenden Sie sich bitte an:
If you have any questions regarding product safety, please contact:

Walter de Gruyter
Genthiner Straße 13
10785 Berlin
productsafety@degruyterbrill.com